Workshop Statistics

Discovery with Data and the Graphing Calculator

Third Edition

Allan J. Rossman
Beth L. Chance
California Polytechnic State University
San Luis Obispo

J. Barr von Oehsen
Clemson University

WILEY
John Wiley & Sons, Inc.

To order books or for customer service, please call 1(800)-CALL-WILEY (225-5945).

Printed in the United States of America.

ISBN-13 978- 0-470-41272-5

10 9 8 7 6 5 4 3 2

About the Authors

Allan Rossman and Beth Chance, professors of statistics at California Polytechnic State University in San Luis Obispo, are coauthors of the *Workshop Statistics* coursebook series and *Investigating Statistical Concepts, Applications, and Methods* (Duxbury). They have received several grants from the National Science Foundation to support curriculum development projects for introductory statistics. They have also served as coeditors of *STATS: The Magazine for Students of Statistics* and the *Proceedings of the Seventh International Conference on Teaching Statistics*. They have published articles on teaching statistics in *The American Statistician* and *The College Mathematics Journal*, and have given scores of statistics education presentations and workshops nationally and internationally. They recently received an award from the MAA's special interest group on statistics for a presentation about assessment in a session about implementing GAISE guidelines. Both have served in various leadership roles for the Advanced Placement program in Statistics.

Allan Rossman received his PhD in statistics from Carnegie Mellon University. Before he came to Cal Poly, he taught for twelve years at Dickinson College in Pennsylvania, where he served a term as department chair. He is president of the International Association for Statistics Education from 2007–2009 and was the Program Chair for the 2007 Joint Statistical Meetings. He has served as chair of the ASA's Section on Statistical Education and of the ASA/MAA Joint Committee on Undergraduate Statistics. He was selected as Fellow of the American Statistical Association in 2001. Allan served as project director for the Mathematical Association of America's NSF-funded STATS (Statistical Thinking with Active Teaching Strategies) project, which conducted workshops for mathematicians who teach statistics.

Beth Chance received her PhD in operations research, with an emphasis in statistics and a minor in education from Cornell University. She taught at University of the Pacific before moving to Cal Poly. She has served on the Test Development Committee for Advanced Placement Statistics and as secretary/treasurer of the ASA Section on Statistical Education. She was the inaugural recipient of the American Statistical Association's Waller Education Award for Excellence and Innovation in Teaching Introductory Statistics in 2002 and received the 2003 Mu Sigma Rho Statistical Education Award. She was selected as a Fellow of the ASA in 2005. Beth's professional interests include development of curricular materials for introductory statistics and research into how students learn statistics, particularly on the role of assessment and technology. She and her collaborators have published in the *Journal of Statistics Education* and the *Statistics Education Research Journal* (*SERJ*), and she has collaborated on several chapters and books aimed at enhancing teacher preparation to teach statistics. She currently serves as the assistant editor for *SERJ*.

James Barr von Oehsen is currently at Clemson University, where he is a Research Assistant Professor in the Department of Mathematical Sciences and a Senior Developer in the Cyberinfrastructure Technology Integration (CITI) group. The goal of CITI is to provide Clemson University with an advanced cyberinfrastructure through the integration of high-performance computing and networks, data perceptualization and visualization, storage architecture, and middleware that offers cyber-environments for e-communities. CITI leads research in areas such as high-performance computing applications, high throughput computing, high-performance networking, data access and interpretation (including visualization), and software environments for cyber-communities. Dr. von Oehsen has a B.S. in Mathematics from Rutgers University and a Ph.D. in Mathematics, with a concentration in Algebraic Topology, from Rutgers University. He is a member of The American Mathematical Society, the Society for Industrial and Applied Mathematics, and the Association for Computing Machinery.

Contents

Annotated Contents

Preface to the Graphing Calculator Version of *Workshop Statistics*

Shorn of all subtlety and led naked out of the protective fold of educational research literature, there comes a sheepish little fact: lectures don't work nearly as well as many of us would like to think.

—George Cobb

Statistics teaching can be more effective if teachers determine what it is they really want students to know and to do as a result of their course, and then provide activities designed to develop the performance they desire.

—Joan Garfield

Guidelines for Introductory Statistics

When *Workshop Statistics: Discovery with Data* appeared in 1996, it was the first book to embrace a constructivist approach for introductory statistics. Rather than describing and explaining statistics through exposition alone, Workshop *Statistics* offered activities designed to lead students to discover statistical concepts, explore statistical principles, and apply statistical techniques. In addition to its focus on active learning, *Workshop Statistics* emphasized developing students' understanding of statistical concepts and their ability to analyze real data, integrating technology to help achieve both of these goals. In these respects, we aimed to provide students and instructors with a resource that supported the 1992 recommendations of the American Statistical Association/Mathematical Association of America (ASA/MAA) Joint Committee on Undergraduate Statistics:

- Teach statistical thinking
- Employ more data and concepts, less theory and fewer recipes
- Foster active learning

Over the next several years, the recommendations that shaped our first edition received even more support from the statistics education community. Reaffirming and expanding on the ASA/MAA recommendations, in 2005, the ASA endorsed Guidelines for Assessment and Instruction in Statistics Education (GAISE):

- Emphasize statistical literacy and develop statistical thinking
- Use real data
- Stress conceptual understanding rather than mere knowledge of procedures
- Foster active learning in the classroom
- Use technology for developing conceptual understanding and analyzing data
- Use assessments to improve and evaluate student learning

These guidelines signal quite a change from the introductory statistics course of previous generations. Most introductory courses now make substantial use of genuine data and technology. Although developing students' statistical thinking and emphasizing conceptual understanding are more challenging goals, more courses are aiming for these goals, and new resources are available to support instructors (and students) in achieving them.

Evolution of the Workshop Approach

Since the publication of the second edition of *Workshop Statistics* in 2000, we have developed new insights into how students learn statistics, both from teaching thousands of students and from interacting with our colleagues who teach statistics and conduct research into how students learn. We have undertaken a substantial revision in this third edition to provide students with a better understanding of statistical practice and scientific investigation, offering additional support for developing their knowledge while maintaining the active-learning focus that distinguished the earlier editions. We also provide instructors with a resource for helping their students understand statistical ideas and apply them to their everyday lives as well as to their academic disciplines. To achieve this, we subscribe to the GAISE recommendations and implement them in the ways we describe here.

Workshop Statistics features current data and a focus on everyday applications of statistics as a way to promote **statistical literacy.** We want to help students develop the skills they need to become thoughtful and critical consumers of the statistics they will encounter in the news and in their everyday life. Toward that end, we have revised this edition to expand significantly the emphasis on always considering the scope of conclusions that can be drawn from a research study. In other words, we want students to develop the habit of asking two questions of any study they encounter:

- *To what group can the study results be generalized?*
- *Do the study findings support a cause-and-effect conclusion?*

Our workshop format gives students the opportunity to think deeply and communicate their statistical thinking in writing. Throughout the book, we frequently ask open-ended questions:

- *What effect would an outlier have on this analysis?*
- *How would your conclusions differ if the sample size had been larger (and all else had remained the same)?*
- *Why do statisticians like to reduce variability in a process?*

We implore students to write well in order to communicate and practice good statistical thinking habits, such as asking how the data were collected and beginning an analysis with graphical and numerical summaries.

Workshop Statistics contains **real data,** some that students collect about themselves, some from commonly available sources such as almanacs, and some from academic research studies. We have come to put a premium on involving students with *authentic studies* that address specific scientific research questions in contexts that will appeal to students. Presenting students with such studies helps them to recognize the power of statistics to answer questions of genuine interest in everyday life and in academic discourse. We have designed many new activities in this edition around studies and data whose underlying research questions we hope students will find engaging. One way that we try to deepen students' statistical thinking capability is to return to the same dataset many times, analyzing it at deeper levels each time.

We strive to help students understand **statistical concepts**—such as sampling, experimentation, bias, and confounding—so they can thoughtfully consider and communicate the conclusions of a study. A solid conceptual understanding is at the center of quantitative literacy: understanding concepts provides the basis for applying statistical methods correctly, transferring knowledge to new types of statistical problems, and interpreting the results appropriately. For example, we realize that students often have difficulty distinguishing between two very different uses of *randomness:* random sampling from a population and random assignment to treatment groups. This distinction is related directly to the scope of conclusion: random sampling enables generalization to the population, and random assignment permits causal conclusions to be drawn. Understanding how the data collection strategy relates to the appropriate scope of conclusion requires studying concepts of bias and confounding. This leads, in turn, to understanding the important processes of statistical thinking. We have completely reordered the topics and rewritten the first unit to consider the scope of conclusion issues and all of the attendant statistical concepts that students must understand to address them thoughtfully. This reordering enables us to revisit these issues throughout the book, giving students repeated exposure to the concepts and additional time to develop their understanding.

Our experience, along with educational research, has taught us that some concepts are particularly challenging for students, including the concept of a variable and sampling distributions. The concept of sampling distributions requires students to think about a sample statistic not simply as a number and not only as a function of the values in a sample, but also as a variable itself that fluctuates from sample to sample. We emphasize variables in the first topic, and most activities throughout the book begin by asking students to describe the observational units and variables in the study. Conceptualizing the long-run behavior of a sample statistic is a daunting task, so we devote several topics, with many activities and assessment items, to helping students meet the challenge. Our discovery-based, collaborative approach helps students to develop their conceptual understanding.

Active learning remains the hallmark of *Workshop Statistics,* and we believe in it as strongly as ever. Students need to experience some level of struggle in order to achieve meaningful learning, and this necessary struggle is given a prominent place within the workshop approach. To make this struggle more beneficial, many conscientious students have asked us for more resources to support their active learning of statistics. Toward that end, we provide much more exposition and guidance in this edition, especially in summarizing the lessons to be learned from the activities.

We have also come to believe all the more strongly in the potential of **technology,** not only for analyzing data but also for facilitating students' explorations of statistical concepts. For example, technology allows students to change a data value to investigate the concept of resistance of the sample mean or median or correlation coefficient. A more involved application of technology is to simulate repeated random sampling to investigate the effect of sample size on a sampling distribution. Dynamic and visual aspects of modern software make it not only easier but also more effective for students to investigate such concepts. Throughout this edition, interactive Java applets accompany many activities. The applets have been class-tested by students for several years and utilize a consistent, easy-to-use layout that is closely tied to the contextual framework of the activity in which they are introduced.

This version of the book assumes that students will use graphing calculator technology. Instructions for using a Texas Instrument calculator, specifically a TI-83 or TI-84, have been integrated into the activities. Accompanying data files and programs for the TI-83, TI-84, and TI-89 are available on the Student CD and Web Resource Center. Although we do not intend this version to be a user's manual for the graphing calculator, the focus is on helping students to use the calculator as an aid in discovering statistical concepts and exploring statistical principles. We strongly recommend that students read the user's manual that accompanies their graphing calculator for more detailed information on applications and features specific to their calculators.

We have also become convinced that **assessment** is the most crucial aspect of teaching and learning. Unfortunately, it is often the most overlooked. We subscribe wholeheartedly to George Cobb's mantra to "judge a book by its exercises, and you cannot go far wrong." We have put considerable thought into what questions to ask and *how* to ask them, in a way that helps introductory students develop the conceptual understanding and data analysis habits that we want them to achieve. When using classroom activities, we strongly encourage instructors to follow up with related assessments so students know the activities are not only fun, but also develop concepts and model statistical thinking. We hope that students' enjoyment of the activities will enhance their learning experience, and we have found that frequent assessments provide a quicker feedback loop to enhance students' conceptual development.

Acknowledgments

We appreciate the very helpful feedback we have received on previous editions from Al Coons, Michael Lacey, Robin Lock, Bill Rinaman, Mark Schilling, and Ned Schillow. We also gratefully acknowledge the informative feedback we received on drafts of this new edition from these reviewers:

- Stanley J. Benkoski, West Valley College, Saratoga, California
- Sean Bradley, Clarke College, Dubuque, Iowa
- Kenneth M. Brown, College of San Mateo, San Mateo, California
- Beth Burns, Bowling Green State University, Bowling Green, Ohio
- Smiley W. Cheng, University of Manitoba, Winnipeg, Manitoba, Canada
- Julie Clark, Hollins University, Roanoke, Virginia
- Sheryl Clayton, Hume-Fogg Magnet High, Nashville, Tennessee
- Jule Connolly, Wake Forest University, Winston-Salem, North Carolina
- Judith Dill, Mid-Pacific Institute, Honolulu, Hawaii
- Peter Flanagan-Hyde, Phoenix Country Day School, Paradise Valley, Arizona
- Dwight Galster, South Dakota State University, Brookings, South Dakota
- Jonathan Graham, The University of Montana, Missoula, Montana
- Todd Hendricks, Georgia Perimeter College, Rockdale Campus, Conyers, Georgia
- James Hoffman, University of Delaware, Newark, Delaware
- Melinda Holt, Sam Houston State University, Huntsville, Texas
- Lifang Hsu, Le Moyne College, Syracuse, New York
- William Josephs, Windward School, Los Angeles, California
- Maryann Justinger, Erie Community College, South Campus, Orchard Park, New York
- James Kiernan, Brooklyn College, Brooklyn, New York
- Gabriel Lampert, New Mexico State University, Las Cruces, New Mexico
- Gary Motta, Lassen Community College, Susanville, California
- Bernard Omolo, University of South Carolina Upstate, Spartanburg, South Carolina
- Marianne Parker, Weber High School, Pleasant View, Utah
- Karen Pikula, Dearborn Center for Math, Science, and Technology, Dearborn Heights, Michigan

- Andrew Pingitore, SUNY Fredonia and Fredonia High School, Fredonia, New York
- William G. Quinn, Frederick Community College, Frederick, Maryland
- Ginger Rowell, Middle Tennessee State University, Murfreesboro, Tennessee
- Robin Schwartz, College of Mt. St. Vincent/Math Confidence, Riverdale, New York
- Rana Singh, Virginia State University, Petersburg, Virginia
- Mark Smith, St. Charles North High School, St. Charles, Illinois
- Maria J. Vlahos, Barrington High School, Barrington, Illinois
- Joan Weinstein, Harvard Extension School, Cambridge, Massachusetts

We especially appreciate the very thorough attention that Kristin Burke and Annie Mac have given to this project. We also thank Patrick Farace for coordinating the review process, LeeAnn Pickrell for her skillful copyediting, and Christa Edwards and Holly Rudelitsch for their work in production.

We appreciate the support and encouragement that we receive from our colleagues at Cal Poly and Clemson University.

We thank Cal Poly student Emily Tietjen for helping to gather datasets, and other students mentioned in the text whose project data we use.

Perhaps most importantly, we thank our many students of introductory statistics over the years for helping us to understand how students learn and inspiring us to become better teachers.

On a personal level, Allan thanks his wife Eileen for her encouragement and support and also his feline friends Eponine and Cosette. Beth thanks her husband, Frank, and son, Ben, for continued inspiration, understanding, and forgiveness. Barr thanks his wife, Shari Prevost, and their two sons, Graham and Evan.

Allan J. Rossman

Beth L. Chance

J. Barr von Oehsen
October 2007

To the Student

How can statistics help identify people who cheat on their tax returns, or ensure a boat can hold the weight of its passengers? What advice can statistics offer to a waitress who wants to earn larger tips? Do people tend to scoop bigger portions of ice cream if they are given bigger bowls? How can we increase our life expectancy—does consuming more candy and sweets actually help?

You will investigate all these questions, and many more, as you learn the basic ideas and techniques of statistics. You will find that statistics has innumerable applications to real-world situations, both in everyday life and in nearly all fields of academic endeavor. The vast majority of examples and exercises concern real data from authentic studies. The issues range from completely silly, such as whether kissing couples tend to lean their head to the left or right, to deadly serious, such as analyzing the results of draft lotteries for serving in the Vietnam War. You will also collect and analyze data about yourself and your classmates.

How to Approach Your Course

You will quickly discover that this is not a typical textbook. This book consists primarily of activities that have been carefully designed to lead you to discover fundamental statistical ideas for yourself, in collaboration with your peers and your instructor. The in-class activities ask a series of guided questions and leave space for you to record your answers directly in the book. We are asking you to construct your own knowledge because we believe this building process, although potentially frustrating at times, overall leads to a deeper understanding and more lasting ability to retain and apply what you learn. The homework activities provide an opportunity to apply what you have learned and assess your understanding.

To support you in this learning process, you will find expository paragraphs throughout the book that present the most important pieces of information and summarize what you should be learning. We encourage you to pay particular attention to what we call "watch out" items, because these points highlight common student misconceptions and provide advice for overcoming and avoiding them. We also urge you to read and periodically review the Wrap-Up section of each topic carefully.

We also encourage you to write as well as you can. One reason is that, to some extent, you are creating your own textbook by responding to the activity questions. Expressing yourself clearly will be helpful as you review material later. More importantly, learning to interpret and communicate the results of statistical studies is an important skill for all educated citizens. You will be expected to learn to construct and analyze numerical arguments, using data to support your statements. In contrast to most mathematics courses, you will be using phrases such as "there is strong evidence that . . ." and "the data suggest that . . ." rather than "the exact answer is . . ." and "it is, therefore, proven that. . . ."

Our advice for your success in learning statistics can be summed up in two words: *think* and *participate.* We hope you will be actively engaged with the material as opposed to reading passively and taking notes. The activities will ask you to think critically and defend your arguments. Do not be afraid to make mistakes, because the most valuable learning can be learning from mistakes. You will be asked to make conjectures, collect data, draw conclusions, write summaries, discuss findings, explore alternatives, investigate scenarios, and more. You must have an open and active mind in order to complete these tasks; in other words, you must accept responsibility for your own learning. We hope your classmates and instructor will help you with this responsibility. Our responsibility has been to provide you with a resource to facilitate this learning process and lead you on the path toward understanding statistics.

Two final words of advice: Have fun! We sincerely hope you will enjoy a dynamic and interactive learning environment as you study statistics.

Student Support Materials

Student resources are available on a Student CD.

Student CD

The Student CD, packaged in the back of every new textbook, contains applets and data files (TI-83/84 Plus, TI-89, Fathom, Excel, Minitab, SPSS, JMP, and tab-delimited text) for all the activities. For your reference, all data files and applets used in the textbook are listed alphabetically in Appendix C.

Organization of *Workshop Statistics*

Workshop Statistics: Discovery with Data, Third Edition, is organized into seven units, each consisting of several topics. Each topic addresses one important idea and comprises these elements:

Overview sections present a motivating example and introduce the topic objectives within the context of material learned in previous topics.

TOPIC 3 • • •

Drawing Conclusions from Studies

There is considerable concern about the issue of young people injuring themselves intentionally. Can you use statistics to better understand the seriousness of this problem, by estimating the proportion of college students who have attempted to injure themselves? On an issue of less societal importance, can you estimate what proportion of people believe that Elvis Presley faked his widely reported death, and would it matter which people you asked? Or consider a different kind of question, which may appear whimsical but may prove important: Do candy lovers live longer than other people? If so, is candy a secret to long life? In this topic, you will begin to study issues related to these questions, focusing on concerns that limit the scope of conclusions you can draw from some statistical studies.

Overview

You have begun to understand that data can be useful for gaining insights into interesting questions. But to what extent can statistics provide answers to these questions? This topic begins your introduction to key concepts that determine the scope of conclusions you can draw from a study. For example, when can you *generalize* the results of a study to a larger group than those used in the study itself? Also, why can't you always conclude that one variable *affects* another when a study shows a relationship between the variables?

As you consider those questions, you will encounter some more fundamental terms, such as population and sample, parameter and statistic, and explanatory and response variables. You will also study the important concepts of bias and confounding, and you will begin to understand why those concepts sometimes limit the scope of conclusions you can draw.

Overview sections present a motivating example and introduce topic objectives.

Preliminaries ask a series of questions designed to get students thinking about the contexts of studies and statistical issues explored in the topic.

Preliminaries

1. What do you think is a typical weight for a male Olympic rower?
2. Take a guess as to the length of the longest reign of a British monarch since William the Conqueror.
3. Do you think states in the eastern or western U.S. tended to have greater population growth (on a percentage basis) in the 1990s?
4. How long (in miles) would you consider the ideal day hike?

Preliminaries engage students with the study contexts and statistical issues in the topic.

In-Class Activities guide students through a series of directed questions to explore the material for the topic within the context of authentic studies and using real data. Each topic contains four to six in-class activities. To facilitate student discovery and learning, in-class activities include these features:

- **Writing** format provides enough space for students to record predictions, interpretations, and explanations in the textbook itself. This form of written self-discovery gives students opportunities to construct their own knowledge and communicate their statistical thinking.
- **Important Terms and Results** are highlighted in shaded boxes. Important terms are defined early and revisited often.
- **Expository Paragraphs** summarize major statistical concepts learned in the activities.
- **Hints** provide occasional advice for carrying out calculations and answering conceptual questions.
- **Watch Out** reminders caution students about common misconceptions and help them develop useful habits.

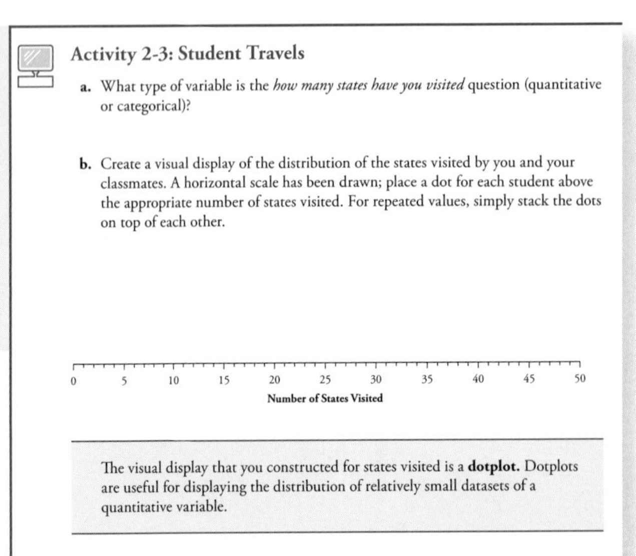

Activity 2-3: Student Travels

a. What type of variable is the *how many states have you visited* question (quantitative or categorical)?

b. Create a visual display of the distribution of the states visited by you and your classmates. A horizontal scale has been drawn; place a dot for each student above the appropriate number of states visited. For repeated values, simply stack the dots on top of each other.

The visual display that you constructed for states visited is a **dotplot.** Dotplots are useful for displaying the distribution of relatively small datasets of a quantitative variable.

In-Class Activities guide students through a series of directed questions within the context of an authentic study and using real data.

√ **Self-Check Examples** appear as the last in-class activity in each topic. These examples cover the key ideas of the topic and include detailed solutions so students can check their work and assess their understanding. Students should attempt self-check examples on their own first and then review the model solutions. Self-check examples are indicated by a checkmark icon.

√ **Activity 3-5: Childhood Obesity and Sleep**

A March 2006 article in the *International Journal of Obesity* described a study involving 422 children aged 5–10 from primary schools in the city of Trois-Rivieres, Quebec, (Chaput, Brunet, and Tremblay, 2006). The researchers found that children who reported sleeping more hours per night were less likely to be obese than children who reported sleeping fewer hours.

a. Identify the explanatory and response variables in this study. Also classify them.

Explanatory: Type:

Response: Type:

b. Is it legitimate to conclude from this study that less sleep caused the higher rate of obesity in Quebec children? If so, explain. If not, identify a confounding variable and explain why its effect on the response is confounded with that of the explanatory variable.

c. Do you think that the study's conclusion (of a relationship between sleep and obesity) applies to children outside of Quebec? Explain.

Solution

a. The explanatory variable is the amount of sleep that a child gets per night. This is a quantitative variable, although it would be categorical if the sleep information were reported only in intervals. The response variable is whether the child is obese, which is a binary categorical variable.

b. This is an observational study because the researchers passively recorded information about the child's sleeping habit. They did not impose a certain amount of sleep on children. Therefore, it is not appropriate to draw a cause-and-effect conclusion that less sleep causes a higher rate of obesity. Children who get less sleep may differ in some other way that could account for the increased rate of obesity. For example, amount of exercise could be a confounding variable. Perhaps children who exercise less have more trouble sleeping, in which case exercise would be confounded with sleep. You have no way of knowing whether the higher rate of obesity is due to less sleep or less exercise, or both, or some other variable that is also related to both sleep and obesity.

Self-Check Examples cover the key ideas of the activity and include detailed solutions.

Wrap-Up sections follow the in-class activities and provide a review of the major ideas studied in the topic, along with useful habits to develop and definitions to remember. The Wrap-Up sections have been greatly expanded to provide more information about what was covered in a topic (with ties back to specific activities) and to facilitate students' review of concepts.

Wrap-Up............

This topic provided you with more techniques for analyzing categorical data. You learned to summarize data from two categorical variables in a two-way table, and you calculated marginal and conditional proportions. You also discovered that segmented bar graphs are useful graphical displays for recognizing a relationship (association) between categorical variables. For example, you analyzed data to investigate a relationship between political viewpoint and opinion about government spending. On the issue of spending on the space program, you found essentially no relationship between the *political viewpoint* and *opinion about federal spending on the space program* variables; in other words, you found that liberals, moderates, and conservatives have very similar distributions of opinions. On the other hand, you found clear evidence of a relationship between *political viewpoint* and *opinion about federal spending on the environment*.

You encountered the notion of relative risk, finding that pregnant HIV-positive women who took AZT were about three times less likely to have an HIV-positive baby than those who took a placebo. You also discovered the phenomenon known as Simpson's paradox, which raises interesting issues with regard to analyzing two-way tables. For instance, you explained why one hospital can have a higher survival rate than another for all types of patients and yet have a lower survival rate overall: The hospital with higher success rates received most of the poor condition patients, who were naturally less likely to survive than those patients in fair condition.

Wrap-Up sections provide a review of major ideas studied in the topic.

Homework Activities appear at the end of each topic. These activities reinforce the material learned and test students' conceptual understanding and ability to apply what they have learned. Lists of **Related Activities** indicate to the student when an activity or dataset is being revisited.

● ● ● **Homework Activities**

Activity 4-6: Rating Chain Restaurants

The July 2006 issue of *Consumer Reports* included ratings of 103 chain restaurants. The ratings were based on surveys that *Consumer Reports* readers sent in after eating at one of the restaurants. The article said, "The survey is based on 148,599 visits to full-service restaurant chains between April 2004 and April 2005, and reflects the experiences of our readers, not necessarily those of the general population."

a. Do you think that the sample here was chosen randomly from the population of *Consumer Reports* readers? Explain.
b. Why do the authors of the article make this disclaimer about not necessarily representing the general population?
c. To what population would you feel comfortable generalizing the results of this study? Explain.

Activity 4-7: Sampling Words

4-1, 4-2, 4-3, 4-4, **4-7**, 4-8, 8-9, 9-15, 14-6

Reconsider Activities 4-1 through 4-4, in which the population of interest was the 268 words in the Gettysburg Address. Now consider the variable *whether the word has at least five letters*.

a. Is this a categorical (also binary) or a quantitative variable?
b. Among the 268 words, 99 contain at least five letters. What proportion of these 268 words contains at least five letters?

Homework Activities test students' conceptual understanding and ability to apply what they have learned.

Datasets and **interactive Java applets** are featured throughout the book. Many activities involve the use of Java applets designed specifically to help students explore a statistical concept. The applets are embedded in the specific context of the activity with step-by-step instructions. A computer screen icon indicates when students can obtain datasets and Java applets from the Web-based Student Resource Center or from their Student CD.

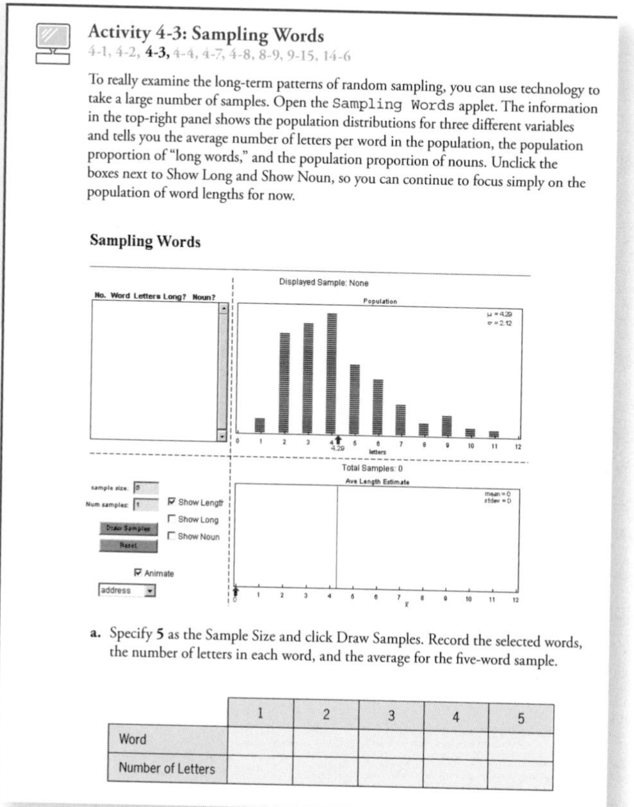

Activity 4-3: Sampling Words
4-1, 4-2, **4-3,** 4-4, 4-7, 4-8, 8-9, 9-15, 14-6

To really examine the long-term patterns of random sampling, you can use technology to take a large number of samples. Open the `Sampling Words` applet. The information in the top-right panel shows the population distributions for three different variables and tells you the average number of letters per word in the population, the population proportion of "long words," and the population proportion of nouns. Unclick the boxes next to Show Long and Show Noun, so you can continue to focus simply on the population of word lengths for now.

Sampling Words

a. Specify **5** as the Sample Size and click Draw Samples. Record the selected words, the number of letters in each word, and the average for the five-word sample.

	1	2	3	4	5
Word					
Number of Letters					

Java applets are designed to help students explore a statistical concept.

Roughly half the activities require the use of a software package or graphing calculator. This version of the book—*Workshop Statistics and the Graphing Calculator*—assumes that students will use graphing calculator technology. Instructions for using a Texas Instruments calculator, specifically a TI-83 or TI-84, have been integrated into the activities. Accompanying data files and programs for the TI-83, TI-84, and TI-89 are available on the Student CD and Web Resource Center.

List of Activities by Application

Business and Economics	Activity Numbers
Cell Phone Fraud	1-3
Credit Card Usage	1-9, 16-12, 17-11, 19-8, 20-10
Got a Tip?	1-10, 5-28, 14-12, 15-17, 22-4, 22-9, 25-3
Penny Thoughts	2-1, 3-10, 16-23
Accumulating Frequent Flyer Miles	3-16
Children's Television Advertisements	6-11, 25-14
Gender-Stereotypical Toy Advertising	6-21, 6-22
Hypothetical Employee Retention Predictions	6-28
Honda Prices	7-10, 10-19, 12-21, 28-14, 28-15, 29-10, 29-11
Hypothetical Manufacturing Processes	7-17
Turnpike Distances	7-20
Wrongful Conclusions	8-6, 16-21, 17-8, 28-25
Salary Expectations	8-26
Hypothetical ATM Withdrawals	9-24, 19-21, 22-25
Digital Cameras	10-5, 26-18, 27-21
Random Cell Phones	11-6
Dice-Generated Ice Cream Prices	11-17, 11-18
Collecting Prizes	11-20
Christmas Shopping	14-3, 14-7, 15-11, 19-1
Cars' Fuel Efficiency	14-11
Magazine Advertisements	16-15, 17-20, 21-21
Charitable Contributions	16-18, 18-7
Hiring Discrimination	17-21, 18-21
Editorial Styles	22-12, 22-13
Cow Milking	23-6
Car Ages	23-7
Comparison Shopping	23-22, 23-23, 26-21
House Prices	8-10, 26-1, 27-5, 28-2, 28-12, 28-13, 29-3
Car Data	26-3, 27-1, 28-10
Airline Maintenance	26-7
Textbook Prices	28-4, 28-5, 29-5
Airfares	28-6, 28-7, 28-8, 28-9, 29-16
Electricity Bills	28-11

Education

Activity Numbers

Entertainment and Media

Food

Food (*continued*)

Government and History

Health and Medicine

Health and Medicine *(continued)*

Learn-the-Concept

Population Statistics *(continued)*

Psychology

Sports

List of Activities Using Student-Generated Data

The following activities ask students to analyze data about themselves and their classmates, which they collect in the Preliminaries section of the topic.

Student Data	1-1, 1-5, 2-7, 2-8, 3-7, 7-8
Penny Thoughts	2-1
Student Travels	2-3
Value of Statistics	2-9
Memorizing Letters	5-5, 7-15, 8-13, 9-19, 10-9, 22-3
Lifetime Achievements	6-3, 6-6, 24-17
Watching Films	6-17, 6-18
Age Guesses	8-20, 20-13
Salary Expectations	8-26
Social Acquaintances	9-8, 9-9, 10-13, 10-14, 19-9, 19-10, 20-12
Candy Colors	13-1
Calling "Heads" or "Tails"	13-10, 17-14, 17-15, 24-18
Flat Tires	17-10, 24-16
Sleeping Times	19-5, 19-12, 19-19, 20-2
Chip Melting	23-2, 26-20, 29-15
Heights, Handspans, and Foot Lengths	26-5, 28-1, 29-12, 29-13, 29-14

In the following activities, students perform simulations on their own and then combine their results with those of their classmates.

Sampling Words	4-1, 4-2
Testing Strength Shoes	5-2
Random Babies	11-1, 11-2, 16-17, 20-16, 20-22
Coin Ages	14-1, 19-15
Penny Activities	16-9, 16-10, 16-11, 18-14
Friendly Observers	21-1

UNIT

1

Collecting Data and Drawing Conclusions

TOPIC

1 • • • Data and Variables

How can statistics be used to help decide the guilt or innocence of a nurse accused of murdering some of her patients? If chief executive officers tend to be taller than average, would this convince you that being tall provides advantages in the business world? Do some students do worse on standardized tests when they are first asked to indicate their race than when they are not, perhaps due to negative stereotypes of their academic ability? In this topic, you will begin your exploration of statistics and start gathering the building blocks and tools needed to address such questions.

Overview

Statistics is the science of reasoning from data, so a natural place to begin your study is by examining what is meant by the term *data*. You will find that data *vary,* and variability abounds in everyday life and in academic study. Indeed, the most fundamental principle in statistics is that of variability. If the world were perfectly predictable and showed no variability, you would not need to study statistics. Thus, you will learn about variables and consider their different classifications. You will also begin to experience the interesting research questions that you can investigate by collecting data and conducting statistical analyses.

Preliminaries

1. Kristen Gilbert, a nurse for a veteran's hospital, was charged with murdering some of her patients by administering fatal doses of a heart stimulant. Some of the evidence presented at her trial was statistical. Researchers combed through the records of all eight-hour shifts at the hospital in the previous eighteen months. What information do you think they recorded about each shift?

2. Telephone companies constantly collect data on cell phone calls in an effort to detect anomalies that might indicate fraud (in other words, that someone other than the rightful owner has been using the phone). What information should you record about each cell phone call, in order to develop a profile for the cell phone owner so that you could then monitor calls for anomalies?

3. Have you used a cell phone so far today?

4. Take a guess for what percentage of your classmates have used a cell phone today.

In-Class Activities

Activity 1-1: Student Data

1-1, 1-5, 2-7, 2-8, 3-7, 7-8

You will encounter some fundamental ideas in statistics by first considering the word *statistics* itself.

a. Write two complete and grammatically correct sentences, explaining your primary reason for taking this course and then describing what the term *statistics* means to you.

b. For each word in your response to part a, record the number of letters in the word:

The numbers that you have recorded are data. Not all numbers are data, however. Data are numbers collected in a particular context. For example, the numbers 10, 3, and 7 do not constitute data in and of themselves. They are data, however, if they refer to the number of letters in the first three words of your response to part a.

c. Did every word in your two sentences contain the same number of letters?

Although it might be obvious that different words contain different numbers of letters, the fact that the same measurement (e.g., "number of letters") can produce different responses illustrates one of the most fundamental ideas in statistics: variability.

A **variable** is any characteristic of a person or thing that can be assigned a number or a category. The person or thing to which the number or category is assigned, such as a student in your class, is called the **observational unit. Data** consist of the numbers or categories recorded for the observational units in a study. **Variability** refers to the phenomenon of a variable taking on different values or categories from observational unit to observational unit.

d. For the data that you recorded in part b, the observational units are the words that you wrote. What is the variable?

> A **quantitative** variable measures a numerical characteristic such as height, whereas a **categorical** variable records a group designation such as gender. **Binary** variables are categorical variables with only two possible categories, for example, male and female.

e. Now consider the students in your class as observational units. Classify each of the following variables as categorical or quantitative. If it is categorical, also indicate whether or not it is binary.

- How many hours you have slept in the past 24 hours
- Whether you have slept for at least 7 hours in the past 24 hours
- How many states you have visited
- Handedness (which hand you write with)
- Day of the week on which you were born
- Gender
- Average study time per week
- Score on the first exam in this course

We will continue to focus your attention on observational units and variables in virtually all of the studies and data that you encounter throughout this book. Along the way, keep the following points in mind:

- This distinction between categorical and quantitative variables is quite important because determining which statistical tools to use for analyzing a given set of data often depends on the type of variables involved.
- Notice how the variable of *sleep time* can be measured either quantitatively (first bullet) or categorically (second bullet). This is true of many variables: The classification of the variable often depends on how the quantity is measured more than on any intrinsic property.
- A variable that takes on numerical values that are really just category labels, such as a zip code, is categorical.

f. Still considering yourself and your classmates as observational units, can *average height of students in the class* be legitimately considered a variable? What about *percentage of students in the class who have used a cell phone today?* Explain.

g. If you record average student height or the percentage of students who have used a cell phone today for all classes taught at your school, are these now variables? What are the observational units in this case?

Watch Out

As the term *variable* suggests, the values assumed by a variable can vary from observational unit to observational unit. Summaries, such as average height in class or percentage of students in class who have used a cell phone, are not variables for the students in the class, because there is just one value for that class. To determine whether something is actually a variable, ask yourself whether or not it represents a question that can be asked of each observational unit and whether the values can potentially vary from observational unit to observational unit. Notice that *average time studying per week* can be legitimately considered a variable because it can be recorded about each student and can vary from student to student. Also notice that *average height* becomes a variable if the observational unit is a class of students rather than an individual student.

Activity 1-2: Variables of State

Suppose that the observational units of interest are the fifty states. Identify which of the following are variables and which are not. Also classify the variables as categorical or quantitative.

a. Gender of the state's current governor

Variable, Categorical, binary

b. Number of states that have a female governor

Statistic

c. Percentage of the state's residents older than 65 years of age

Variable, quantitative

d. Highest speed limit in the state

Variable, quantitative

e. Whether or not the state's name contains one word

Variable, categorical, binary

f. Average income of the adult residents of the state

Variable, quantitative

g. How many states were settled before 1865

Statistic

Activity 1-3: Cell Phone Fraud

Lambert and Pinheiro (2006) describe a study in which researchers try to identify characteristics of cell phone calls that suggest the phone is being used fraudulently. For each cell phone call, the researchers recorded information on its direction (incoming or outgoing), location (local or roaming), duration, time of day, day of week, and whether the call took place on a weekday or weekend.

a. Identify the observational units in this study.

Cell phone calls

b. Identify the categorical variables mentioned in the preceding paragraph. Indicate which are binary and which are not.

Direction - binary
Location - binary
weekday/weekend - binary
duration - quantitative
Day of week - quantitative
time - quantitative

c. Identify the quantitative variables mentioned in the preceding paragraph.

Activity 1-4: Studies from *Blink*

1-4, 5-14

The following studies are all described in the popular book *Blink: The Power of Thinking Without Thinking,* by Malcolm Gladwell (2005). For each study, identify the observational units and variables. Also, classify each variable as quantitative or categorical.

a. An economist suspects that chief executive officers (CEOs) of American companies tend to be taller than the national average height of 69 inches, so she takes a random sample of 100 CEOs and records their heights.

Observational units:

Variable: Type:

b. A psychologist shows a videotaped interview of a married couple to a sample of 150 marriage counselors. Each counselor is asked to predict whether the couple will still be married five years later. The psychologist wants to test whether marriage counselors make the correct prediction more than half the time.

Observational units:

Variable: Type:

c. A psychologist gives an SAT-like exam to 200 African-American college students. Half of the students are randomly assigned to use a version of the exam that asks them to indicate their race, and the other half are randomly assigned to use a version of the exam that does not ask them to indicate their race. The psychologist suspects that those students who are not asked to indicate their race will score significantly higher on the exam than those who are asked to indicate their race.

Observational units:

Variable 1: Type:

Variable 2: Type:

d. An economist randomly assigns four actors to go to ten different car dealerships each and negotiate the best price they can for a particular model of car. The four people are all the same age, dressed similarly, and tell the car salespeople that they have the same occupation and neighborhood of residence. One of the actors is a white male, one is a black male, one is a white female, and one is a black female.

The economist wants to test whether or not the average prices differ significantly among these four types of customers.

Observational units:

Variable 1: Type:

Variable 2: Type:

Variable 3: Type:

This activity reveals that variables are the building blocks of interesting research questions that can be addressed with data. The next activity asks you to think of some questions that can be addressed with data collected on yourself and your classmates.

Activity 1-5: Student Data

1-1, **1-5,** 2-7, 2-8, 3-7, 7-8

Recall the variables listed in part e of Activity 1-1 (page 5) that could be measured concerning yourself and your classmates. Some research questions that you could investigate with data on those variables include

- Do female students tend to study more than male students?
- Do left-handers and right-handers differ with regard to sleeping times?
- Do students who study more tend to score higher on exams?

a. Suggest two other questions that you could investigate from the variables listed in part e.

Question 1: Are more female or males left handed?

Question 2: Do people who travel more sleep less?

b. Suggest four additional variables that you could record about yourself and your classmates, and then propose two research questions that you could address using those variables. *Hint:* Be sure to distinguish the variables from the research questions; remember a variable is some characteristic that can be recorded for each student and can vary from student to student.

Variable 1: Variable 2:

Variable 3: Variable 4:

Question 1:

Question 2:

√

Activity 1-6: A Nurse Accused

1-6, 3-20, 6-10, 25-23

Statistical evidence played an important role in the murder trial of Kristen Gilbert, a nurse who was accused of murdering hospital patients by giving them fatal doses of a heart stimulant (Cobb and Gerlach, 2006). Hospital records for an eighteen-month period indicated that of the 257 eight-hour shifts that Gilbert worked, a patient died on 40 of those shifts (15.6%). But during the 1384 eight-hour shifts that Gilbert did not work, a patient died on only 34 of those shifts (2.5%). (You will learn how to analyze such data in Topics 6 and 21.)

a. Identify the observational units in this study. *Hint:* The correct answer here is more subtle than most students expect.

b. Identify the two variables mentioned in the preceding paragraph. Classify each as categorical (possibly binary) or quantitative.

Variable 1: Type:

Variable 2: Type:

Solution

a. The observational units are the eight-hour shifts.

b. One variable is whether or not Gilbert worked on the shift. This variable is categorical and binary. The other variable is whether or not a patient died on the shift. This variable is also categorical and binary.

Watch Out

It's tempting to call the patients the observational units, but that is not consistent with the data reported. The data indicate what happened on each *shift,* not what happened to each patient. The variables, therefore, need to refer to something that can be recorded about each shift, namely whether Gilbert worked that shift or not and whether a patient died on that shift or not. Notice that we are asking these variables as questions to be posed to each shift. Another way to spot the observational units is to focus on how many data values are in the study; in this case, there are 257 + 1384, or 1641, shifts, not 1641 patients.

Some common errors in reporting variables include

- Providing a summary, such as "the total number of patient deaths" or "the percentage who died on Gilbert's shifts"
- Giving an ambiguous answer, such as "patient deaths"
- Stating the research question rather than a variable, such as "Did patients die at a higher rate on Gilbert's shifts?"
- Describing a subset of the observational units, such as "the patients who died on Gilbert's shifts"

Wrap Up . . . .

You can use statistics to address interesting research questions that help you better understand the world and whatever academic discipline you study. You have seen that statistics played an important role in the murder trial of Kristen Gilbert, and that statistics enabled researchers to answer questions such as whether CEOs are taller than average and whether thinking about their race causes African-American students to do worse on standardized exams.

Because statistics is the science of **data,** this topic has given you a sense of what data are and a glimpse of what data analysis entails. Data are not mere numbers: Data are collected for some purpose and have meaning in some context. For example, the numbers 5.25 and 37 are not data until you learn that they represent the number of hours slept last night and the number of states that a person has visited.

You encountered the most fundamental concept of statistics: **variability.** This concept will be central throughout the course. How long your classmates slept last night varies from student to student, as does the day of the week on which your classmates were born. One key idea to learn quickly is that of a variable. Correctly identifying and classifying variables will serve you well throughout this course and help you determine which statistical tools to apply to the data.

Some useful definitions to remember and habits to develop from this topic include

- Always consider data in context and anticipate reasonable values for the data collected and analyzed.
- A **variable** is a *characteristic* that varies from person to person or from thing to thing. The person or thing is called an **observational unit.**
- Variables can be classified as **categorical** or **quantitative,** depending on whether the characteristic is a categorical designation (such as gender) or a numerical value (such as height). A categorical variable with only two possible categories is also called **binary.**

In the next topic, you will deepen your understanding of variables by studying the concept of distribution. You will also discover two simple graphical techniques for displaying data, and you will investigate the concept of a statistical tendency.

● ● ● Homework Activities

Activity 1-7: Miscellany

Classify each of the following variables as categorical (possibly binary) or quantitative. Also identify the observational unit for each variable. (You will encounter each of these variables later in the book.)

a. Whether a spun penny lands heads or tails
b. Whether or not a person washes his/her hands before leaving a public restroom
c. Number of calories in a fast-food sandwich
d. Life expectancy in a country
e. Whether an American household owns a cat or does not own a cat
f. Year in which a college was founded
g. The comprehensive fee charged by a college
h. Candidate an American voted for in the 2004 presidential election
i. Whether or not a newborn baby tests HIV-positive
j. Running time of an Alfred Hitchcock movie
k. Age of an American penny
l. Weight of an automobile
m. Whether an automobile is foreign or domestic to the United States

n. Classification of an automobile as small, midsize, or large
o. Whether an applicant for graduate school is accepted
p. Ratio of backpack weight to body weight for a college student
q. Whether a person classifies him/herself as a political conservative, moderate, or liberal
r. Whether a college student has abstained from the use of alcohol for the past month
s. Whether a participant in a sport suffers an injury in a given year
t. A sport's injury rate per 1000 participants
u. A state's rate of automobile thefts per 1000 residents
v. Amount of liquid poured into a glass by a bartender
w. Number of "close friends" a person reports having
x. Age of a bride on the wedding day
y. Whether the bride is older, younger, or the same age as the groom in a couple getting married
z. The difference in ages (groom's age minus bride's age, which could be negative) of a couple getting married

Activity 1-8: Top 100 Films

In 1998, the American Film Institute selected the 100 greatest American films of all time. The following variables use these films as observational units. Identify each of the following variables by type:

a. Box office revenue
b. Number of years since production
c. Decade produced
d. Whether the film was produced before 1960
e. Whether the film won an Academy Award for Best Picture
f. Whether you have seen the film
g. The number of people in your class who have seen the film

Activity 1-9: Credit Card Usage

1-9, 16-12, 17-11, 19-8, 20-10

The Nellie Mae organization conducts an extensive annual study of credit card usage by college students. The observational units are the college students contacted by the organization. Some of the variables recorded include

- Year in school (freshman, sophomore, ...)
- Whether the student has a credit card
- Outstanding balance on the credit card
- Whether the outstanding balance exceeds $1000
- Source for selecting a credit card (direct mail, parent referral, ...)
- Region of the country

a. Classify each of these variables as categorical (possibly binary) or quantitative.
b. List two research questions that could be investigated using these variables.

Activity 1-10: Got a Tip?

1-10, 5-28, 14-12, 15-17, 22-4, 22-9, 25-3

Suppose that a waitress wants to investigate some factors that might affect the amount of her tips. She decides to record the amount of the tip she gets from each table that she waits on, along with other variables that might affect the amount she receives.

a. Suggest three variables that you would encourage her to record because they might be related to the amount of tip she receives.
b. Suggest some specific research questions that you could investigate with the resulting data.

Activity 1-11: Proximity to the Teacher

1-11, 5-23, 27-9, 29-6

Suppose that you want to investigate whether or not students who sit closer to the teacher tend to score higher on quizzes than students who sit farther away.

a. Identify the observational units in this study.
b. Identify the variables in this study, and classify each variable as categorical or quantitative.

Activity 1-12: Emergency Rooms

Suppose that the observational units in a study are the patients arriving at an emergency room in a given day. For each of the following, indicate whether it can legitimately be considered a variable or not. If it is a variable, classify it as categorical (and if it is binary) or quantitative. If it is not a variable, explain why not.

a. Blood type
b. Waiting time
c. Mode of arrival (ambulance, personal car, on foot, other)
d. Whether men have to wait longer than women
e. Number of patients who arrive before noon
f. Whether the patient is insured
g. Number of stitches required
h. Whether stitches are required
i. Which patients require stitches
j. Number of patients who are insured

Activity 1-13: Candy Colors

1-13, 2-19, 13-1, 13-2, 15-7, 15-8, 16-4, 16-22, 24-15, 24-16

Suppose that you are given a sample of 25 Reese's Pieces™ candies and asked to record the color of each candy.

a. What are the observational units?
b. What is the variable? Is it categorical (also binary) or quantitative?

Now suppose that your classmates are each given a sample of 25 Reese's Pieces candies. Also suppose that students are asked to report the proportion of orange candies in their sample of 25 candies.

c. Now what are the observational units?
d. Now what is the variable? Is it categorical (also binary) or quantitative?

Activity 1-14: Natural Light and Achievement

1-14, 5-17

A recent study by the Heschong Mahone Group, based near Sacramento, California, found that students who received instruction in classrooms with more natural light scored as much as 25 percent higher on standardized tests than other students in the same school district.

a. Identify the observational units in this study.

b. Identify the two primary variables of interest. *Hint:* Be sure to state each of these variables as characteristics that change from observational unit to observational unit.

c. Indicate whether you believe that the variables in part b were measured as categorical or quantitative data.

Activity 1-15: Children's Television Viewing

1-15, 19-16, 20-4, 22-14, 22-15

Researchers at Stanford University studied whether or not reducing children's television viewing might help to prevent obesity (Robinson, 1999). Third- and fourth-grade students at two public elementary schools in San Jose were the subjects. One of the schools incorporated a curriculum designed to reduce watching television and playing video games, whereas the other school made no changes to its curriculum. At the beginning and end of the study, a variety of variables were measured on each child. These included body mass index, tricep skinfold thickness, waist circumference, waist-to-hip ratio, weekly time spent watching television, and weekly time spent playing video games.

a. Identify the observational units in this study.

b. Identify some of the variables measured on these units, making sure to phrase them as variables. You should identify at least one quantitative and at least one categorical variable.

Activity 1-16: Nicotine Lozenge

1-16, 2-18, 5-6, 9-21, 19-11, 20-15, 20-19, 21-6, 22-8

An article in the June 10, 2002, issue of the *Archives of Internal Medicine* reported on a study of the effectiveness of a nicotine lozenge for helping smokers to quit smoking. Smokers who participated in the study were randomly assigned to receive either the nicotine lozenge being tested or a placebo (no active ingredient) lozenge. Several variables were recorded on each subject at the beginning of the study, such as weight, gender, number of cigarettes smoked per day, and whether or not the person had made previous attempts to quit smoking. At the end of the study, the primary variable recorded was whether or not the subject had successfully refrained from smoking.

a. Identify the observational units in this study.

b. Identify the categorical variables mentioned in the preceding paragraph.

c. Identify the quantitative variables mentioned in the preceding paragraph.

d. Did you include *type of lozenge assigned* as one of the variables? Which type of variable is this?

Activity 1-17: Oscar Winners and Super Bowls

As you begin to develop your own research questions, a key first step is to define measurable variables on the observational units of interest.

a. Consider all movies that have won the Academy Award (Oscar) for Best Picture as the observational units for a study. Provide examples of two categorical variables and two quantitative variables that you could examine for these observational units. *Hint:* State these clearly as variables. Ask yourself, "What question could I ask each of these movies?" supposing that movies could answer.

b. Now consider as observational units all of the Super Bowl games that have been played. Provide examples of two categorical variables and two quantitative variables that you could examine for these observational units.

TOPIC

2 • • •

Data and Distributions and the Graphing Calculator

How often do people wash their hands after using a public restroom? Are men or women more likely to wash their hands? Are people in some parts of the country, or in some types of public places, more likely to wash their hands than others? These might not be life-or-death issues, but they have been analyzed in an extensive national study. A more serious issue is whether states can encourage drivers to wear seatbelts by passing laws to require seatbelt use. Are people in states with stricter laws more likely to buckle up than those in states with less strict laws? You will investigate these issues in this topic, as you begin to learn about comparing distributions of data across groups.

Overview

This topic continues your study of the ideas of variable and variability, as you begin to explore the notion of the distribution of a set of data measuring a particular variable. One way to examine the distribution of a set of data is to create a graph, which also leads you to a fundamental principle of data analysis: Start any analysis by looking at a graph of the data. By examining graphs and comparing distributions, you will discover the idea of a statistical tendency, which will arise throughout the course. As in Topic 1, you will find that context is crucial in statistics and that data can address interesting research questions.

Preliminaries

1. Take a guess for the percentage of people who wash their hands after using a public restroom.

60%

2. Do you suspect that men or women are more likely to wash their hands after using a public restroom?

Women

3. Guess the percentage of adult Americans who regularly wear a seatbelt in a car.

90%

4. If the decision were entirely up to you, would you favor abolishing the penny, making the nickel the lowest denomination coin?

5. Guess what percentage of American adults favor abolishing the penny.

6. Count how many states you have visited (or lived in or even just driven through) by looking through the list in Activity 2-4 on page 20.

7. Guess the number of different states that a typical student at your school has visited (or lived in). Also, guess what the largest and smallest numbers of states visited by the students in your class will be.

Smallest: 5 Typical: 10 Largest: 28

In-Class Activities

Activity 2-1: Penny Thoughts

2-1, 3-10, 16-23

Consider the question about whether you favor retaining or abolishing the penny.

a. Is this a categorical (also binary) or quantitative variable?

Categorical

b. Count how many students responded to the penny question and how many of them voted to retain the penny. Calculate the proportion who voted to retain the penny by dividing that count by the total count. (Note that a proportion is a number between 0 and 1. Report your answer to three decimal places.)

60%- No
.6

c. How many and what proportion of the respondents voted to abolish the penny?

40% .4 Yes

d. Create a visual display of these responses by drawing rectangles with heights corresponding to the proportions voting for each option. Be sure to give labels to the horizontal and vertical axes.

The visual display that you have constructed is called a **bar graph.** Bar graphs display the distribution of a categorical variable. The **distribution** of a variable refers to its pattern of variation. With a categorical variable, distribution means the variable's possible categories and the proportion of responses in each.

e. Write a sentence or two describing what your analysis reveals about the attitudes of students in this class toward the penny.

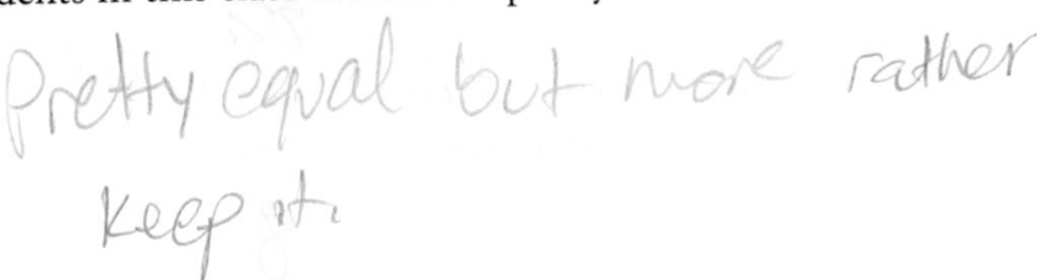

Activity 2-2: Hand Washing

2-2, 21-16, 25-16

In August 2005, researchers for the American Society for Microbiology and the Soap and Detergent Association monitored the behavior of more than 6300 users of public restrooms. They observed people in public venues such as Turner Field in Atlanta and Grand Central Station in New York City. They found that 2393 of 3206 men washed their hands, compared to 2802 of 3130 women.

a. What proportion of the men washed their hands? What proportion of the women washed their hands?

b. Are these proportions consistent with the following pair of bar graphs?

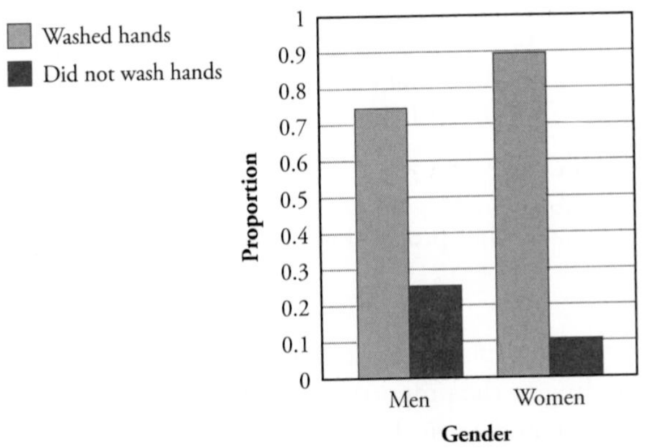

c. Comment on what your calculations and the bar graph reveal about whether one gender is more likely to wash their hands after using a public restroom.

Women more likely

The following bar graphs reveal the distribution of the *washed hands* variable based on the city in which the person was observed:

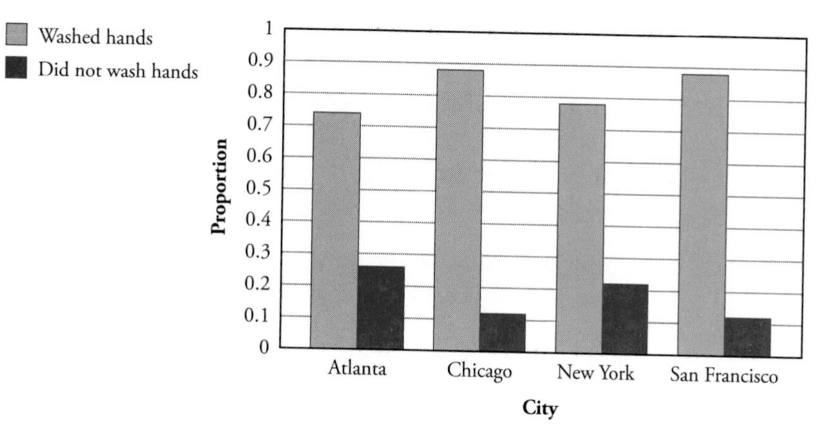

d. For each city, estimate the proportion of people who washed their hands as accurately as you can from the graph.

Atlanta: Chicago: New York: San Francisco:

e. Comment on what the bar graphs reveal about how these cities compare with regard to hand washing.

Activity 2-3: Student Travels

a. What type of variable is the *how many states have you visited* question (quantitative or categorical)?

b. Create a visual display of the distribution of the states visited by you and your classmates. A horizontal scale has been drawn; place a dot for each student above the appropriate number of states visited. For repeated values, simply stack the dots on top of each other.

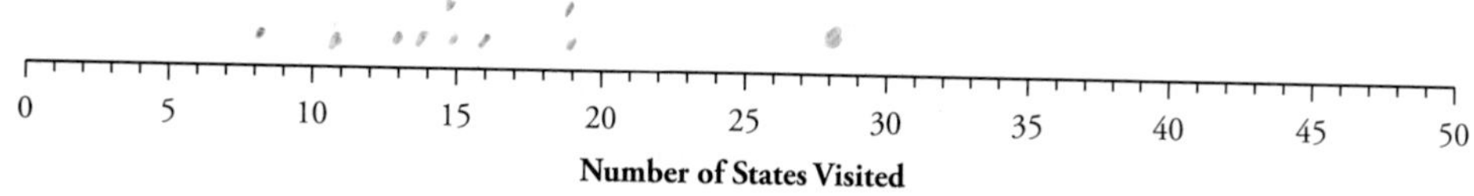

The visual display that you constructed for states visited is a **dotplot.** Dotplots are useful for displaying the distribution of relatively small datasets of a quantitative variable.

Watch Out

- Each dot in a dotplot represents a different observational unit.
- Develop the habit of always labeling the axis with the name of the variable.
- Dotplots can be cumbersome to create by hand with larger datasets, so you can often use technology to create the graph.
- Dotplots pertain to a quantitative variable; bar graphs, to a categorical variable.

c. Circle your own value on the dotplot you just created. Where do you seem to fall in relation to your peers with regard to states visited?

d. Based on this display, comment on the accuracy of your guesses in the Preliminaries section.

e. Write a paragraph describing various features of the distribution of states visited. Imagine that you are describing this distribution to someone who cannot see the display and has absolutely no idea how many states people visit. *Advice:* Here and throughout the course, *always* relate your comments to the context. Remember that data are not just numbers, but numbers with a context: in this case, the number of states visited by you and your classmates.

f. Enter the states visited data into your graphing calculator. To do this, follow these steps:

1. Turn on your calculator.
2. Press the 2ND key (this accesses the commands written in blue on the TI-84 Plus and the commands written in yellow on the TI-83 Plus) followed by [{]. You should now see a left curly bracket { on your screen. Enter the number of states visited by each student, using the comma key to separate the numbers. Once you have entered all the numbers, press 2ND [}] to close the list with a right curly bracket }.
3. Press the STO▸ key to store your data in a named list. To name the list, you will use the ALPHA key to access the green letters printed above your calculator keys. Press the ALPHA key before selecting each character and type the name STATE (a list name can only contain five letters). Your Home screen should now look similar to the following:

```
{1,5,2,10,3,1,1,
4,35,2,47}→STATE
```

Note: The numbers shown in the screen are example data; the data you enter will most likely be different.

4. Now press the [ENTER] key.

g. Download the DOTPLOT program to your graphing calculator. You can either use the USB cable that links a computer to your calculator (this option requires that the TI-Connect software be installed on your computer), or download it from another calculator (of the same type), using the cable that comes with the calculator. You will use one (possibly both) of these methods to download all subsequent programs and files into your calculator.

h. Use the DOTPLOT program to re-create a dotplot of the states visited data:

1. Run the program by pressing the [PRGM] key. Select DOTPLOT from the EXEC menu, and then press the [ENTER] key twice to start the program.
2. Select **One Plot** and then enter the STATE list at the prompt. You can find the STATE list by pressing [2ND] [LIST] and then [STAT] to access the list directory. Use the up or down arrow keys to move the cursor until you locate **STATE.** *TI hint:* To jump to the names of lists that begin with *S*, press the [ALPHA] key and the [LN] key (note that the letter *S* is printed above this key).
3. Once you have selected the STATE list, press the [ENTER] key. Your screen should look like the following:

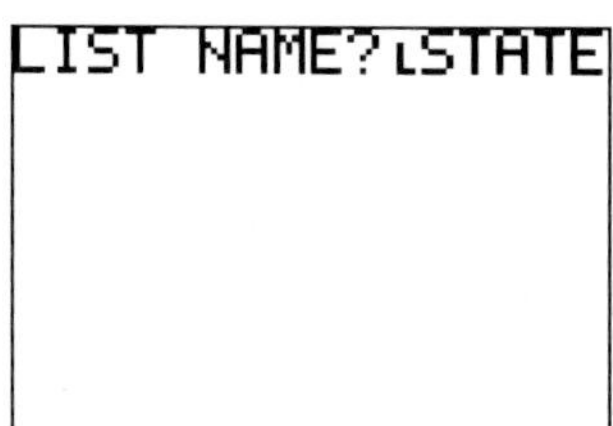

Your calculator will display the dotplot after you press [ENTER]. The granularity is a number that falls between 0 and 1, inclusive, and allows you to adjust the grouping within the dotplot. Zero will plot the true data (no grouping), and 1 will do the most grouping (depending on your data). Zero is a great setting for small datasets. The larger the dataset, the larger you might want to set the granularity. Press the [TRACE] key to show information about your plot.

4. Try different granularities (start with 0) on your data to see how the settings affect your plot. *TI note:* You might not notice any difference in the plots for small datasets.

In Topic 1, we mentioned that classifying a variable as categorical or quantitative is extremely important because that classification helps determine the type of statistical analysis to perform. You have already begun to experience this because a bar graph is the simplest graphical display of a categorical variable and a dotplot is the simplest graphical display of a quantitative variable. In other words, by identifying the type of variable, you restrict your choice of graphical displays immediately.

Activity 2-4: Buckle Up!

2-4, 3-21, 8-5

The National Highway Traffic Safety Administration (NHTSA) reports the percentage of residents in each state who regularly wear a seatbelt in a car and also whether the state has a primary or secondary type of seatbelt law. A primary law means that motorists can be stopped based solely on belt usage, whereas a secondary law means that the motorist can be stopped only for another reason. The 2005 data appear in the next table (s = secondary, p = primary, and * = not known):

State	Law	% Usage	State	Law	% Usage	State	Law	% Usage
Alabama	p	81.8	Louisiana	p	77.7	Ohio	s	78.7
Alaska	s	78.4	Maine	s	75.8	Oklahoma	p	83.1
Arizona	s	94.2	Maryland	p	91.1	Oregon	p	93.3
Arkansas	s	68.3	Massachusetts	s	64.8	Pennsylvania	s	83.3
California	p	92.5	Michigan	p	92.9	Rhode Island	s	74.7
Colorado	s	79.2	Minnesota	s	82.6	South Carolina	s	69.7
Connecticut	p	81.6	Mississippi	s	60.8	South Dakota	s	68.8
Delaware	p	83.8	Missouri	s	77.4	Tennessee	p	74.4
Florida	s	73.9	Montana	s	80.0	Texas	p	89.9
Georgia	p	81.6	Nebraska	s	79.2	Utah	s	86.9
Hawaii	p	95.3	Nevada	s	94.8	Vermont	s	84.7
Idaho	s	76.0	New Hampshire	none	*	Virginia	s	80.4
Illinois	p	86.0	New Jersey	p	86.0	Washington	p	95.2
Indiana	p	81.2	New Mexico	p	89.5	West Virginia	s	84.9
Iowa	p	85.9	New York	p	85.0	Wisconsin	s	73.3
Kansas	s	69.0	North Carolina	p	86.7	Wyoming	s	*
Kentucky	s	66.7	North Dakota	s	76.3			

a. What are the observational units for these data?

States

b. Classify each of the variables in the table as categorical (also binary) or quantitative.

usage - quantitative
law - categorical, binary

The following dotplots reveal the distribution of seatbelt usage percentages in a state, separated by type of law.

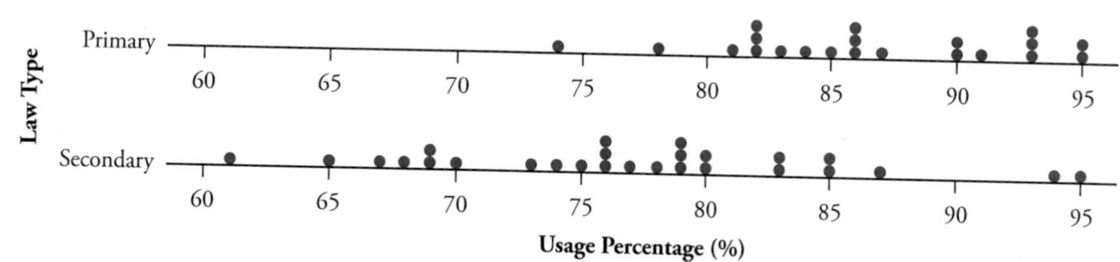

c. What would you estimate is a typical usage percentage for a state with a primary-type seatbelt law? How about a state with a secondary-type law? (Do not perform any calculations; base your answers on a casual reading of the dotplots.)

Primary: 86 Secondary: 76

d. Does a state with a primary law *always* have a higher usage percentage than a state with a secondary law? Explain. If not, identify a pair of states for which the state with a primary law has a lower usage percentage than the state with a secondary law.

Arizona & Indiana

Nevada & Tennessee

e. Do states with a primary law *tend to* have higher usage percentages than states with a secondary law? Explain how you can tell from the dotplots.

Dots are above 70%

f. Do the data seem to support the contention that tougher (primary) laws lead to more seatbelt usage? Can you draw this conclusion definitively? Explain.

Yes b/c of the law

Dotplots and bar graphs are usually most illuminating when used to *compare* the distribution of a variable between two or more groups. Questions c–f point to the concept of a **statistical tendency:** States with primary laws *tend to* have higher seatbelt usage percentages than states with secondary laws, but there are many exceptions to this general "rule." Another way to say this is that states with primary laws have higher seatbelt usage percentages *on average* than states with secondary laws. Try to get in the habit of using phrases such as "tend to" or "on average" to express statistical tendencies. You will consider whether you can draw a cause-and-effect conclusion (which would justify the use of the phrase "lead to" in part f) between tougher laws and increased usage in Topic 3.

Activity 2-5: February Temperatures

2-5, 8-19, 9-7

The following dotplots display the distributions of daily high temperatures for the month of February 2006 in Lincoln, Nebraska; San Luis Obispo, California; and Sedona, Arizona.

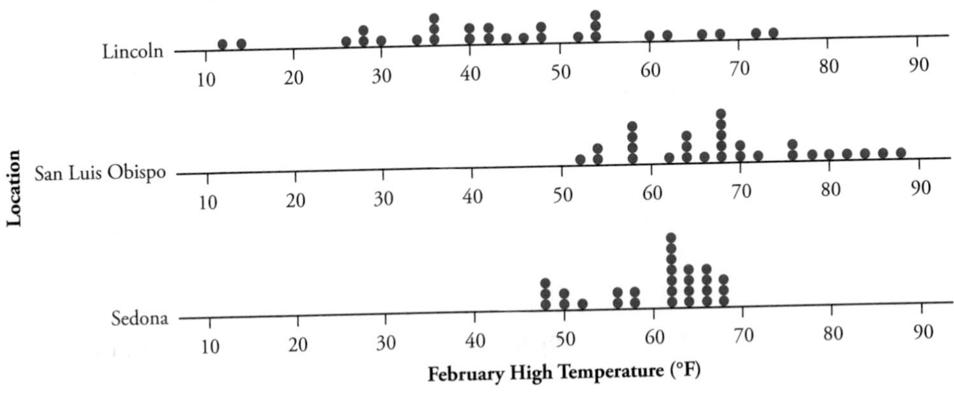

a. Which city tended to have the highest temperatures that month? Which city tended to have the lowest temperatures? Explain how you can tell.

Highest: San Luis Obispo Lowest: Lincoln

Explanation:

b. Which city had the most day-to-day consistency in its high temperatures that month? Which city had the least? Explain how you can tell.

Most: Sedona Least: Lincoln

Explanation:

Spread of Dots on graph

When comparing distributions of a quantitative variable, in addition to tendency, look for **consistency.** Tendency refers to the *center* of the distribution, consistency to the *spread* (or variability) in the distribution.

Activity 2-6: Sporting Examples

2-6, 3-4, 8-14, 10-11, 22-26

A statistics professor conducted a study in which he taught two sections of the same introductory class, with one difference between the sections. In one section, he used only examples taken from sports applications, and in the other section, he used examples taken from a variety of application areas (Lock, 2006). The sports-themed section was advertised as such, so students knew which type of section they were enrolling in. The sports-themed section was also scheduled earlier in the day than the "regular" section. At the end of the semester, the professor compared student performance in the two sections using course grades and total points earned in the course.

a. Identify the observational units in this study.

Students in Stat. classes

b. Identify three variables mentioned in the preceding paragraph. Classify each as categorical (also binary) or quantitative.

which Section: categorical, binary
Quantitative: Time of class

The following dotplots compare the distributions of total points earned in the course (out of 400 possible points) between the two sections.

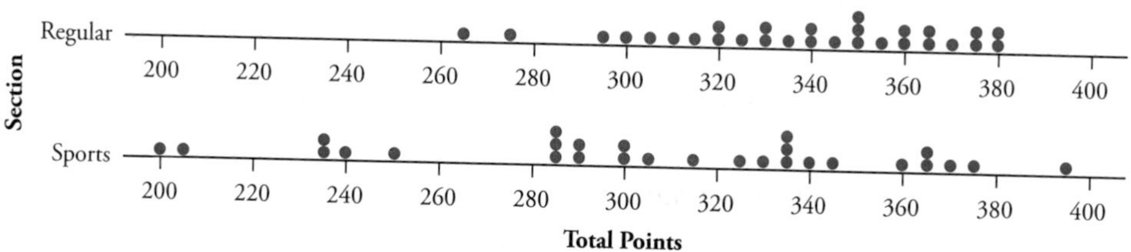

c. Compare and contrast the point distributions between the two sections. Comment on both tendency and consistency.

d. Would you say that one section tended to score more points than the other? Justify your answer.

e. Did every student in one section score more points than every student in the other section? If not, explain what a statistical tendency means in this context.

Suppose that you categorize student performance as "good" for those students earning at least 85% of the possible points, "fair" for those students earning at least 70% but less than 85%, and "poor" for those students earning less than 70%. Then the results are

Regular section:	16 good	11 fair	2 poor	29 total
Sports section:	7 good	15 fair	6 poor	28 total

f. For each section, calculate the proportion of students falling into each of the three performance categories.

g. Construct bar graphs of the distributions of performance categories for each section.

h. Comment on what these bar graphs reveal about student performance in the two sections. Are your findings consistent with your analysis of the quantitative variable (*total points*) in questions c and d?

The regular section scored higher but also had more points for the total. More scored in the good section but lower in the fair and poor sections

i. Do you think it's reasonable to conclude that the sports examples caused student performance to decrease in that section? Put more positively, do you think it's reasonable to conclude that the regular section's assortment of examples caused student performance to improve in the course? Explain.

No, we can not put causation to the testing scores

Solution

a. The observational units are the students enrolled in one or the other of these sections.

b. The variables are *section* (categorical, binary), *grade* (categorical, not binary), and *total points earned* (quantitative).

c. Students in the regular section tended to score more points than those in the sports section. Scores in the regular section appear to be centered around 340 (85% of the possible points), whereas those in the sports section are centered around 310–320 points (a bit less than 80% of the possible points). Scores in the sports section are more spread out than those in the regular section. Students in the sports section had the six lowest scores, all less than 260 points, but that section also had the highest overall score, greater than 390 points.

d. Students in the regular section tended to score more points than those in the sports section. Most students in the regular section scored between 300–380 points, with a center of approximately 340 points. In contrast, many students in the sports section scored less than 300 points, and the center was approximately 310–320 points.

e. No, some sports students scored more points than some regular students. The statistical tendency means that a typical student in the regular section scored more points than a typical student in the sports section.

f. The proportions are found by dividing the counts by 29 for the regular section and by 28 for the sports section. These proportions are

Regular section:	.552 good	.379 fair	.069 poor
Sports section:	.250 good	.536 fair	.214 poor

g. The bar graphs follow:

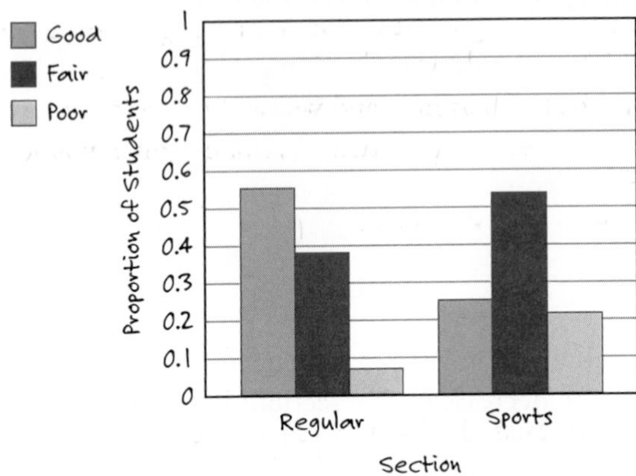

h. The bar graphs reveal similar results to the dotplots: Students in the regular section tended to score higher than those in the sports section. More than half of the regular students were in the good category, compared to only one-fourth of the students in the sports section. At the other extreme, only 6.9% of the regular students did poor work, compared to 21.4% of sports students.

i. You cannot draw a cause-and-effect conclusion between the type of section and student performance. You will study these issues again in the next topic, but one key is that students self-selected which section to take. Perhaps those who chose to take the sports section had lower academic aptitude to begin with than those who selected the regular section, or perhaps students were sleepier in the sports section because it met earlier in the day.

Watch Out

- A statistical tendency is not a hard-and-fast rule for either a quantitative or a categorical variable. Be careful with your language on this issue. For example, it's better to say that "men tend to be taller than women" than to say that "men are taller than women."
- Always label axes in bar graphs and dotplots. The label on the horizontal axis is the variable of interest. Including units (such as minutes or hours) with the label is also a good habit.
- When asked for a proportion, be sure to report a number between 0 and 1, possibly including 0 and 1. Do not confuse this with a percentage, which is a number between 0% and 100% (inclusive). Multiplying a proportion by 100 converts it to a percentage.
- Bar graphs are easier to compare if you present the vertical axis as a proportion (or percentage) rather than a raw count. For example, if a hand-washing study includes 100 men and 1000 women, then displaying the proportion in each gender who wash their hands would be much more informative than presenting the number in each gender who wash their hands.
- Be careful about drawing cause-and-effect conclusions even when you find a big difference between two groups. You will study this issue in more detail in Topics 3 and 5, and it will arise throughout the course.
- Always relate your comments to the context of the data and, ideally, to the research question of interest. One way to check yourself is to ask whether someone who only reads your comments would know that you are talking about, for example, total points earned in a statistics course or weights of football players.

Wrap Up . . . .

Graphical displays are a useful first step in analyzing the **distribution** of a variable. A **bar graph** is the simplest graph to create for a categorical variable, such as whether a person washes his/her hands when leaving a public restroom. Comparing bar graphs between two groups (such as men and women) or across several groups (such as location) can reveal interesting tendencies. For example, you learned that women tend to wash their hands more often than men, and sports fans at Atlanta's Turner Field are less likely to wash their hands than travelers or commuters at Grand Central Station.

A **dotplot** is the simplest graph to create for a quantitative variable. Dotplots can reveal not only interesting tendencies but also information about consistency (or spread). For example, you found that states with strict seatbelt laws tend to have higher percentages of people who buckle up than states with less strict laws. Note also that states with strict laws also have more consistency (less spread) in seatbelt usage percentages than states with less strict laws.

Some useful definitions to remember and habits to develop from this topic include

- Always begin to analyze data by creating a graph.
- The simplest graphs to construct and interpret are bar graphs (for categorical data) and dotplots (for quantitative data).
- Always label graphs clearly. Indicate the variable on the horizontal axis and the count, proportion, or percentage on the vertical axis.
- A **statistical tendency** refers to observational units in one group being more likely to be in a certain category (for a categorical variable) or to have higher values (for a quantitative variable) than those in another group.
- **Consistency** refers to how variable, or spread out, the values in a dataset are for a quantitative variable.
- When describing the distribution of a quantitative variable, refer to both *center* (tendency) and *spread* (consistency).
- When you describe the distribution of a variable as revealed in a graph, always relate your comments to the context of the data.

In the next topic, you will begin to consider the crucial question of what conclusions can be drawn from analyzing data, depending on how the data were collected in the first place.

Homework Activities

Activity 2-7: Student Data

1-1, 1-5, **2-7,** 2-8, 3-7, 7-8

Reconsider the variables listed in part e of Activity 1-1 on page 6. For each variable, indicate whether a bar graph or dotplot would be the appropriate graphical display.

Activity 2-8: Student Data

1-1, 1-5, 2-7, **2-8,** 3-7, 7-8

Explain what the term *statistical tendency* means as it relates to the issues mentioned in Activity 1-5.

a. Do female students tend to study more than male students?
b. Do students who study more tend to score higher on exams?

Activity 2-9: Value of Statistics

2-9, 2-10, 9-3

This activity asks you to consider the value of statistics in society.

a. How would you rate the value of statistics in society, on a scale of 1 (completely useless) to 9 (extremely important)?

b. Tally the responses of students in your class by counting how many students answered 1, how many answered 2, and so on. Create a table similar to the following to organize the data:

Rating	1	2	3	4	5	6	7	8	9
Tally (Count)	0	0	0	0	1	1	6	0	1

c. Is there one value that students chose more than any other? If so, what is it?

d. How many and what proportion of students gave a response (strictly) greater than 5? Less than 5?

e. Based on these data, write a paragraph interpreting how your class generally seems to feel about the value of statistics in society. Specifically, comment on the degree to which these students seem to be in agreement. Also address whether students seem to be generally optimistic, pessimistic, or undecided about the value of statistics.

Activity 2-10: Value of Statistics

2-9, **2-10,** 9-3

Reconsider Activity 2-9. Examine the following frequency (tally) table for hypothetical classes A–E. (Consider all empty cells counts of zero.) For each of the following descriptions, identify which of the classes fits the description best. *Hint:* Use each class letter once and only once.

Rating	1	2	3	4	5	6	7	8	9
Class A Count	1	4	7	1		1	3	5	2
Class B Count	2	3	6	5	2	1	2	3	
Class C Count						1	19	4	
Class D Count		2	3		2	5	6	2	4
Class E Count	2	3	1	3	2	4	5	1	3

i. The class is in considerable agreement that statistics is useful.

ii. The class feels generally that statistics is useful, but to varying degrees and with a few disagreements.

iii. The class displays a wide range of opinions, with a slight preference toward feeling that statistics is useful.

iv. The class is sharply divided on the issue.

v. The class feels generally that statistics is not useful but displays a range of opinions.

Activity 2-11: Quiz Scores

Suppose that 20 students in a class take four quizzes. Each quiz is worth 10 points, so a student can obtain any integer score between 0 and 10, inclusive. For each of the following descriptions, create a frequency distribution (as in Activity 2-10) of scores that fits the description. Also produce a dotplot for each distribution that you create.

a. Quiz 1: Most students scored quite well, but a few did very badly.
b. Quiz 2: All students did a mediocre job, with very little variability (much consistency) from student to student.
c. Quiz 3: Students' scores show very little consistency, running the gamut from very poor to very good and everything in between.
d. Quiz 4: Approximately half of the students did very well, but the other half did very poorly.

Activity 2-12: Responding to Katrina

2-12, 4-10, 16-13

A poll conducted by CNN, *USA Today*, and Gallup on September 8–11, 2005, asked people whether they thought race was a factor in the government's slow response to Hurricane Katrina in New Orleans. The following bar graphs compare the responses of blacks and whites.

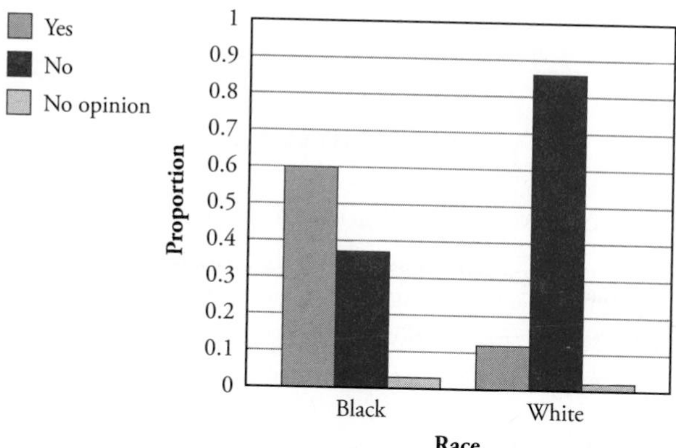

Comment on what these bar graphs reveal about the opinions of whites and blacks on whether race was a factor in the slow response to Katrina.

Activity 2-13: Backpack Weights

2-13, 10-12, 19-6, 20-17

A group of undergraduate students at Cal Poly (Mintz, Mintz, Moore, and Schuh, 2002) conducted a study examining how much weight students carry in their backpack. They weighed the backpacks of 100 students and compared the distributions of backpack weights between men and women.

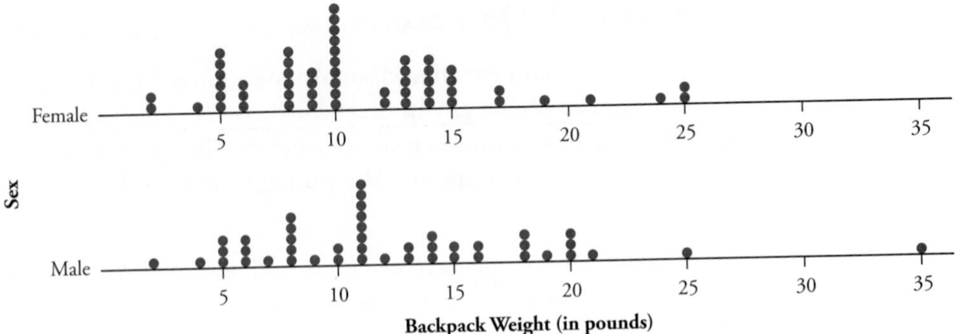

a. Compare and contrast the distribution of backpack weights between male students and female students.

b. Does one sex tend to carry more weight in their backpack than the other sex? Explain how you can tell from the dotplots.

The student researchers also recorded the body weight of each student and computed the ratio of backpack weight to body weight for each student. Here are the dotplots of these ratios.

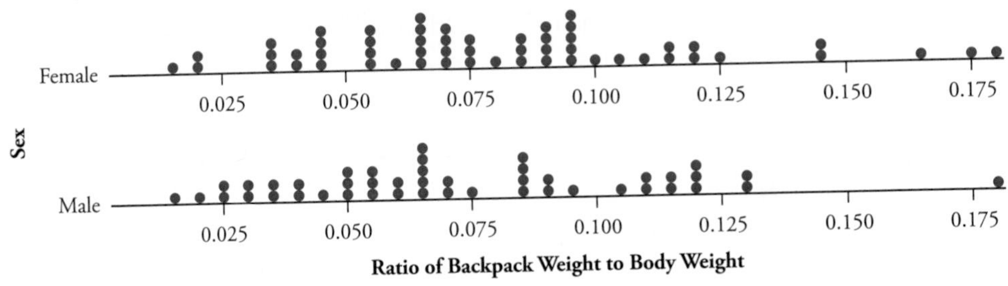

c. Compare and contrast the distribution of ratios between male students and female students.

d. Does one sex tend to carry a higher ratio of their body weight in their backpack than the other sex? Explain how you can tell from the dotplots.

e. Explain why your answers to part b and part d differ.

Activity 2-14: Broadway Shows

2-14, 26-10, 26-11, 27-15

The following dotplots pertain to the 29 Broadway shows in production for the week of June 19–25, 2006. The shows are categorized as plays and musicals.

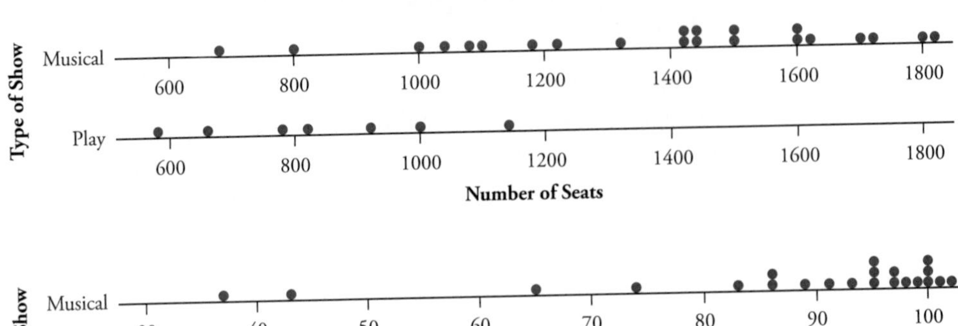

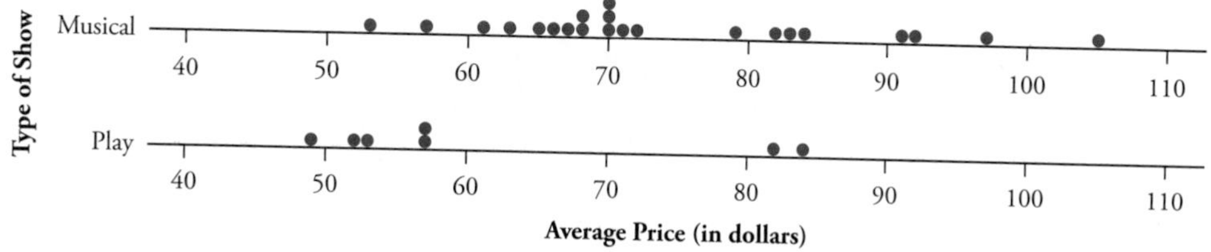

For each of these three variables (*number of seats in the theater, percentage of the theater's seats that were occupied,* and *average ticket price paid*), write a paragraph comparing and contrasting its distribution between plays and musicals.

Activity 2-15: Highest Peaks

The following dotplots display the distribution of highest points (elevation of tallest peak) in each state, separated into states that are east vs. west of the Mississippi River.

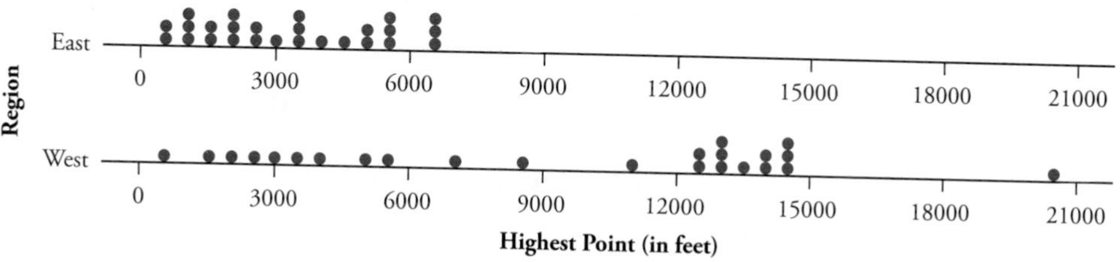

Write a paragraph commenting on the similarities and differences in these distributions of states' highest points for the two parts of the country.

Activity 2-16: Pursuit of Happiness

2-16, 3-25, 13-17, 25-1, 25-2, 25-4

The Pew Research Center conducted a survey in February of 2006 in which American adults were asked: "Generally, how would you say things are these days in your life—would you say that you are very happy, pretty happy, or not too happy?" Of the 3014 respondents, 34% answered "very happy," 50% said "pretty happy," 15% responded "not too happy," and 1% replied that they did not know. Construct a well-labeled bar graph to display these results.

Activity 2-17: Roller Coasters

2-17, 10-2

The Roller Coaster DataBase (`www.rcdb.com`) contains extensive data on roller coasters around the world. Some of the variables for which data are provided include height, length, speed, number of inversions, type (steel or wooden), and design (sit down, stand up, inverted).

a. What are the observational units for these data?
b. Classify each of the variables listed in the previous paragraph as categorical (also binary) or quantitative.

The following dotplots reveal the distributions of heights, separated by type, for 145 roller coasters in the United States.

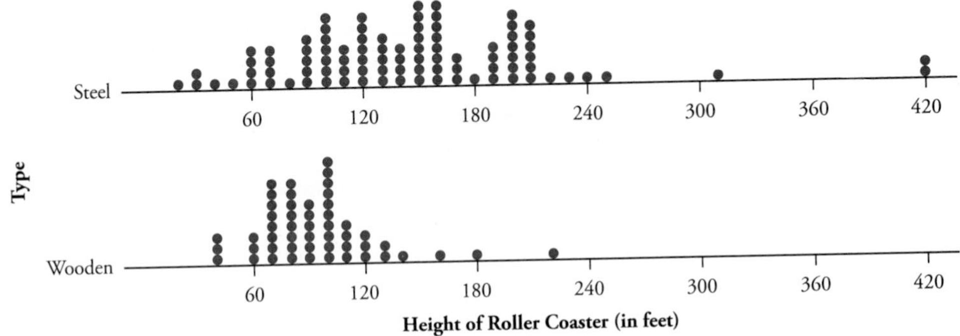

c. Comment on how the distributions of roller coaster heights compare between steel and wooden coasters.

d. What is a typical height for a steel roller coaster? What is a typical height for a wooden coaster? (Do not perform any calculations; base your answers on a casual reading of the dotplots.)

e. Does one type of roller coaster (steel or wooden) *tend to* be taller than the other type? Explain.

f. Is one type of roller coaster *always* taller than the other type? Explain.

Activity 2-18: Nicotine Lozenge

1-16, **2-18,** 5-6, 9-21, 19-11, 20-15, 20-19, 21-6, 22-8

A study reported in the *Archives of Internal Medicine* reported on the effectiveness of a nicotine lozenge for smokers who wanted to quit (Shiffman et al., 2002). One group of smokers took the nicotine lozenge and a control group took a placebo lozenge with no active ingredient. The following bar graphs display the breakdown of quitting successfully, comparing the nicotine and placebo groups, after 6 weeks and again after 52 weeks.

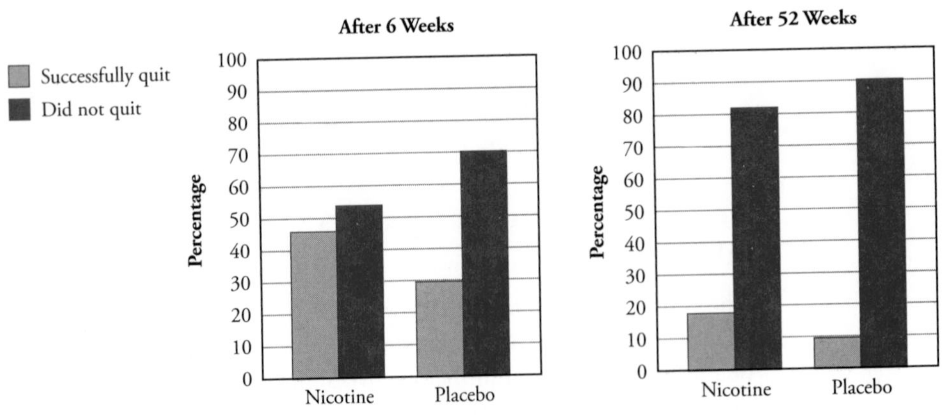

Write a paragraph or two commenting on what these graphs reveal. Be sure to address two questions: Do smokers using the nicotine lozenge tend to quit more often than those using the placebo, and how do the success rates compare after 6 weeks and after 52 weeks?

Activity 2-19: Candy Colors

1-13, **2-19,** 13-1, 13-2, 15-7, 15-8, 16-4, 16-22, 24-15, 24-16

A class of statistics students examined the distribution of colors in a bag of Reese's Pieces. They found 273 orange, 154 brown, and 114 yellow candies.

a. Identify the observational units and variable. Also classify the variable as categorical (also binary) or quantitative.

b. Produce the relevant graph (bar graph or dotplot) to display the distribution of colors observed by these students. Also comment on what the graph reveals.

c. Examine the distribution of colors in another bag of Reese's Pieces. Create a graph to compare your distribution to the one reported in part b. Comment on how similar or different the distributions are.

TOPIC

3 • • • Drawing Conclusions from Studies

There is considerable concern about the issue of young people injuring themselves intentionally. Can you use statistics to better understand the seriousness of this problem, by estimating the proportion of college students who have attempted to injure themselves? On an issue of less societal importance, can you estimate what proportion of people believe that Elvis Presley faked his widely reported death, and would it matter which people you asked? Or consider a different kind of question, which may appear whimsical but may prove important: Do candy lovers live longer than other people? If so, is candy a secret to long life? In this topic, you will begin to study issues related to these questions, focusing on concerns that limit the scope of conclusions you can draw from some statistical studies.

Overview

You have begun to understand that data can be useful for gaining insights into interesting questions. But to what extent can statistics provide answers to these questions? This topic begins your introduction to key concepts that determine the scope of conclusions you can draw from a study. For example, when can you *generalize* the results of a study to a larger group than those used in the study itself? Also, why can't you always conclude that one variable *affects* another when a study shows a relationship between the variables?

As you consider those questions, you will encounter some more fundamental terms, such as population and sample, parameter and statistic, and explanatory and response variables. You will also study the important concepts of bias and confounding, and you will begin to understand why those concepts sometimes limit the scope of conclusions you can draw.

Preliminaries

1. Do you believe that Elvis Presley faked his death on August 16, 1977?

2. Take a guess for the percentage of adult Americans who believe that Elvis Presley faked his death.

3. Guess the percentage of American college students who have ever injured themselves intentionally.

35%

4. Would you say that you consume candy rarely, sometimes, or often?

Sometimes

In-Class Activities

Activity 3-1: Elvis Presley and Alf Landon

3-1, 3-6, 16-5

Elvis Presley is reported to have died in his Graceland mansion on August 16, 1977. On the 12th anniversary of this event, a Dallas record company wanted to learn the opinions of all adult Americans on the issue of whether Elvis was really dead.

But of course they could not ask every adult American this question, so they sponsored a national call-in survey. Listeners of more than 100 radio stations were asked to call a 1-900 number (at a charge of $2.50) to voice an opinion concerning whether Elvis was really dead. It turned out that 56% of the callers thought that Elvis was alive.

This scenario is very common in statistics: wanting to learn about a large group based on data from a smaller group.

The **population** in a study refers to the *entire* group of people or objects (observational units) of interest. A **sample** is a (typically small) *part* of the population from whom or about which data are gathered to learn about the population as a whole. If the sample is selected carefully, so it is **representative** of (has similar characteristics to) the population, you can learn useful information. The number of observational units (people or objects) studied in a sample is the **sample size.**

a. Identify the population and sample in this study.

Population: Adult Americans

Sample: Radio Station Listeners & Callers

b. Do you think that 56% accurately reflects the opinions of all Americans on this issue? If not, identify some of the flaws in the sampling method.

In 1936, *Literary Digest* magazine conducted the most extensive (to that date) public opinion poll in history. They mailed out questionnaires to over 10 million people whose names and addresses they had obtained from telephone books and vehicle registration lists. More than 2.4 million people responded, with 57% indicating they would vote for Republican Alf Landon in the upcoming presidential election. (Incumbent Democrat Franklin Roosevelt won the actual election, carrying 63% of the popular vote.)

c. Identify the population of interest and the sample actually used to study that population in this poll.

Population: Voters

Sample: People in Phone books or who had vehical registration

d. Explain how *Literary Digest*'s prediction could have been so much in error. In particular, comment on why its sampling method made it vulnerable to overestimating support for the Republican candidate.

In both the Elvis study and the *Literary Digest* presidential election poll, the goal was to learn something about a very large population (all American adults and all American registered voters, respectively) by studying a sample. However, both studies used a poor method of selecting the sample from the population. In neither case was the sample representative of the population, so you could not accurately infer anything about the population of interest from the sample results. This is because the sampling methods used were *biased.*

A sampling procedure is said to have **sampling bias** if it tends systematically to overrepresent certain segments of the population and to underrepresent others.

These scenarios also indicate some common problems that produce biased samples. Both are **convenience samples** to some extent because they reach those people most readily accessible (e.g., those listening to the radio station or listed in the phone book). Another problem is **voluntary response,** which refers to samples collected in such a way that members of the population decide for themselves whether or not to participate in the study. For instance, radio stations asked listeners to call in if they wanted to participate. The related problem of **nonresponse** can arise even if an unbiased sampling method is used (e.g., those who are not home when the survey is conducted may have longer working hours than those who participate). Furthermore, the **sampling frame** (the list used to select the subjects) in the *Literary Digest* poll was not representative of the population in 1936 because the wealthier segment of society was more likely to have vehicles and telephones, which overrepresented those who would vote Republican.

A **parameter** is a number that describes a population, whereas a **statistic** is a number that describes a sample.

Note: To help keep this straight, notice that population and parameter start with the same letter, as do sample and statistic.

e. Identify each of the following as a parameter or a statistic:

- The 56% of callers who believed that Elvis was alive

  Statistic

- The 57% of voters who indicated they would vote for Alf Landon

  Statistic

- The 63% of voters who actually voted for Franklin Roosevelt

  Parameter

f. Consider the students in your class as a sample from the population of all students at your school. Identify each of the following as a parameter or a statistic:

- The proportion of students in your class who use instant-messaging or text-messaging on a daily basis

  Statistic

- The proportion of students at your school who use instant-messaging or text-messaging on a daily basis

  Parameter

- The average number of hours students at your school spent watching television last week

  Parameter

- The average number of hours students in your class slept last night

  Statistic

g. Identify each of the following as a parameter or a statistic. If you need to make an assumption about who or what the population of interest is in a given case, explain that.

- The proportion of voters who voted for President Bush in the 2004 election

  Statistic

- The proportion of voters surveyed by CNN who voted for John Kerry in the 2004 election

  Statistic

- The proportion of voters among your school's faculty members who voted for Ralph Nader in the 2004 election

  Statistic

- The average number of points scored in a Super Bowl game

  Parameter

h. What type of variable leads to a parameter or statistic that is a proportion? What type of variable leads to a parameter or statistic that is an average? *Hint:* Review your answers to the last few questions and look for a pattern.

Proportion: Categorical Average: quantitative

Watch Out

- A categorical variable leads to a parameter or statistic that is a proportion, whereas a quantitative variable usually leads to a parameter or statistic that is an average.
- Many students confuse a parameter with a population and a statistic with a sample. Remember that parameters and statistics are *numbers,* whereas populations and samples are groups of observational units (people or objects).
- If you believe that a sampling method is biased, suggest a likely direction for that bias. Do not simply say "there is bias" or "the results will be off" or "the results could be higher or lower." Remember, bias is a systematic tendency to err in a particular direction, not just to err.

Activity 3-2: Self-Injuries

An article published in the June 6, 2006, issue of the journal *Pediatrics* describes the results of a survey on the topic of college students injuring themselves intentionally (Whitlock, Eckenrode, and Silverman, 2006). Researchers invited 8300 undergraduate and graduate students at Cornell University and Princeton University to participate in the survey. A total of 2875 students responded, with 17% of them saying that they had purposefully injured themselves.

a. Identify the observational units and variable in this study. Also, classify the variable as categorical (also binary) or quantitative.

Observational units:

Variable: Type:

b. Identify the population and sample.

Population: Sample:

c. What is the sample size in this study?

d. Is 17% a parameter or a statistic? Explain.

e. Do you think it likely this sample is representative of the population of all college students in the world? What about of all college students in the U.S.? Explain.

Activity 3-3: Candy and Longevity

3-3, 21-27

Newspaper headlines proclaimed that chocolate lovers live longer, following the publication of a study titled "Life Is Sweet: Candy Consumption and Longevity" in the *British Medical Journal* (Lee and Paffenbarger, 1998). In 1988, researchers sent a health questionnaire to men who entered Harvard University as undergraduates between 1916 and 1950. They then obtained death certificates for those who died by the end of 1993.

a. With regard to learning about the health habits in the population of all adult Americans, do you consider this a representative sample? Explain.

Can't be, all men

Researchers found that 3312 of the respondents said that they almost never consumed candy, whereas 4529 did consume candy.

b. Determine the proportion of respondents who consumed candy. Is this a parameter or a statistic?

57%, statistic

Of the 3312 nonconsumers of candy, 247 had died by the end of 1993, compared to 267 of the 4529 consumers of candy.

c. Calculate the proportion of deaths in each group.

Nonconsumers: Consumers:

The variable whose effect you want to study is the **explanatory variable.** The variable that you suspect is affected by the other variable is the **response variable.**

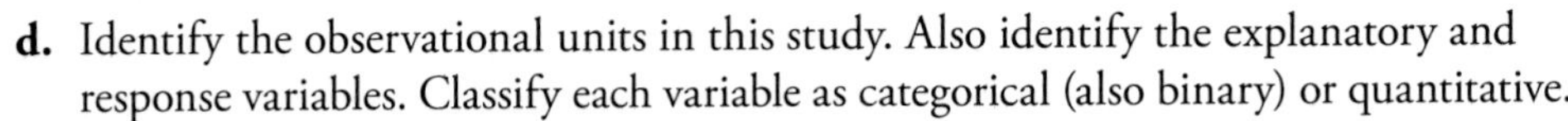

d. Identify the observational units in this study. Also identify the explanatory and response variables. Classify each variable as categorical (also binary) or quantitative.

Observational units:

Explanatory: Type:

Response: Type:

The researchers went on to show that this difference in proportions is too large to have reasonably occurred by chance. They also used more sophisticated analyses to estimate that candy consumers in this study would enjoy 0.92 added years of life, compared with nonconsumers.

e. Even if you focus on this group of males who attended Harvard and not some larger population, it is not reasonable to conclude that candy consumption caused the lower death rate and the higher longevity. Provide an alternative explanation for why candy consumers might live longer than nonconsumers.

In part e, you identified a possible second difference between the candy consumers and the nonconsumers, which often happens when the individuals self-select into the explanatory variable groups. Whenever a second variable changes between the explanatory variable groups, you cannot conclude that the explanatory variable causes an effect on the response variable. You have no way of knowing whether the explanatory variable or some other variable led to the different response variable outcomes in the two groups.

In an **observational study,** researchers passively observe and record information about observational units. An observational study may establish an association or relationship between the explanatory and response variables, but you cannot draw a cause-and-effect conclusion between the explanatory and response variables from an observational study.

The researchers provided data on more health-related variables in this study. Among the 3312 nonconsumers of candy, 1201 had never smoked, compared to 1852 who had never smoked among the 4529 consumers.

f. Among the nonconsumers, calculate the proportion who never smoked. Then do the same for the candy consumers.

Nonconsumers: Consumers:

g. Comment on what the calculations in part f reveal, and how they might help to explain why candy consumers in this study tended to live longer than nonconsumers.

An observational study does not control for the possible effects of variables that are not considered in the study but could affect the response variable. These unrecorded variables are called **lurking variables.** Lurking variables can have effects on the response variable that are confounded with those of the explanatory variable. A **confounding variable** is a lurking variable whose effects on the response variable are indistinguishable from the effects of the explanatory variable.

When confounding variables are present, even if you observe a difference in the response variable between treatment groups, you have no way of knowing which variable (explanatory or confounding), or some combination of the two, is responsible because the treatment groups differ in more ways than simply the explanatory variable. Thus, you cannot draw cause-and-effect conclusions from observational studies because *confounding* might provide an alternative explanation for any observed relationship.

In this study, the person's smoking status is confounded with his candy consumption because those who consumed candy were less likely to smoke than those who did not consume candy. Because smoking status is known to be associated with longevity, you cannot say whether it was the lack of smoking or the candy consumption, or both (or something else entirely) that led to a tendency to live longer in the candy-consuming group.

In order to conduct a study in which you can draw a cause-and-effect conclusion, you have to impose the explanatory variable on subjects (i.e., assign the subjects to "treatment groups") in such a way that the groups are nearly identical except for the explanatory variable (eating candy or not, for example). Then, if the groups are found to differ substantially on the response variable as well, you can attribute that difference to the explanatory variable. You will study strategies for designing such a study and assigning subjects to treatment groups in Topic 5.

Note that the issue of generalizing results from a sample to a larger population is a completely different issue than drawing a cause-and-effect conclusion. In this study, it might be reasonable to generalize the finding about a relationship between candy and longevity to all males who attend Ivy League colleges. But it might not be safe to generalize to all males because those who attend Harvard probably have access to better healthcare and other advantages. It would certainly be risky to generalize this finding to women, as their bodies may respond differently to candy than men's bodies do. In Topic 4, you will learn how to select a sample from a population so that it is likely to be representative.

Activity 3-4: Sporting Examples

2-6, **3-4,** 8-14, 10-11, 22-26

Recall from Activity 2-6 that a statistics professor compared academic performance between two sections of students: one taught using sports examples exclusively and the other taught using a variety of examples. The sections were clearly advertised, and students signed up for whichever section they preferred. The sports section was offered at an earlier hour of the morning than the regular section. The professor found that the students taught using sports examples exclusively tended to perform more poorly than students taught with a variety of examples.

a. Identify the observational units and explanatory and response variables. Also classify the variables' type.

Observational units: Students

Explanatory variable: examples Sports (non sports) used — Type: categorical binary

Response variable: Performance — Type: categorical

b. Explain why this is an *observational* study.

Cause & effect conclusion can not be drawn

- Students had the choice to pick either class

c. Is it legitimate to conclude that the sports examples *caused* the lower academic performance from students? If so, explain. If not, identify a potential confounding variable and explain why it is confounded with the explanatory variable. *Hint:* Describe how the confounding variable provides an alternative explanation for the observed difference in academic performance between the two groups. Be sure to explain the connection of your proposed confounding variable to both the explanatory variable and the response variable.

√

Activity 3-5: Childhood Obesity and Sleep

A March 2006 article in the *International Journal of Obesity* described a study involving 422 children aged 5–10 from primary schools in the city of Trois-Rivieres, Quebec, (Chaput, Brunet, and Tremblay, 2006). The researchers found that children who reported sleeping more hours per night were less likely to be obese than children who reported sleeping fewer hours.

a. Identify the explanatory and response variables in this study. Also classify them.

Explanatory: hours of sleep Type: quant.

Response: obesity Type: categ.

b. Is it legitimate to conclude from this study that less sleep caused the higher rate of obesity in Quebec children? If so, explain. If not, identify a confounding variable and explain why its effect on the response is confounded with that of the explanatory variable.

- No b/c it was only a regional study & Observational study

Confounding variables: diet, extracurriculars family history, what grade they are in, growth spurt

c. Do you think that the study's conclusion (of a relationship between sleep and obesity) applies to children outside of Quebec? Explain.

Perhaps, even though it's regional

Can't apply to other places different lifestyle

Solution

a. The explanatory variable is the amount of sleep that a child gets per night. This is a quantitative variable, although it would be categorical if the sleep information were reported only in intervals. The response variable is whether the child is obese, which is a binary categorical variable.

b. This is an observational study because the researchers passively recorded information about the child's sleeping habit. They did not impose a certain amount of sleep on children. Therefore, it is not appropriate to draw a cause-and-effect conclusion that less sleep causes a higher rate of obesity. Children who get less sleep may differ in some other way that could account for the increased rate of obesity. For example, amount of exercise could be a confounding variable. Perhaps children who exercise less have more trouble sleeping, in which case exercise would be confounded with sleep. You have no way of knowing whether the higher rate of obesity is due to less sleep or less exercise, or both, or some other variable that is also related to both sleep and obesity.

c. The population from which these children were selected is apparently all children aged 5–10 in primary schools in the city of Trois-Rivieres. These Quebec children might not be representative of all children in this age group worldwide, so you should be cautious about generalizing that a relationship between sleep and obesity exists for children around the world.

Watch Out

- Confounding is a tricky concept to grasp. Many students find it especially hard to express a confounding variable as a legitimate variable, and many also neglect to explain its connection both to the explanatory variable and to the response variable. For example, in the candy and longevity study, a confounding variable is a man's smoking habit/status. It is not enough to say that this variable is confounding because smokers tend not to live as long; you also need to note that candy consumers are less likely to smoke than nonconsumers.
- After learning that cause-and-effect conclusions cannot be drawn from observational studies, some students overreact and believe that observational studies are useless. On the contrary, an observational study can still be interesting and important by establishing a relationship between two variables, even though you cannot draw a cause-and-effect conclusion. For example, although the sleep and obesity study does not establish a cause-and-effect relationship between sleep and obesity, the connection is still interesting and important for children and parents to know about. Of course, there may be a cause-and-effect relationship between sleep and obesity; the point is simply that an observational study cannot establish this conclusion.

Wrap Up ....

This topic introduced you to two sets of key terms that will recur throughout this course: population and sample, parameter and statistic. The population of interest in the self-injury study might well be all college students in the U.S., but the sample consisted only of those students at Cornell and Princeton who responded to a survey. The parameter was then the proportion of students who had tried to injure themselves among all college students, whereas the statistic was the proportion of students who admitted to having injured themselves among those Cornell and Princeton students who responded to the survey. Another set of important terms is explanatory and response variables. For example, whether or not a person consumed candy regularly is an explanatory variable, and whether or not the person survived the five-year study period is a response variable whose behavior you think might be *explained* by the candy-consumption habits. Although we have introduced many new terms in this topic, it is important to learn them quickly as they recur throughout the course.

You have also learned about two things that can prevent you from drawing certain conclusions, bias and confounding. Sampling bias occurs when the sampling method tends to produce samples that are not representative of the population, in which case you cannot *generalize* findings in the sample to the larger population. The *Literary Digest*'s sampling method was biased: It overrepresented support for the Republican challenger Alf Landon by sampling only those (wealthier) Americans with a telephone or vehicle, and also because the sample only consisted of those who voluntarily chose to respond. Confounding can always occur with observational studies, precluding you from drawing cause-and-effect conclusions because the groups you are comparing might differ in more ways than just the explanatory variable. For example, smoking status was a confounding variable in the candy study, because smokers were less likely to eat candy than nonsmokers were. Thus, you cannot tell whether the increased longevity of candy lovers is due to the candy or to not smoking. This is an issue even if you are not trying to generalize the results to a larger population.

Most importantly, you have begun to consider two key questions to ask of statistical studies:

- To what population can you reasonably generalize the results of a study?
- Can you reasonably draw a cause-and-effect connection between the explanatory variable and the response variable?

The answer to the first question depends on how the sample was selected. The answer to the second question depends on whether or not the explanatory variable was assigned to the observational units.

Some useful definitions to remember and habits to develop from this topic include

- The **population** is the entire collection of observational units (people or objects) of interest; the **sample** is the part of the population on which you gather data.
- A **parameter** is a number that describes a population; a **statistic** is a number that describes a sample. Notice that these are numbers, not people or objects as population and sample are.
- An **explanatory variable** is one whose effect you study; a **response variable** is the outcome you record.
- **Sampling bias** is the systematic tendency of a sampling method to overrepresent some parts of the population and underrepresent others.
- An **observational study** is one in which researchers record information passively, without attempting to impose the explanatory variable on the observational units.
- A **confounding variable** is one whose effect on the response variable cannot be distinguished from the effect of the explanatory variable. To properly identify a confounding variable, discuss how it is related to both the explanatory and the response variables.
- You cannot draw cause-and-effect conclusions from observational studies because other factors (confounding variables) might differ between the groups.

In the next two topics, you will learn how to answer two key scope-of-conclusion questions, generalizability and causation. You will also learn how to design studies well in the first place, so that you can make generalizations and/or draw causal conclusions. More specifically, you will discover how to select a sample from the population of interest (Topic 4) and how to assign subjects to treatment groups (Topic 5). Both of these design considerations involve the concept of randomness, but always keep in mind that these are two separate issues.

Homework Activities

Activity 3-6: Elvis Presley and Alf Landon

3-1, **3-6**, 16-5

The question of whether Elvis Presley faked his death in 1977 has also been asked on an Internet site called `misterpoll.com`, where anyone can post a poll question and get responses from whoever sees the site and chooses to respond.

a. Is this sampling method unbiased for estimating the proportion of adults who believe that Elvis faked his death in 1977? If not, identify the direction in which you expect the sample to be biased. Explain your answer.

Of the 2032 responses posted by August 1, 2006, 50% responded that Elvis had faked his death.

b. Is this number a parameter or a statistic? Explain.
c. Do you want to rethink your answer to part a in light of this sample result? Explain.
d. Identify the sample size. Would taking a larger sample in the same manner help reduce bias? Explain.

Activity 3-7: Student Data

1-1, 1-5, 2-7, 2-8, **3-7,** 7-8

Consider your class of students as a sample from the population of all students at your school.

a. For each of the following variables, indicate whether or not your class would likely constitute a *representative* sample from this population with regard to that variable. Explain your answers.

- Grade point average
- Hours slept last night
- Number of siblings
- Whether you prefer to call or to send a text/instant message to friends
- Political viewpoint
- Gender
- Total income that you will receive over your lifetime

b. Think of two additional variables for which you believe your class is a representative sample from this population. Explain your choices.
c. Think of two additional variables for which you believe your class is not a representative sample from this population. Explain your choices.

Activity 3-8: Generation M

3-8, 4-14, 13-6, 16-1, 16-3, 16-7, 18-1, 21-11, 21-12

The generation of children raised in the 1990s and 2000s has been dubbed "Generation M" because of the impact of media on their lives. The Kaiser Family Foundation conducted an extensive study to quantify how much time teenagers spend with various types of media. They gathered data from what they describe as a "nationally representative sample" of 2032 teenagers. Identify each of the following as a parameter or a statistic:

a. The proportion of all American teenagers who have a television in their bedrooms
b. The proportion of these 2032 teenagers who have a television in their bedrooms
c. The proportion of boys in this sample who have a television in their bedrooms
d. The average number of hours per week the teenagers in this study spent reading
e. The average number of hours per week of recreational computer use among all American teenagers
f. The proportion in this study who rated their level of contentedness as high
g. The proportion who rated their contentedness as high among all American teenagers

h. The average amount of weekly media exposure among those surveyed who rate their level of contentedness as high

Activity 3-9: Community Ages

Suppose that you want to estimate the average age of the residents in your community.

a. Is this number a parameter or a statistic? Explain.
b. Suppose that you take a sample of people attending a local church on Sunday morning. Would you expect this sampling method to be biased for estimating the average age of community residents? If so, in which direction? Explain.
c. Suppose that you take a sample of drivers as they arrive at a local daycare facility. Would you expect this sampling method to be biased for estimating the average age of community residents? If so, in which direction? Explain.

Activity 3-10: Penny Thoughts

2-1, **3-10,** 16-23

On July 15, 2004, the Harris Poll released the results of a survey asking whether people favored or opposed abolishing the penny. Of a national sample of 2136 adults, 59% opposed abolishing the penny. Explain what is wrong with each of the following statements:

a. The population is the 2136 adults contacted by the Harris Poll.
b. The sample is the 59% who oppose abolishing the penny.
c. The variable is the 59% who oppose abolishing the penny.
d. The observational units in this survey are pennies.
e. The parameter consists of all American adults.
f. The statistic is the average number of adult Americans who oppose abolishing the penny.

Activity 3-11: Class Engagement

Suppose that you want to investigate the conjecture that statistics students at your school are more actively engaged in class than calculus students. You regularly attend classes and rate the level of student engagement based on the number of questions asked by students and the number of students who offer responses to instructor questions. You do this for Professor Newton's 8 AM calculus class and for Professor Fisher's 11 AM statistics class.

a. If the statistics class has a substantially higher level of student engagement than the calculus class, can you attribute the difference to the subject matter? Explain, based on how the study was conducted.
b. Identify two confounding variables in this study, being sure to identify them clearly as variables.

Activity 3-12: Web Addiction

The August 23, 1999, issue of the *Tampa Tribune* reported on a study involving data volunteered by 17,251 users of the `abcnews.com` Web site. Users of the site were asked to respond to questions including whether or not they used the Internet to escape problems, tried unsuccessfully to cut back their Internet use, and found themselves preoccupied with the Internet even when not using it. Almost 6% of those responding confessed to some sort of addiction to the Internet.

a. Identify the population and sample used in this study.
b. The number 6% applies to the sample and so is a statistic. Identify, in words, the corresponding parameter of interest in this study.
c. Do you think that 6% is a reasonable estimate of the parameter you defined in part b? If not, indicate whether you think this estimate is too high or too low. Explain.

Activity 3-13: Alternative Medicine

In a spring 1994 issue, *Self* magazine reported that 84% of its readers who responded to a mail-in poll indicated that they had used a form of alternative medicine, such as acupuncture, homeopathy, or herbal remedies. Comment on whether or not this sample result is representative of the population of all adult Americans. Do you suspect that the sampling method is biased? If so, is the sample result likely to overestimate or underestimate the proportion of all adult Americans who have used alternative medicine? Explain your answers.

Activity 3-14: Courtroom Cameras

An article appearing in the October 4, 1994, issue of *The Harrisburg Evening-News* reported that Judge Lance Ito (who was trying the O.J. Simpson murder case) had received 812 letters from around the country on the subject of whether or not to ban cameras from the courtroom. Of these 812 letters, 800 expressed the opinion that cameras should be banned.

a. What proportion of this sample supports a ban on cameras in the courtroom? Is this number a parameter or a statistic?
b. Do you think that this sample represents the population of all American adults? Comment on the sampling method.

Activity 3-15: Junior Golfer Survey

In the "Golf Plus" section of its August 21, 2006, issue, *Sports Illustrated* presented the results of its Junior Golfer Survey. Participants in the survey were the 72 golfers, 36 boys and 36 girls aged 13–18, who had played in a recent American Junior Golf Association Tournament.

a. Would you consider this to be a representative sample from the population of all American teenagers? Explain why or why not.
b. One of the questions asked was, "If you could vote for President, would you be more likely to vote for a Democrat or a Republican?" Would you expect this sampling procedure to be biased with regard to this variable? If so, do you think the bias will overestimate support for Democrats or Republicans? Explain.
c. In response to the question posed in part b, 13 responded "Democrat," compared to 48 for "Republican," 4 for "neither," and 7 for "don't know." Construct an appropriate graph of these responses, and describe the distribution. Also, comment on whether you want to rethink your answer to part b in light of these data.

d. Two other questions asked of these golfers were

- How many hours are you online during a typical week?
- Do you have your own cell phone?

Would you expect this sampling procedure to be biased with regard to this variable? If so, indicate the direction of the bias, and explain your answers.

/Activity 3-16: Accumulating Frequent Flyer Miles

Frequent flyers can accumulate their miles to receive free flights and other benefits. This practice has become so widespread that consumers can now accumulate miles through means other than flying, for example by using a credit card affiliated with an airline's frequent flyer program. On September 28, 2005, the Web site msnbc.com published an article about how consumers can redeem frequent flyer miles from airlines. Within the article was a link to participate in an online poll that asked, "Do you use a credit card to accumulate airline miles?" Of the 1935 responses that had been received when we checked, 83% had answered "yes."

a. Identify the observational units and variable. Also, classify the variable as categorical or quantitative.
b. Is 83% a parameter or a statistic? Explain.
c. Comment on whether this sampling method is unbiased for estimating the proportion of all American adults who use a credit card to accumulate airline miles. If you think this sampling method is biased, indicate whether you think it will overestimate or underestimate the population proportion. Explain your answer.
d. Identify the sample size, and comment on whether, and if so how, it affects your answer to part c.

Activity 3-17: Foreign Language Study

3-17, 5-11

Studies often show that students who study a foreign language in high school tend to perform better on the verbal portion of the SAT exam than students who do not study a foreign language.

a. Do you think these are observational studies? Explain.
b. Is it legitimate to conclude that foreign language study causes an improvement in students' verbal abilities? If so, explain. If not, identify a confounding variable and explain why it is confounded with the explanatory variable. *Hint:* Describe how the confounding variable provides an alternative explanation for the observed difference in SAT scores between the two groups.

Activity 3-18: Smoking and Lung Cancer

3-18, 3-19

Many early studies on the relationship between smoking and lung cancer in the 1950s found that smokers were about 13 times more likely to die from lung cancer than nonsmokers. Still, people argued against a cause-and-effect conclusion, citing numerous possible confounding variables. Suppose a student argues that these studies are not convincing evidence because the researchers did not record the diets of the individuals. What more does the student need to say in order to provide a complete explanation of why diet is a potentially confounding variable?

Activity 3-19: Smoking and Lung Cancer

3-18, **3-19**

One of the early studies of the relationship between smoking and lung cancer was conducted by Hammond and Horn (1958). They and an extensive set of volunteers tracked 187,766 men over a 44-month period, noting their smoking habits and whether or not they died of lung cancer.

a. Identify the explanatory and response variables in this study.
b. Is this an observational study? Explain.
c. Would you have any qualms about generalizing the results of this study to the population of all adults? Explain.

Activity 3-20: A Nurse Accused

1-6, **3-20,** 6-10, 25-23

Recall from Activity 1-6 that nurse Kristin Gilbert was accused of murdering hospital patients. Hospital records showed that of 257 eight-hour shifts on which Gilbert worked, a patient died on 40 of those shifts (15.6%). But of 1384 eight-hour shifts on which Gilbert did not work, a patient died on only 34 of those shifts (2.5%).

a. Identify the observational units, explanatory variable, and response variable.
b. Is this an observational study? Explain.
c. Based on the type of study, can you legitimately conclude that Gilbert caused the higher death rate on her shifts?
d. Put yourself in the role of the defense attorney. Suggest a confounding variable that offers an alternative explanation (other than Gilbert's guilt) for the observed relationship between her shifts and a higher death rate. Also explain, as if to a jury, how this confounding variable provides an alternative explanation.

Activity 3-21: Buckle Up!

2-4, **3-21,** 8-5

In Activity 2-4, you found that states with tougher seatbelt laws tended to have a higher proportion of residents who complied with the law to wear a seatbelt.

a. Is this an observational study? Explain how you know.
b. Can you conclude from these data that the tougher seatbelt laws cause a higher proportion of residents to comply?
c. Even if no cause-and-effect conclusion can be drawn, do the data suggest a potential benefit of tougher seatbelt laws? Explain.

Activity 3-22: Yoga and Middle-Aged Weight Gain

Researchers studied 15,500 healthy middle-aged adults, asking them to complete extensive surveys detailing their physical activities and weight gain/loss between the ages of 45 and 55. They found that those who practiced yoga had substantially lower weight gain than those who did not practice yoga.

a. Identify and classify the explanatory and response variables in this study.
b. Is this an observational study? Explain how you can tell.
c. Does this type of study allow for drawing a cause-and-effect conclusion between practicing yoga and gaining less weight? Explain.
d. Suggest a confounding variable that provides an alternative explanation for the observed relationship between practicing yoga and gaining less weight. (State this

as a variable, and make a plausible argument for how this confounding variable is related to both the explanatory and response variables.)

Activity 3-23: Pet Therapy

3-23, 5-13

Suppose that you want to study whether having a pet is beneficial to patients recovering from a heart attack. You obtain records of such patients from a local hospital and select a sample for your study. For each patient, you record whether he/she has a pet, and then you follow the patients for five years and see which patients survive.

a. Is this an observational study? Explain.

b. Identify the explanatory and response variables. Also classify them as categorical (also binary) or quantitative.

c. If you find that pet owners survive at a higher rate than non–pet owners, can you conclude that pet ownership leads to therapeutic benefits for heart attack patients? Explain.

Activity 3-24: Winter Heart Attacks

Studies conducted in New York City and Boston have noted that more heart attacks occur in December and January than in all other months. Some people have tried to conclude that holiday stress and overindulgence cause the increased risk of heart attack.

a. Identify a confounding variable whose effect on heart attack rate might be confounded with that of the month variable, providing an alternative explanation for the increased risk of heart attack in December and January.

A more recent study in Los Angeles revealed a similar finding.

b. Identify a potential confounding variable in the Boston and New York City studies that was eliminated from consideration in the Los Angeles study.

c. Identify another confounding variable that still pertains to the Los Angeles study. *Hint:* You might think of a variable that would be eliminated by conducting the study in the southern hemisphere.

Activity 3-25: Pursuit of Happiness

2-16, **3-25,** 13-17, 25-1, 25-2, 25-4

Reconsider Activity 2-16. The Pew Research Center conducted a survey in February of 2006 in which American adults were asked how happy they were with their lives. Among those with a family income of less than $30,000, only 24% responded that they were very happy. This percentage increased to 33% in the $30–$75,000 income group, 36% in the $75–$100,000 income group, and 49% in the greater than $100,000 income group. Do these study results establish a causal connection between income and happiness? Explain based on statistical principles.

Activity 3-26: Televisions, Computers, and Achievement

Researchers compared achievement scores of 348 Chicago third-graders, based on whether or not the child had a television in his/her bedroom and whether or not there was a computer in the home (Hancox, Milne, and Poulton, 2005). Researchers found that those children with a television in their bedrooms scored substantially lower on the mathematics portion of the test than those without a television in their bedrooms. Also, those with a computer in their homes scored substantially higher on the language arts portion of the test than those without a computer.

a. Identify the two explanatory variables and two response variables. Also classify each as categorical (also binary) or quantitative.
b. Is this an observational study? Explain.
c. Can you conclude that a bedroom television is harmful to a child's academic development? Can you conclude that a home computer is helpful to a child's academic development? Explain.
d. Suggest a potentially confounding variable in this study, and explain why it is confounding.
e. Identify the sample in this study.
f. To what population would you feel comfortable generalizing the results of this study? Explain.

Activity 3-27: Parking Meter Reliability

In 1998, for her sixth-grade science project, Ellie Lammer selected 50 parking meters along Solano and Shattuck Avenues in Berkeley, California. The *Los Angeles Times* reported that she put in one hour's worth of coins in each meter and used three stopwatches to see how long the meters actually lasted. She found that only three meters provided the correct amount of time. Would you be willing to generalize Ellie's results to all Berkeley parking meters? To all California parking meters? Explain.

Activity 3-28: Night Lights and Nearsightedness

Studies have shown children who sleep with a night light are more likely to become nearsighted than those who sleep in a dark room (Quinn, Shin, Maguire, and Stone, 1999).

a. Is it legitimate to conclude that sleeping with a night light causes a higher rate of nearsightedness? Explain.
b. Suppose that a student is asked to identify a confounding variable and responds, "Genetics, because nearsighted parents tend to have nearsighted children." Explain why this argument is incomplete and then complete it.

TOPIC

4 • • • Random Sampling

In the mid-1990s, public information campaigns urged parents to put babies on their backs to sleep, in an effort to reduce the occurrence of sudden infant death syndrome (SIDS). How can statistics help to determine whether or not these campaigns were successful? Can statistics help to resolve authorship of literary works whose origins are hotly debated? And what do these two research questions have in common? In both cases, it is not feasible to interview every member of the population; instead, you turn to sampling, which you will investigate more thoroughly in this topic.

Overview

Much of statistics concerns generalizing about observed findings from a sample to a larger population. For example, in the previous topic, you saw that a radio station wanted to estimate the proportion of all listeners who believe Elvis is alive by asking them to phone in with their opinions. However, you saw that generalizing information from the sample to the larger population of interest can be very misleading when the sampling method is biased, giving you a sample that is not representative of the population. This topic explores methods for selecting a sample so that trends and patterns observed in the sample can be reasonably generalized to the larger population of interest.

Preliminaries

1. Guess how many words are in the Gettysburg Address.

2. Guess the proportion of American infants who sleep on their backs.

3. Think about and then record your own answers to the following questions:
    - Have you sent or read an e-mail so far today?
    - Have you spoken on a cell phone so far today?

- Is it more common for you to call your friends or to text/instant message them?

text Text

4. Collect data from your classmates on their answers to the three preceding questions.

In-Class Activities

Activity 4-1: Sampling Words

4-1, 4-2, 4-3, 4-4, 4-7, 4-8, 8-9, 9-15, 14-6

Consider the population of 268 words in Lincoln's Gettysburg Address, which follows.

> Four score and seven years ago, our fathers brought forth upon this continent a new nation, conceived in liberty, and dedicated to the proposition that all men are created equal.
>
> Now we are engaged in a great civil war, testing whether that nation, or any nation so conceived and so dedicated, can long endure. We are met on a great battlefield of that war.
>
> We have come to dedicate a portion of that field as a final resting place for those who here gave their lives that that nation might live. It is altogether fitting and proper that we should do this.
>
> But, in a larger sense, we cannot dedicate, we cannot consecrate, we cannot hallow this ground. The brave men, living and dead, who struggled here have consecrated it, far above our poor power to add or detract. The world will little note, nor long remember, what we say here, but it can never forget what they did here.
>
> It is for us the living, rather, to be dedicated here to the unfinished work which they who fought here have thus far so nobly advanced. It is rather for us to be here dedicated to the great task remaining before us, that from these honored dead we take increased devotion to that cause for which they gave the last full measure of devotion, that we here highly resolve that these dead shall not have died in vain, that this nation, under God, shall have a new birth of freedom, and that government of the people, by the people, for the people, shall not perish from the earth.

a. Select a sample by circling 10 words that you believe to be representative of this population of words.

b. Record which words you selected and the number of letters in each word:

Word					
Number of Letters					

Word					
Number of Letters					

c. Create a dotplot of your sample results (number of letters in each word). Also indicate what the observational units and variable are in this dotplot. Is the variable categorical or quantitative?

Dotplot:

Observational units:

Variable:

Type:

d. Calculate the average (mean) number of letters per word in your sample. Is this number a parameter or a statistic? *Hint:* Add up the number of letters in each word and divide by ten.

e. Combine your sample average with those of your classmates to produce a dotplot of sample averages. Be sure to label the horizontal axis appropriately.

f. Indicate what the observational units and variable are in the dotplot in part e. *Hint:* To identify the observational units, ask yourself what each dot on the plot represents. The answer is different than the answer for part c.

Observational units:

Variable:

One conceptual challenge here is realizing that the observational units are no longer the individual words, but the *samples* of ten words. Each dot in the plot in part e comes from a sample of ten words, not from an individual word.

g. The average number of letters per word in the population of all 268 words is 4.29 letters. Mark this value on the dotplot in part e. How many students produced a sample average greater than the actual population average? What proportion of the students does this represent?

h. Would you say that this sampling method (asking people simply to circle ten representative words) is biased? If so, in which direction? Explain how you can tell this from the dotplot.

i. Suggest some reasons why this sampling method turns out to be biased.

j. Consider a different sampling method: Close your eyes and point to the page ten times in order to select the words for your sample. Explain why this method would also be biased toward overestimating the average number of letters per word in the population.

Most people tend to choose larger words, perhaps because these words are more interesting or convey more information. Therefore, the first sampling method is biased toward *over*estimating the average number of letters per word in the population. Some samples might not overestimate this population average, but samples chosen with this method *tend* to overestimate the population mean. Closing your eyes does not eliminate the bias because you are still more likely to select larger words because those words take up more space on the page. In general, human judgment is not very good at selecting representative samples from populations.

k. Would using this same sampling method but with a larger sample size (say, 20 words) eliminate the sampling bias? Explain.

l. Suggest how you might employ a different sampling method that would be unbiased.

Activity 4-2: Sampling Words

4-1, **4-2,** 4-3, 4-4, 4-7, 4-8, 8-9, 9-15, 14-6

Now you will discover and investigate a better procedure for sampling words from the Gettysburg Address and for taking samples from populations in general.

One way to avoid a biased sampling method is to give every member of the population the same chance of being selected for the sample. Moreover, your selection method should ensure that every possible sample (of the desired sample size) has an equal chance of being the sample ultimately selected. Such a sampling design is called **simple random sampling.**

Although the principle of simple random sampling is probably clear, random sampling is by no means simple to implement. One method is to rely on physical mixing: Write each word from the Gettysburg Address on an individual piece of paper, put them all into a hat, mix them thoroughly, and draw them out one at a time until the sample is complete. Unfortunately, this method is fraught with the potential for hidden biases, such as different sizes of paper pieces and insufficient mixing.

A better alternative for selecting a simple random sample (hereafter to be abbreviated SRS) is to use a computer-generated **table of random digits.** Such a table is constructed so that each position is equally likely to be occupied by any one of the digits 0, 1, 2, 3, 4, 5, 6, 7, 8, and 9, and so that the value in any one position has no impact on the value in any other position. A table of random digits can be found in the back of the book (see Table I).

The first column in the random digit table gives a line number for you to refer to. The other columns list the random digits in groups of five. It is often convenient to read across a line, but you can begin anywhere on a line and move in any direction. If you need more digits, just continue to the next line.

But before you use the table, you must produce a numbered list of the words in the population:

001	Four	046	nation	091	live	136	to	181	have	226	we
002	score	047	so	092	It	137	add	182	thus	227	here
003	and	048	conceived	093	is	138	or	183	far	228	highly
004	seven	049	and	094	altogether	139	detract	184	so	229	resolve
005	years	050	so	095	fitting	140	The	185	nobly	230	that
006	ago	051	dedicated	096	and	141	world	186	advanced	231	these
007	our	052	can	097	proper	142	will	187	It	232	dead
008	fathers	053	long	098	that	143	little	188	is	233	shall
009	brought	054	endure	099	we	144	note	189	rather	234	not
010	forth	055	We	100	should	145	nor	190	for	235	have
011	upon	056	are	101	do	146	long	191	us	236	died
012	this	057	met	102	this	147	remember	192	to	237	in
013	continent	058	on	103	but	148	what	193	be	238	vain
014	a	059	a	104	in	149	we	194	here	239	that
015	new	060	great	105	a	150	say	195	dedicated	240	this
016	nation	061	battlefield	106	larger	151	here	196	to	241	nation
017	conceived	062	of	107	sense	152	but	197	the	242	under
018	in	063	that	108	we	153	it	198	great	243	God
019	liberty	064	war	109	cannot	154	can	199	task	244	shall
020	and	065	We	110	dedicate	155	never	200	remaining	245	have
021	dedicated	066	have	111	we	156	forget	201	before	246	a
022	to	067	come	112	cannot	157	what	202	us	247	new
023	the	068	to	113	consecrate	158	they	203	that	248	birth
024	proposition	069	dedicate	114	we	159	did	204	from	249	of
025	that	070	a	115	cannot	160	here	205	these	250	freedom
026	all	071	portion	116	hallow	161	It	206	honored	251	and
027	men	072	of	117	this	162	is	207	dead	252	that
028	are	073	that	118	ground	163	for	208	we	253	government
029	created	074	field	119	The	164	us	209	take	254	of
030	equal	075	as	120	brave	165	the	210	increased	255	the
031	Now	076	a	121	men	166	living	211	devotion	256	people
032	we	077	final	122	living	167	rather	212	to	257	by
033	are	078	resting	123	and	168	to	213	that	258	the
034	engaged	079	place	124	dead	169	be	214	cause	259	people
035	in	080	for	125	who	170	dedicated	215	for	260	for
036	a	081	those	126	struggled	171	here	216	which	261	the
037	great	082	who	127	here	172	to	217	they	262	people
038	civil	083	here	128	have	173	the	218	gave	263	shall
039	war	084	gave	129	consecrated	174	unfinished	219	the	264	not
040	testing	085	their	130	it	175	work	220	last	265	perish
041	whether	086	lives	131	far	176	which	221	full	266	from
042	that	087	that	132	above	177	they	222	measure	267	the
043	nation	088	that	133	our	178	who	223	of	268	earth
044	or	089	nation	134	poor	179	fought	224	devotion		
045	any	090	might	135	power	180	here	225	that		

a. Use the Random Digits Table to select a simple random sample of five words from the Gettysburg Address. Do this by starting in the table at any point (it does not have to be at the beginning of a line) and reading off the first 5 three-digit numbers that you come across. (When you get a number greater than 268, skip it and move on. If you happen to get repeats, keep going until you have five different numbers between 001 and 268.) Record the random digits that you selected, the corresponding words, and the lengths (number of letters) of the words:

	1	2	3	4	5
Random Digits	086	096	106	172	248
Word	lives	and	larger	to	birth
Word Length	5	3	6	2	5

b. Determine the average length in your sample of five words.

4.2

c. Again, combine your sample mean with those of your classmates to produce a dotplot. Be sure to label the horizontal axis appropriately.

d. Comment on how the distribution of sample averages from these random samples compares to those of your "circle ten words" samples.

e. Do the sample averages from the random samples tend to over- or underestimate the population average, or are they roughly split evenly on both sides?

A statistic is said to provide **unbiased** estimates of a population parameter if values of the statistic from different random samples are centered at the actual parameter value.

4

f. Does random sampling appear to have produced unbiased estimates of the average word length in the population, which is 4.29 letters?

Activity 4-3: Sampling Words

4-1, 4-2, **4-3,** 4-4, 4-7, 4-8, 8-9, 9-15, 14-6

To really examine the long-term patterns of random sampling, you can use technology to take a large number of samples. Open the `Sampling Words` applet. The information in the top-right panel shows the population distributions for three different variables and tells you the average number of letters per word in the population, the population proportion of "long words," and the population proportion of nouns. Unclick the boxes next to Show Long and Show Noun, so you can continue to focus simply on the population of word lengths for now.

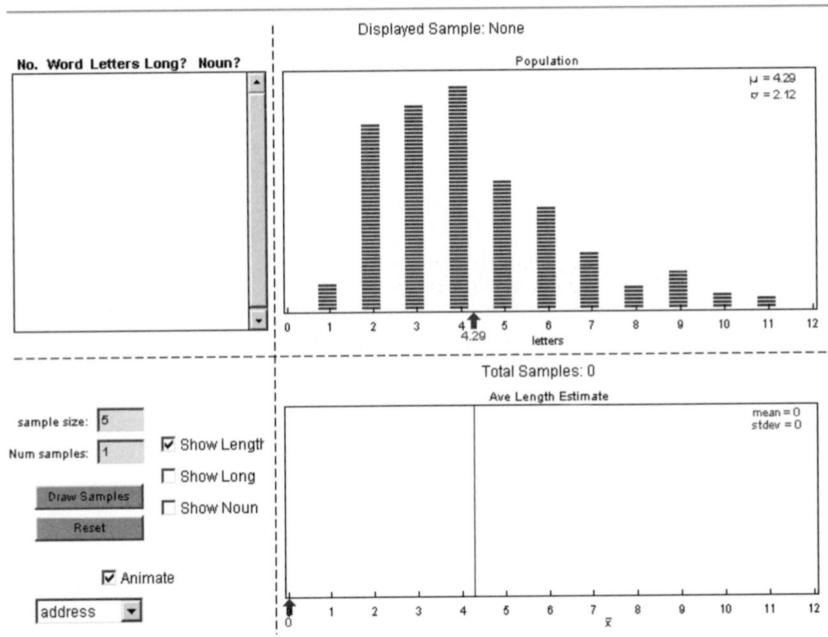

a. Specify **5** as the Sample Size and click Draw Samples. Record the selected words, the number of letters in each word, and the average for the five-word sample.

	1	2	3	4	5
Word					
Number of Letters					

Average number of letters:

b. Click Draw Samples again. Did you obtain the same sample of words this time? Did your sample have the same average length?

An important statistical property known as **sampling variability** refers to the fact that the values of sample statistics vary from sample to sample.

This simple idea is actually a powerful one, for it lays the foundation for many statistical inference techniques that you will study in Units 4–6.

c. Change the number of samples from **1** to **498** in the Num Samples field. Click the Animate button. Then click the Draw Samples button. The applet now takes 498 more simple random samples from the population (for a total of 500) and adds the resulting sample averages to the graph in the lower-right panel. The red arrow indicates the average of the 500 sample averages. Record this value here.

d. If the sampling method is unbiased, the sample averages should be centered "around" the population average of 4.29 letters. Does this appear to be the case?

e. What do you suspect would happen to the distribution of sample averages if you took a random sample of 20 words rather than 5? Explain briefly. *Hint:* Comment on center (tendency) and spread (consistency).

f. Change the sample size in the applet to **20,** and take **500** random samples of **20** words each. Summarize how this distribution of average word lengths compares to the distribution when the sample size was 5 words per sample.

g. Which of the two distributions (sample size 5 or sample size 20) has less variability (more consistency) in the values of the sample average word length?

h. In which case (sample size 5 or sample size 20) is the result of a single sample more likely to be close to the actual population value?

> The **precision** of a sample statistic refers to how much the values vary from sample to sample. Precision is related to sample size: Sample statistics from larger samples are more precise and closer together than those from smaller samples. Statistics from larger random samples, therefore, provide a more *accurate* estimate of the corresponding population parameter.

i. Would taking a larger sample using a biased sampling method tend to reduce the bias? Explain.

Two caveats: First, even with random sampling, you can still get the occasional "unlucky" sample whose results are not close to the population, although this is less likely with large sample sizes. Second, the sample size means little if the sampling method is not random. Remember that the *Literary Digest* poll (Activity 3-1) had a huge sample of 2.4 million people, yet their results were not close to the population value.

Watch Out

- Try not to confuse sample size (n) with the number of samples. *Sample size* refers to how many observational units are in the sample. For example, if there were 25 students in your class for the start of the Gettysburg Address activity, then the sample size would have been 10 (number of words per sample) and the number of samples would have been 25. In an actual study, you would only take one sample from a given population, but as a learning tool, you have taken many samples from the same population to study how sample results vary from sample to sample. (This idea is a key one that will come up again and again later in the course, so try to get comfortable with it now.)

Activity 4-4: Sampling Words

4-1, 4-2, 4-3, **4-4,** 4-7, 4-8, 8-9, 9-15, 14-6

Now that you have explored the effect of sample size on sampling variability, you will investigate the effect of population size.

a. Return to the `Sampling Words` applet. Again, ask for **500** samples of **5** words each, and look at the distribution of average word lengths in those 500 samples. Draw a rough sketch of the resulting distribution.

b. In the lower-left corner of the applet, change the setting from **Address** to **Fouraddresses.** Now the population consists of four copies of the 268 words, for a total population size of 1072 words. Once again, ask for **500** samples of **5** words each, and examine the distribution of average word lengths in those 500 samples. Comment on features of the resulting distribution, and compare this distribution to the one from part a.

c. Do these distributions seem to have similar variability (consistency)?

d. Did, in fact, much change at all when you sampled from the larger population?

Although the role of sample size is crucial to assessing how a sample statistic varies from sample to sample, the size of the population does not affect this sampling variability. As long as the population is large relative to the sample size (at least 10 times as large), the precision of a sample statistic depends on the sample size, not on the population size.

Activity 4-5: Back to Sleep

4-5, 6-5, 21-2

In June of 1992, the American Academy of Pediatrics (AAP) issued a recommendation that healthy infants be placed on their backs or on their sides, rather than on their stomachs, to sleep. This recommendation was based on mounting evidence that sleeping on their stomachs might be related to occurrences of sudden infant death syndrome (SIDS). The recommendation was strengthened in 1994 and accompanied by a national public education campaign called "Back to Sleep." The National Infant Sleep Position Study was launched in 1992 to determine how widely this recommendation was adopted for American infants.

The researchers took a random sample of households with infants younger than eight months, obtained from what they described as a "nationally representative list" generated from public birth records, infant photography companies, and infant formula companies. In 1992, they made 2068 calls and completed 1002 interviews. The primary reasons that some calls did not result in interviews were that the number was not a household's, the household's eligibility could not be determined, or the respondent declined to be interviewed.

a. Identify the population of interest, the sampling frame (see page 36 of Topic 3), and the sample.

Population: birth records, Photo company, formula company

Sampling frame:

Sample: infants participated

b. What is the sample size?

1002

c. Explain how this sampling method differs from a simple random sample.

We chose 3 companies

The report (Willinger et. al., 1998) mentions that compared to birth statistics compiled by the National Center for Health Statistics, the sample underrepresented black mothers, young (younger than age 20) mothers, and low birth-weight infants.

d. Suggest some reasons for why the sampling method might have led to underrepresentation of these groups.

Young mothers wouldn't want babies photographed
not enough money

e. Do these comparisons to birth statistics pertain to the issue of bias or to the issue of precision? Explain.

bias

Some of the findings from the study were that the proportion of infants placed on their stomachs (prone) to sleep fell from 70% in 1992 to 24% in 1996, and the proportion of infants placed on their backs (supine) rose from 13% in 1992 to 35% in 1996.

f. Are these numbers parameters or statistics? Explain how you know this.

Statistic

g. Describe an advantage of using such a large sample size for this study.

closer to real population

Researchers were also interested in estimating the proportion of infants sleeping in various positions for subgroups based on factors such as the infant's gender and the mother's age.

h. Would the sample results for these subgroups be more precise, less precise, or just as precise as for the entire group of infants? Explain.

less Precise b/c subgroup is smaller

4

Solution

a. The population of interest is all infants younger than eight months in the United States in those years. The sampling frame is the list of households with such infants, generated from birth records, infant photography companies, and infant formula companies. The sample consists of the infants in the 1002 households whose mothers (or other caregivers) participated in the interview.

b. The sample size is 1002. (Actually, a total of 1015 infants were in the sample because some households had twins.)

c. The researchers did not technically obtain a simple random sample of infants. One reason is that the sampling frame did not include the entire population. Another reason is that more than half of the numbers called did not lead to an interview. Infants who were not included in the sampling frame or whose mothers declined to participate might differ systematically in some ways from those who were included. Nevertheless, the researchers did use randomness to select their sample, and they probably obtained as representative a sample as reasonably possible.

d. Perhaps mothers in those groups were in a lower economic class and, therefore, less likely to have phones in the first place, or perhaps they had to work so their children were in daycare.

e. These comparisons address the issue of bias, not precision. The sampling method was slightly biased with regard to the mother's race and age and the infant's birth weight.

f. These percentages are statistics because they are based on the samples.

g. The large sample size produces high precision. This means that the sample statistics are likely to be close to their population counterparts. For example, the population proportion of infants who sleep on their backs should be close to the sample proportion who sleep on their backs.

h. The sample size for subgroups is smaller than for the whole group, so the sample results would be less precise.

Watch Out

- Do not use the term *random* in a nontechnical sense. When you use that term in this course, you must really mean it. A random sample is one chosen with an impersonal mechanism such as a random digit table or a calculator or a computer. For example, if you take as your sample the students who you happen to see on campus, that is *not* a random sample of students at your school and, consequently, cannot be trusted to be representative.
- When a study involves a sample that is not drawn randomly from a population, ask yourself whether something in the sampling method suggests that the results from many samples are likely to be biased in one direction or the other. Depending on the variable being examined, a nonrandom sample could still be representative of the population, but you should be much less confident of that than with a random sample. Conversely, whenever a truly random sample is selected from the population of interest, you can assume that it is reasonable to generalize results from this sample to that population.
- Many students confuse descriptions of the sample, the population, the variable, and the parameter. For example, some common incorrect statements are to say that the sample is those who think Elvis is still alive, or the variable is whether most people think Elvis is alive, or the parameter is those who think Elvis is still alive (from Activity 3-1). The sample here, however, is those people who heard about the poll and called in their votes (including those who said that they do not think Elvis is alive as well as those who do). The variable records the response of each individual to whether he or she believes Elvis is alive. The parameter is the proportion of adult Americans who believe that Elvis is still alive. Notice that the parameter has to describe a number, in this case a proportion, not a subset of observational units.
- Many people don't sufficiently distinguish between the terms *bias* and *precision,* which often leads to the mistaken belief that taking a larger sample reduces bias. But if the sampling method is biased, then taking a larger sample does not reduce the bias. In fact, taking a larger sample exacerbates the problem by producing a more precise estimate that is still not close to the population value. The infamous *Literary Digest* poll of 1936 (from Activity 3-1) is a prime example of this phenomenon: Despite its enormous sample size, that poll produced a very misleading result due to its biased sampling method.

Wrap Up . . . .

In this topic, you continued to study the problem of sampling bias, which arose when you chose nonrandom samples from the Gettysburg Address. That self-selection method tends to favor longer words rather than shorter ones. You learned that simple random sampling achieves the goal of selecting a sample from a population, so it is likely to be representative (having the same characteristics as) of that population. In fact, taking a random sample of five words probably produced better results (sample statistics tended to be closer to the population parameter) than did the self-selected sample of ten words.

You also began to study properties of random sampling, such as the obvious (but important) property known as sampling variability. You learned that random sampling produces sample statistics that are **unbiased** estimates for their population parameter counterparts, and you found that larger samples produce more precise estimates. But remember that the *Literary Digest* survey involved a huge sample size and yet horribly underestimated support for the incumbent Democratic president. Therefore, you will often design an unbiased sampling method before you consider the level of precision. You also discovered that although the sample size has a substantial effect on sampling variability, population size does not.

Some useful definitions to remember and habits to develop from this topic include

- Take care when selecting a sample from a population if you hope to generalize results from the sample to describe the population.
- Sampling bias occurs when a sampling method favors certain outcomes over others. Bias is a property of the sample selection method, not simply of a single sample.
- **Simple random sampling** eliminates sampling bias by giving every observational unit in the population an equal chance of being selected. You discovered this property by hypothetically taking multiple simple random samples and seeing that the sample statistics *center* around the population parameter rather than systematically tending to over- or underestimate the population parameter.
- **Sampling variability** means that different samples produce different results (i.e., different values of sample statistics). Even with simple random samples, you expect results to vary from sample to sample, but with a predictable long-term pattern.
- Increasing the sample size reduces sampling variability. In other words, larger samples produce more **precision,** less variation in results from sample to sample. But if the sampling method is biased in the first place, then taking a larger sample does not reduce bias.
- The size of the *population* is not relevant to the issues of sampling bias or precision.

Whereas this topic indicated how to address the problem of sampling bias introduced in Topic 3, the next topic tackles the problem of confounding. One of the keys will again be the deliberate introduction of randomness into the data-collection process, but it entails an entirely different type and application of randomness.

Homework Activities

Activity 4-6: Rating Chain Restaurants

The July 2006 issue of *Consumer Reports* included ratings of 103 chain restaurants. The ratings were based on surveys that *Consumer Reports* readers sent in after eating at one of the restaurants. The article said, "The survey is based on 148,599 visits to full-service restaurant chains between April 2004 and April 2005, and reflects the experiences of our readers, not necessarily those of the general population."

a. Do you think that the sample here was chosen randomly from the population of *Consumer Reports* readers? Explain.

b. Why do the authors of the article make this disclaimer about not necessarily representing the general population?

c. To what population would you feel comfortable generalizing the results of this study? Explain.

Activity 4-7: Sampling Words

4-1, 4-2, 4-3, 4-4, **4-7,** 4-8, 8-9, 9-15, 14-6

Reconsider Activities 4-1 through 4-4, in which the population of interest was the 268 words in the Gettysburg Address. Now consider the variable *whether the word has at least five letters.*

a. Is this a categorical (also binary) or a quantitative variable?

b. Among the 268 words, 99 contain at least five letters. What proportion of these 268 words contains at least five letters?

c. Is your answer to part b a parameter or a statistic? Explain.
d. If you take a simple random sample of five words, would you expect the sample proportion that contained at least five letters to equal this value from part b? Is it even possible? Explain.

Activity 4-8: Sampling Words

4-1, 4-2, 4-3, 4-4, 4-7, **4-8,** 8-9, 9-15, 14-6

Reconsider the previous activity.

a. Use the `Sampling Words` applet to take 500 random samples of 5 words each. Look at the distribution of the sample proportions that contain at least 5 letters. (Make sure that the Show Long box is checked.) Comment on whether or not this distribution appears to be centered at the population value.
b. Now change the sample size to **20,** and take **500** random samples. Again, examine the distribution of the sample proportions that contain at least 5 letters. Comment on how this distribution compares to that from part a.
c. Explain why it makes intuitive sense that the two distributions in part a and part b differ as they do. Comment on both center and spread.

Activity 4-9: Sampling Senators

4-9, 4-18

Suppose you want to take a sample of 5 from the population of 100 current members of the U.S. Senate in order to estimate the average number of years that a current senator has served.

a. Identify (in words) the observational units, variable, population, sample, parameter, and statistic.
b. Suppose that you ask each of your classmates to select a sample of five senators by thinking of the five who come to mind most readily. This sampling method would likely be biased for estimating the average years of service. Indicate and explain the direction of the bias.
c. Would asking your classmates to think of a sample of 10 senators help to avoid this bias? Explain.
d. Provide detailed instructions for how to use the table of random digits to take a simple random sample of five current senators.
e. There are 435 members of the U.S. House of Representatives. Describe, in detail, how you would have to change your instructions in part d to take a simple random sample of ten current representatives.

Activity 4-10: Responding to Katrina

2-12, **4-10,** 16-13

Recall Activity 2-12, which presented results of a national survey asking about perceptions of whether or not race was a factor in the government's slow response to Hurricane Katrina. The survey involved a random sample of 262 black adults and 848 non-Hispanic white adults in the U.S. Of the blacks sampled, 60% said that race was a factor, compared with 12% of whites. Based on the sample sizes, which group's sample responses probably come closer to reflecting that group's population values? Explain.

Activity 4-11: Rose-y Opinions

Suppose you want to investigate the opinions of American sports fan about the famous baseball player Pete Rose. The Gallup Organization conducted such a poll on January 9–11, 2004. They asked a random sample of 1000 individuals, "Next, we'd like to get your overall opinion of some people in the news. As I read each name, please say whether you have a favorable or unfavorable opinion of these people—or whether you have never heard of them. How about . . . Pete Rose?"

a. One proposal is to ask the first 1000 people leaving a Los Angeles Lakers' basketball game. Identify the observational units and the variable of interest. Is this a quantitative or a categorical variable?
b. Define the population of interest and the sample being considered.
c. Is this a reasonable sampling plan? Explain.

An alternative proposal is to select 1000 people from each National Basketball Association game in the U.S. that weekend. Now the sample size is about 20,000 people.

d. Will this approach address the problems you identified in part c?
e. Suppose you had a list of all subscribers to *Sports Illustrated*. Explain how you could use this list to select a random sample. Also, describe the population that would be represented by this random sample.
f. Gallup's poll found that 49% of the respondents had an unfavorable opinion of Pete Rose. Define the parameter and the statistic for this study.
g. If Gallup had selected another random sample of 1000 people and interviewed them, which of your answers to part f would change? Explain.

Activity 4-12: Sampling on Campus

For each of the following situations, describe how you would take a simple random sample of observational units. Also identify the observational units, variable, population, sample, and parameter in each situation.

a. You want to study whether or not college freshmen really tend to gain an average of 15 pounds during their first term at college.
b. You want to estimate the average price paid for textbooks by a student.
c. You want to estimate the average number of words on a page of your history textbook.
d. You want to study the political party registrations of college faculty.

Activity 4-13: Sport Utility Vehicles

Suppose you want to estimate the proportion of vehicles on the road in your hometown that are sport utility vehicles (SUVs). You decide to stand at the intersection closest to your home one morning between 7 and 8 AM, observing how many vehicles go by and how many are SUVs.

a. Identify the observational units, variable, population, sample, parameter, and statistic (all in words) in this study.
b. Explain why this sampling method is not likely to be unbiased.

Now suppose that you go to a local car dealership and obtain a list of all vehicles sold within the past month. You take a simple random sample from this list.

c. Identify the sampling frame.
d. Is this sampling plan likely to be unbiased for estimating the proportion of SUVs among vehicles on the road in your hometown? Explain.

Activity 4-14: Generation M

3-8, **4-14,** 13-6, 16-1, 16-3, 16-7, 18-1, 21-11, 21-12

Recall that in the Preliminaries section you gathered data from yourself and your classmates concerning e-mail, cell phone, and text-messaging usage.

a. If your goal is to learn about media usage among all students at your school, do your classmates form a population or a sample? Explain.
b. What proportion of students in your class had sent or read at least one e-mail message so far that day? Is this number a statistic or a parameter? Explain.
c. Repeat part b for the other two questions, and summarize your findings.
d. Do you and your classmates constitute a random sample of the students at your school? Explain.
e. Do you think that you and your classmates form a representative sample of students at your school with regard to media usage? Explain.
f. If the population were all Americans in your age group, do you think that your class data would be representative in terms of media usage? Explain. If not, in which direction would you expect your sample results to be biased?

Activity 4-15: Emotional Support

4-15, 18-19

In the mid-1980s, Shere Hite undertook a study of women's attitudes toward relationships, love, and sex by distributing 100,000 questionnaires in women's groups. Of the 4500 women who returned the questionnaires, 96% said that they gave more emotional support than they received from their husbands or boyfriends.

a. Comment on whether Hite's sampling method is likely to be biased in a particular direction. Specifically, do you think that the 96% figure overestimates or underestimates the proportion who give more support in the population of all American women?

An ABC News/*Washington Post* poll surveyed a random sample of 767 women, finding that 44% claimed to give more emotional support than they received.

b. Which poll surveyed the larger number of women?
c. Which poll's results do you think are more representative of the population of all American women? Explain.

Activity 4-16: College Football Players

Following is the opening day roster for the 2006 Cal Poly football team:

No.	Name	Pos	Class	Wt	No.	Name	Pos	Class	Wt
1	Tredale Tolver	WR	So.	175	9	Kevin Van Gorder	WR	Sr.	198
2	Courtney Brown	CB	Sr.	205	10	Joseph Wighton	FS	Sr.	185
3	Matt Brennan	QB	So.	205	11	Ramses Barden	WR	So.	227
4	Nick Emmons	QB	Jr.	188	11	Tyler Mariucci	QB	Jr.	190
5	Mike Anderson	QB	Fr.	180	12	Ernie Cooper	WR	So.	185
6	Fred Hives II	RB	So.	210	13	Clint Autry	SS	Jr.	190
7	Pat Johnston	QB	So.	195	14	Wes Pryor	LB	Jr.	195
8	Cordel Webb	SS	Sr.	185	15	Keoni Akina	QB	Jr.	185

(continued)

No.	Name	Pos	Class	Wt	No.	Name	Pos	Class	Wt
16	Michael Chou	QB	Fr.	190	56	Josh Mayfield	OL	Sr.	240
17	Kyle Shotwell	LB	Sr.	235	57	Bobby Best	DE	Fr.	245
18	Tim Chicoine	P	Jr.	190	58	Carlton Gillespie	DE	R-Fr.	240
19	Kenny Chicoine	FS	Sr.	205	59	Justen Peek	LB	Sr.	225
20	Anthony Randolph	CB	Sr.	220	60	Brock Daniels	OL	Jr.	275
21	Jeremy Konaris	RB	Sr.	155	61	Kenny Calderone	OL	Fr.	285
22	Phil Johnston	WR	Fr.	185	62	Perris Kelly	OL	Jr.	285
23	Randy Samuel	SS	Sr.	174	63	Alex Lee	DE	R-Fr.	220
24	Scottie Cordier	DB	Fr.	170	64	Patrick Koligian	OL	So.	250
25	Mark Cordes	LB	Jr.	180	65	Dylan Roddick	OL	Jr.	280
26	Andre Thomas	CB	Jr.	195	66	Ryan Fink	OL	R-Fr.	255
27	James Noble	RB	So.	180	67	Arturo Munoz	OL	Fr.	260
28	Eric Jaso	DB	Fr.	175	67	Jeremy Sutherland	DL	R-Fr.	265
29	Drew Robinson	RB	Jr.	195	68	Lucas Trily	DL	Fr.	235
30	Martin Mohamed	FS	Fr.	210	69	Eric Blank	OL	Fr.	245
31	Xavier Gardner	WR	R-Fr.	170	70	Julai Tuua	OL	Sr.	275
32	Mark Restelli	LB	So.	185	71	Justin Reece	OL	R-Fr.	245
33	David Elmerick	WR	Sr.	185	72	Will Hames III	OL	Sr.	265
33	Martin Mares	DB	Jr.	185	73	Eric White	OL	So.	267
34	Chris Edwards	CB	Fr.	195	75	Daniel Bradley	OL	Jr.	291
34	Jono Grayson	WR	Fr.	170	76	William Mitchell	DL	Fr.	280
35	Nick Coromelas	K	Sr.	200	77	Kevin O'Flaherty	OL	Fr.	275
36	Andrew Breaux	RB	Fr.	205	78	Stephen Field	OL	So.	280
36	Brandon Williamson	CB	Fr.	180	79	Mike Porter	OL	So.	275
37	Aris Borjas	FS	Jr.	200	81	Favi Diaz	WR	So.	195
38	Gene Grant	CB	So.	190	82	Jon Hall	HB	R-Fr.	230
39	Matt Hill	DB	R-Fr.	190	83	Ryan Galloway	DE	Fr.	220
40	Mike Montero	LB	Fr.	210	84	Louis Shepherd	DE	Jr.	250
41	Matt Kirschner	RB	So.	175	85	Kyle Maddux	LB	Fr.	210
42	Vince Freitas	WR	So.	215	86	Donald McCormack	PK	Fr.	150
43	Jaymes Thierry	RB	R-Fr.	200	87	Michael Chowtham	WR	Jr.	185
44	Ryan Shotwell	DE	R-Fr.	220	88	Anthony Gutierrez	DB	R-Fr.	195
46	Tommy Pace	HB	R-Fr.	220	89	Eric Gardley	WR	Fr.	180
47	David Fullerton	SS	R-Fr.	195	90	Kava Tyrell	HB	Fr.	230
48	Michael Maye	CB	Jr.	185	91	Ian Masterson	LB	R-Fr.	205
49	Kevin Spach	HB	Jr.	220	92	Adam Torosian	DE	Jr.	255
50	Hal Kelley	OL	Fr.	260	94	Travis Harwood	DL	Sr.	260
51	Danny Rohr	LB	R-Fr.	220	95	Kevin Okarski	P	R-Fr.	185
52	Alex Bynum	LB	R-Fr.	230	96	Sean Lawyer	DL	So.	270
53	James Chen	OL	Fr.	240	97	Jorge Vazquez	OL	Fr.	220
54	Travis Smith	LB	So.	206	99	Chris White	DL	Sr.	275
55	Jason Relyea	LB	Jr.	220					

a. Identify each of the following variables as quantitative or categorical: number, position, weight, and class.

b. Use the Random Digits Table to take a simple random sample of 15 players. (Be sure to think carefully about how to use the sampling frame and to list the line number you used in the table.) Calculate the average weight of these players. How do you think this sample mean weight compares to the average weight of all 99 players? Explain.

Activity 4-17: Phone Book Gender

4-17, 16-16, 18-11

The authors of your text were curious about the proportion of women living in San Luis Obispo County. They selected a page randomly from the 1998–1999 SLO phone book (page 40) and then randomly selected columns 1 and 4 on that page (this is an example of a *multistage* sampling design). They found the following results: 36 listings had both male and female names, 77 had male names, 14 had female names, 34 had initials only, and 5 had pairs of initials.

a. Identify the parameter and the statistic in this study.

b. Do you believe that this sampling technique will give an unbiased estimate for the proportion of women living in San Luis Obispo? If not, explain whether you think the statistic will be an overestimate or an underestimate of the population parameter.

Activity 4-18: Sampling Senators

4-9, **4-18**

Suppose you took simple random samples repeatedly from a population and computed the proportion of Democrats and then examined how much those proportions varied from sample to sample. Arrange the following four situations from the one that would produce the most variability to the one that would produce the least variability in these sample proportions, and explain your reasoning.

a. A sample of size 20 from the population of U.S. senators

b. A sample of size 1000 from the population of New York residents

c. A sample of size 100 from the population of New York residents

d. A sample of size 500 from the population of Wyoming residents

Activity 4-19: Voter Turnout

4-19, 18-10

In the 1998 General Social Survey, a random sample of 2613 adult Americans were asked whether or not they had voted in the 1996 presidential election, and 1783 said yes.

a. What proportion of these people said that they had voted?

b. Is this number a parameter or a statistic? Explain.

c. Create a bar graph to display the proportions who claimed to have voted and not.

The Federal Election Commission reported that 49.0% of those eligible to vote in the 1996 election had actually voted.

d. Is this number a parameter or a statistic? Explain.

e. Does the sample result seem to be consistent with the actual percentage who voted in the election?

f. Do you suspect that the difference between the sample and population values is simply the result of sampling variability? Explain.

g. Suggest a concern with doing reliable survey work that this example illustrates.

Activity 4-20: Nonsampling Sources of Bias

a. Suppose that simple random samples of adult Americans are asked to complete a survey describing their attitudes toward the death penalty. Suppose that one group is asked, "Do you believe that the U.S. judicial system should have the right to call for executions?" whereas another group is asked, "Do you believe that the death penalty should be an option in cases of horrific murder?" Would you anticipate that the proportions of "yes" responses might differ between these two groups? Explain.

b. Suppose that simple random samples of students on this campus are questioned about a proposed policy to ban smoking in all campus buildings. If one group is interviewed by a person wearing a t-shirt and jeans and smoking a cigarette, whereas another group is interviewed by a nonsmoker wearing a business suit, would you expect that the proportions declaring agreement with the policy might differ between these two groups? Explain.

c. Suppose that an interviewer knocks on doors in a suburban community and asks the person who answers whether he or she is married. If the person is married, the interviewer proceeds to ask, "Have you ever engaged in extramarital sex?" Would you expect the proportion of "yes" responses to be close to the actual proportion of married people in the community who have engaged in extramarital sex? Explain.

d. Suppose that simple random samples of adult Americans are asked whether or not they approve of the president's handling of foreign policy. If one group is questioned prior to a nationally televised speech by the president on his or her foreign policy and another is questioned immediately after the speech, would you be surprised if the proportions of people expressing approval differed between these two groups? Explain.

e. List four sources of bias that can affect sample survey results even if the sampling procedure used is indeed a random one. Base your list on the preceding four questions.

Activity 4-21: Prison Terms and Car Trips

a. Suppose you want to estimate the average length of a prisoner's sentence at a local prison, for the population of all prisoners sentenced to serve in that prison over the past five years. You decide to take a random sample of prisoners currently serving time and calculate the average length of sentence for those prisoners. Explain why this sampling method is likely to be biased toward overestimating the average length of a sentence in the population. *Hint:* Consider which prisoners are more likely to be in the sample.

b. Explain how this same principle applies to estimating the average length of a car trip by taking a sample of cars on the road at any one time.

c. Describe a third situation in which this same principle would apply and thereby produce a biased sampling method.

TOPIC 5

Designing Experiments

Do strength shoes (modified athletic shoes with a 4 cm platform attached to the front half of the sole) really help a person jump farther? How would you design a study to investigate this claim? What factors in a memory experiment affect how many letters a person can memorize correctly? Can a nicotine lozenge help smokers who want to quit? One common aspect of these questions is an interest in finding out whether or not one variable has an effect on another variable. The question is not simply whether those who take a nicotine lozenge tend to quit smoking, but whether you can say that their quitting is because of the lozenge and not some other factor. This topic teaches you how to design studies so they produce data that can answer such questions.

Overview

In the previous topic, you studied the idea of random sampling as a fundamental principle by which to gather information about a population. Sometimes, however, your goal is not to describe a population but to investigate whether one variable has an effect on another variable. This topic introduces you to the design of controlled experiments for this purpose, contrasting them with the observational studies that you explored in earlier topics. You will discover principles for designing controlled experiments and learn how to avoid the problem of confounding variables. You will also explore properties of random assignment and see how this process differs from random sampling, which you studied in Topic 4.

Preliminaries

1. Your instructor will distribute a sequence of letters and give you twenty seconds to memorize as many as you can. Record how many you remember correctly, in the right order.

2. Record these memory scores for your classmates, where the score is the number of letters they remember in the correct order before the first mistake. Also keep track of which version of the letters each student had.

In-Class Activities

Activity 5-1: Testing Strength Shoes

5-1, 5-2, 5-3

The strength shoe is a modified athletic shoe with a 4 cm platform attached to the front half of the sole. Its manufacturer claims that this shoe can increase a person's jumping ability.

a. If your friend who wears strength shoes can jump much farther than another friend who wears ordinary shoes, would you consider that compelling evidence that strength shoes really do increase jumping ability? Explain.

Anecdotal evidence results from situations that come to mind easily and is of little value in scientific research. Much of the practice of statistics involves designing studies and collecting data so people do not have to rely on anecdotal evidence.

Now suppose that you take a random sample of individuals, identify who does and does not wear strength shoes, and then compare their jumping ability.

b. Identify the explanatory and response variables in this study. Also classify each as categorical (also binary) or quantitative.

Explanatory: Type:

Response: Type:

c. Even if the strength shoe group tends to jump much farther than the other group, can you conclude legitimately that strength shoes cause longer jumps? Explain.

The problem with this study, as with all observational studies, is that you do not know whether the two groups might differ in more ways than simply the explanatory variable. For example, subjects who choose to wear the strength shoes could be more athletic to begin with than those who opt to wear the ordinary shoes.

When investigating whether one variable causes an effect on another, researchers create a comparison group and assign subjects to the explanatory variable groups.

An **experiment** is a study in which the experimenter *actively imposes* the **treatment** (explanatory variable group) on the subjects. Ideally, the groups of subjects are identical in all respects other than the explanatory variable, so the researcher can then see the explanatory variable's direct effects on the response variable.

A 1993 study published in the *American Journal of Sports Medicine* investigated the strength shoe claim with a group of 12 intercollegiate track and field participants (Cook et al., 1993). Suppose you also want to investigate this claim, and you recruit 12 of your friends to serve as subjects. You plan to have 6 people wear strength shoes and the other 6 wear ordinary shoes and then measure their jumps.

d. How might you assign subjects to these two groups in an effort to balance out potentially confounding variables?

Random assignment is the preferred method of assigning subjects to treatments (explanatory variable groups) in an experiment: Each subject has an equal chance of being assigned to any of the treatment groups. Such a study is called a **randomized comparative experiment.**

e. Describe in detail how you might implement the process of randomly assigning subjects to treatments.

Activity 5-2: Testing Strength Shoes

5-1, **5-2,** 5-3

In this activity, you will explore properties of random assignment.

Reconsider the experiment described in the previous activity. Suppose that your 12 subjects are listed in the following table. You record their gender and height (in inches) because you suspect that these variables might be related to jumping ability:

Name	Gender	Height	Name	Gender	Height	Name	Gender	Height
Anna	female	61	Kyle	male	71	Patrick	male	70
Audrey	female	67	Mary	female	66	Peter	male	69
Barbie	female	63	Matt	male	73	Russ	male	68
Brad	male	70	Michael	male	71	Shawn	male	67

a. Take 12 index cards, and write each subject's name on a different card. Shuffle the cards and randomly deal out 6 for the strength shoe group and 6 for the ordinary shoe group. Record the names assigned to each group in this table, along with their genders and heights:

Strength Shoe Group			Ordinary Shoe Group		
Name	Gender	Height	Name	Gender	Height

b. Calculate and report the proportion of men in each group. Also subtract these two proportions (taking the strength shoe group's proportion minus the ordinary shoe group's proportion).

Strength shoe group: Ordinary shoe group:

Difference (*strength – ordinary*):

c. Calculate and report the average height in each group. Also subtract these two averages (taking the strength shoe group's average minus the ordinary shoe group's average).

Strength shoe group: Ordinary shoe group:

Difference (*strength – ordinary*):

d. Are the two groups identical with regard to both of these variables? Are they similar?

e. Combine your results with those of your classmates. Produce a dotplot of the differences in proportions of men. Also produce a dotplot of the differences in average heights. Be sure to label the axes of the dotplots clearly, and also identify the observational units in those plots.

f. Do both dotplots appear to be centered around zero? Explain why this indicates that random assignment is effective.

Activity 5-3: Testing Strength Shoes

5-1, 5-2, **5-3**

Reconsider the previous activity. Now you will explore properties of random assignment further by using an applet to repeat this randomization process many more times.

a. Open the `Randomization of Subjects` applet. You should see a card for each subject, with the person's height recorded and gender color-coded. Click Randomize, and the applet will randomize the 12 subjects between the two groups. (Click Show Tables to see the lists of names.)

Randomizing Subjects

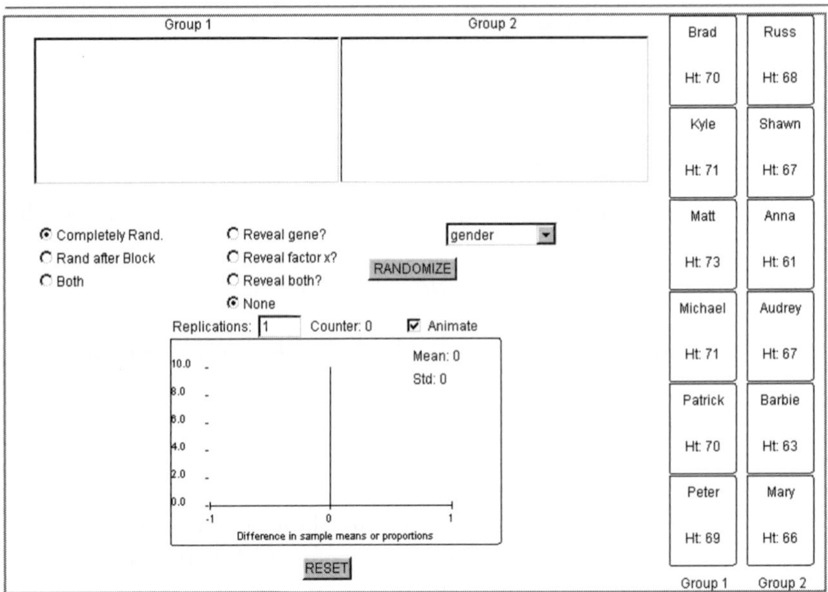

Report the proportion of men in each group and the difference in proportions for this particular random assignment outcome.

Group A: Group B:

Difference:

b. Click Randomize again. Did you get the exact same assignment of subjects to groups? Is the difference in proportions of men the same as before?

c. Before you ask for more randomizations, predict what the dotplot of differences in proportions will look like. In particular, where do you think that distribution will be centered? Explain.

d. Click Reset, change the number of replications to **200,** and uncheck the Animate box. Click the Randomize button. The applet will display the distribution of the 200 differences in proportions of men between the two groups. Where is this distribution centered? Is this what you predicted?

e. Does random assignment *always* balance out the gender variable exactly between the two treatment groups? Does it *tend to* balance out the gender variable in the long run? Explain.

f. Now use the pull-down menu to change Gender to Height. The applet now displays the distribution of 200 differences in average heights between the two treatment groups. Describe this distribution. Also, address whether or not random assignment tends to balance out the heights between the two groups, and explain the evidence for your conclusion.

g. Now suppose that there are two more variables related to jumping ability that you had not considered or could not measure. Would you expect random assignment to balance out these variables between the two treatment groups as well? Explain.

h. Click Reveal Both, and the applet reports values for a categorical genetic trait and a quantitative *x* factor. Then using the pull-down menu, select each of these in turn. Comment on what the dotplots reveal about whether random assignment is effective for unrecorded or unseen variables.

i. Now suppose you conduct this randomized experiment and find that the strength shoe group jumps substantially farther, on average, than the ordinary shoe group. Would you be comfortable concluding that the strength shoes caused the increased jumping distance? Explain how you would argue that no confounding variable was responsible for this increase.

Experimenters try to assign subjects to groups so that lurking and potentially confounding variables tend to balance out between the two groups. These activities demonstrate that **random assignment** generally achieves the goal of creating treatment groups that are similar in all respects except for the treatment imposed. Thus, if the groups turn out to differ substantially on the response variable, you can draw a cause-and-effect conclusion between the explanatory and response variables. As you can see, a randomized comparative experiment has the potential to establish a cause-and-effect relationship between two variables.

Activity 5-4: Botox for Back Pain

5-4, 6-19, 6-20, 21-7

In a recent study, 15 patients suffering from severe back pain were assigned to receive the drug botulinum (Botox), and 9 of these patients reported a substantial decrease in pain (Foster et al., 2001).

a. Would you conclude that Botox is an effective treatment for back pain? Explain.

An alternative design might be to randomly assign 15 subjects to receive Botox and 15 subjects not to receive a treatment and then compare the pain reduction between the two groups.

b. Identify two ways in which this is an improvement over the original design.

c. Can you identify a flaw in this design that would still prevent researchers from concluding that Botox causes a reduction in back pain, even if its group experiences much more pain reduction than the no-treatment group? *Hint:* Identify a confounding variable that exists even in this randomized comparative experiment!

d. Suggest how to eliminate this confounding variable, thereby creating a more effective study of whether Botox is effective for treating back pain.

Comparison groups are especially important in medical studies because subjects often respond positively simply to being given a treatment, whether or not the treatment is actually effective. This phenomenon is known as the **placebo effect.** Experimenters control for this by administering a **placebo** (a "sugar pill" with no active ingredient) to subjects in a control group or by comparing a new treatment against a standard treatment. Naturally, subjects must be **blind** as to which treatment they actually receive or else that knowledge could affect their responses. When possible, experiments should be **double-blind** so the person evaluating the subjects is also unaware of which subjects receive which treatment. In this way, the evaluator's judgment is not influenced (consciously or subconsciously) by any hidden biases.

The comparison group and use of random assignment make the study in part b preferable to the one in part a. But if the comparison group is not given any treatment, then a confounding variable is whether the subject believed him/herself to be getting a treatment and, therefore, expected to improve. A better plan for studying this issue is to use a placebo treatment for the comparison group.

Activity 5-5: Memorizing Letters

5-5, 7-15, 8-13, 9-19, 10-9, 22-3, 23-3

Recall the data collected in the Preliminaries section about how many letters you could memorize in 20 seconds. Every person received the same sequence of 30 letters, but they were presented in two different groupings. One group received

JFK-CIA-FBI-USA-SAT-GPA-GRE-IBM-NBA-CPR

and the other received

JFKC-IAF-BIU-SASA-TGP-AGR-EIB-MN-BAC-PR

Similar experiments have shown that those receiving the letters already organized in familiar chunks are able to memorize more than those with the less memorable groupings.

a. Explain why this study is an experiment and not an observational study.

b. Identify and classify the explanatory and response variables in this study.

Explanatory: Type:

Response: Type:

c. Explain how random assignment was implemented and why it was important in this study.

d. Explain how blindness was implemented and why it was important in this study.

e. Create dotplots of the memory scores, comparing the two treatment groups. *Hint:* Be sure to label the horizontal axis.

f. Comment on whether these experimental data appear to support the conjecture that those who receive the letters in convenient three-letter chunks tend to memorize more letters.

g. If the "JFK" group does substantially better than the "JFKC" group, could you legitimately conclude that the grouping of letters *caused* the higher scores? Explain how you would respond to the argument that perhaps the good memorizers were in the JFK group and the poor memorizers in the JFKC group.

Activity 5-6: Nicotine Lozenge

1-16, 2-18, **5-6,** 9-21, 19-11, 20-15, 20-19, 21-6, 22-8

An article in the June 10, 2002, issue of the *Archives of Internal Medicine* reported on a study of the effectiveness of a nicotine lozenge for helping smokers quit smoking (Shiffman et al., 2002). Newspaper advertisements sought volunteers who were smokers interested in quitting. Those volunteers selected to participate in the study were randomly assigned to receive either the nicotine lozenge or a placebo lozenge (with no active ingredient).

a. Is this an observational study or an experiment? Explain.

b. Who or what are the observational units in this study?

c. Identify and classify the explanatory and response variables.

The article reports on many background variables, such as age, weight, gender, smoking amount, and whether the person made a previous attempt to quit smoking. It shows that the two groups (nicotine lozenge and placebo lozenge) had similar distributions for these variables at the start of the study.

d. Why do the researchers report this information? Do you think they were pleased that the distributions for these variables were similar between the two groups? Explain how this is helpful to the kind of conclusion they were hoping to draw.

e. The researchers found that the proportion of subjects who successfully abstained from smoking was substantially higher in the nicotine lozenge group than in the placebo group. Is it legitimate to conclude that the nicotine lozenge was responsible for this higher rate of quitting? Explain.

Solution

a. This is an experiment because the researchers imposed the nicotine lozenge (or placebo) on subjects.

b. The experimental units are the smokers interested in quitting who volunteered to participate in the study.

c. The explanatory variable is whether the smoker was given a nicotine lozenge or a placebo. This variable is categorical and binary. The response variable is whether the smoker successfully quit smoking by the end of the study. This variable is also categorical and binary.

d. This information validates that the random assignment achieved its goal of balancing out all of these variables, which could potentially be related to a smoker's ability to quit, between the nicotine lozenge and placebo groups. Thus, if the nicotine lozenge group has a higher proportion who quit smoking, then the researchers can attribute that to the lozenge and not to any of these background variables because they were similar between the groups.

e. Yes. Because this was a randomized experiment, it is legitimate to conclude that the nicotine lozenge caused the increase in the proportion who successfully quit smoking.

Watch Out

- Keep in mind that random *assignment* is very different from random *sampling.* They are different processes with different objectives.
- Random sampling aims to produce a sample that is representative of a population, so meaningful inferences can be made about the population based on the sample. In other words, random sampling eliminates sampling bias.
- Random assignment (or randomization) aims to produce treatment groups that are similar in all respects except for the treatment imposed. Then, if the groups differ substantially in the response variable, you can conclude that the explanatory variable caused that difference. In other words, random assignment eliminates confounding.
- Although you learned in Topic 3 that you should not draw cause-and-effect conclusions from observational studies, do not go overboard and believe that no study can ever produce a cause-and-effect conclusion. A well-designed, randomized comparative experiment can indeed lead to cause-and-effect conclusions.
- When you are asked to design an experiment to investigate some issue, include lots of details so someone else can conduct the experiment based solely on your description. Such a description is called the **protocol** of the study.

Wrap Up ....

This topic introduced the need for randomized comparative experiments to be able to establish cause-and-effect relationships between variables. Random assignment of subjects to experimental groups (treatments) eliminates the problem of confounding by, on average, balancing out other variables and creating treatment groups that are similar in all ways except for the explanatory variable being studied. For example, randomly assigning subjects to wear a strength shoe or an ordinary shoe serves to balance out factors such as gender and athleticism between the two groups. If those in the strength shoe group then out-perform those in the control group, you can attribute the better performance to the strength shoes because any other major differences between the two groups can be assumed to have been eliminated.

You also studied the importance of using a comparison group in an experiment. For example, it is not very informative simply to see whether patients report a reduction in back pain after being given an injection of a drug. Even if the drug were not effective, patients might have felt better naturally as time passed, or they might have responded positively to receiving any kind of injection (**placebo effect**). Blindness is another technique for ensuring that the treatment groups are as similar as possible. Subjects should not know which treatment they are receiving. If you had known you were in the group expected to do well in the memory study, that knowledge could have affected your performance. Ideally, experiments are also double-blind, so those interacting with and evaluating the subjects are also unaware of who is in which group. These levels of blindness provide additional controls against potential confounding variables, such as the placebo effect or inconsistent measurement of the response variable.

Some useful definitions to remember and habits to develop from this topic include

- An **experiment** is distinguished from an observational study by the experimenters deliberately imposing the explanatory variable on the subjects, rather than passively observing and recording information.
- A **randomized comparative experiment** is the "gold standard" for establishing a cause-and-effect relationship between two variables. By randomly assigning subjects to treatment groups, you guard against the possibility that the groups differ systematically in another way (i.e., you guard against confounding).
- **Random assignment,** or randomization, is a different use of randomness than is random sampling. Although random assignment allows you to potentially draw cause-and-effect conclusions, be cautious about generalizing the results of randomized experiments to a larger population unless the subjects were randomly sampled from the population of interest.
- **Blindness** refers to the subjects in an experiment being unaware of which treatment they receive. **Double-blindness** refers to the evaluators (not the researchers so much as those determining the response variable outcome) also being unaware of which group received which treatment.
- The scope of conclusions that you can draw from a study depends primarily on how the data were collected. Random *sampling* allows for *generalizing* from a sample to the larger population. Random *assignment* allows for drawing a *cause-and-effect* conclusion between the explanatory and response variables if, in fact, the experimental groups differ substantially with regard to the response variable at the end of the study. When describing study conclusions, address generalization and causation separately.

Now that you have learned sound strategies for collecting data, the next unit will introduce you to basic ideas related to analyzing data.

Homework Activities

Activity 5-7: An Apple a Day

Suppose you want to investigate whether or not the expression "an apple a day keeps the doctor away" has any validity. In other words, you want to investigate whether eating apples has any health benefit. For each of the following three approaches, identify whether the approach is an experiment, an observational study, or an anecdote.

a. You recall that your Uncle Joe loved apples and was never sick a day in his life, whereas your Uncle Tom despised apples and was often ill.
b. You take a random sample of individuals, identify who does and does not eat apples regularly, and then follow them for six months to see who requires a visit to a doctor and who does not.
c. You take a random sample of individuals, randomly assign half to eat an apple every day and the other half not to. You then follow them for six months to see who requires a visit to a doctor and who does not.

Activity 5-8: Treating Parkinson's Disease

Spheramine is an experimental cell therapy treatment for patients with advanced stages of Parkinson's disease. Brain surgery is required to implant Spheramine in the patient. The following statement comes from an announcement (appearing on the National Institutes of Health `clinicaltrials.gov` Web site) seeking volunteers to participate in an experiment: "This double-blind, randomized, placebo-controlled study will evaluate the safety and efficacy of bilateral Spheramine® implantation into the brain compared to sham surgery in patients with Parkinson's Disease."

a. What do you think sham surgery means?
b. Explain what double-blind means and why it is important in this study.
c. Explain what randomized means and why it is important in this study.
d. Explain what placebo-controlled means and why it is important in this study.

Activity 5-9: Ice Cream Servings

5-9, 22-19, 22-20

Researchers conducted a study in which they invited 85 nutrition experts to an ice cream social (Wansink et al., 2006). These experts were randomly given either a 17- or a 34-ounce bowl, and they were also randomly given either a 2- or a 3-ounce ice cream scoop. They were then invited to serve themselves ice cream. The data revealed that those with larger bowls served and ate substantially more ice cream than those with smaller bowls, as did those with larger scoops.

a. Identify and classify the explanatory and response variables in this study.
b. Is this an observational study or an experiment? Explain.
c. Explain why random assignment was important in this study.
d. Describe how and why this study made use of blindness.
e. From this study, can you draw a cause-and-effect conclusion between the size of the bowl or scoop and the size of the serving? Explain.
f. How would you respond to the argument that perhaps the people with bigger appetites tend to eat more and so you can't attribute the bigger servings to the bigger bowls and bigger spoons?

Activity 5-10: Spelling Errors

Describe how you might design a study to investigate whether college students tend to do a better job of reducing spelling errors when proofreading a research paper if they use a computer's spell-checker than if they do not use a spell-checker.

Activity 5-11: Foreign Language Study

3-17, **5-11**

Suppose that students in your high school who study a foreign language tend to score higher on the verbal portion of the SAT than those who do not study a foreign language.

a. Can you conclude that foreign language study improves your verbal skills? Explain.
b. Describe the design of an experiment to investigate the question of whether foreign language study improves your verbal skills.
c. Explain why it would not be very feasible to conduct such an experiment.

Activity 5-12: AZT and HIV

5-12, 6-2, 21-13

In 1993, one of the first studies aimed at preventing maternal transmission of AIDS to infants gave the drug AZT to pregnant, HIV-infected women (Connor et al., 1994). Roughly half of the women were randomly assigned to receive the drug AZT, and the others received a placebo. The HIV-infection status was then determined for 363 babies, 180 from the AZT group and 183 from the placebo group. Of the 180 babies whose mothers had received AZT, 13 were HIV-infected, compared to 40 of the 183 babies in the placebo group.

a. Identify and classify the explanatory and response variables in this study.
b. Is this an observational study or an experiment? Explain.
c. Explain how and why the study makes use of comparison.
d. Explain how and why the study incorporates random assignment.
e. Explain how and why the study makes use of blindness.

Activity 5-13: Pet Therapy

3-23, **5-13**

Suppose you want to study whether pets provide a therapeutic benefit for their owners. Specifically, you decide to investigate whether heart attack patients who own a pet tend to recover more often than those who do not. You randomly select a sample of heart attack patients from a large hospital and then follow them for five years. You compare the sample proportions who survived for those five years between those who had a pet and those who did not.

a. Identify and classify the explanatory and response variables in this study.
b. Is this study a controlled experiment or an observational study? Explain.
c. Does this study make use of a comparison group? Explain.
d. Does this study make use of random assignment? Explain.
e. Does the study design enable you to conclude that owning a pet does indeed have a therapeutic benefit for heart attack survivors? Explain.
f. Describe, in principle, how you could design a controlled experiment to address this issue.
g. Is such an experiment feasible in practice? Explain.

Activity 5-14: Studies from *Blink*

1-4, **5-14**

Recall the four studies from the book *Blink* described in Activity 1-4 (page 7).

a. Identify the explanatory and response variables for those studies.
b. Identify which of the studies are observational and which are experimental.
c. Comment on the scope of conclusions (generalizability, causation) that can potentially be drawn from each of those studies.

Activity 5-15: Reducing Cold Durations

The November 1, 1999, issue of *USA Today* reported on a study of the effectiveness of a zinc nasal spray for reducing the duration of a common cold. Researchers recruited 104 subjects who agreed to report to their lab within 24 hours of getting cold symptoms. Each subject was randomly assigned to one of three groups: One group received a full dosage of the zinc spray; another received a low dosage; and a third received a placebo spray. The cold symptoms lasted an average of 1.5 days for the full dosage group, 3.5 days for the low dosage group, and 10 days for the placebo group.

a. Identify the observational/experimental units in this study.
b. Identify the explanatory and response variables.
c. Is this an observational study or an experiment? Explain.
d. Why did the researchers use a placebo spray, as opposed to just providing no treatment for that group of subjects?

Activity 5-16: Religious Lifetimes

The August 9, 1999, issue of *USA Today* reported that a national study of 3617 adult Americans concluded that people who attend religious services at least once per month live substantially longer than those who do not.

a. Identify the explanatory and response variables.
b. Is this an experiment or an observational study? Explain.
c. Can you conclude from the study that attendance at religious services causes people to live longer? If not, identify a confounding variable that provides a plausible alternative explanation.
d. Because over 300 million people live in the United States, is a sample size of 3617 adequate for describing the population? Explain. *Hint:* Remember what you learned about the effect of population size in Activity 4-4 on page 62.

Activity 5-17: Natural Light and Achievement

1-14, **5-17**

Recall the study by the Heschong Mahone group, based near Sacramento, California, and discussed in Activity 1-14 on page 12, which found that students who took their lessons in classrooms with more natural light scored as much as 25% higher on standardized tests than other students in the same school district.

a. Describe the design of a controlled experiment that could determine whether or not natural lighting in classrooms improves test scores.
b. Explain briefly why carrying out such an experiment would be difficult.
c. John B. Lyons, an Educational Department official, commented that this was "one of the first studies that shows a clear correlation" between daylight and achievement (*The Tribune*, Dec. 21, 1999). How could he reword his conclusion if this study were conducted as a controlled experiment?

Activity 5-18: SAT Coaching

Suppose you want to study whether an SAT coaching program actually helps students to score higher on the SATs, so you gather data on randomly selected students who have attended the program. You find that 95% of the sample scored higher on the SATs after attending the program than before attending the program. Moreover, you calculate that the sample mean of the improvements in SAT scores was a substantial 120 points.

a. Identify the explanatory and response variables in this study.
b. Is the SAT coaching study, as described, a controlled experiment or an observational study?
c. Explain why you cannot legitimately conclude that the SAT coaching program caused these students to improve on the test. Suggest some other reasons for their improvement.

Activity 5-19: Capital Punishment

Suppose that you want to study whether the death penalty acts as a deterrent against homicide, so you compare the homicide rates between states that have the death penalty and states that do not.

a. Is this study an experiment? Explain.
b. If you find a large difference in the homicide rates between these two types of states, can you attribute that difference to the deterrent effect of the death penalty? Explain.
c. If you find no difference in the homicide rates between the two types of states, can you conclude that the death penalty has no deterrent effect? Explain.

Activity 5-20: Literature for Parolees

The October 6, 1993, issue of *The New York Times* reported on a study in which 32 convicts were given a course in great works of literature. To be accepted for the program, the convicts had to be literate and to convince a judge of their intention to reform. After 30 months of parole, only 6 of these 32 had committed another crime. This group's performance was compared against a similar group of 40 parolees who were not given the literature course; 18 of these 40 had committed a new crime after 30 months.

a. What proportion of the literature group committed a crime within 30 months of release? What proportion of this group did not commit a crime?
b. What proportion of the control group committed a crime within 30 months of release? What proportion of this group did not commit a crime?
c. Which fundamental principles of control does this experiment lack? Comment on how this lack hinders the conclusion of a cause-and-effect relationship in this case.

Activity 5-21: Therapeutic Touch
5-21, 17-19

Practitioners of a controversial medical practice known as *therapeutic touch* claim that they can manipulate a person's human energy field in order to provide healing powers, without actually touching the patient's body. Emily Rosa, an 11-year-old in Colorado, recruited 21 practitioners of therapeutic touch in a study she conducted in 1996 (Rosa et al., 1998). Her study consisted of placing a screen between a subject's eyes and hands. For several trials, she hovered her hand over one of theirs, flipping a coin each time to decide which hand to hold hers over. She then asked the practitioner to decide which of their hands her hand was near to see whether the practitioner could detect Emily's energy field.

a. Was this an observational study or experiment? Explain.
b. Explain how the principle of random assignment was employed in this study.
c. Was the study double-blind? Explain.
d. Would you be willing to generalize the results Emily obtained for these 21 practitioners to all practitioners? Explain.
e. If these 21 practitioners do show a strong tendency to correctly identify which hand Emily is holding hers over, would you be willing to attribute this tendency to detection of Emily's energy field? Be sure to discuss the design of the study in your answer.

Activity 5-22: Prayers, Cell Phones, School Uniforms

Select one of the following issues, and describe in detail the design of an experiment to address it. Also explain why observational studies investigating the issue could not settle the question by describing at least one potential confounding variable.

a. Is the use of cell phones by automobile drivers a hazard?
b. Does prayer help to reduce suffering?
c. Does requiring students to wear uniforms in school lead to better academic achievements?

Activity 5-23: Proximity to the Teacher

1-11, **5-23,** 27-9, 29-6

Suppose that you want to investigate whether students who sit close to the teacher tend to perform better on quizzes than those who sit farther away.

a. Identify the observational units, explanatory variable, and response variable.
b. Describe how you could perform an experiment to address this research question.
c. Identify an advantage of the experiment. Explain.
d. Identify an advantage of the observational study. Explain.

Activity 5-24: Smoking While Pregnant

Many studies have shown that children of mothers who smoked while pregnant tend to be less well-developed physically and intellectually than children of mothers who did not smoke while pregnant.

a. Do you suspect that these studies are observational or experimental? Explain.
b. Is it feasible or ethical to conduct experiments to investigate whether smoking while pregnant has harmful effects for the child's development? Explain.

Activity 5-25: Dolphin Therapy

In a recent study conducted in Honduras and reported in the *British Medical Journal,* subjects suffering from mild to moderate depression were taken off their medication and assigned to one of two groups (Antonioli and Reveley, 2005). One group swam with dolphins every day, whereas the other group swam and snorkeled on their own around a coral reef every day. Researchers found that the patients who swam with dolphins showed a greater improvement in depression symptoms than the other group.

a. Is this an observational study or an experiment? Explain.
b. Identify and classify the explanatory and response variables.
c. Does this study permit a cause-and-effect conclusion between swimming with dolphins and depression symptoms improving? Explain.
d. Were the subjects in this study blind as to which treatment they received? Could they have been, considering the research question? Explain.

Activity 5-26: Cold Attitudes

5-26, 25-21

In a study published in the July 2003 issue of the journal *Psychosomatic Medicine,* researchers reported that people who tend to think positive thoughts catch a cold less often than those who tend to think negative thoughts (Cohen et al., 2003). The scientists recruited over 300 initially healthy volunteers, and they first interviewed them over two weeks to gauge their emotional state, eventually assigning them a numerical score for positive emotions and a numerical score for negative emotions. Then the researchers injected rhinovirus, the germ that causes colds, into each subject's nose. The subjects were then monitored for the development of cold-like symptoms. Subjects scoring in the bottom third for positive emotions were three times more likely to catch a cold than those scoring in the top third.

a. Identify the explanatory and response variables in this study. Classify each as categorical (also binary) or quantitative.
b. Is this an observational study or an experiment? Explain. *Hint:* Ask yourself whether the explanatory variable was assigned by the researchers.
c. Based on this study and the substantial differences in likelihood of catching a cold between these groups, can you draw a cause-and-effect conclusion between emotional outlook and likelihood of catching a cold? Explain.

Activity 5-27: Friendly Observers

5-27, 21-1, 23-11

Psychology researchers investigated a conjecture that using an observer with a vested interest would decrease subjects' performance on a skill-based task (Butler and Baumeister, 1998). Subjects were given time to practice playing a video game that required them to navigate an obstacle course as quickly as possible. They were then told to play the game one final time with an observer present. Subjects were assigned to one of two groups: One group was told that the participant and observer would each win $3 if the participant beat a certain threshold time, and the other group was told only that the participant would win the prize if the threshold were beaten.

a. Is this an observational study or an experiment? Explain.
b. Identify the observational/experimental units in this study.
c. Identify the explanatory and response variables in this study.
d. Does this study make use of blindness? Explain.

Activity 5-28: Got a Tip?

1-10, **5-28,** 14-12, 15-17, 22-4, 22-9, 25-3

Suppose that a college student who works as a waitress wants to examine factors that might influence how much her customers tip. Some of the factors that she considers testing are

- Whether she introduces herself by name
- Whether she stands throughout or squats at the table
- Whether she wears a flower in her hair

a. Are these explanatory variables or response variables?
b. How would you measure the response variable in this study?
c. Is it possible to conduct this study as an experiment rather than an observational study? Explain.
d. Describe an advantage of conducting this study as an experiment.
e. Describe how she could use random assignment in conducting this study as an experiment. Address the question of whether the random assignment would be performed on a customer-to-customer level or a shift-by-shift level.

UNIT 2 •••
Summarizing Data

TOPIC

6 • • •

Two-Way Tables

Do men and women differ with regard to career aspirations? Do political conservatives and liberals have different opinions about how the federal government should spend taxpayer dollars? Is AZT an effective drug for preventing mother-to-child transmission of AIDS? Did the "Back to Sleep" promotional campaign help convince more parents to put babies on their backs to sleep? You will investigate these and other questions in this topic as you discover statistical methods for exploring relationships between categorical variables.

Overview

After you have collected data, a useful first step in analyzing the data is to summarize their distribution. This often involves creating both graphical displays and numerical summaries. In this unit, you will examine strategies and methods for summarizing data, starting with categorical variables. In this topic, you will learn to construct and analyze two-way tables that allow you to compare the data for categorical variables. Although the mathematical operations you will perform are no more complicated than adding and calculating proportions, you will acquire some powerful analytical tools for addressing interesting research questions. You will also find that the lessons you learned in the first unit about collecting data and the types of conclusions you can draw from a study continue to apply.

Preliminaries

1. Do you think a person's political inclination has any bearing on his or her opinion about how much the federal government spends on the environment?

2. Is there a difference between the proportion of American men who are U.S. senators and the proportion of U.S. senators who are American men? If so, which proportion is greater? Yes, Men

3. If you could choose one of the following accomplishments for your life, which would you choose:

   - To win an Olympic gold medal

- To win a Nobel Prize
- To win an Academy Award

4. Record the responses of your classmates and yourself to the previous "lifetime achievement" question, along with each student's gender.

In-Class Activities

Activity 6-1: Government Spending

6-1, 25-7, 25-8, 25-24

The General Social Survey (GSS) is a large-scale survey conducted every two years on a nationally representative sample of adult Americans. Some of the questions asked in the 2004 survey concerned political viewpoint and opinion about the federal government's spending priorities. The following two-way table presents a summary of responses to a question about political inclination and how much the federal government spends on the environment (notice that row and column totals are also provided):

	Liberal	Moderate	Conservative	Total
Too Little	127	158	113	398
About Right	27	80	91	198
Too Much	1	17	32	50
Total	155	255	236	646

This table is called a **two-way table** because it classifies each person according to two categorical variables. The columns include the explanatory variable categories, and the rows include the response variable categories.

In particular, this is a 3 × 3 table; the first number represents the number of categories of the row variable (*opinion about federal spending on the environment*), and the second number represents the number of categories of the column variable (*political viewpoint*). Note that the entries of a two-way table are *counts* and, therefore, integers.

a. What proportion of the survey respondents said that the federal government spends too little on the environment? About the right amount? Too much? *Hint:* Take advantage of the totals provided in the table to calculate your responses.

Too little: 62% About right: 31% Too much: 7%

The **marginal distribution** of a categorical variable is the proportional breakdown of its categories. The marginal distribution identifies the individual categories and the proportion of observations in each category.

The word *marginal* indicates that this distribution is calculated from the totals in the margins of the table. In part a, you have calculated the marginal distribution of the *opinion about federal spending on the environment* variable. When analyzing two-way tables, you start typically by considering the marginal distribution of each variable by itself before moving on to explore possible relationships between the two variables. As you saw in Topic 2, marginal distributions can be represented in bar graphs.

b. Construct a bar graph to display the marginal distribution of the *opinion about federal spending on the environment* variable.

c. Comment on what your calculations and bar graph reveal about Americans' opinions regarding federal spending on the environment. For example, do Americans tend to believe that the government spends too little or too much or about the right amount?

d. Now restrict your attention (for the moment) to only the respondents who classify themselves as "liberal." What proportion of these liberal respondents say that the federal government spends too little on the environment?

e. What proportion of the liberal respondents say that the federal government spends about the right amount on the environment?

f. What proportion of the liberal respondents say that the federal government spends too much on the environment?

g. Do these three proportions add up to one?

> A **conditional distribution** is the distribution (i.e., which outcomes and how often they occur) of the response variable for a *particular category* of the explanatory variable.

You have now calculated in parts d–f the conditional distribution of the *opinion about federal spending on the environment* variable for the liberal category of the *political viewpoint* variable. You could also calculate the conditional distributions for the moderate and conservative viewpoints.

h. The following table reports the conditional distribution of the *opinion about federal spending on the environment* for the "moderate" and "conservative" political viewpoints. Record the conditional distribution that you calculated in the "liberal" column of the table:

	Liberal	Moderate	Conservative
Too Little	.819	.619	.479
About Right	.174	.314	.385
Too Much	.1	.067	.136
Total	1.000	1.000	1.000

Conditional distributions can be represented visually with **segmented bar graphs.** Each category of the explanatory variable is represented by a rectangle with a total height of 100%, but each rectangle contains *segments* whose lengths correspond to the conditional proportions.

i. Complete this segmented bar graph by constructing the conditional distribution of spending opinion for those with a liberal political viewpoint.

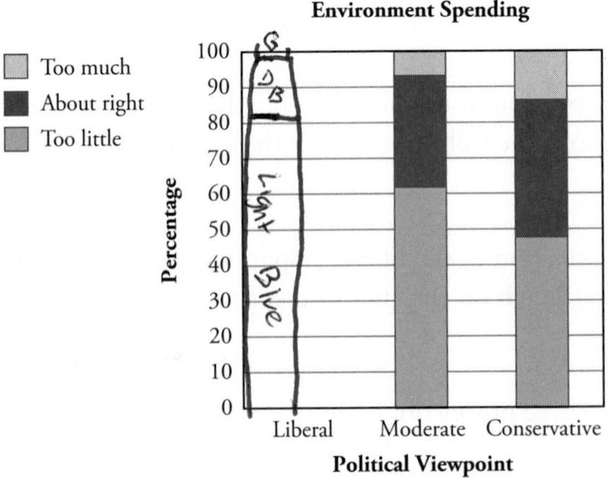

Note: The segmented bar graph is an alternative to presenting multiple bar graphs, one for each category of the explanatory variable, as you did in Activity 2-2 to analyze hand-washing data comparing men and women. Segmented bar graphs make the comparison across groups a bit easier to see visually.

j. Based on the conditional proportions and segmented bar graph, does the distribution of this opinion seem to differ among the three political groups? If so, describe key features of the differences.

k. The following segmented bar graph pertains to similar data from the 2004 GSS when people were asked about their opinion regarding the federal government's spending on the space program.

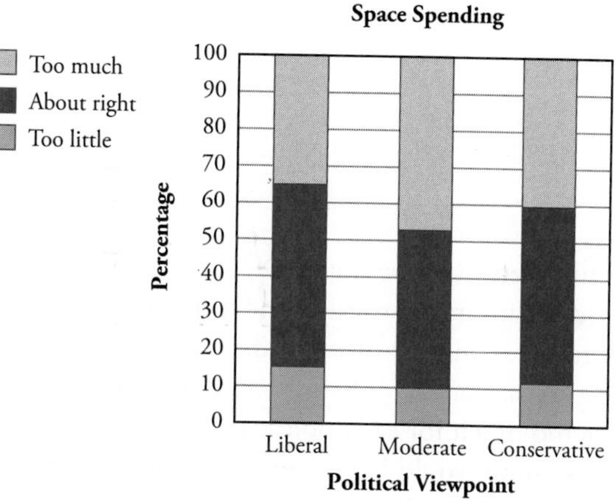

l. Based on the graph, approximately what proportion of liberals believe that the government spends too much on the space program?

35%

m. Judging from the graph, is there a big difference in how the three political groups feel about government spending on the space program? Explain.

no.

Two categorical variables are said to be **independent** if the conditional distributions of one variable are identical for every category of the other variable. Otherwise, the variables have an **association,** or a relationship.

Note: With sample data, you do not expect conditional distributions to be completely identical even if the variables are essentially independent in the population. In Unit 6, you will learn how to assess the level of dependence between the variables.

n. With which issue does opinion about government spending come closer to being independent of political viewpoint—environment or space program? Explain how you can tell.

o. Sketch what a segmented bar graph might look like for an issue where opinion about government spending is perfectly independent of political viewpoint.

Watch Out

Some people mistakenly believe that independence requires that each of the three opinion categories appear one-third of the time for each political viewpoint group. But independence requires only that the conditional distributions be the same for each of the viewpoint groups. For example, if the breakdown was 10% too little, 60% about right, and 30% too much for each political viewpoint group, then the conditional distribution of opinion would have been identical regardless of viewpoint, and so opinion would have been independent of viewpoint even though some opinions occur more than others.

Two final questions reveal a common problem with reading two-way tables:

p. Among those who say that the government spends too little on the environment, what proportion identify themselves as liberal?

127 of 398 32%

q. Is this close to the proportion of liberals who say that the government spends too little on the environment (as you found in part d)?

No, 81.9% vs. 32%

not even close

Watch Out

In dealing with conditional proportions, it is important to remember which category is the one being conditioned on. For example, the proportion of *American males* who are U.S. senators is very small (less than 1/1,000,000), yet the proportion of U.S. *senators* who are American males is very large (86/100 in the year 2007).

Activity 6-2: AZT and HIV

5-12, **6-2,** 21-13

Recall that Activity 5-12 described one of the first studies aimed at preventing maternal transmission of AIDS to infants. The drug AZT was given to pregnant, HIV-infected women in 1993 (Connor et al., 1994). Roughly half of the women were randomly assigned to receive the drug AZT, and the others received a placebo. The HIV-infection status was then determined for 363 babies, 180 from the AZT group and 183 from the placebo group. Of the 180 babies whose mothers had received AZT, 13 were HIV-infected, compared to 40 of the 183 babies in the placebo group.

a. Is this an observational study or an experiment? Explain how you can tell.

b. Identify the explanatory and response variables.

Explanatory: Response:

c. Organize the information provided into the following 2 × 2 table:

	Placebo	AZT	Total
HIV-infected			
Not HIV-infected			
Total			

d. Within each treatment group, calculate the conditional proportion of babies who were HIV-infected. Also, verify that your calculations are consistent with the following segmented bar graph.

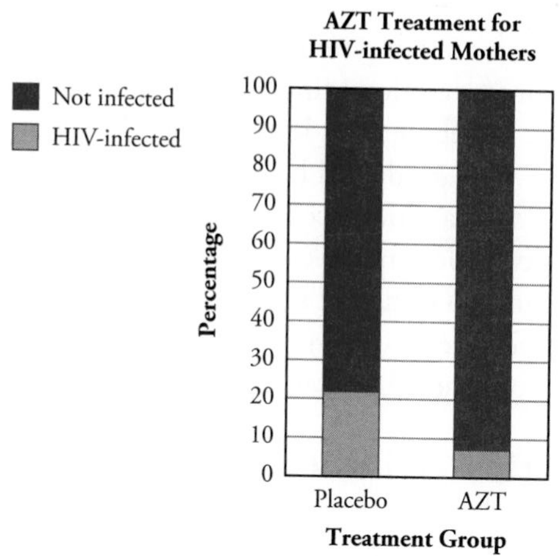

e. Calculate the difference between these two conditional proportions. Does this difference appear to be a large one?

f. Calculate the ratio of the two proportions in part d. Also write a sentence interpreting what this number represents. *Hint:* The convention is to put the larger proportion in the numerator.

If the response variable categories are incidence and nonincidence of a disease, then the **relative risk** is the ratio of the proportions having the disease between the two groups of the explanatory variable. Relative risk is especially useful for situations in which the conditional proportions are fairly small.

Even though the difference in the HIV-positive proportions between the two treatment groups was "small," the relative risk allows us to see that the rate of HIV-infection was three times larger for the placebo group, a much more revealing comparison.

g. Can you legitimately conclude that AZT is the *cause* of the three-fold reduction in HIV-infection rate compared to the placebo group? Explain based on the type of study conducted.

Watch Out
Randomization should balance out potential confounding variables between the two groups, which eliminates confounding as an explanation for why the AZT group outperformed the placebo group. In principle, though, there could be another explanation for why the AZT group did better than the placebo group: Perhaps randomness itself led to such a difference in recovery rates. The researchers could have been victims of one of those unlucky random assignments that created dissimilar groups. This issue is related to statistical significance, and you will learn methods for assessing statistical significance in Unit 4. In particular, you will learn that in this study, the AZT group did perform *significantly* better than the placebo group.

Activity 6-3: Lifetime Achievements

6-3, 6-6, 24-18

Consider the two variables on which you collected class data in question 4 of the Preliminaries: *gender* and *preferred lifetime achievement*.

a. Classify each variable as either categorical (also binary) or quantitative.

Gender:

Preferred lifetime achievement:

b. Which of these variables would you consider the explanatory variable and which the response variable?

Explanatory: Response:

c. Determine the marginal distribution of the *preferred lifetime achievement* variable. Display its distribution with a bar graph, and comment on what it reveals about your classmates' preferences.

d. Count how many students fall into each of the six possible pairs of responses to these questions. Record these counts in the appropriate cells of the following table. For example, the number that you place in the upper-left cell of the table should be the number of male students who classify themselves as preferring an Olympic medal. *Hint:* You might want to start by putting tally marks in the appropriate cell as you work through the list of responses.

	Male	Female
Olympic Medal		
Nobel Prize		
Academy Award		

e. Suppose that a friend in another class tells you that there were 15 male students and 21 female students in that class, and that 16 preferred an Olympic medal, 12 a Nobel Prize, and 8 an Academy Award. Explain why this would not be enough information to re-create the two-way table.

Watch Out

In order to construct a two-way table, you need to know more than just the total counts for each variable separately. You need information on both variables *simultaneously* in order to know how many people of each gender chose a particular achievement.

For example, the following two-way tables could both be created from the marginal totals given in part e:

	Male	Female	Total
Olympic Medal	12	4	16
Nobel Prize	3	9	12
Academy Award	0	8	8
Total	15	21	36

	Male	Female	Total
Olympic Medal	7	9	16
Nobel Prize	5	7	12
Academy Award	3	5	8
Total	15	21	36

Both tables have the same marginal distributions for each variable individually (15 male students and 21 female students; 16 Olympic medals, 12 Nobel prizes, and 8 Academy Awards), but the tables reveal a very different relationship between gender and achievement preference. The table on the left shows dissimilar choices between male students and female students, whereas the table on the right reveals similar choices between the genders.

Activity 6-4: Hypothetical Hospital Recovery Rates

6-4, 6-25

The following two-way table classifies hypothetical hospital patients according to the hospital that treated them and whether they survived or died:

	Survived	Died	Total
Hospital A	800	200	1000
Hospital B	900	100	1000

a. Calculate the proportion of hospital A's patients who survived and the proportion of hospital B's patients who survived. Which hospital saved the higher percentage of its patients?

A: B:

Suppose that when you further categorize each patient according to whether they were in fair condition or poor condition prior to treatment, you obtain the following two-way tables:

Fair Condition

	Survived	Died	Total
Hospital A	590	10	600
Hospital B	870	30	900

Poor Condition

	Survived	Died	Total
Hospital A	210	190	400
Hospital B	30	70	100

b. Convince yourself that when you combine the "fair" and "poor" condition patients, the totals are indeed those given in the table in part a.

c. Among those patients who were in *fair* condition, compare the recovery rates for the two hospitals. Which hospital saved the greater percentage of its patients who had been in fair condition?

Hospital A: Hospital B:

d. Among those patients who were in *poor* condition, compare the recovery rates for the two hospitals. Which hospital saved the greater percentage of its patients who had been in poor condition?

Hospital A: Hospital B:

> This phenomenon is called **Simpson's paradox,** which refers to the fact that aggregate proportions can reverse the direction of the relationship seen in the individual pieces.

In this case, hospital B has the higher recovery rate (proportion) overall, yet hospital A has the higher recovery rate for each type of patient.

e. Write a few sentences explaining (arguing from the data given) how hospital B can have the higher recovery rate overall, yet hospital A can have the higher recovery rate for each type of patient. *Hint:* Do fair or poor patients tend to survive more often? Does one hospital tend to treat more of one type of patient? Is there any connection here?

f. Which hospital would you rather go to if you were ill? Explain.

Watch Out

Two conditions need to be met for Simpson's paradox to occur. For example, the survival rate among fair- and poor-condition patients differed, and the proportion of fair-condition patients at each hospital differed (more went to hospital B). In general, for Simpson's paradox to occur, a third variable needs to be related to the other two. In the hospital example, the third variable was the condition of the patients, which was differentially related to both the *survival* variable and the *hospital* variable. The figure gives you a visual representation:

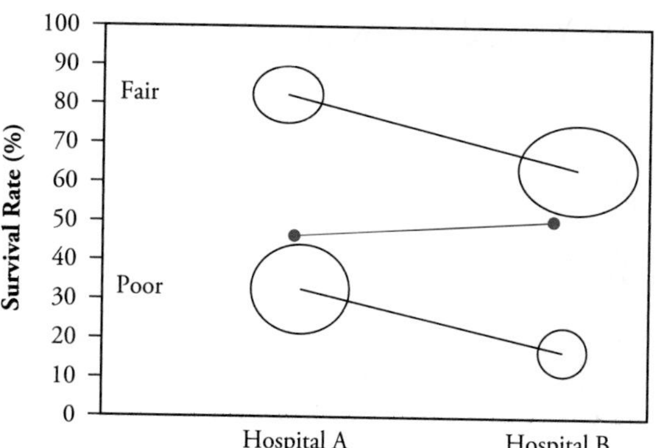

The sizes of the ovals represent the sample sizes in each group, and the centers of the ovals represent the overall survival rate of that group. The fair-condition patients have higher survival rates than the poor-condition patients at both hospitals. However, most of the fair-condition patients are at hospital B, and most of hospital B's patients are in fair condition. (Perhaps hospital B is a small rural hospital.) So when you look at the overall survival rate for hospital B (the solid circle), it is closer to the survival rate of the fair-condition patients. But the overall survival rate of hospital A (perhaps a large urban hospital), weighted by the larger proportion of poor-condition patients there, will be closer to the lower survival rate of the poor-condition patients. In this case, the overall survival rate for hospital A was pulled below the overall survival rate for hospital B.

Activity 6-5: Back to Sleep

4-5, **6-5,** 21-2

Recall from Activity 4-5 that the National Infant Sleep Positioning Study began conducting national surveys in 1992 to examine how well parents were heeding recommendations to place sleeping babies on their backs or sides (Willinger et al., 1998). This recommendation had been issued in 1992, and a promotional campaign called "Back to Sleep" was launched in 1994. For five consecutive years, from 1992 through 1996, researchers interviewed a random sample of between 1002 and 1008 parents of infants less than eight months old. The following table gives the percentages responding for each of the three sleep positions:

	1992	1993	1994	1995	1996
Stomach	70%	58%	43%	29%	24%
Side	15%	22%	27%	32%	39%
Back	13%	17%	27%	38%	35%
No Regular Position	2%	3%	3%	1%	2%

a. Are these percentages parameters or statistics? Explain.

b. Explain how you can tell that these are conditional percentages.

Conditional : which year

c. Identify the explanatory and response variables.

Explanatory: Years Response:

d. Construct a segmented bar graph to display how the conditional distribution of sleep position changed over these five years.

See Aswer Next Page

e. Summarize what the data reveal about whether, and if so how, the conditional distribution of sleep position changed over these five years. Do the data suggest that the recommendation and "Back to Sleep" promotional campaign had an impact? Explain. *Hint*: Remember to consider the design of the study.

The intention of babies

Solution

a. These are statistics because they pertain to the sample of parents interviewed in the study. They are not calculated based on the population of all American parents with infants less than eight months old.

b. These are conditional distributions. Each year has its own percentages sleeping in each of the positions, which sum to 100%.

c. The explanatory variable is *year*; the response variable is *infant's sleep position*.

d. The segmented bar graph is shown here:

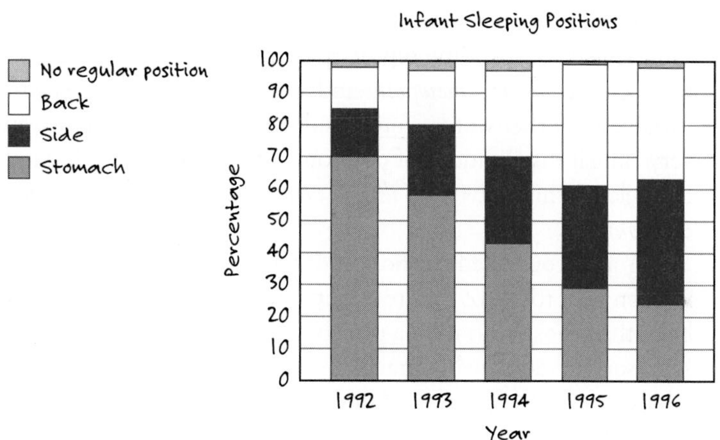

e. The segmented bar graph reveals that the distribution of infant sleep positions changed considerably over this five-year period. The percentage of parents who placed infants on their stomachs declined dramatically, from more than two-thirds (70%) to less than one-fourth (24%) in these five years. The proportions of infants placed on their backs and sides both increased during this period. Because this is an observational study and not an experiment, you cannot say that the recommendation or promotional campaign caused these changes, but it was nonetheless heartening to find that parents were generally changing their habits and placing infants in safer positions to sleep.

Watch Out

- When constructing a two-way table, put the explanatory variable in columns and the response variable in rows. The segmented bar graph is easier to interpret if you also follow that same pattern in creating the graph: Each bar should correspond to a category of the explanatory variable, not the response variable.
- The methods learned in this topic apply only to categorical variables, not to quantitative variables. All of the data that you have analyzed in this topic is categorical.

- The design of the study determines the scope of conclusions you can draw. The AZT study is an experiment, so it allows for a cause-and-effect conclusion (provided the difference proves to be statistically significant). But the sleep position study is observational, so it does not justify a cause-and-effect conclusion, even though the relationship between *year* and *infant's sleep position* is quite strong. However, because the respondents were randomly selected in the sleep position study, we are willing to believe these percentages are representative of all parents of infants. With the AZT study, we would be more cautious in generalizing from the pregnant women who volunteered for that study to all pregnant HIV-positive women.

Wrap-Up...........

This topic provided you with more techniques for analyzing categorical data. You learned to summarize data from two categorical variables in a two-way table, and you calculated marginal and conditional proportions. You also discovered that segmented bar graphs are useful graphical displays for recognizing a relationship (association) between categorical variables. For example, you analyzed data to investigate a relationship between political viewpoint and opinion about government spending. On the issue of spending on the space program, you found essentially no relationship between the *political viewpoint* and *opinion about federal spending on the space program* variables; in other words, you found that liberals, moderates, and conservatives have very similar distributions of opinions. On the other hand, you found clear evidence of a relationship between *political viewpoint* and *opinion about federal spending on the environment.*

You encountered the notion of relative risk, finding that pregnant HIV-positive women who took AZT were about three times less likely to have an HIV-positive baby than those who took a placebo. You also discovered the phenomenon known as Simpson's paradox, which raises interesting issues with regard to analyzing two-way tables. For instance, you explained why one hospital can have a higher survival rate than another for all types of patients and yet have a lower survival rate overall: The hospital with higher success rates received most of the poor condition patients, who were naturally less likely to survive than those patients in fair condition.

Some useful definitions to remember and habits to develop from this topic include

- A **two-way table** of counts organizes data from two categorical variables. Put the explanatory variable in columns and the response variable in rows. When you are collecting such data, keep the outcomes of both variables together for each observational unit, not just the totals for the two variables separately.
- **Segmented bar graphs** are informative graphical displays for two-way tables. These graphs display the conditional proportions of the response variable for each category of the explanatory variable.
- The most relevant numerical summaries for analyzing two-way tables are **conditional distributions. Relative risk** is the ratio of conditional proportions between two groups, which is especially useful when the conditional proportions are small.
- Two categorical variables are **independent,** meaning that they have no **association** or relationship, if the conditional proportions of the response variable are identical across all categories of the explanatory variable.
- **Simpson's paradox** is an interesting phenomenon by which the direction of conditional proportions can reverse when you take into account (and disaggregate the data according to) a third variable.

Some habits we have mentioned before and that you will see often include

- Distinguish between explanatory and response variables.
- Summarize data first through graphical displays and numerical summaries.
- Remember the data collection method (for example, observational study or randomized experiment, random sample from population or not) determines the scope of conclusion you can draw.

Whereas this topic focused exclusively on *categorical* variables, in the next several topics you will learn how to summarize data from quantitative variables using both graphical displays and numerical summaries.

Homework Activities

Activity 6-6: Lifetime Achievements

6-3, **6-6,** 24-18

Reconsider the class data on preferred lifetime achievement and gender. Refer to the two-way table that you constructed in Activity 6-3.

a. Determine the conditional distribution of preferred achievement for each gender.
b. Construct a segmented bar graph to display these conditional distributions.
c. Comment on whether the data suggest that male and female students differ with regard to their choices for lifetime achievement.

Activity 6-7: "Hella" Project

6-7, 21-8

Three students at Cal Poly conducted a class project to investigate their conjecture that students from northern California are more likely to use the word *hella* in their everyday conversation than are students from southern California (Mead, Olerich, Selenkow, 2005). (*Hella* is used to add emphasis, similar to "very.") They collected a sample of first-year students by randomly selecting room numbers in two first-year residence halls. They then asked the occupants of those rooms where they were from and whether they used *hella* in their everyday vocabulary. The responses of the 40 subjects were

(south, yes)	(south, no)	(south, no)	(north, yes)
(north, no)	(north, yes)	(south, yes)	(south, no)
(south, no)	(south, no)	(south, no)	(south, no)
(south, no)	(north, yes)	(north, no)	(south, no)
(north, yes)	(south, no)	(north, yes)	(south, yes)
(north, no)	(north, yes)	(south, no)	(south, no)
(south, no)	(south, no)	(north, no)	(south, no)
(south, no)	(south, no)	(north, yes)	(north, yes)
(north, yes)	(south, no)	(south, no)	(south, no)
(south, no)	(north, yes)	(north, no)	(south, no)

a. Is this an observational study or an experiment?
b. Identify the explanatory and response variables.
c. Organize these responses into a two-way table.
d. Calculate the conditional proportion of students who use the word *hella* in each region.
e. Create a segmented bar graph to display these data.
f. Comment on whether the data seem to support the students' conjecture.

Activity 6-8: Suitability for Politics

6-8, 6-9, 21-15, 25-19, 25-20

Another question asked on the 2004 General Social Survey was whether the respondent generally agreed or disagreed with the assertion that "men are better suited emotionally for politics than women." The following two-way tables summarize the results, one by the respondent's *political viewpoint* and one by *gender.*

	Liberal	Moderate	Conservative
Agree	40	68	96
Disagree	168	225	215

	Male	Female
Agree	109	103
Disagree	276	346

Analyze these data to address the question of whether there is a relationship between political inclination and reaction to the statement, and/or between gender and reaction to the statement. Write a paragraph or two describing your findings, and include appropriate calculations and displays in your answer.

Activity 6-9: Suitability for Politics

6-8, **6-9,** 25-15, 25-19, 25-20

Refer to the previous activity. The GSS has asked this same question for several decades. The following two-way table summarizes the responses over the years:

	1970s	1980s	1990s	2000s
Agree	2398	2563	1909	802
Disagree	2651	4597	6427	2609

Analyze these data to address the question of whether reactions to this statement appear to have changed over the decades. Include appropriate calculations and graphical displays in your answer.

Activity 6-10: A Nurse Accused

1-6, 3-20, **6-10,** 25-23

Reconsider the murder case against nurse Kristen Gilbert, as described in Activity 1-6. Hospital records for an 18-month period indicated that of 257 eight-hour shifts on which Gilbert worked, a patient died during 40 of those shifts. But of 1384 eight-hour shifts on which Gilbert did not work, a patient died during only 34 of those shifts.

a. Organize these data into a 2 × 2 two-way table. Be sure to put the explanatory variable in columns.
b. For each type of shift, calculate the conditional proportion of shifts on which a patient died.
c. Calculate the relative risk of a patient dying, comparing Gilbert shifts to non-Gilbert shifts.
d. Write a sentence interpreting the relative risk calculation from part c.

Activity 6-11: Children's Television Advertisements

6-11, 25-14

Outley and Taddese (2006) conducted an analysis of food advertisements shown during children's television programming on three networks (Black Entertainment Television (BET), Warner Brothers (WB), and Disney Channel) in July 2005. They classified each commercial according to the type of product advertised as well as the network it appeared on. Their data are summarized in the following two-way table:

	BET	WB	Disney	Total
Fast Food	61	32	0	93
Drinks	66	9	5	80
Snacks	3	0	2	5
Cereal	15	16	4	35
Candy	17	26	0	43
Total	162	83	11	256

a. What are the dimensions of this table? *Hint*: In other words, say that this is a ? × ? table.
b. What proportion of the food advertisements on BET were for fast food?
c. What proportion of the fast-food advertisements were on BET?
d. For each network, calculate the conditional distribution of the types of food commercials shown.
e. Use these conditional distributions to construct a segmented bar graph.
f. Comment on what your calculations and graph reveal about possible differences in the types of advertisements shown on children's programs across these three networks.

Activity 6-12: Female Senators

6-12, 18-4, 18-18

The 2007 U.S. Senate consists of 49 Democrats, 49 Republicans, and 2 Independents. (Both Independents associate themselves with Democrats.) There are 16 women in the 2007 U.S. Senate, 11 Democrats and 5 Republicans.

a. What proportion of the senators are women?
b. What proportion of the senators are Democrats?
c. Is it fair to say that most Democratic senators are women? Support your answer with an appropriate calculation.

d. Is it fair to say that most women senators are Democrats? Support your answer with an appropriate calculation.

Activity 6-13: Weighty Feelings

6-13, 25-15, 26-13

One of the questions in the 2003–2004 National Health and Nutrition Examination Survey (NHANES) asked respondents whether they felt their current weight was about right, overweight, or underweight. Responses by gender are summarized in the following table:

	Female	Male	Total
Underweight	116	274	390
About Right	1175	1469	2644
Overweight	1730	1112	2842
Total	3021	2855	5876

a. Identify the explanatory and response variables in this table.
b. Determine and graph the marginal distribution of the *feeling about one's weight* variable.
c. Determine and graph the conditional distribution of weight feelings for each gender.
d. Comment on whether the two genders appear to differ with regard to feelings about their weight.

Activity 6-14: Preventing Breast Cancer

6-14, 6-15, 21-3, 21-18, 25-22

The Study of Tamoxifen and Raloxifene (STAR) enrolled more than 19,000 postmenopausal women who were at increased risk for breast cancer. Women were randomly assigned to receive one of the drugs (tamoxifen or raloxifene) daily for five years. Initial results released in April 2006 revealed that 163 of 9726 women in the tamoxifen group had developed invasive breast cancer, compared to 167 of the 9745 women in the raloxifene group.

a. Is this an observational study or an experiment? Explain how you know.
b. Identify the explanatory and response variables.
c. Organize the data in a two-way table.
d. For each treatment group, calculate the conditional proportion who developed invasive breast cancer.
e. Construct a segmented bar graph to compare these conditional proportions.
f. Calculate the relative risk of developing invasive breast cancer, comparing the raloxifene group to the tamoxifen group.
g. Summarize what your analysis reveals about the relative effectiveness of these two drugs.

Activity 6-15: Preventing Breast Cancer

6-14, **6-15,** 21-3, 21-18, 25-22

Refer to the previous activity. The study hoped to show that raloxifene would reduce the risk of dangerous side effects involving blood clots, as compared to tamoxifen (with a similar cancer incidence rate). The initial report also revealed that 53 of the 9726 women in the tamoxifen group had a blood clot in a major vein, compared to 65 of the 9745 women taking raloxifene. Further, 54 of the 9726 women taking tamoxifen developed a blood clot in the lung, compared to 35 of the 9745 women taking raloxifene. Analyze these data for each of these side effects (blood clot in major vein, blood clot in lung). Be sure that your analysis includes

a. A two-way table
b. Conditional proportions
c. A segmented bar graph
d. Relative risk
e. A summary of conclusions (including interpreting your calculations and considering the scope of conclusions; in particular, can you conclude that raloxifene is the cause of the increased risk of these blood clots? Explain.)

Activity 6-16: Flu Vaccine

6-16, 21-14

Toward the end of 2003, there were many warnings that the flu season would be especially severe, and many more people chose to obtain a flu vaccine than in recent years. In January 2004, the Centers for Disease Control and Prevention published the results of a study that looked at workers at Children's Hospital in Denver, Colorado. Of the 1000 people who had chosen to receive the flu vaccine (before November 1, 2003), 149 still developed flu-like symptoms. Of the 402 people who did not get the vaccine, 68 developed flu-like symptoms.

a. Identify the observational units and the two variables of interest in this study. Comment on the type of variables involved. Which variable do you consider the explanatory variable and which the response variable?
b. Create a two-way table to summarize the results of this study.
c. Produce numerical and graphical summaries of the results of this study. Write a paragraph describing what these summaries reveal, remembering to place your statements in context and being careful how you talk about which group is being conditioned on.
d. Does this study provide evidence that the flu vaccine helped reduce the occurrence of a flu-like illness? *Hint*: Is a cause-and-effect conclusion appropriate? Explain.

Activity 6-17: Watching Films

6-17, 6-18

Find a listing of the 30 movies from last year that generated the most box office revenue in the U.S. (You might try looking at www.boxofficemojo.com.) Review the list and check off the movies that you saw. Then ask a friend to go through the list and do the same.

a. Create a 2 × 2 table for which the column variable is *whether you saw the movie,* and the row variable is *whether your friend saw the movie.*
b. Construct graphical and numerical summaries to investigate whether you and your friend appear to have similar movie-watching habits. Write a paragraph summarizing your findings. *Hint:* Ask whether the likelihood of your friend having watched a movie seems to differ depending on whether you saw it.

Activity 6-18: Watching Films

6-17, **6-18**

Reconsider the previous activity.

a. Construct a hypothetical example of a 2 × 2 table for which you and your friend's movie-watching habits are completely independent.
b. Construct a hypothetical example of a 2 × 2 table for which you and your friend have very similar movie-watching habits.
c. Construct a hypothetical example of a 2 × 2 table for which you and your friend have very dissimilar (almost opposite) movie-watching habits.

Activity 6-19: Botox for Back Pain

5-4, **6-19,** 6-20, 21-17

The study described in Activity 5-4 was actually a randomized, double-blind study to investigate whether botulinum toxin A (botox) is effective for treating patients with chronic low-back pain (Foster et al., 2001). At the end of eight weeks, 9 of 15 subjects in the botox group had experienced substantial relief, compared to 2 of 16 in the group that had received a normal saline injection.

a. Explain what *randomized* and *double-blind* mean in this study.
b. Organize the study results in a two-way table.
c. Calculate conditional proportions for patients who experienced substantial relief and create a segmented bar graph to display the study results.
d. How many times more likely was a subject to experience pain reduction if he or she was in the botox group as opposed to the saline group?
e. Summarize what your analysis reveals about whether botox appears to have been effective for treating back pain.

Activity 6-20: Botox for Back Pain

5-4, 6-19, **6-20,** 21-17

Reconsider the previous activity. Suppose that the study had involved ten times as many patients, and that the results had turned out to be identical proportionally.

a. Multiply every entry in the 2 × 2 table by 10.
b. Describe how the segmented bar graph would change (if at all).
c. Calculate the relative risk for this new table. Describe how it has changed (if at all).

Now suppose that the original study had turned out differently: Suppose that all 15 in the botox group, and 8 of 16 in the saline group, experienced relief.

d. Calculate the conditional proportions of patients who experienced substantial pain relief, and calculate the difference between them.
e. How does the difference in proportions in part d compare to that from the actual study results?
f. Calculate the relative risk of experiencing pain relief for these hypothetical data. How does it compare to the relative risk from the actual study results?
g. Summarize what these calculations reveal about how sample size affects relative risk.

Activity 6-21: Gender-Stereotypical Toy Advertising

6-21, 6-22

To study whether toy advertisements tend to picture children with toys considered typical of their gender, researchers examined pictures of toys in a number of children's catalogs. For each picture, they recorded whether the child pictured was a boy or girl. (For this activity, we will ignore ads in which boys and girls appeared together.) They also recorded whether the toy pictured was a traditional "male" toy (such as a truck or a toy soldier) or a traditional "female" toy (such as a doll or a kitchen set) or a "neutral" toy (such as a puzzle or a toy phone). Their results are summarized in the following two-way table:

	Boy Shown	Girl Shown
Traditional "Male" Toy	59	15
Traditional "Female" Toy	2	24
Neutral Gender Toy	36	47

a. Calculate the marginal totals for the table.
b. What proportion of the ads showing boys depicted traditionally male toys? Traditionally female toys? Neutral toys?
c. Calculate the conditional distribution of toy types for ads showing girls.
d. Construct a segmented bar graph to display these conditional distributions.
e. Based on the segmented bar graph, comment on whether the researchers' data seem to suggest that toy advertisers do indeed tend to present pictures of children with toys stereotypical of their gender.

Activity 6-22: Gender-Stereotypical Toy Advertising

6-21, **6-22**

Reconsider the previous activity. For this activity, we will refer to ads that show boys with traditionally "female" toys and ads that show girls with traditionally "male" toys as "crossover" ads.

a. What proportion of the ads under consideration are crossover ads?
b. What proportion of the crossover ads depict girls with traditionally male toys?
c. What proportion of the crossover ads depict boys with traditionally female toys?
d. When toy advertisers do defy gender stereotypes, in which direction does their defiance tend?

Activity 6-23: Baldness and Heart Disease

6-23, 25-9

To investigate a possible relationship between heart disease and baldness, researchers asked a sample of 663 male heart patients to classify their degree of baldness on a 5-point scale (Lasko et al., 1993). They also asked a control group (not suffering from heart disease) of 772 males to do the same baldness assessment. The results are summarized in the following table:

	None	Little	Some	Much	Extreme
Heart Disease	251	165	195	50	2
Control	331	221	185	34	1

a. What proportion of these men identified themselves as having little or no baldness?
b. Of those who had heart disease, what proportion claimed to have some, much, or extreme baldness?
c. Of those who declared themselves as having little or no baldness, what proportion were in the control group?
d. Construct a segmented bar graph to compare the distributions of baldness ratings between subjects with heart disease and those from the control group.
e. Summarize your findings about whether a relationship seems to exist between heart disease and baldness.
f. Consider the "none" or "little" baldness categories as one group and the other three categories as another group. Calculate the relative risk of heart disease between these two groups.

Activity 6-24: Gender and Lung Cancer

Researchers in New York investigated a possible gender difference in incidence of lung cancer among smokers. They screened 1000 people who were smokers of age 60 or older. They found 459 women and 541 men, with 19 of the women and 10 of the men suffering from lung cancer.

a. Identify the explanatory and response variables in this study.
b. Construct a 2×2 table for these data. Also include the marginal totals in the table.
c. Calculate the conditional distributions of lung cancer for each gender.
d. Construct a segmented bar graph to display these distributions.
e. Calculate the relative risk of having lung cancer between women and men.
f. Write a sentence or two interpreting the relative risk value in this context.

Activity 6-25: Hypothetical Hospital Recovery Rates

6-4, **6-25**

Reconsider the hypothetical data in Activity 6-4 on page 104. Calculate and interpret the relative risk of dying between hospital A and hospital B for patients

a. Overall **b.** In fair condition **c.** In poor condition

Activity 6-26: Graduate Admissions Discrimination

6-26, 21-5

The University of California at Berkeley was charged with having discriminated against women in their graduate admissions process for the fall quarter of 1973 (Bickel et al., 1975). The following table identifies the number of acceptances and denials for both men and women applicants in each of the six largest graduate programs at the institution at that time.

	Men Accepted	Men Denied	Women Accepted	Women Denied
Program A	511	314	89	19
Program B	352	208	17	8
Program C	120	205	202	391
Program D	137	270	132	243
Program E	53	138	95	298
Program F	22	351	24	317
Total	1295	1486	559	1276

a. Start by ignoring the program distinction, collapsing the data into a two-way table of gender by admission status. To do this, find the total number of men accepted and denied and the total number of women accepted and denied. Construct a table such as the one shown here:

	Admitted	Denied	Total
Men	1295	1486	2781
Women	559	1276	1835
Total	1854	2762	4616

b. Consider, for a moment, just the men applicants. Of the men who applied to one of these programs, what proportion were admitted? Now consider the women applicants; what proportion of them were admitted? Do these proportions seem to support the claim that men were given preferential treatment in admissions decisions?

c. To try to isolate the program or programs responsible for the mistreatment of women applicants, calculate the proportion of men and the proportion of women within each program who were admitted. Record your results in a table such as the one here:

	Proportion of Men Admitted	Proportion of Women Admitted
Program A		
Program B		
Program C		
Program D		
Program E		
Program F		

d. Does it seem as if any particular program is responsible for the large discrepancy between men and women in the overall proportions admitted?
e. Reason from the data given to explain how it happened that men had a much higher admission rate overall, even though women had higher rates in most programs, and no program favored men very strongly.

Activity 6-27: Softball Batting Averages

Construct your own hypothetical data to illustrate Simpson's paradox in the following context. Show that it is possible for one softball player (Amy) to have a higher proportion of hits than another (Barb) in June and in July and yet have a lower proportion of hits for the two months combined. To get started: Suppose that Amy has 100 at-bats in June and 400 in July, and suppose that Barb has 400 at-bats in June and 100 in July. Make up how many hits each player had in each month so the previous statement holds. See how great the differences can be in the players' proportions of hits. (The proportion of hits is the number of hits divided by the number of at-bats.)

	June	July	Combined
Amy's Hits			
Amy's At-bats	100	400	500
Amy's Proportion of Hits			
Barb's Hits			
Barb's At-bats	400	100	500
Barb's Proportion of Hits			

Activity 6-28: Hypothetical Employee Retention Predictions

Suppose an organization is concerned about the number of new employees who leave the company before they finish one year of work. In an effort to predict whether a new employee will leave or stay, they develop a standardized test and apply it to 100 new employees. After one year, they note what the test had predicted (stay or leave) and whether the employee actually stayed or left. They then compile the data into the following table:

	Predicted to Stay	Predicted to Leave	Total
Actually Stayed	63	21	84
Actually Left	12	4	16
Total	75	25	100

a. Of those employees predicted to stay, what proportion actually left?
b. Of those employees predicted to leave, what proportion actually left?
c. Is an employee predicted to stay any less likely to leave than an employee predicted to leave?

d. Considering your answers to parts a–c, does the standardized test provide any helpful information about whether an employee will leave or stay?

e. Is the employee's action *independent* of the test's prediction? Explain.

f. Create a segmented bar graph to display these conditional distributions.

g. Sketch what the segmented bar graph would look like if the test were perfect in its predictions.

h. Sketch what the segmented bar graph would look like if the test were very useful but not quite perfect in its predictions.

Activity 6-29: Politics and Ice Cream

Suppose that 500 college students are asked to identify their preferences in political affiliation (Democrat, Republican, or Independent) and ice cream flavor (chocolate, vanilla, or strawberry). Fill in the following table in such a way that the variables *political affiliation* and *ice cream preference* turn out to be completely independent. In other words, the conditional distribution of ice cream preference should be the same for each political affiliation, and the conditional distribution of political affiliation should be the same for each ice cream flavor.

	Chocolate	Vanilla	Strawberry	Total
Democrat	108			240
Republican		72	27	
Independent		32		80
Total	225			500

Activity 6-30: Your Choice

6-30, 12-22, 19-22, 21-28, 22-27, 26-23, 27-22

Consider a pair of categorical variables whose relationship you would be interested in exploring. Describe the variables in as much detail as possible, including any definitions you need to clarify how the variables would be measured, and indicate how you would present the data in a two-way table.

TOPIC

7 • • • Displaying and Describing Distributions

How much do Olympic rowers typically weigh, and how much variability do these weights reveal? Why does one member of the rowing team weigh so much less than all of his teammates? How long do British monarchs tend to remain on the throne? Did eastern or western states tend to experience more population growth in the 1990s? These are some of the questions you will investigate in this topic that introduces you to analyzing quantitative data.

Overview

In the previous topic, you examined ways to summarize categorical data both graphically and numerically by creating two-way tables and segmented bar graphs and calculating conditional proportions and relative risk. In this topic and in the next several, you will turn your attention to summarizing *quantitative* data. Previously in Topic 2, you learned about using dotplots for displaying the distribution of relatively small datasets of a quantitative variable, and you began to comment on the center and spread of the distribution. In this topic, you will discover some more key features of a distribution and also become familiar with two new visual displays: stemplots and histograms.

Preliminaries

1. What do you think is a typical weight for a male Olympic rower?

2. Take a guess as to the length of the longest reign of a British monarch since William the Conqueror.

3. Do you think states in the eastern or western U.S. tended to have greater population growth (on a percentage basis) in the 1990s?

4. How long (in miles) would you consider the ideal day hike?

In-Class Activities

Activity 7-1: Matching Game

7-1, 8-3

Consider the following seven variables:

A. Point values of letters in the board game Scrabble
B. Prices of properties on the Monopoly game board
C. Jersey numbers of Cal Poly football players in 2006
D. Weights of rowers on the 2004 U.S. men's Olympic team
E. Blood pressure measurements for a sample of healthy adults
F. Quiz percentages for a class of statistics students (quizzes were quite straightforward for most students)
G. Annual snowfall amounts for a sample of cities around the U.S.

a. What type of variables are these: categorical or quantitative?

The following dotplots display the distributions of these variables, but the variables are not shown in the same order as they are listed. Moreover, the scales have been intentionally left off the axes!

b. For each dotplot, try to identify the variable displayed (by letter, from the previous list). Also, provide a brief explanation of your reasoning in each case. Note: You might make different matches than other students; be prepared to justify your choices.

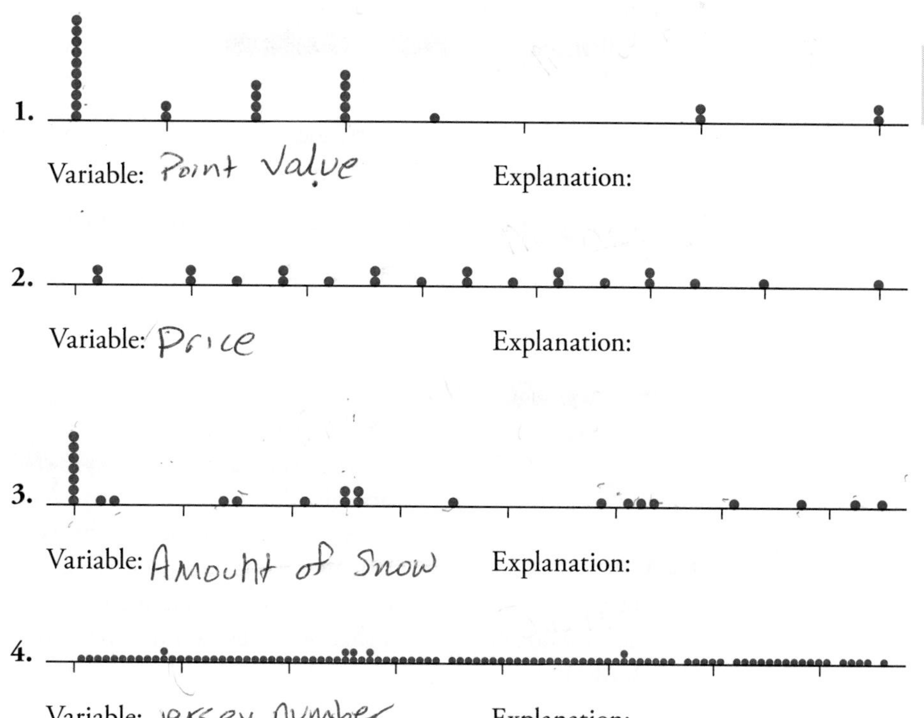

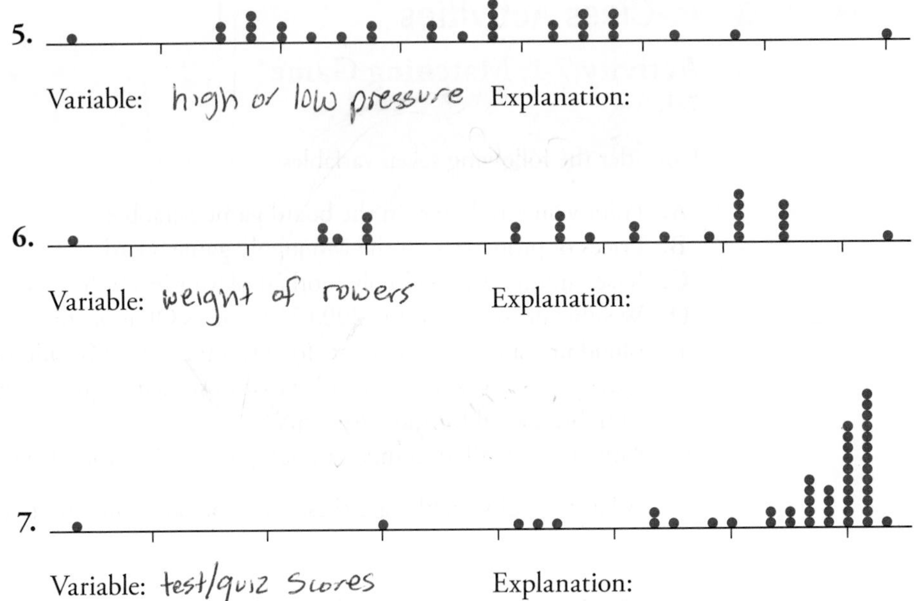

One of the goals of this matching game is to illustrate that you can anticipate what the distribution of a set of data might look like by considering the context of the data. Another goal is to help you develop a checklist of features to look for when describing the distribution of a quantitative variable. You encountered the two most important features previously in Topic 2:

1. The **center** of a distribution is usually the most important aspect of a set of data to notice and describe.
2. A distribution's **spread,** or **variability** (or consistency), is a second important feature.

Of course, with this activity, these two features are not helpful because no scales are given on the axes.

c. What do dotplots 1 and 3 have in common, as opposed to 6 and 7 especially?

The **shape** of a distribution is the third important feature. Although distributions come in a limitless variety of shapes, certain shapes arise often enough to have their own names. A distribution is said to be **symmetric** if the left side is roughly a mirror image of the right side. A distribution is **skewed to the right** if its tail extends toward larger values (on the right side of the axis) and **skewed to the left** if its tail extends toward smaller values. Another aspect of shape is whether the distribution contains clusters or gaps. The following drawings illustrate these shapes, but these are idealizations in that real data would never display such perfectly smooth distributions.

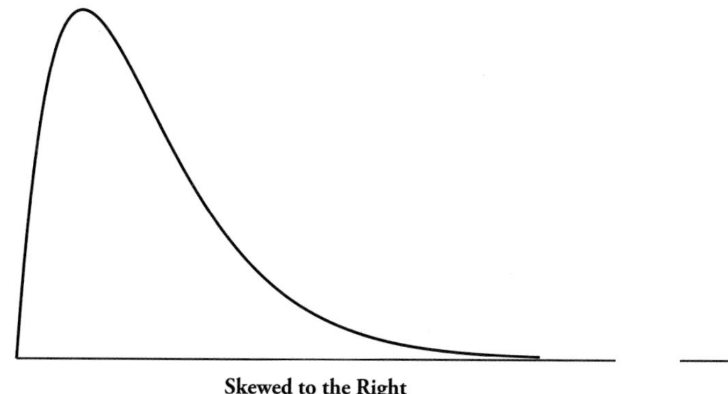
Skewed to the Right

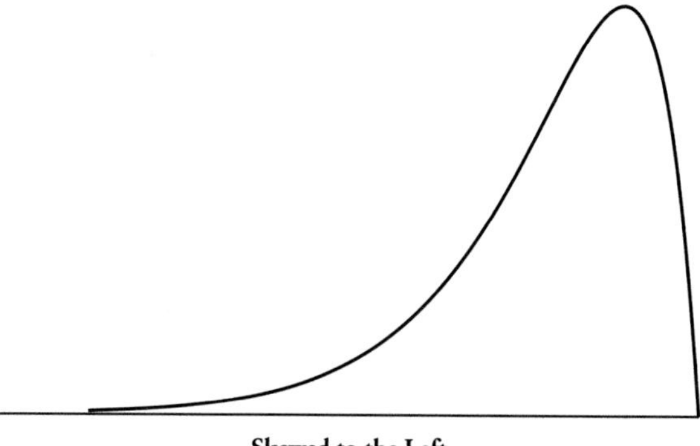
Skewed to the Left

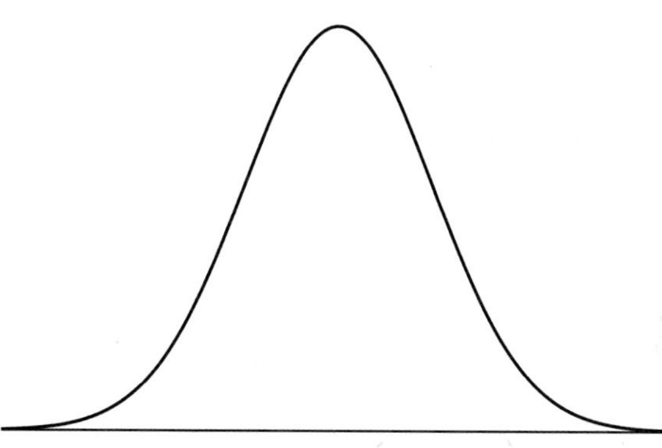
Symmetric, Mound-Shaped

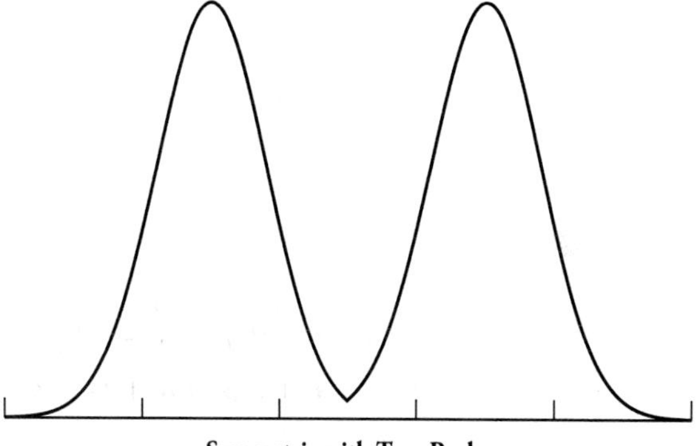
Symmetric with Two Peaks

When describing any distribution of a quantitative variable, you should always comment on center, spread, and shape. Also, note any unusual features revealed in the distribution, such as outliers. **Outliers** are data values that differ markedly from the pattern established by the vast majority of data. Outliers warrant close examination.

Keep in mind that although you have created a checklist of important features, every dataset is unique and may have some interesting aspects not covered by this list.

Activity 7-2: Rowers' Weights

7-2, 8-4, 10-21

The following table records the weight (in pounds) of each rower on the 2004 U.S. Olympic men's rowing team. A dotplot of these weights follows.

Name	Weight	Event	Name	Weight	Event
Abdullah	185	Double sculls	Holbrook	195	Quad sculls
Ahrens	215	Eight	Hoopman	185	Eight
Allen	210	Eight	Klugh	205	Four
Beery	215	Eight	Moser	210	Four
Cipollone	120	Eight	Nuzum	210	Double sculls
Deakin	200	Eight	Read	180	Eight
DuRoss	210	Quad sculls	Ruckman	155	LW double sculls
Hansen	210	Eight	Samsonov	185	Pair

(*continued*)

Name	Weight	Event	Name	Weight	Event
Schroeder	229	Four	Volpenhein	215	Eight
Smack	195	Quad sculls	Walton	180	Pair
Smith	160	LW four	Warner	160	LW four
Teti	160	LW four	Wherley	215	Four
Todd	154	LW four	Wilkinson	190	Quad sculls
Tucker	153	LW double sculls			

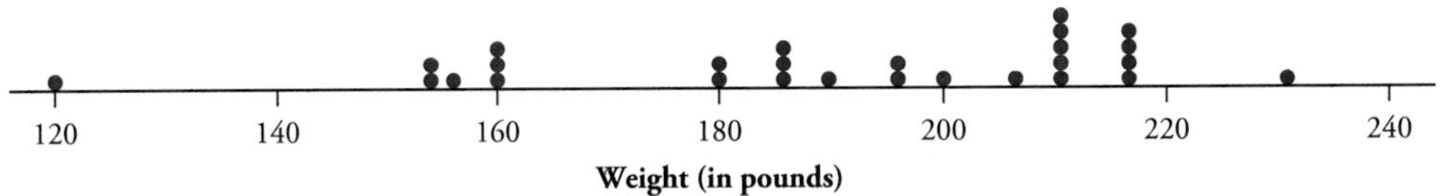

a. Write a paragraph describing key features of this distribution of rowers' weights. *Hint:* Remember the checklist, and relate your comments to the data's context.

- Light weight boat
- The center is
- Spread is pretty ranged (varied)
- Many guys between 180-218
- Outlier is 120 (coxswain)
- Symetric w/ 2 peaks

b. Give the name of the team member whose weight makes him an apparent outlier. Suggest an explanation for why his weight differs so substantially from the others.

- Outlier is 120 = Coxswain

↳ Cipollone

c. Suggest an explanation for the clusters and gaps apparent in the distribution. *Hint:* Consider the events in which the cluster of less heavy rowers competed.

Clusters: light weight vs. heavy weight

Activity 7-3: British Monarchs' Reigns

7-3, 7-9

The following table lists the years on the throne (calculated as *year leaving throne* minus *year taking throne*) for the monarchs of Great Britain, beginning with William the Conqueror in 1066:

Ruler	Reign (in years)	Ruler	Reign (in years)	Ruler	Reign (in years)	Ruler	Reign (in years)
William I	21	Edward III	50	Edward VI	6	George I	13
William II	13	Richard II	22	Mary I	5	George II	33
Henry I	35	Henry IV	13	Elizabeth I	44	George III	59
Stephen	19	Henry V	9	James I	22	George IV	10
Henry II	35	Henry VI	39	Charles I	24	William IV	7
Richard I	10	Edward IV	22	Charles II	25	Victoria	63
John	17	Edward V	0	James II	3	Edward VII	9
Henry III	56	Richard III	2	William III	13	George V	25
Edward I	35	Henry VII	24	Mary II	6	Edward VIII	1
Edward II	20	Henry VIII	38	Anne	12	George VI	15

a. Who is the current monarch? Why do you think his/her reign is not represented here (as of December 2006)?

• Elizabeth II
• It is not over

b. How many years was the longest reign? Which monarch was on the throne the longest?

• 63 for Victoria

c. How many years was the shortest reign? What do you think this value really means? Which monarch spent the shortest time on the throne?

• 0 for Edward V

You can create a **stemplot** of a distribution of quantitative data by separating each data value into two pieces: a "stem" and a "leaf." When the data consist primarily of two-digit numbers, you can make the tens digit the stem and the ones digit the leaf. For example, for William I's reign of 21 years, 2 is the stem and 1 is the leaf; for Richard III's reign of 2 years, 0 is the stem and 2 is the leaf.

Consider the following stemplot of the length of monarchs' reigns.

```
0 | 0123566799
1 | 0023333579 — William I
2 | 012224455
3 | 355589
4 | 4
5 | 069
6 | 3
```

Scewed to the Right
Center is somewhere in 20's
Varied 1–63 years

d. How many years on the throne does 6|3 represent? Which monarch reigned for that many years?

e. How many monarchs reigned for 13 years? Without looking at the original data, explain how you can tell from the stemplot.

4

f. Based on this stemplot, describe the *shape* (symmetric, skewed to the left, skewed to the right) of the distribution of length of monarchs' reigns.

The stemplot is a simple but useful visual display. Its principal virtues are that it is easy to construct by hand (for relatively small sets of data) and it retains the actual numerical values of the observations. Stemplots also conveniently sort the data values from least to greatest.

g. Determine a value such that half of these 40 monarchs reigned for more years and half reigned for fewer years.

h. Determine a value such that one-fourth of the monarchs reigned for fewer years. Then find a value such that one-fourth of the monarchs reigned for more years.

Watch Out

Be sure to sort the leaves from smallest to largest in the rows (stems) of a stemplot.

Activity 7-4: Population Growth

7-4, 8-15, 9-18, 10-10

The population of the U.S. grew by 13.2% between 1990 and 2000. The following table reports the percentage growth in population for each of the fifty states during the 1990s. States are classified by whether they are east or west of the Mississippi River:

Western State	%	Western State	%	Eastern State	%	Eastern State	%
Alaska	14.0	Montana	12.9	Alabama	10.1	New Hampshire	11.4
Arizona	40.0	Nebraska	8.4	Connecticut	3.6	New Jersey	8.9
Arkansas	13.7	Nevada	66.3	Delaware	17.6	New York	5.5
California	13.8	New Mexico	20.1	Florida	23.5	North Carolina	21.4
Colorado	30.6	North Dakota	0.5	Georgia	26.4	Ohio	4.7
Hawaii	9.3	Oklahoma	9.7	Illinois	8.6	Pennsylvania	3.4
Idaho	28.5	Oregon	20.4	Indiana	9.7	Rhode Island	4.5
Iowa	5.4	South Dakota	8.5	Kentucky	9.7	South Carolina	15.1
Kansas	8.5	Texas	22.8	Maine	3.8	Tennessee	16.7
Louisiana	5.9	Utah	29.6	Maryland	10.8	Vermont	8.2
Minnesota	12.4	Washington	21.1	Massachusetts	5.5	Virginia	14.4
Missouri	9.3	Wyoming	8.9	Michigan	6.9	West Virginia	0.8
				Mississippi	10.5	Wisconsin	9.6

To investigate whether eastern or western states tend to have higher population growths, you can produce a visual display to compare the two groups.

> In a **side-by-side stemplot,** a common set of stems (for example, the tens digit) is placed in the middle of the display with leaves (for example, the ones digit) branching out in either direction to the left and right. The convention is to order the leaves from the middle out from least to greatest.

a. Complete the following side-by-side stemplot to compare the distributions of population growth between eastern and western states. *Hint:* Use the stems provided here. To view the data with an appropriate level of detail, truncate the data by ignoring the tenths digit after the decimal point. For example, report Alabama's percentage growth as 1|0 and Connecticut's as 0|3. We have already put ordered leaves on the stemplot for all states except for the eight that begin with the letter *N*. An underscore (_) indicates where you should fill in a leaf value.

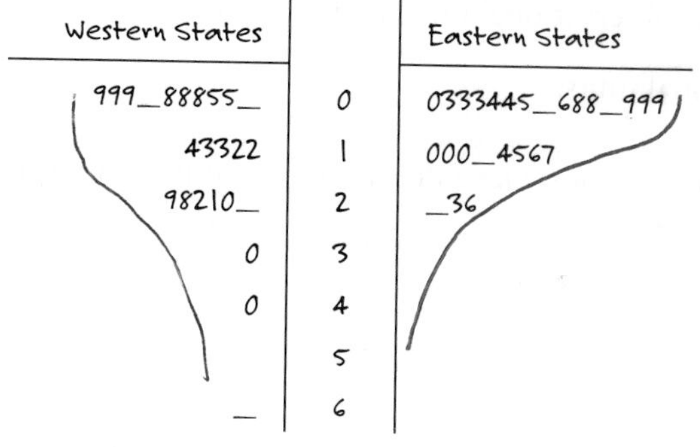

leaf unit = one percentage point

b. Based on this side-by-side stemplot, compare and contrast the distributions of population growth between eastern and western states. *Hint:* Remember to comment on center, spread, shape, and any other apparent features, such as outliers.

East:
Center: teens (lower)
Shape: skewed to right
not varied: 1 → 26

West:
Center: teens (higher)
Shape: skewed to right
Varied: 0 → 66

Activity 7-5: Diabetes Diagnoses

7-5, 10-15

The National Health and Nutrition Examination Survey (NHANES) is a large-scale study conducted annually by the National Center for Health Statistics. It involves over 10,000 Americans, randomly selected according to a multistage sampling plan. All sampled subjects are asked to complete a survey and take a physical examination.

One of the questions asked in the 2003–2004 NHANES survey pertained only to subjects who had been diagnosed with diabetes. Subjects were asked to indicate the age at which they were first diagnosed with diabetes by a health professional. Responses for the 548 subjects with diabetes are summarized in the following frequency table:

Age	1	2	3	4	5	6	7	8	9	10	11	12	13	15	16	17	18
Tally	1	3	7	5	5	2	4	5	1	5	1	2	1	2	3	1	4
Age	19	20	21	24	25	26	27	28	29	30	31	32	33	34	35	36	37
Tally	3	1	3	2	1	4	4	2	1	6	5	3	8	5	15	4	7
Age	38	39	40	41	42	43	44	45	46	47	48	49	50	51	52	53	54
Tally	10	9	15	9	8	9	5	12	8	7	11	12	25	9	9	7	8
Age	55	56	57	58	59	60	61	62	63	64	65	66	67	68	69	70	71
Tally	22	13	8	11	16	19	5	16	9	10	10	9	6	12	7	13	4
Age	72	73	74	75	76	77	78	79	80	81	82	83	84	85	87	88	
Tally	4	9	4	8	7	5	6	3	2	3	2	1	2	1	1	1	

a. Identify the observational units and variable.

Observational units:

Variable:

b. Is it practical to construct a dotplot or a stemplot to display this distribution of ages? Explain.

A **histogram** is a graphical display similar to a dotplot or stemplot, but histograms are more feasible with very large datasets; histograms also permit more flexibility than stemplots. You construct a histogram by dividing the range of data into subintervals (**bins**) of equal length, counting the number (**frequency**) of observational units in each subinterval, and constructing bars whose heights correspond to the frequency in each subinterval. The bar heights can also correspond to the proportions (**relative frequencies**) of observational units in the subintervals.

The following histogram displays the distribution of ages at which the 548 survey subjects were diagnosed with diabetes. Note that the midpoints of the subintervals are reported. For example, the first subinterval indicates that four people were diagnosed with diabetes at age 2 or younger, and the second subinterval reveals that 23 people were diagnosed with diabetes between the ages of 3 and 7, inclusive.

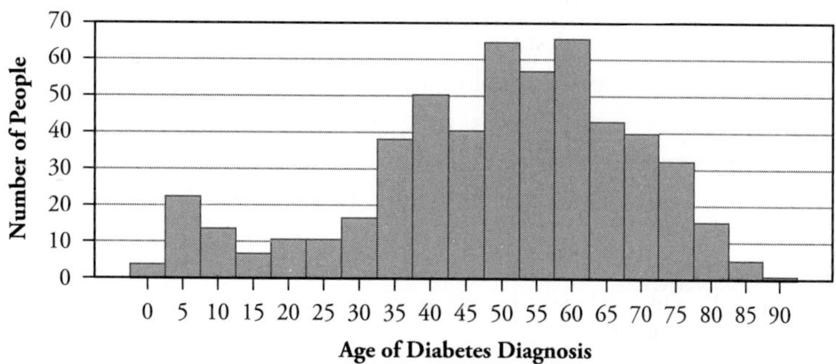

c. How many and what proportion of the 548 people were diagnosed with diabetes before the age of 18?

d. How many and what proportion of the 548 people were diagnosed with diabetes after the age of 62?

e. Comment on the shape, center, and spread of the distribution of these ages of diagnosis.

Watch Out

- A histogram is not the same as a bar graph. A histogram displays the distribution of a quantitative variable, whereas a bar graph displays the distribution of a categorical variable. Note that the horizontal axis of a histogram is a numerical scale. In fact, all of the graphs that you have studied in this topic (dotplot, stemplot, and histogram) apply only to quantitative variables.
- The choice of subintervals (bins) can have a substantial effect on the visual impression conveyed by a histogram. You might want to try several choices to see which subintervals provide the most informative display.
- Histograms are much easier to construct with the graphing calculator than by hand.
- Some students mistakenly think that the frequencies in the previous data table (1, 3, 7, 5, 5, and so on) are the actual data. In this case, the data are the *ages* at which the subjects were diagnosed with diabetes. The frequencies provide a convenient way to report the ages without having to type, for example, the age of 50 a total of 25 times. When you look at a histogram, make sure you are clear on the total number of observational units involved.

The TI-83 Plus and TI-84 Plus graphing calculators have the capability to create visual displays of statistical data from the Stat Plot screen. Before attempting to use your calculator to display a histogram (or any other statistical display), make sure all graphing functions are turned off (cleared). To access the **STAT PLOTS** menu, press [2ND] [STATPLOT]. Your calculator screen should look similar to the example shown here, but may have a different set-up depending on your calculator.

In the screen shot shown above, you can see that **Plot1** is on and set up to draw a histogram of a list named AGE and to plot the frequency in a list named TALLY; **Plot2** is off and set up to draw a boxplot of a list named PROP; and **Plot3** is off and set up to draw a scatterplot of list Y versus list X (you will learn about boxplots and scatterplots later). To turn off all of the plots, press the [4] key and then press [ENTER].

f. Download the **Diabetes** list into your calculator.

g. Use your graphing calculator to re-create the histogram shown following part b.

1. Press [2ND] [STAT PLOT] and [ENTER] or the [1] key to begin setting up **Plot1.**

You have six choices of graphical displays: scatterplot, xy-line, histogram, modified boxplot, boxplot, and normal probability plot. You also have the capability to select any list and frequency (in this case, you will select the TALLY list).

2. Make **Plot1** a histogram by arrowing down to the **Type** line and to the right to highlight the histogram icon. Press the ENTER key. Arrow down to **Xlist** and enter the list name. *TI hint:* Above many of the keys are letters. To access these letters, press the ALPHA key before pressing each letter. You know you are in the Alphanumeric mode when the cursor flashes the letter *A*.
3. Next, you must set the viewing window for your histogram. To do this, press the WINDOW key and enter the following numbers:

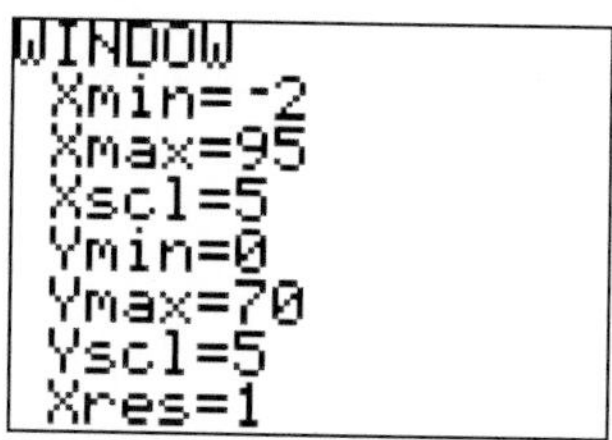

TI tip: Make sure you use the (-) key to express the negative sign, not the subtraction key.

4. Now press the GRAPH key, followed by the TRACE key and then the right and left arrows to obtain information about your histogram.

h. Now change the **Xscl** setting to **10** in the Window screen (this reduces the number of subintervals to 10), and then reduce the number of subintervals to **5.** Finally, increase the number of subintervals to **30.** Comment on the different appearances of the distribution (of the same age data) revealed by these histograms. Which histogram do you think provides the most informative display? Explain.

i. Suppose that someone creates the following graphs. How would you characterize the shape of the weight distribution for each graph? Does either convey appropriate information about shape, center, and spread of the distribution of weights? Explain.

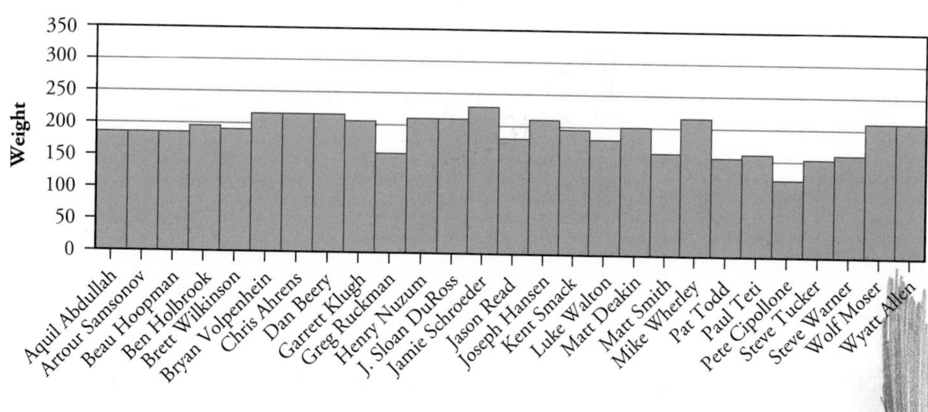

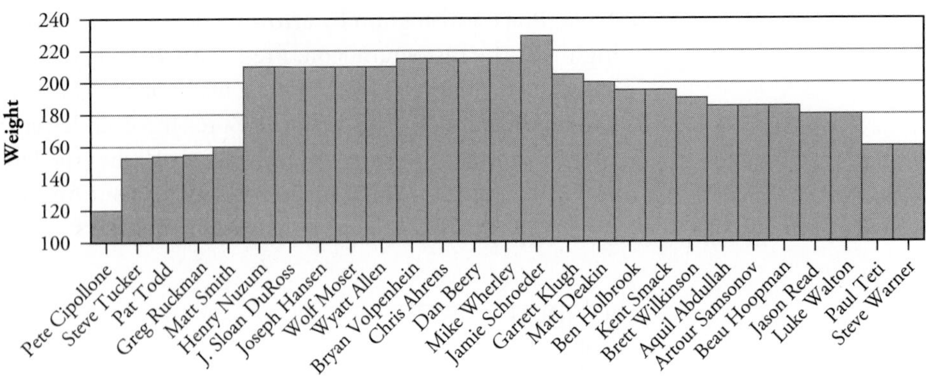

Watch Out

Neither of these graphs is a legitimate histogram, and neither is helpful for examining the distribution (shape, center, spread) of male Olympic rower weights. Remember that the variable is weight, and so weight is what needs to be presented on the horizontal axis. Both of these graphs present the observational units (the rowers) on the horizontal axis and present very misleading information about the shape, center, and spread of the distribution of weights. For example, someone interpreting these graphs might consider the shape of the weight distribution to be symmetric. Further, the outlier and clusters that we found before are not readily seen.

Activity 7-6: Go Take a Hike!

7-6, 7-14

The following data are the distances (in miles) of the 72 hikes described in the book *Day Hikes in San Luis Obispo County* (Stone, 2000), listed in the order that the hikes appear in the book:

0.6	1.0	2.5	4.0	2.5	2.0	3.8	4.6	1.5	1.0	5.8	2.2	1.5
2.5	4.0	3.5	2.6	1.8	2.0	2.5	3.2	1.5	1.0	2.0	9.5	2.6
4.6	6.0	2.1	5.5	4.0	3.4	5.0	3.0	2.5	3.2	4.5	3.0	1.5
2.0	0.8	3.0	3.0	5.5	2.8	5.6	7.0	6.0	3.2	2.6	4.0	5.0
6.0	7.4	2.0	5.6	2.0	1.0	1.5	1.5	2.0	1.8	3.4	5.0	1.5
7.0	1.0	3.0	4.0	6.0	2.2	2.0						

a. Create a stemplot of this distribution.

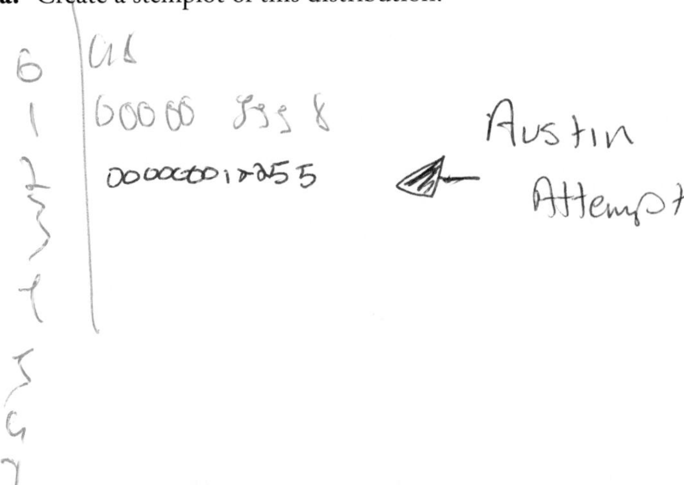

b. Write a paragraph describing the distribution of these hike distances.

Scewed right (tail toward big #)

Center: ≈ 2

Spread: pretty spread out .6 - 9.5

Solution

a. The stemplot is shown here.

```
0 | 68
1 | 00000555555588
2 | 000000001225555556668      leaf unit = .1 mile
3 | 000002224458
4 | 00000566
5 | 00055668
6 | 0000
7 | 004
8 |
9 | 5
```

b. The distribution of hike distances is sharply skewed to the right, indicating there are many hikes on the short side and only a few longer hikes. A typical hike is between 2 and 3 miles. Most hikes are between 1 and 6 miles, but two hikes are less than a mile and a few are more than 6 miles. The longest hike is 9.5 miles, which is a bit unusual and could be considered an outlier because this hike is more than two miles longer than the next longest hike (7.4 miles). Many hikes have a reported distance that is a multiple of a whole number or a half number of miles.

Watch Out

- It's easy to confuse "skewed to the left" and "skewed to the right." Remember the direction of the skew is indicated by the longer tail. You can also think of "skewed to the right" as being "piled up on the left" and vice versa.
- Pay attention to what measurement units are represented in a stemplot. In the hike stemplot, the leaf unit is 0.1 miles, but it's easy to forget that and say, for example, that the longest hike is 95 miles. Include a scale to remind the reader what the units represent (9|5 = 9.5 miles, or leaf unit = tenths of a mile).
- Remember context! In this case, you're describing hike distances. It should be clear to anyone reading your description that the stemplot shows a distribution of hike distances, not lengths of British monarchs' reigns or rowers' weights!

Wrap-Up...........

In this topic, you returned to the study of quantitative variables and learned more about distributions of data. You found that thinking about the context of the data can help you to anticipate what the distribution will look like. This was the point of the first activity, helping you to think about how the distribution of snowfall amounts would differ from quiz percentages, which in turn would differ from blood pressure measurements or property prices on the Monopoly game board.

You created a checklist of various features to consider when describing a distribution, including **center, spread, shape,** and **outliers.** You learned about the three shapes that distributions often follow: **symmetric, skewed to the left,** and **skewed to the right.** For example, you found that the distribution of length of British monarchs' reigns is skewed to the right, indicating that most reigns are short and only a few reigns last a long time, whereas the distribution of Olympic rower weights is skewed to the left, revealing that most rowers are heavier, with only a few "lightweight" rowers. Returning to the rowing example, these features reveal that a rowing team includes rowers for lightweight events, in addition to the person who calls out instructions and, therefore, needs to weigh less and so might be an outlier.

You also learned two new techniques for displaying a distribution of quantitative data: **stemplots** and **histograms.** For comparing distributions between two groups, you used **side-by-side stemplots.** Finally, you began to see how useful the graphing calculator can be for producing such graphs.

Some useful definitions to remember and habits to develop from this topic include

- When describing the distribution of a quantitative variable, comment on at least center, spread, shape, and outliers. Provide enough information about these features so someone not looking at the graph can understand the distribution simply from your description. Your comments should always relate to the data's context.
- Pay particular attention to outliers. Identify them, and investigate possible explanations for their occurrence. In particular, make sure they are not simply typographical errors. Similarly, you should note and try to explain any clusters or gaps in the distribution.
- Remember to label (describe) and scale (number) the axes on dotplots and histograms. Consider the observational units and the variable being measured.
- Examine several different types of graphs. Each type has advantages and disadvantages, and creating several types of graphs of the same data will enhance your understanding of the data.
- Anticipate features of the data by considering the nature of the variable involved.

You now have a checklist of distribution features; however, we have been somewhat vague about specifying the two most important aspects of a distribution: center and spread. The next two topics will remedy this ambiguity by introducing you to specific numerical measures of center and spread. In addition to learning how to perform these calculations, you will investigate the properties of these measures and examine their utility and limitations.

Homework Activities

Activity 7-7: Newspaper Data

Consider the following variables on which you could find data in your local newspaper. For each, indicate whether you would expect the shape of its distribution to be skewed to the right, skewed to the left, or roughly symmetric. Explain your reasoning. Also sketch a histogram, labeling and scaling the horizontal axis clearly, to illustrate what you expect each distribution to look like.

a. Ages at death for deceased persons in the obituary notices
b. Asking prices of houses listed in real estate section
c. Final stock prices of the day
d. High temperatures for a January day in major cities across the U.S.
e. High temperatures for a July day in major cities across the U.S.

Activity 7-8: Student Data

1-1, 1-5, 2-7, 2-8, 3-7, **7-8**

Consider the following variables on which you could gather data from students at your school. For each, indicate whether you would expect the shape of its distribution to be skewed to the right, skewed to the left, or roughly symmetric. Explain your reasoning. Also sketch a histogram, labeling and scaling the horizontal axis clearly, to illustrate what you expect each distribution to look like.

a. Amount paid on the student's last visit to a grocery store
b. Last digits of cell phone number
c. Number of siblings
d. Price paid for most recent haircut
e. Number of pairs of shoes owned
f. Number of t-shirts owned
g. Distance from place of birth
h. Age
i. Duration of most recent phone conversation

Activity 7-9: British Monarchs' Reigns

7-3, **7-9**

Reconsider the data from Activity 7-3 on length of reigns for British monarchs, and then examine the following graph.

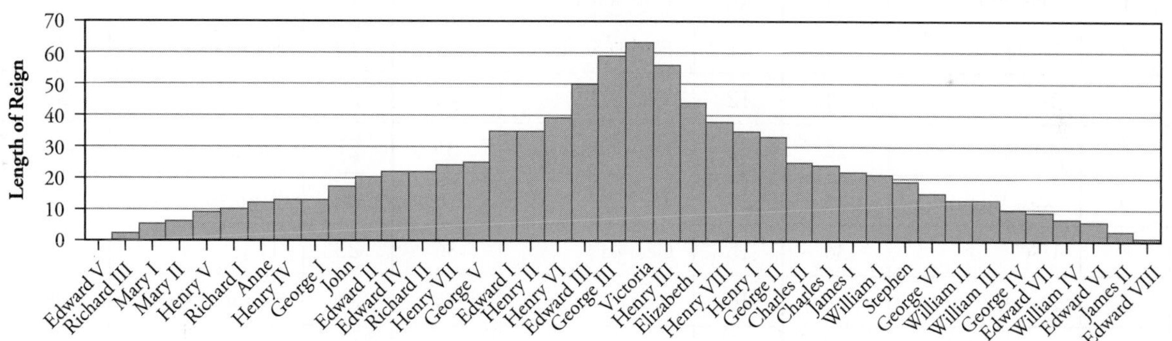

a. Is this a legitimate histogram of the distribution of lengths of reign for British monarchs? Explain.
b. This graph suggests that the distribution of lengths of British monarchs' reigns is quite symmetric. Is this an accurate impression? Explain.

Activity 7-10: Honda Prices

7-10, 10-19, 12-21, 28-14, 28-15, 29-10, 29-11

The following dotplots pertain to data collected on used Honda Civics for sale on the Internet. The three variables represented here are the car's price, mileage, and year of manufacture. Make an educated guess as to which dotplot represents which variable, and explain your reasoning.

a.

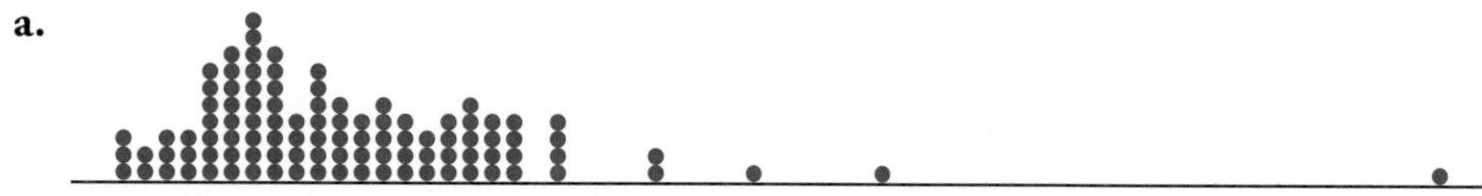

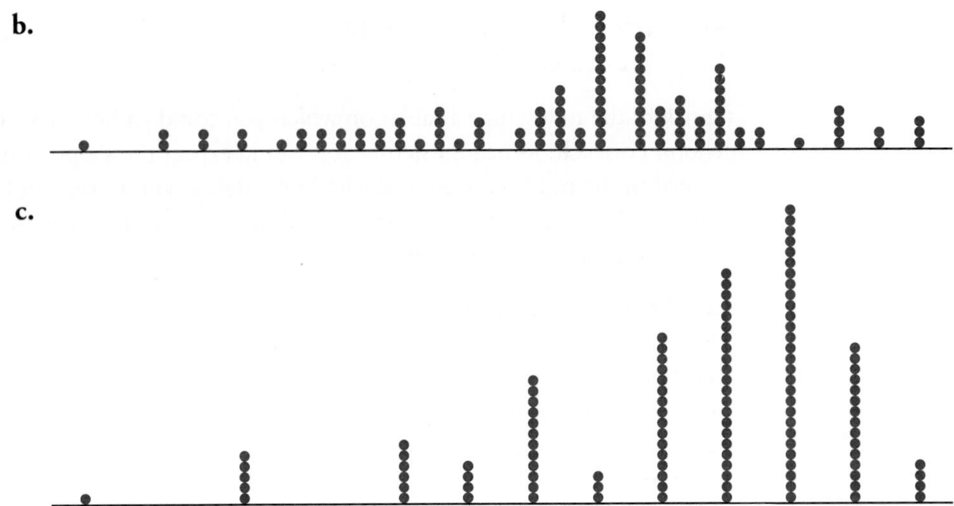

Activity 7-11: College Football Scores

The following data are the margins of victory in college football games involving Top 25 teams for the first weekend of the 2006 season:

Rank	Team	Margin	Rank	Team	Margin	Rank	Team	Margin
1	Ohio State	23	10	Oklahoma	7	18	Clemson	48
2	Notre Dame	4	11	Florida State	3	19	Penn State	18
3	Texas	49	12	Miami (FL)	–3	20	Nebraska	39
4	Auburn	26	13	Louisville	31	21	Oregon	38
5	West Virginia	32	14	Michigan	20	22	Tennessee	17
6	USC	36	15	Georgia	36	23	TCU	10
7	Florida	27	16	Iowa	34	24	Arizona State	35
8	Louisiana State	42	17	Virginia Tech	38	25	Texas Tech	32
9	Cal	–17						

Note: All of the teams in the Top 25 won, except for the two games in which Top 25 teams played each other, where the loser's margin is reported as negative. Most top teams try to schedule a game against a much weaker opponent for the first week of the season.

a. Create a stemplot of these data.
b. Write a paragraph describing the distribution of these margins of victory.
c. How would you expect this distribution to differ later in the season, when teams generally play stronger opponents within their conference schedule?

Activity 7-12: Hypothetical Exam Scores

7-12, 8-22, 8-23, 9-22, 10-22, 27-8

Create a side-by-side stemplot of hypothetical exam scores for students studying under two different professors, with the property that Professor Cobb's students tend to score higher than Professor Moore's, but Cobb's students' scores also display more variability than Moore's students' scores.

Activity 7-13: Hypothetical Commuting Times

7-13, 22-2, 22-6, 22-7

Suppose you want to compare commuting times for two different routes (A and B) to school. Create histograms of this hypothetical data using the same scale, with the property that route A tends to take a bit longer on average, but the time it takes to travel route A is more consistent than for route B.

Activity 7-14: Go Take a Hike!

7-6, **7-14**

Recall Activity 7-6, which detailed data on day hikes in San Luis Obispo county. Some of the other quantitative variables reported on those hikes include elevation gain and expected time to complete the hike. Use the graphing calculator (the data are stored in the grouped file DAYHIKES containing the lists **DIST, TIME,** and **ELEV,** respectively) to produce graphical displays of these variables. Comment on key features of the distribution of each variable.

Activity 7-15: Memorizing Letters

5-5, **7-15,** 8-13, 9-19, 10-9, 22-3, 23-3

Reconsider the class data collected from the memory experiment in Topic 5, discussed in Activity 5-5. Subjects were randomly assigned to see convenient three-letter groupings of letters (the "JFK" group) or less convenient, more irregular groupings of letters (the "JFKC" group).

a. Create a side-by-side stemplot to compare the distributions of scores between the two groups. *Hint:* You might want to consider a variation of a stemplot that repeats each stem and then puts leaves from 0–4 on the first line of the stem and leaves from 5–9 on the second line of the stem. This produces more stems, which elongates the graph but might provide a more appropriate level of detail and, therefore, a more informative display.

b. Comment on what this stemplot reveals about the comparative performance of the two groups on this memory task.

Activity 7-16: Placement Exam Scores

7-16, 9-4

Dickinson College's Department of Mathematics and Computer Science gives an exam each fall to freshmen who intend to take calculus; scores on the exam are used to determine which level of calculus a student should be placed in. The exam consists of 20 multiple-choice questions. Scores for the 213 students who took the exam in 1992 are tallied in the following table, and the raw scores are stored in the data file MATHPLACEMENT, which contains the list **SCORE**).

Score	1	2	3	4	5	6	7	8	9	10
Count	1	1	5	7	12	13	16	15	17	32
Score	11	12	13	14	15	16	17	18	19	
Count	17	21	12	16	8	4	7	5	4	

a. Use the graphing calculator to produce a dotplot and a histogram of this distribution. Try a few different histograms, with varying subintervals, to find the one that you consider the best summary of the distribution.

b. Write a few sentences describing the shape, center, and spread of this distribution.

Activity 7-17: Hypothetical Manufacturing Processes

A manufacturing process strives to make steel rods with diameters of 12 centimeters, but the actual diameters vary slightly from rod to rod; however, rods with diameters within ± 0.2 centimeters of the target value are considered to be within specifications (i.e., acceptable). Suppose that 50 rods are collected for inspection from each of four processes and that the dotplots of their diameters are shown here.

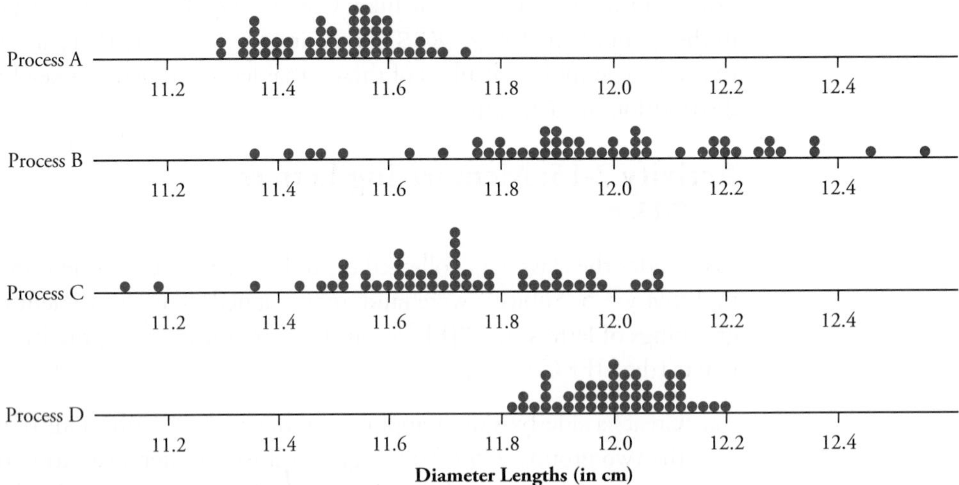

Write a paragraph describing each of these distributions, concentrating on the center and variability of the distributions. Also address each of the following questions in your paragraph:

- Which process is the best?
- Which process is the most stable, i.e., has the least variability in rod diameters?
- Which process is the least stable?
- Which process produces rods with diameters that are generally farthest from the target value?

Activity 7-18: Hitchcock Films

The following table lists the running times (in minutes) of the videotape versions of 22 movies directed by Alfred Hitchcock:

Film	Time	Film	Time
The Birds	119	*Psycho*	108
Dial M for Murder	105	*Rear Window*	113
Family Plot	120	*Rebecca*	132
Foreign Correspondent	120	*Rope*	81
Frenzy	116	*Shadow of a Doubt*	108
I Confess	108	*Spellbound*	111
The Man Who Knew Too Much	120	*Strangers on a Train*	101
Marnie	130	*To Catch a Thief*	103
North by Northwest	136	*Topaz*	126
Notorious	103	*Under Capricorn*	117
The Paradine Case	116	*Vertigo*	128

a. Construct a stemplot of the distribution of the films' running times.
b. Comment on key features of this distribution.
c. One of these movies is unusual in that all of the action took place in one room and Hitchcock filmed it without editing. Explain how you might be able to identify this unusual film based on the distribution of the films' running times.

Activity 7-19: *Jurassic Park* Dinosaur Heights

In the blockbuster movie *Jurassic Park,* dinosaur clones run amok on a tropical island intended to become the world's greatest theme park. In Michael Crichton's novel on which the movie was based, the examination of dotplots of dinosaur heights provides the first clue that the dinosaurs are not as under control as the park's creator would like to believe. Here are reproductions of two dotplots presented in the novel.

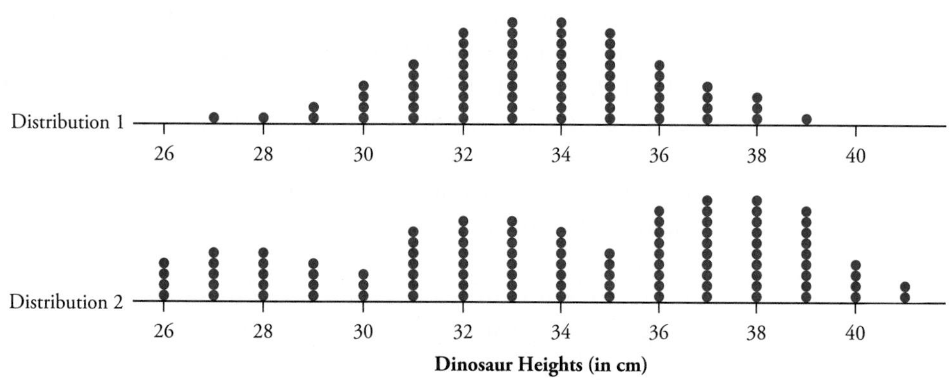

a. Comment briefly on the most glaring difference in these two distributions of dinosaur heights.
b. The cynical mathematician Ian Malcolm (a character in the novel) argues that one of these distributions is characteristic of a normal biological population, whereas the other distribution is what you would expect from a controlled population that had been introduced in three separate batches (as these dinosaurs had). Identify which distribution corresponds to which type of population.

c. Take a closer look at distribution 1. Something about it suggests that it does not come from real data but rather from the mind of an author. Can you identify its suspicious quality?

Activity 7-20: Turnpike Distances

The Pennsylvania Turnpike extends from Ohio in the west to New Jersey in the east. The distances (in miles) between its exits (as of 1996) as you travel west to east are listed in this table.

Exit	Name	Miles	Exit	Name	Miles
1	Ohio Gateway	*	16	Carlisle	25.0
1A	New Castle	8.0	17	Gettysburg Pike	9.8
2	Beaver Valley	3.4	18	Harrisburg West Shore	5.9
3	Cranberry	15.6	19	Harrisburg East	5.4
4	Butler Valley	10.7	20	Lebanon–Lancaster	19.0
5	Allegheny Valley	8.6	21	Reading	19.1
6	Pittsburgh	8.9	22	Morgantown	12.8
7	Irwin	10.8	23	Downingtown	13.7
8	New Stanton	8.1	24	Valley Forge	14.3
9	Donegal	15.2	25	Norristown	6.8
10	Somerset	19.2	26	Fort Washington	5.4
11	Bedford	35.6	27	Willow Grove	4.4
12	Breezewood	15.9	28	Philadelphia	8.4
13	Fort Littleton	18.1	29	Delaware Valley	6.4
14	Willow Hill	9.1	30	Delaware River Bridge	1.3
15	Blue Mountain	12.7			

a. To prepare for constructing a histogram to display this distribution of distances, count how many values fall into each of the subintervals: 0.1–5.0, 5.1–10.0, 10.1–15.0, 15.1–20.0, 20.1–25.0, 25.1–30.0, 30.1–35.0, 35.1–40.0.

b. Construct (by hand) a histogram of this distribution, clearly labeling both axes.

c. Comment on key features of this distribution.

d. Find a value such that half of the exits are greater than this distance from one another and half are less than this distance apart. Also explain why such a value is not unique.

e. If a person has to drive between consecutive exits and has only enough gasoline to drive 20 miles, is she likely to make it? Assume you do not know which exits she is driving between and explain your answer.

f. Repeat part e supposing she has only enough gasoline to drive 10 miles.

Activity 7-21: Exam Scores

The following stemplot displays the scores of 62 students who took the first exam in a statistics course:

```
4 | 9
5 | 578
6 | 344568
7 | 0224444456677799
8 | 001112222333455577788999
9 | 000001112223345
```

leaf unit = 1 point

a. How many students received a score of 77 on the exam?
b. How many and what proportion of students scored 90 or greater?
c. How many and what proportion of students scored less than 70?
d. Which score appears most often?
e. There are only two values between 75 and 95 for which no one obtained those particular scores. What are the two values?

Activity 7-22: Tennis Simulations

7-22, 8-21, 9-16, 22-18

As part of a study investigating alternative scoring systems for tennis, researchers analyzed computer simulations of tennis matches (Rossman and Parks, 1993). For 100 simulated tennis games, researchers recorded the number of points played in each game. (A game of tennis is won when one player wins four points, with a margin of at least two points above their opponent's score.) The data are tallied in the following table. From the table, you can see that 12 games ended after just 4 points and 2 games required 18 points to complete.

Points in Game	4	5	6	8	10	12	14	16	18
Tally for Standard System	12	21	34	18	9	2	1	1	2

a. Create a visual display of these data, and comment on key features of the distribution.
b. Describe and explain the unusual gaps that the distribution exhibits.

The study also simulated tennis games using the "no-ad" scoring system, which awards the game to the first player to reach four points. These data are tallied here.

Points in Game	4	5	6	7
Tally for No-ad System	13	22	33	32

c. Create a graph of these data using the same scaling as your graph in part a. Comment on how the distribution of game lengths compares between the two scoring systems.

Finally, the study also simulated tennis games using a "handicap" scoring method that uses no-ad scoring and also awards weaker players bonus points at the start of a game. The lengths of 100 simulated games using this handicap scoring system are tallied here:

Points in Game	1	2	3	4	5	6	7
Tally for Handicap System	3	4	12	18	28	25	10

d. Create a graph of these data using the same scaling as in parts a and c. Comment on how the distributions of game lengths compare among all three scoring systems.

Activity 7-23: Blood Pressures

The grouped file BLOODPRESSURES contains the lists **SYSBP** (systolic blood pressure), **DIABP** (diastolic blood pressure), and **PLRT** (pulse rate), respectively. These lists contain data from the NHANES study on systolic and diastolic blood pressure measurements. Produce graphical displays of these distributions, and describe their key features.

TOPIC

8 • • • Measures of Center

How much money does the typical contestant on a television game show win? Is this amount the same as the average amount won by contestants on television game shows? Is it possible for most contestants to win less than the average amount? What about real estate—how can most houses cost less than the average price? Averages such as these are commonly reported in news articles and product advertisements, but these averages can be misleading and misunderstood. For example, would comparing averages be a meaningful way to investigate whether cancer pamphlets are written at an appropriate level to be read and understood by cancer patients? You will investigate properties of averages and other measures of center in this topic.

Overview

In the previous topic, you examined distributions of quantitative data, representing them graphically with stemplots and histograms and describing their key features such as center, spread, shape, clusters, and outliers. When analyzing distributions of quantitative data, it is also handy to use a single numerical measure to summarize a certain aspect of a distribution, such as its center. In this topic, you will encounter some common measures of the center of a distribution, investigating their properties, and applying these measures to some real data, while exposing some of their limitations.

Preliminaries

1. Who would you expect to get the most sleep—college students with a 7 AM statistics class, an 8 AM statistics class, or an 11 AM statistics class? Which section's students would you expect to get the least sleep?

   Most: 11:00 AM　　　　Least: 7:00 AM

2. Estimate the typical winning amount on the television game show *Deal or No Deal.*

   1,000

3. Estimate the age of this course's instructor (or, if your instructor prefers, another teacher or administrator at your school).

4. Record these age estimates for yourself and your classmates.

5. How much do you expect to make for your annual salary in your first year of full-time employment?

   45,000

6. Record these salary expectation values for yourself and your classmates.

In-Class Activities

Activity 8-1: Sleeping Times

8-1, 19-4, 19-5, 19-12, 19-19, 20-2, 20-7

The following dotplots display the distributions of sleeping times for students in three sections of a statistics course. Section 1 met at 7 AM, Section 2 at 8 AM, and Section 3 at 11 AM.

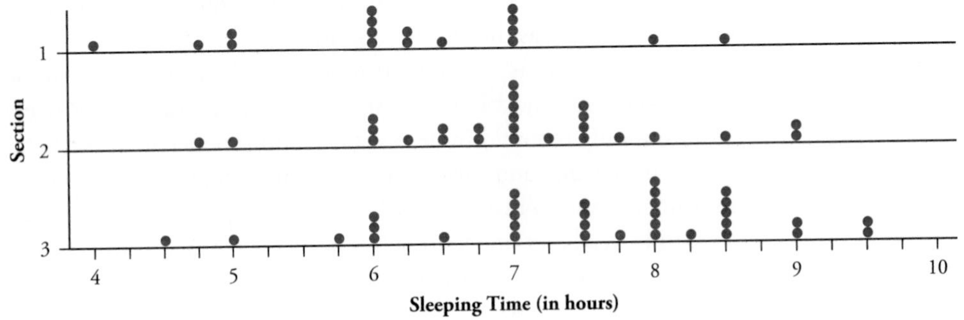

a. Identify the observational units in this graph. Also identify and classify the presumed explanatory and response variables.

Observational units:

Explanatory variable: Class time   Type: Cat.

Response variable: hrs of sleep   Type: Quant.

b. How would you compare the *centers* of these three distributions of sleep times? Would you say that they are all similar, or do some centers differ noticeably from others?

c. Concentrate on Section 1 for the moment. If you were asked to select a single number to represent the center of this distribution of hours of sleep, what number would you choose? Briefly explain how you arrived at this choice.

Up to this point, we have been quite vague about how to determine the center of a distribution of quantitative data. Now, however, we will introduce two common ways to measure center: the mean and the median.

The **mean** is the ordinary arithmetic average, found by adding up the values for each of the observational units and dividing by the number of values. You can think of the mean as the balance point of the distribution.

The **median** is the value of the middle observation (once the n values have been arranged in order). The median of an odd number of data values is located in position $(n + 1)/2$. The median of an even number of values is defined to be the average (mean) of the middle two values, on either side of position $(n + 1)/2$.

d. Calculate the mean of the sleep times in Section 1. In the dotplot shown at the beginning of this activity, mark the mean value with an "x" on the axis for Section 1. *Hint:* There are 17 values in Section 1. When you read the values from the dotplot, note that all values are multiples of 0.25 hours.

e. Determine the median of the sleep times in Section 1. Explain how you arrived at this number.

f. The mean sleep times in the other two sections are 7.523 hours and 7.000 hours. Which is the mean for Section 2, and which is the mean for Section 3? Explain how you know without performing the calculations.

Section 2 mean: Section 3 mean:

Explanation:

g. Calculate the median sleep times for Section 2 and Section 3. *Hint*: There are 26 student values in Section 2 and 33 in Section 3.

Section 2 median: Section 3 median:

h. Do the mean and median of a dataset always equal each other?

Watch Out

- These measures of center (mean and median) apply only to *quantitative* variables, such as *hours spent sleeping*. Because *section* is a categorical variable, it makes no sense to calculate the mean section or median section.
- The mean and median can be either parameters or statistics, depending on whether the data constitute a population or a sample. We will use μ to represent a population mean and $\bar{x}$ to represent a sample mean.

- There is a third quantity that is sometimes used as a measure of center: The **mode** is the numerical value that appears most often in a distribution of data. For example, in Section 3, the mode is 8 hours of sleep. The mode is often not useful, however, because the values might not repeat or because the most common value might not be near the center of the distribution. The mode does apply to categorical as well as quantitative variables, but the notion of "center" does not often make sense with a categorical variable.

i. What is the mode of the *section* variable? Interpret this value?

You will usually rely on your calculator to compute measures of center. Moreover, becoming aware of the properties associated with interpreting the mean and median of a dataset is at least as important as being able to calculate those values. The next few activities introduce you to some of these properties and to the use of your calculator to compute the mean and median of a distribution of quantitative data.

Activity 8-2: Game Show Prizes

On the TV game show *Deal or No Deal,* a contestant chooses from among 26 suitcases, each containing a dollar amount. Those 26 amounts (as of April 2007) are listed here:

.01	1	5	10	25	50	75	100	200
300	400	500	750	1000	5000	10,000	25,000	50,000
75,000	100,000	200,000	300,000	400,000	500,000	750,000	1,000,000	

a. Determine the median prize amount.

$M = \frac{(750 + 1000)}{2} = 875$

b. Without doing the calculation, how do you expect the mean to compare to the median prize amount? Explain.

Very different because of the wide range of $'s

c. Use your calculator to compute the median and mean of these prize amounts. You will enter, by hand, the data from the table above into a list named **PRIZE.** You will then use the **1-Var Stats** option from the **CALC** menu. To access this feature, press STAT and then arrow right to select CALC. Press ENTER or 1 to select **1-Var Stats.** Your screen should now look like:

```
1-Var Stats
```

At the prompt, choose the PRIZE list from your LIST menu. You screen should now look like this:

```
1-Var Stats LPRI
ZE
```

Now press ENTER and arrow up and down to move your cursor through the information.

Mean: 131477.5388 Median: 875

d. Did your calculation confirm your prediction in part b? Are the mean and median close to each other? Explain why this makes sense.

Yes, no because there is such a huge difference in the #'s

e. How many and what proportion of the prize amounts are greater than the mean?

6/26 23%

f. How many and what proportion of the prize amounts are greater than the median?

20/26 = 77%

g. If the producers of the show want to advertise either the mean or median prize amount in order to give the impression that contestants win huge amounts, which (mean or median) should they advertise? Explain.

mean - it is much higher than the median

This activity shows you that the mean and median are not always similar values. Therefore, it is important to understand their properties and limitations. Here, the mean represents the average amount won by a contestant when this game is played many times. The median represents the value that he or she would exceed half the time.

Activity 8-3: Matching Game

7-1, **8-3**

Reconsider Activity 7-1, in which you matched up seven variables with their dotplots. Concentrate on three of these variables:

- Prices of properties on the Monopoly game board
- Annual snowfall amounts for a sampling of U.S. cities
- Quiz percentages for a class of statistics students

a. Use your calculator to re-create the dotplots for only these three variables (the data are stored in the grouped file MATCHING, which contains the lists **BOFFC** (box office), **BPRSS** (blood pressure), **JNUM** (jersey numbers), **MONOP** (Monopoly prices), **QUIZ** (quiz percent), **RWGHT** (rowers' weights), **SNOW** (snowfall amounts), and **TLPNT** (title points), respectively. When running the DOTPLOT program, select the **COMPARE PLOTS** option. Which dotplot is roughly symmetric, which is skewed to the left, and which is skewed to the right?

Symmetric:

Skewed to the left:

Skewed to the right:

b. Use the **1-Var Stats** feature on your calculator to compute the mean and median for each of these variables. Record them in the next table:

	Monopoly Prices	Snowfall Amounts	Quiz Percentages
Mean			
Median			

c. Summarize what these data and calculations reveal about how the shape of the distribution (symmetric or skewed to the left or right) relates to the relative location of the mean and median.

These data reveal that the mean is close to the median with symmetric distributions, whereas the mean is greater than the median with distributions that are skewed to the right and the mean is less than the median with distributions that are skewed to the left. In other words, when one tail of the distribution is longer, the mean follows the tail.

Watch Out

These are not hard and fast rules. Do not look at the relative positions of the mean and median to decide about the shape of the distribution. Rather, look at the shape of the distribution to predict how the mean and median will compare to each other.

Activity 8-4: Rowers' Weights

7-2, **8-4,** 10-21

Refer to Activity 7-2 on page 123, in which you examined the weights of rowers on the 2004 U.S. men's Olympic rowing team.

a. Use your calculator to re-create the dotplot of rowers' weights (the data are stored in the file ROWERS04, which contains a list named WGHT). Based on this dotplot, estimate the values of the mean and median of the rowers' weights.

Mean (estimate): Median (estimate):

b. Use the **1-Var Stats** function to compute the mean and median. Record them in the first column of the following table. Why does it make sense that the mean and median differ in this direction? *Hint:* Think about more than the shape of the distribution.

Explanation:

	Whole Team	Without Coxswain	With Max Weight at 329	With Max Weight at 2229
Mean				
Median				

c. Predict what will happen to the mean if you remove the coxswain (Cipollone, the outlier whose job is to call out rowing instructions and keep the rowers in synch) from this analysis. Also predict what will happen to the median. Explain.

d. Remove the coxswain Cipollone and recalculate the mean and median.

1. To remove data from a list, you will need to use the List Editor screen. Press the [STAT] key and then the [5] key. Your screen should look like the following:

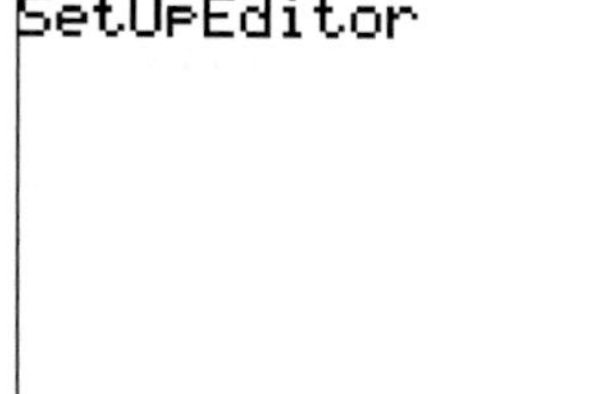

2. Now insert the name of the list after **SetUpEditor.** *TI hint:* You can add as many lists as you want in the List Editor screen by entering the list names after **SetUpEditor** separated by commas. Your Home screen should now look like this:

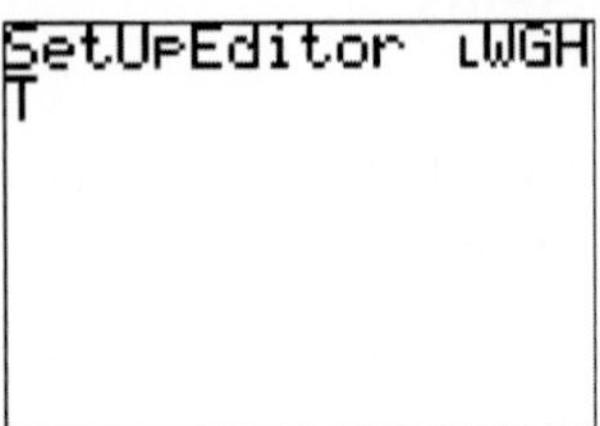

3. Press [STAT] and then press [ENTER] and [1]. You can now edit any of the entries using your cursor to highlight a value. To delete an entry, press the [DEL] key.

e. Record the values in the second column of the table. Did these values change as you predicted?

f. Now consider what would happen if you increased the weight of the heaviest rower (do not forget to re-enter the coxswain Cipollone's weight in the list. *TI note:* You can enter Cipollone's weight at the end of the list). How would this change the mean (if at all)? How would it affect the median? Explain.

g. Increase Schroeder's weight from 229 to **329.** Recalculate the mean and median, and record them in the third column of the previous table. Did they change (or not) as you predicted?

h. Suppose the typist's fingers slipped when recording Schroeder's weight and entered 2229. Predict how this error will affect the mean and median.

i. Increase Schroeder's weight to **2229** and re-calculate the mean and median. Record them in the last column of the table.

A measure whose value is relatively unaffected by the presence of outliers in a distribution is said to be **resistant.**

j. Based on these calculations, is the mean and/or the median resistant? Explain why this conclusion makes sense; base your argument on how each measure is calculated.

k. In principle, is there any limit to how much the mean can increase or decrease simply by changing *one* of the values in the distribution?

Activity 8-5: Buckle Up!

2-4, 3-21, **8-5**

Reconsider Activity 2-4, in which you compared seatbelt compliance percentages between states with strict enforcement laws and states with less strict laws.

a. Use your calculator to re-create dotplots for comparing the distributions of seatbelt compliance percentages between the two types of states. (The data are stored in a grouped file `SEATBELTUSAGE05`, containing the lists **PRMRY** and **SCNDY**, respectively) Does one type of state tend to have higher compliance percentages than the other?

b. Use your calculator to determine the mean and median of the compliance percentages in the two types of states. How do the two types of states compare on compliance percentages, according to these two measures of center? By how many percentage points is the compliance percentage greater (on average) in states with strict enforcement laws?

c. Do such large differences in mean and median between the two types of states enable you to draw a causal conclusion between stricter laws and the higher percentage of people who wear seatbelts? Explain.

Activity 8-6: Wrongful Conclusions

8-6, 16-21, 17-8, 28-25

For each of the following arguments, explain why the conclusion drawn is invalid. Also, include a simple hypothetical example illustrating that the conclusion drawn need not follow from the information given.

a. A real-estate agent notes that the mean housing price for an area is $250,000 and concludes that half of the houses in the area cost more than that.

b. A businesswoman calculates that the median cost of the five business trips she took in a month is $600 and concludes that the total cost must have been $3000.

c. A company executive concludes that an accountant must have made a mistake because she prepared a report stating that 90% of the company's employees earn less than the mean salary.

d. A restaurant owner decides that more than half of her customers prefer chocolate ice cream because chocolate is the mode when customers choose among chocolate, vanilla, and strawberry ice cream.

√

Activity 8-7: Readability of Cancer Pamphlets

8-7, 8-27

Researchers in Philadelphia investigated whether pamphlets containing information for cancer patients are written at a level that the cancer patients can comprehend (Short et al., 1995). They applied tests to measure the reading levels of 63 cancer patients and also the readability levels of 30 cancer pamphlets (based on such factors as sentence length and number of polysyllabic words). These measurements correspond to grade levels, but cancer patient reading levels below grade 3 and above grade 12 were not determined exactly.

The following tables indicate the number of patients at each reading level and the number of pamphlets at each readability level:

Patients' Reading Level	<3	3	4	5	6	7	8	9	10	11	12	>12
Count	6	4	4	3	3	2	6	5	4	7	2	17

Pamphlets' Readability Level	6	7	8	9	10	11	12	13	14	15	16
Count	3	3	8	4	1	1	4	2	1	2	1

a. Explain why the form of the data does not allow you to calculate a cancer patient's mean reading skill level.

b. Determine a cancer patient's median reading level. *Hint:* Consider the counts, and remember there are 63 patients.

c. Determine a pamphlet's median readability level.

d. How do these medians compare? Are the values fairly close?

e. Does the closeness of these medians indicate the pamphlets are well matched to the patients' reading levels?

f. What proportion of the patients are not at the reading skill level necessary to understand even the simplest pamphlet in the study? Should you reconsider your answer to part e?

Solution

a. To calculate the mean, you need to know *all* of the actual data values, and you do not know the values for those patients with a reading level below grade 3 or above grade 12.

b. There are 63 patients, so the median is the ordered value in position (63 + 1)/2, or the 32nd value. If you start counting from the low end, you find 6 patients read below grade 3, 10 patients at grade 3 or below, 14 patients at grade 4 or below, 17 patients at grade 5 or below, 20 patients at grade 6 or below, 22 patients at grade 7 or below, 28 patients at grade 8 or below, and 33 patients at grade 9 or below. The 32nd value is therefore at grade level 9, which is the median patient reading level.

c. There are 30 pamphlets, so the median readability level is the average of the 15th and 16th pamphlets. Counting in a similar way, the 15th and 16th readability values are both located at grade level 9, so a pamphlet's median readability level is grade 9.

d. These medians are identical.

e. No. The centers of the distributions (as measured by the medians) are well matched, but you need to look at both distributions in their entirety and consider all values. The problem is that many patients read at a level below that of the simplest pamphlet's readability level. Seventeen patients read at a level below grade 6, which is the lowest readability level of a pamphlet.

f. 17/63 = .27, so 27% of the patients have a reading level below that of the simplest pamphlet.

Watch Out

- Center is a property; mean and median are two ways to measure center. Neither mean nor median is synonymous with center. The mean and the median have their own properties and strengths.
- Do not ignore the counts in a frequency table. A common error in analyzing the patient reading skill levels is to claim that the median reading level of a patient is 7.5, because there are six grade levels below 7.5 and six grade levels above 7.5. But this value ignores the fact that there are different numbers of people at the various grade levels.

- Center is only one aspect of a distribution of data. Measures of center do not tell the whole story; other important features are spread, shape, clusters, and outliers.

Wrap-Up

You explored the use of mean and median as measures of the center of a distribution. You found that the mean and median are often similar but can be quite different. Deciding which to use depends on the question involved. For example, if you want to know how much a typical buyer pays for a house, then you calculate the median. But if you want to know the total amount spent on houses in a given month, then you calculate the mean.

You also discovered many properties of these measures, for example, that the median is resistant to outliers but the mean is not, as you saw when examining rowers' weights. You learned that the mean and median are similar for symmetric distributions, whereas the mean is pulled in the direction of the longer tail for skewed distributions.

Perhaps most importantly, you learned that these measures only summarize *one* aspect of a distribution and that you must combine these numerical measures with what you already know about displaying distributions visually and summarizing them. When analyzing data on cancer pamphlet readability, for instance, the medians did not tell researchers who wanted to compare pamphlet readability to patient reading levels the whole story.

Some useful definitions to remember and habits to develop from this topic include

- The **median** is the middle value in a distribution and, therefore, can be considered a "typical" value. The **mean** is the arithmetic average, or the balance point of the distribution.
- Because the median reports the middle value of a distribution, it is **resistant** to (i.e., not affected much by) outliers. The mean is not resistant to outliers; one extreme value can influence the mean considerably.
- In many cases, you should report both the mean and the median. In other cases, the research question determines which measure is more appropriate.
- Mean and median only pertain to the *center* of a distribution. Neither mean nor median conveys any information about the variability or shape of a distribution. Center is often the most important aspect of a distribution, but many research questions require looking at additional features.

In the next topic, you will discover and explore measures of the spread, or variability, of a distribution of data.

Homework Activities

 Activity 8-8: Human Ages

Think about all of the human beings alive at this moment.

a. Which value do you think is greater: the *mean* age or the *median* age of all human beings alive at this moment? Explain.
b. Which value do you think is greater: the mean age of all human beings alive at this moment, or the mean age of all *Americans* alive at this moment? Explain.
c. Estimate the values of the quantities in parts a and b and also estimate the median age of all Americans alive at this moment.

Activity 8-9: Sampling Words

4-1, 4-2, 4-3, 4-4, 4-7, 4-8, **8-9,** 9-15, 14-6

Reconsider the distribution of lengths of the 268 words in the Gettysburg Address. The frequency distribution is given here:

Word Length	1	2	3	4	5	6	7	8	9	10	11
Frequency	7	49	54	59	34	27	15	6	10	4	3

a. Without performing any calculations, how do you expect the mean and median to compare? Explain.
b. Determine the mean and median word length. Explain how you performed these calculations. Also, comment on whether the calculations verified your prediction in part a.
c. In analyzing the mean and median word length in the Gettysburg Address, are you calculating parameters or statistics? Explain.
d. Suppose a student mistakenly calculates the mean to be 24.36 letters. Explain why this value cannot possibly be correct.
e. What error did the hypothetical student in part d make to arrive at the number 24.36? *Hint:* Look at the second row of the previous table.

Activity 8-10: House Prices

8-10, 26-1, 27-5, 28-2, 28-12, 28-13, 29-3

The median house price in San Luis Obispo County (California) in June 2006 was $620,540. Would you expect the mean house price to be greater than, less than, or very close to the median house price? Explain.

Activity 8-11: Supreme Court Justices

As of October 2006, the nine justices of the U.S. Supreme Court and their years of appointment and service are given here:

Justice	Stevens	Scalia	Kennedy	Souter	Thomas	Ginsberg	Breyer	Roberts	Alito
Year Appointed	1975	1986	1988	1990	1991	1993	1994	2005	2006
Years of Service	31	20	18	16	15	13	12	1	0

a. Calculate the mean and median of the justices' years of service.
b. What proportion of the justices have served for longer than the mean service length?
c. If Justice Stevens had served for 51 years rather than 31 years, how would that increase affect the mean? How would that increase affect the median?
d. If the same nine justices were all to serve for an additional five years, how would the mean change? How would the median change?

Activity 8-12: Planetary Measurements

8-12, 10-20, 19-20, 27-13, 28-20, 28-21

The following table lists the average distance from the Sun (in millions of miles), diameter (in miles), and period of revolution around the Sun (in Earth days) for the nine planets of our solar system (as of the year 2005):

Planet	Distance (in millions of miles)	Diameter (in miles)	Period (in Earth days)
Mercury	36	3,030	88
Venus	67	7,520	225
Earth	93	7,926	365
Mars	142	4,217	687
Jupiter	484	88,838	4,332
Saturn	887	74,896	10,760
Uranus	1,765	31,762	30,684
Neptune	2,791	30,774	60,188
Pluto	3,654	1,428	90,467

a. Calculate (by hand) the median value of the variables of *distance, diameter,* and *period.*

b. A classmate uses the $(n + 1)/2$ formula and obtains a median diameter of 88,838 miles. What is the most likely cause of his or her mistake? *Hint:* Notice that this is Jupiter's diameter.

c. Without doing any calculations, do you expect the mean values of these variables to be close to, less than, or greater than the medians? Explain.

d. In August of 2006, the International Astronomical Union decided that Pluto would no longer be considered a planet. Before doing the calculations, predict how the medians of these variables will change once Pluto is removed.

e. Remove Pluto, and for the remaining eight planets, recalculate the medians of these three variables.

f. Would you expect Pluto's removal to have a similar effect on the means, a less extreme effect, or a more extreme effect? Explain.

g. Use your calculator to compute the means for these three variables with and without Pluto. (The data are stored in the file `PLANETS`.) Do you need to rethink your answer to part f? Explain.

Activity 8-13: Memorizing Letters

5-5, 7-15, **8-13,** 9-19, 10-9, 22-3, 23-3

Reconsider the data collected in the memory experiment in Activity 5-5.

a. Calculate the mean and median of the scores for the JFK and JFKC treatment groups.

b. Are these parameters or statistics? Explain.

c. Comment on how these measures of center compare between the two groups. Does the difference in values appear to be substantial?

d. Comment on what your analysis reveals about the question of whether the grouping of letters affects memory performance.

Activity 8-14: Sporting Examples

2-6, 3-4, **8-14,** 10-11, 22-26

Reconsider the data presented in Activity 2-6 comparing two sections of introductory statistics students, one of which used exclusively sports examples in the class (Lock, 2006). The data are stored in the grouped file SPORTSEXAMPLES, containing lists named **RGLR** (regular points) and **SPRTS** (sports points), respectively.

a. Calculate and compare the mean and median total points between the two sections.
b. By approximately how many points, on average, did students in the regular section outperform students in the sports-themed section?
c. Even if students in the regular section outperformed students in the sports-themed section by a whopping margin, could you legitimately conclude that the sports examples caused the lower performance? Explain.

Activity 8-15: Population Growth

7-4, **8-15,** 9-18, 10-10

Reconsider the data from Activity 7-4 on population growth in the 1990s for the 50 states.

a. Calculate the median percentage change for the eastern states and for the western states. Comment on how the medians compare.
b. Based on the shapes of the distributions for population growth in eastern and western states, how would you expect the means to compare to the medians? Explain.
c. How would you expect the mean percentage change in the western states to change if you removed Nevada from the analysis? Explain.

Activity 8-16: Zero Examples

Suppose that the sum of the values in a dataset equals zero.

a. Does it follow that the *mean* of the values in the dataset must equal zero? If so, explain why. If not, provide a counterexample (i.e., make up 10 numerical values where the sum equals zero but the mean does not).
b. Does it follow that the *median* of the values in the dataset must equal zero? If so, explain why. If not, provide a counterexample.

Activity 8-17: Marriage Ages

8-17, 9-6, 16-19, 17-22, 23-1, 23-12, 26-4, 29-17, 29-18

A student investigated whether people tend to marry spouses of similar ages and whether husbands tend to be older than their wives. The student gathered age data from a sample of 24 couples, taken from marriage licenses filed in Cumberland County, Pennsylvania, in June and July of 1993:

Couple	Husband's Age	Wife's Age	Difference	Couple	Husband's Age	Wife's Age	Difference
1	25	22		5	38	33	
2	25	32		6	30	27	
3	51	50		7	60	45	
4	25	25		8	54	47	

(*continued*)

Couple	Husband's Age	Wife's Age	Difference	Couple	Husband's Age	Wife's Age	Difference
9	31	30	1	17	26	27	-1
10	54	44	10	18	31	36	-5
11	23	23	0	19	26	24	2
12	34	39	-5	20	62	60	2
13	25	24	1	21	29	26	3
14	23	22	1	22	31	23	8
15	19	16	3	23	29	28	1
16	71	73	-2	24	35	36	-1

a. Create a side-by-side stemplot to display and compare the distribution of ages between husbands and wives.
b. Comment on how the age distributions compare.
c. Calculate and compare the median age for husbands and wives.
d. Now consider the difference in ages (subtracting the wife's age from the husband's age) for each couple. Calculate the median of these age differences.
e. Does the difference between the median age of the husbands and median age of the wives equal the median of the differences?
f. The data are stored in the file `MARRIAGEAGES`, containing the lists **HSBND, WIFE,** and **DFFR,** respectively). Use your calculator to check whether the difference in mean ages is equal to the mean of the differences. Summarize your findings.

Activity 8-18: Quiz Percentages

Consider the following dotplot of quiz percentage scores for students in an introductory statistics course.

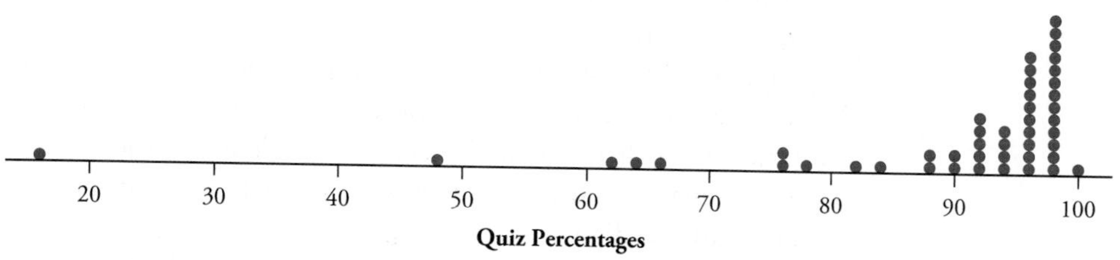

a. Based on this dotplot (without doing any calculations), estimate the mean and median values of these scores.
b. The data are stored in the file `QUIZPCT` with the list named **QZPCT.** Use your calculator to compute the mean and median.
c. How many and what proportion of the students scored greater than the mean value? Explain why this proportion turns out to be so large.
d. Remove the outlier (the student who scored very poorly), and recalculate the mean and median. Comment on how the mean and median changed.

Activity 8-19: February Temperatures

2-5, **8-19,** 9-7

Reconsider the data from Activity 2-5 on February temperatures in three locations: Lincoln, Nebraska; San Luis Obispo, California; and Sedona, Arizona. The data are stored in the grouped file FEBTEMPS, with lists **NEB, CAL, ARZ,** respectively.

a. Use your calculator to compute the mean and median temperatures for each location.

b. Use your calculator to convert the data to the Celsius scale by subtracting 32 and multiplying by 5/9. To convert the Lincoln, Nebraska, data on your calculator, subtract **32** from the list **NEB** and then multiply by **5/9.** Your screen should look like the following (the converted data is stored in a new list called **NEBF**):

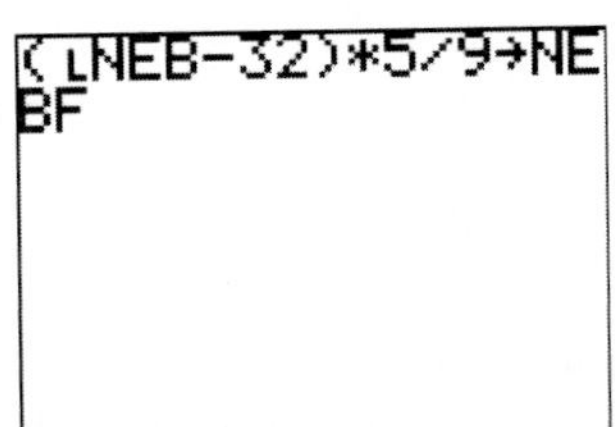

Then, calculate the mean and median temperature on the Celsius scale.

c. How do the mean and median on the Celsius scale compare to the mean and median on the Fahrenheit scale? Can you find an equation that relates the measures of center on the two scales?

Activity 8-20: Age Guesses

8-20, 20-13, 20-14

a. Create a dotplot of the guesses for your instructor's age, which you collected in the Preliminaries section.

b. Write a paragraph describing the distribution of these estimates.

c. If your instructor is willing to reveal her or his actual age, determine how many estimates are correct, greater than, and less than the correct age. Comment on the accuracy of the estimates.

d. Based on this distribution, do you expect the mean to be greater than the median, less than the median, or very close to the median?

e. Calculate (either by hand or using your calculator) the mean and median of these age estimates. Did you confirm your prediction about the relative values of the mean and median?

f. If the data contain any obvious outliers, remove them and recalculate the mean and median. Then comment on how much these values changed.

g. If your instructor is willing to reveal her or his actual age, comment on how well the mean and median of the age guesses approximate the actual value.

Activity 8-21: Tennis Simulations

7-22, **8-21,** 9-16, 22-18

Reconsider Activity 7-22, in which you analyzed simulation results for three different tennis scoring systems (Rossman and Parks, 1993). The data (lengths of games, measured by number of points played) are summarized in the following frequency table, and the raw data are stored in the grouped file TENNISSIM, containing the lists **PSTND** (standard), **PNOAD** (no-ad), and **PHNDC** (handicap), respectively:

Points in Game	1	2	3	4	5	6	7	8	10	12	14	16	18
Tally (Standard)				12	21	34		18	9	2	1	1	2
Tally (No-ad)				13	22	33	32						
Tally (Handicap)	3	4	12	18	28	25	10						

a. Determine the mean and median of the game lengths for all three scoring systems.
b. Comment on how these measures of center compare across the three scoring systems.

Activity 8-22: Hypothetical Exam Scores

7-12, **8-22,** 8-23, 9-22, 10-22, 27-8

For each of the following properties, construct a dataset of ten hypothetical exam scores that satisfies the property. Also create a dotplot of your hypothetical dataset in each case. Assume that the exam scores are integers between 0 and 100, inclusive, repeats allowed. (Feel free to use your calculator to automate the calculations.)

a. Ninety percent of the scores are greater than the mean.
b. The median is less than two-thirds of the mean.
c. The mean does not equal the median, and none of the scores are between the mean and the median.

Activity 8-23: Hypothetical Exam Scores

7-12, 8-22, **8-23,** 9-22, 10-22, 27-8

Suppose an instructor is teaching two sections of a course and she calculates the mean exam score to be 60 for Section 1 and 90 for Section 2.

a. Do you have enough information to determine the mean exam score for the two sections combined? Explain.
b. What can you say with certainty about the value of the overall mean for the two sections combined?
c. Without seeing all of the individual students' exam scores, what information would you need to calculate the overall mean?
d. Suppose Section 1 contains 20 students and Section 2 contains 30 students. Calculate the overall mean exam score. Is the overall mean closer to 60 or to 90?
e. Give an example of sample sizes for the two sections for which the overall mean is less than 65.
f. If you do not know the numbers of students in the sections but you do know that the same number of students are in each section, can you determine the overall mean? If not, explain. If you can, calculate the overall mean.
g. Explain how a student could transfer from Section 1 to Section 2 and cause the mean score for each section to decrease. What would need to be true about that student's score?

Activity 8-24: Class Sizes

Suppose a college is offering five sections of an introductory statistics course. The enrollments for the course are given here:

Section #	1	2	3	4	5
Enrollment	200	35	35	20	10

a. What are the observational units? What is the variable?
b. Calculate the mean enrollment size per section.
c. Now consider the 300 students taking introductory statistics at this college to be the observational units, and the variable to be the number of students in the student's class. Calculate the mean of this variable. *Hint:* Notice that the 200 students in section 1 will all report that there are 200 students in their class.

Activity 8-25: Sports Averages

Averages abound in sports. Some examples include

- Average yards per carry in football
- Average runs per game in baseball
- Average driving distance in golf
- Average goals scored by opponents in hockey
- Average points per game in basketball
- Average speed of a serve in tennis

Choose any two of these sports averages and answer the following questions:

a. Identify the observational units.
b. Create a hypothetical example (make up, say, ten values) and illustrate how to calculate the average. (Better yet, find ten real values from a player or team, and use them to illustrate the calculation.)
c. If the median were calculated instead of the average (mean), would you expect the value to be greater than, less than, or about the same as in part b? Explain your reasoning.

Activity 8-26: Salary Expectations

Reconsider the data collected in the Preliminaries questions about how much students expect to earn the first year of their career.

a. Calculate the mean and median of these salary expectations.
b. Suppose that another student in the class expected to begin her career with a salary of $1,000,000. Perform the appropriate calculation to determine how much this student's expectations affect the mean and median.
c. Is the mean or the median affected more by the inclusion of the student in part b? Explain why.
d. Suppose that every student's estimate was $5000 more than he/she actually answered (each student's answer is $5000 greater than it was in Topic 7). Without

doing any calculations, how would you expect this change to affect the mean and median?

e. Add $5000 to each student's answer, and recalculate the mean and median. How do these values compare to the original mean and median? Comment on whether your answer to part d is correct.

Activity 8-27: Readability of Cancer Pamphlets

8-7, **8-27**

Reconsider the data on cancer pamphlets from Activity 8-7 on page 153. Construct graphical displays of both distributions (readability of pamphlets and reading level of patients). Comment on the usefulness of these graphs for comparing the two distributions and determining if the pamphlets are well matched to the patients' reading levels.

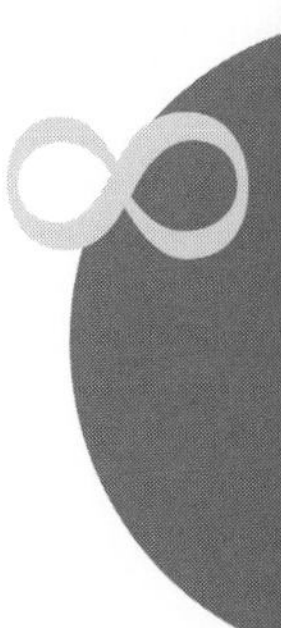

TOPIC

9 • • • Measures of Spread

If two teams of baseball players have the same average age, does that imply that all of the baseball players on both teams are the same age? Suppose you were choosing a city to live in based solely on temperature: Would you consider two cities with the same average annual temperature equivalent? What other issue related to temperature might you consider? How can college admissions officers compare the qualifications of applicants who have taken different entrance exams (for example, the ACT versus the SAT)? The answers to these questions lie in studying the spread—or consistency—of distributions. In this topic, you will even discover how the spread of a distribution relates to the midnight ride of Paul Revere!

Overview

In the previous topic, you explored the mean and median, important numerical measures of the center of a distribution. In this topic, you will build on your knowledge of measures of center by investigating similar numerical measures of a distribution's spread, or variability, namely the interquartile range and the standard deviation. You will examine these measures' properties and investigate some common misconceptions about variability. Studying these measures will also lead you to an important result (the empirical rule) and an important technique (standardization, or computing *z*-scores)—two concepts that will appear throughout this course.

Preliminaries

1. Which would you consider more impressive—a score of 1740 on the SAT or a score of 30 on the ACT?

2. Take a guess for the typical age at which a bride gets married.

3. Do you expect that husbands tend to be older than their wives? If so, by approximately how many years on average?

4. In his book *The Tipping Point,* Malcolm Gladwell describes an exercise in which he presents people with a list of about 250 surnames, all taken at random from

the Manhattan phone book. He asks people to read through the list and see how many acquaintances they know with a surname on the list. Gladwell defines "know" very broadly: If you sat down next to that person on a train, you would know their name if they introduced themselves to you, and they would know your name. Multiple names count: If the name is Johnson, in other words, and you know three Johnsons, you get three points. Read through Gladwell's list and see how many of your acquaintances have a surname on the list:

> Algazi, Alvarez, Alpern, Ametrano, Andrews, Aran, Arnstein, Ashford, Bailey, Ballout, Bamberger, Baptista, Barr, Barrows, Baskerville, Bassiri, Bell, Bokgese, Brandao, Bravo, Brooke, Brightman, Billy, Blau, Bohen, Bohn, Borsuk, Brendle, Butler, Calle, Cantwell, Carrell, Chinlund, Cirker, Cohen, Collas, Couch, Callegher, Calcaterra, Cook, Carey, Cassell, Chen, Chung, Clarke, Cohn, Carton, Crowley, Curbelo, Dellamanna, Diaz, Dirar, Duncan, Dagostino, Delakas, Dillon, Donaghey, Daly, Dawson, Edery, Ellis, Elliott, Eastman, Easton, Famous, Fermin, Fialco, Finklestein, Farber, Falkin, Feinman, Friedman, Gardner, Gelpi, Glascock, Grandfield, Greenbaum, Greenwood, Gruber, Garil, Goff, Gladwell, Greenup, Gannon, Ganshaw, Garcia, Gennis, Gerard, Gericke, Gilbert, Glassman, Glazer, Gomendio, Gonzalez, Greenstein, Guglielmo, Gurman, Haberkorn, Hoskins, Hussein, Hamm, Hardwick, Harrell, Hauptman, Hawkins, Henderson, Hayman, Hibara, Hehmann, Herbst, Hedges, Hogan, Hoffman, Horowitz, Hsu, Huber, Ikiz, Jaroschy, Johann, Jacobs, Jara, Johnson, Kassel, Keegan, Kuroda, Kavanau, Keller, Kevill, Kiew, Kimbrough, Kline, Kossoff, Kotzitzky, Kahn, Kiesler, Kosser, Korte, Leibowitz, Lin, Liu, Lowrance, Lundh, Laux, Leifer, Leung, Levine, Leiw, Lockwood, Logrono, Lohnes, Lowet, Laber, Leonardi, Marten, McLean, Michaels, Miranda, Moy, Marin, Muir, Murphy, Marodon, Matos, Mendoza, Muraki, Neck, Needham, Noboa, Null, O'Flynn, O'Neill, Orlowski, Perkins, Pieper, Pierre, Pons, Pruska, Paulino, Popper, Potter, Purpura, Palma, Perez, Portocarrero, Punwasi, Rader, Rankin, Ray, Reyes, Richardson, Ritter, Roos, Rose, Rosenfeld, Roth, Rutherford, Rustin, Ramos, Regan, Reisman, Renkert, Roberts, Rowan, Rene, Rosario, Rothbart, Saperstein, Schoenbrod, Schwed, Sears, Statosky, Sutphen, Sheehy, Silverton, Silverman, Silverstein, Sklar, Slotkin, Speros, Stollman, Sadowski, Schles, Shapiro, Sigdel, Snow, Spencer, Steinkol, Stewart, Stires, Stopnik, Stonehill, Tayss, Tilney, Temple, Torfield, Townsend, Trimpin, Turchin, Villa, Vasillov, Voda, Waring, Weber, Weinstein, Wang, Wegimont, Weed, Weishaus

5. Record these data on number of acquaintances for yourself and your classmates.

In-Class Activities

Activity 9-1: Baseball Lineups

9-1, 9-2, 9-13, 9-17

When the New York Yankees played the Detroit Tigers on August 30, 2006, those teams had the two best records in the American League. The 2006 Yankees were expected to be a good team, but the 2006 Tigers' success was a surprise to most people. One baseball fan conjectured that the Tigers had achieved their success based on youth and the Yankees on experience. To investigate this conjecture, he gathered data on the ages of the two teams' starting lineups for that particular game:

	Yankees		Tigers	
Position	Name	Age	Name	Age
C	Posada	35	I. Rodriguez	34
1B	Wilson	29	Casey	32
2B	Cano	23	Perez	33
3B	A. Rodriguez	31	Inge	29
SS	Jeter	32	Guillen	30
LF	Cabrera	22	Monroe	29
CF	Damon	32	Granderson	25
RF	Abreu	32	Gomez	28
DH	Giambi	35	Young	32
P	Wang	26	Robertson	28

a. Identify the observational units. Also identify the explanatory variable and the response variable. Classify each variable as categorical (also binary) or quantitative.

Observational units:

Explanatory variable: Type:

Response variable: Type:

b. Create comparative dotplots, using the axes shown here, for comparing the ages between the two teams. (Be sure to label which dotplot represents which team.) Comment on how the age distributions compare.

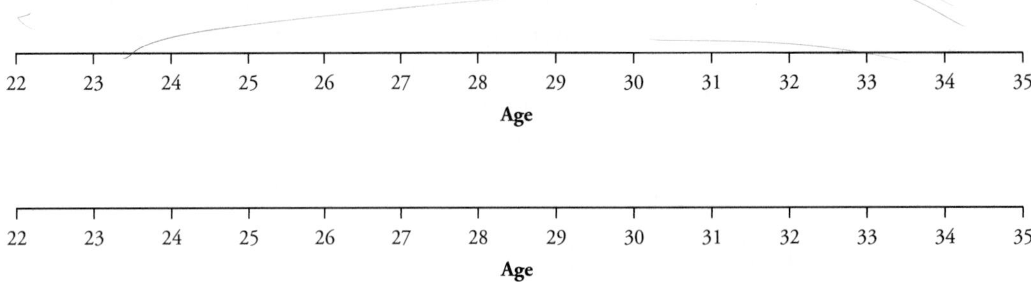

Comment:

c. Calculate the mean and median age of each team's lineup. *Hint:* Employ the fact that the Yankees' ages add up to 297 years and the Tigers' ages add up to 300 years. Are the centers of these two distributions reasonably similar?

Yankees Mean: Median:

Tigers Mean: Median:

d. Because the centers of these age distributions are so close, can you reasonably conclude that there is no difference between the age distributions of these lineups? Explain.

e. Which team's lineup appears to have more variability in its ages?

In the previous topic, you learned that the mean and median are two ways to measure the center of a distribution; you will now learn several ways to measure the spread, or variability, of a distribution.

f. What is the age of the oldest player in the Yankees lineup? The youngest? What is the difference in age between the oldest and youngest players?

Oldest: Youngest: Difference:

g. Repeat part f for the Tigers' lineup.

Oldest: Youngest: Difference:

A very simple, but not particularly useful, measure of variability is the **range,** calculated as the difference between the maximum and minimum values in a dataset.

Another measure of variability is the **interquartile range (IQR),** which is the difference between the upper quartile and the lower quartile of a distribution. The **lower quartile** (or the 25th percentile, abbreviated Q_L) is the value such that 25% of the values in a dataset are less than that value and 75% are greater than it, whereas the **upper quartile** (or 75th percentile, abbreviated Q_U) is the value such that 75% of the values in a dataset are less than that value and 25% are greater than it. Thus, the IQR is the range of the middle 50% of the data.

To find the lower quartile, you first find the median (the middle value) of those observations below the location of the actual median. Similarly, the upper quartile is the median of those observations above the location of the actual median. (When there is an odd number of observations, you do *not* include the actual median in either group.)

For the Yankees, the sorted ages are

22 23 26 29 31 32 32 32 35 35
↑ median

The lower quartile is the median of the five youngest ages (22, 23, 26, 29, 31), which is 26 years, the third youngest age. On the other side of the median, the upper quartile is the third oldest age, which is 32 years. The interquartile range of Yankees' ages is therefore 32 – 26 or 6 years.

h. Determine the lower and upper quartiles of the ages for the Tigers. Then find the IQR of the Tigers' ages. *Hint:* Put the ages in order first, perhaps by using the dotplot you created in part b.

Lower quartile: Upper quartile: IQR:

i. Which team has the greater age range? Which has the greater IQR? Are these values consistent with your answer to part e?

j. Based on this analysis, summarize how the age distributions differ between the Yankees and Tigers (shape, center, and spread).

Activity 9-2: Baseball Lineups

9-1, **9-2,** 9-13, 9-17

Other measures of variability examine how far the data values fall or deviate from the mean of the distribution.

a. The mean age for Detroit's lineup is 30 years. Complete the missing entries for Casey and Granderson in the "deviation from mean" column of the following table by calculating the differences between their ages and the mean age:

Player	Age	Deviation from Mean	Absolute Deviation	Squared Deviation
I. Rodriguez	34	34 – 30 = 4	4	16
Casey	32	32 - 30 = 2	2	4
Perez	33	3	3	9
Inge	29	–1	1	1
Guillen	30	0	0	0
Monroe	29	–1	1	1
Granderson	25	-5	5	25
Gomez	28	–2	2	4
Young	32	2	2	4
Robertson	28	–2	2	4
Total	300	0	22	68

b. Add the values in the "deviation from mean" column. Explain why the value of this sum makes sense.

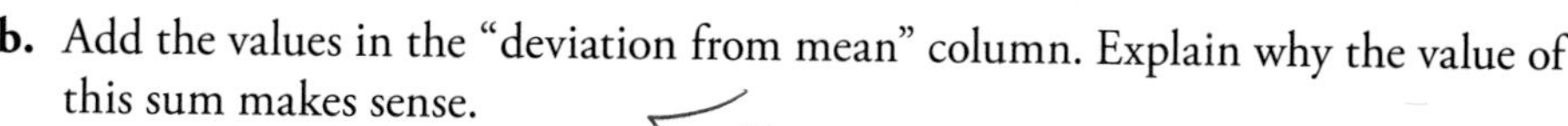

Because a measure of spread is concerned with *distances* from the mean rather than *direction* from the mean, you could work with the absolute values of these deviations.

c. Complete the missing entries in the "absolute deviation" column of the table. Then add those absolute deviations.

d. Calculate the average (mean) of these absolute deviations. Also report the measurement units (for example, inches, hours, dollars, . . . ?) of this calculation.

2.2

The measure of spread you have just calculated is the **mean absolute deviation (MAD)**. It is an intuitively sensible but not widely used measure of spread.

An alternative to working with absolute deviations is to square the deviations from the mean.

e. Complete the missing entries in the "squared deviation" column of the table. Then add those squared deviations.

f. Divide the sum of squared deviations by 9 (one less than the sample size).

g. To convert back to the original units (years of age), take the square root of your answer to part f.

The standard deviation is a widely used measure of variability. To compute the standard deviation, you calculate the difference between the mean and each data value and then square the difference: (data value − mean)^2. Add these squared terms, and divide the sum by $n - 1$ (one less than the sample size). The standard deviation (denoted by s) is the square root of the result:

$$s = \sqrt{\frac{\Sigma(x_i - \bar{x})^2}{n - 1}}$$

The standard deviation can loosely be interpreted as the typical distance that a data value in the distribution deviates from the mean.

You will usually rely on your calculator to compute standard deviations.

h. Enter the Tigers' ages into your graphing calculator. Use the **1-Var Stats** feature to verify your computation of the standard deviation for the Tiger lineup's ages. Then use your calculator to compute the standard deviation for the Yankee lineup's ages. Which is greater? Is this what you expected?

i. Now suppose that I. Rodriguez's age had been incorrectly listed as 43 years rather than 34. Would you expect this change to have much/any effect on the value of the range? How about the IQR? How about the standard deviation? Explain.

Yes, the range is more variable

j. Similar to Activity 8.4, part d, use the **SetUpEditor** feature on your list of Tiger's ages to change I. Rodriguez's age from **34** to **43.** Then use your calculator to recalculate the range, IQR, and standard deviation. Record these, along with the original values, in the following table.

k. Now suppose that I. Rodriguez's age had been incorrectly listed as 134 rather than 34. Use your calculator to recalculate the measures of spread in this case. Record them in the table as well. Do not forget to change I. Rodriguez's age back to **34** after you compute the measures of spread.

	Original Data	With Large Outlier (43)	With Huge Outlier (134)
Range			
Interquartile Range			
Standard Deviation			

l. Which measures of spread are resistant to outliers and which are not? Explain how you decide.

Watch Out

- Be aware that various software packages and calculators might use slightly different rules for calculating quartiles. Because of these differences, your by-hand calculations may be correct and still differ a bit from the value calculated by your software program or calculator.
- It can be tempting to regard range and IQR as an interval of values, but they should each be reported as a single number that measures the spread of the distribution. For example, it is technically wrong to say that the range of Yankee players' ages is 22 to 35. The correct range of the Yankee ages is 13 years, calculated as 35 – 22.
- Remember that these measures of spread apply only to quantitative variables, not to categorical ones.

Activity 9-3: Value of Statistics

2-9, 2-10, **9-3**

Consider hypothetical ratings from five classes of the value of statistics on a 1–9 scale. The data are given in the following frequency table (and stored in the grouped data file VALUESFJ, containing the lists CLSSF, CLSSG, CLSSH, CLSSI, and CLSSJ, respectively) and displayed in the following histograms:

Rating	1	2	3	4	5	6	7	8	9
Class F Count	0	3	1	5	7	2	4	2	0
Class G Count	1	2	3	4	5	4	3	2	1
Class H Count	1	0	0	0	22	0	0	0	1
Class I Count	12	0	0	0	1	0	0	0	12
Class J Count	2	2	2	2	2	2	2	2	2

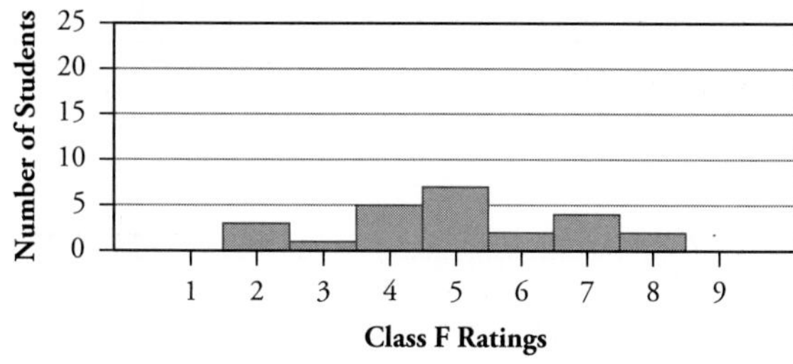

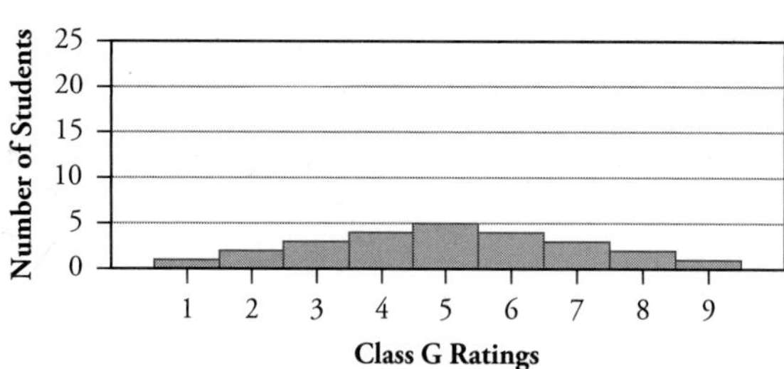

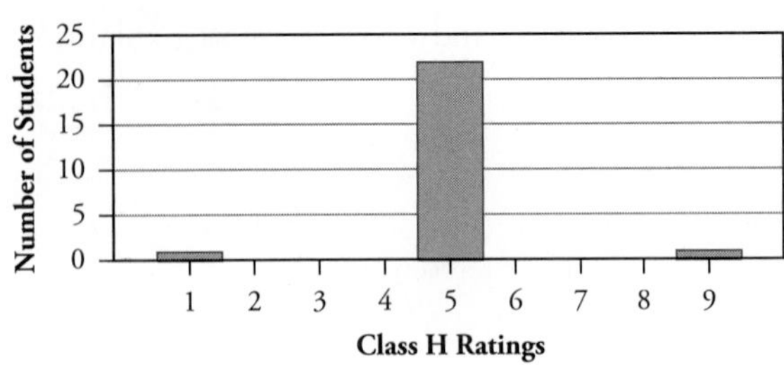

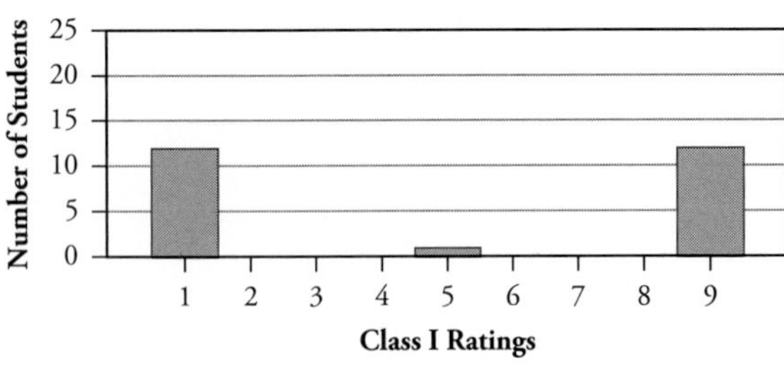

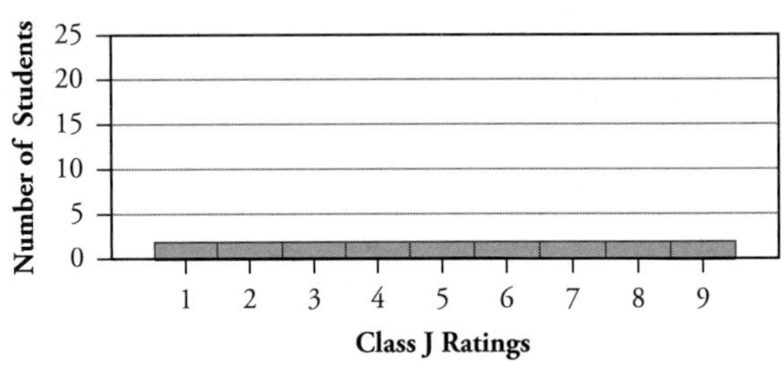

a. Judging from the tables and histograms, make a prediction as to which class's ratings have more variability: class F or class G. Explain your reasoning.

b. Judging from the tables and histograms, which class has the most variability in ratings among classes H, I, and J? Which class has the least variability in ratings? Explain your reasoning.

Most variability: Least variability:

c. Use your calculator to compute the range, interquartile range, and standard deviation of the ratings for each class. Record the results in the table:

	Class F	Class G	Class H	Class I	Class J
Range	6	8	4	8	8
Interquartile Range	2.5	1.5	0	8	4
Standard Deviation	1.76	2.04	1.18	4	2.65

d. Judging from these measures of spread, does class F or G have more variability? Was your prediction in part a correct?

e. Judging from these measures of spread, which class, H, I, or J, has the most variability? Which has the least? Was your prediction in part b correct?

f. Between classes F and G, which has more *bumpiness* or unevenness in its histogram? Does that class have more or less variability than the other class?

g. Among classes H, I, and J, which distribution has the greatest number of distinct values? Does that class have the most variability of the three classes?

h. Based on the previous two questions, does either "bumpiness" or "variety" relate directly to the concept of variability? Explain.

i. Create your own hypothetical example with 10 ratings (each between 1 and 9, inclusive, with repeats allowed) so that the standard deviation is as *small* as possible. Use your calculator to compute the standard deviation, and report its value along with your ten ratings.

j. Create another hypothetical example with 10 ratings (each between 1 and 9, inclusive, with repeats allowed) so that the standard deviation is as *large* as possible. Use your calculator to compute the standard deviation, and report its value along with your ten ratings.

Watch Out

Variability can be a very tricky concept to grasp, but it is absolutely fundamental to working with data and understanding statistics. A common misconception is that a "bumpier" histogram indicates a more variable distribution, but this is not the case. Make sure you focus on the variability in the horizontal values (the variable) and not the heights (the frequencies). Similarly, the number of distinct values represented in a histogram does not necessarily indicate greater variability. Consider how far the values fall from the center more than the variety of their exact numerical values.

As stated earlier, the standard deviation can be loosely interpreted as the typical distance by which a data value deviates from the mean of that distribution. In the next two activities, you will see some other interpretations and applications of standard deviation.

Activity 9-4: Placement Exam Scores

7-16, **9-4**

Recall Activity 7-16, which presented scores on a mathematics placement exam consisting of 20 questions taken by 213 students. The frequency table is reproduced here. The mean score on this exam is $\bar{x} = 10.221$ points. The standard deviation of the exam scores is $s = 3.859$ points. The frequency of each score and a histogram of this distribution follow.

Score	1	2	3	4	5	6	7	8	9	10
Count	1	1	5	7	12	13	16	15	17	32
Score	11	12	13	14	15	16	17	18	19	
Count	17	21	12	16	8	4	7	5	4	

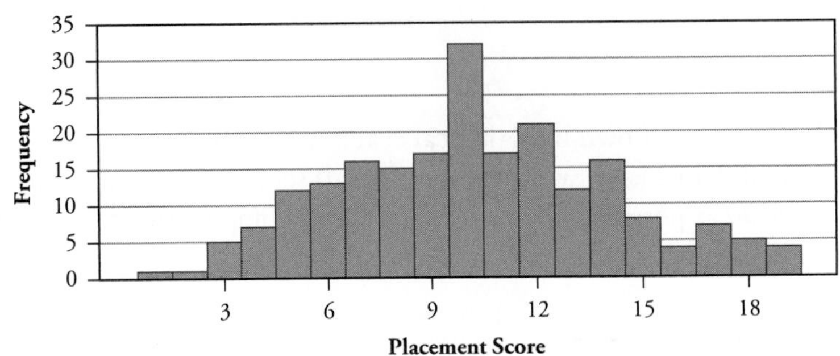

a. Does this distribution appear to be roughly symmetric and mound-shaped?

Yes

b. Consider the question of how many scores fall within one standard deviation of the mean (denoted by $\bar{x} \pm s$). First, determine the upper endpoint of this interval by adding the value of the standard deviation to that of the mean. Then determine the interval's lower endpoint by subtracting the value of the standard deviation from that of the mean.

$\bar{x} + s =$

$\bar{x} - s =$

c. Refer back to the table of tallied scores to determine how many of the 213 scores fall within *one* standard deviation of the mean. What proportion of the 213 scores does this account for?

68.5%

d. Determine how many of the 213 scores fall within *two* standard deviations of the mean, which turns out to be between 2.503 and 17.939 points. What proportion of scores does this represent?

e. Determine how many of the 213 scores fall within three standard deviations of the mean, which turns out to be between −1.356 and 21.798 points. What proportion of scores does this represent?

You have discovered what is called the **empirical rule.** With mound-shaped distributions, approximately 68% of the observations fall within one standard deviation of the mean, approximately 95% fall within two standard deviations of the mean, and virtually all of the observations fall within three standard deviations of the mean.

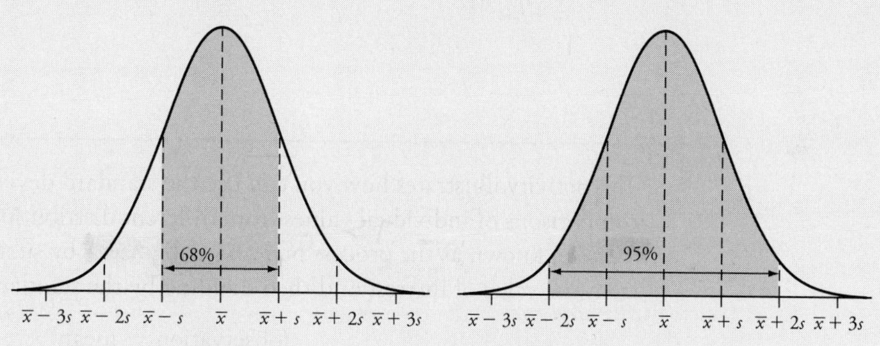

Note that the empirical rule applies to mound-shaped distributions but not necessarily to distributions of other shapes. This rule provides a way of interpreting the value of a distribution's standard deviation as half the width of the middle 68% of the (symmetric) distribution. Consider how the standard deviation compares with the IQR, which reveals the range of the middle 50% of the data, and with the quartiles themselves, which specify the values between which that middle 50% of the data falls.

Activity 9-5: SATs and ACTs

9-5, 12-11

Suppose a college admissions office needs to compare scores of students who take the Scholastic Aptitude Test (SAT) with those who take the American College Test (ACT). Among the college's applicants who take the SAT, scores have a mean of 1500 and a standard deviation of 240. Among the college's applicants who take the ACT, scores have a mean of 21 and a standard deviation of 6.

a. If applicant Bobby scored 1740 on the SAT, how many points above the SAT mean did he score?

b. If applicant Kathy scored 30 on the ACT, how many points above the ACT mean did she score?

c. Is it sensible to conclude that because your answer to part a is greater than your answer to part b, Bobby outperformed Kathy on the admissions test? Explain.

d. Determine how many standard deviations above the mean Bobby scored by dividing your answer to part a by the standard deviation of the SAT scores.

e. Determine how many standard deviations above the mean Kathy scored by dividing your answer to part b by the standard deviation of the ACT scores.

WTF

This activity illustrates how you can use the standard deviation to make comparisons of individual values from different distributions. You calculate a ***z*-score,** known as the process of **standardization,** by subtracting the mean from the value of interest and then dividing by the standard deviation:

$$z = \frac{(\text{observation} - \text{mean})}{\text{standard deviation}}$$

These *z*-scores indicate how many standard deviations above or below the mean a particular value falls.

In other words, the standard deviation serves as a ruler by which you can fairly compare different quantities. For example, the following sketches show that an ACT score of 30 is better (has a higher *z*-score) than an SAT score of 1740. You should use *z*-scores only when working with mound-shaped distributions, however, as with the standard deviation.

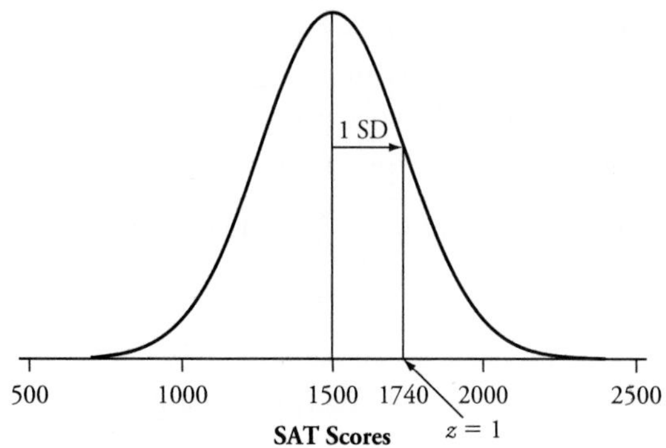

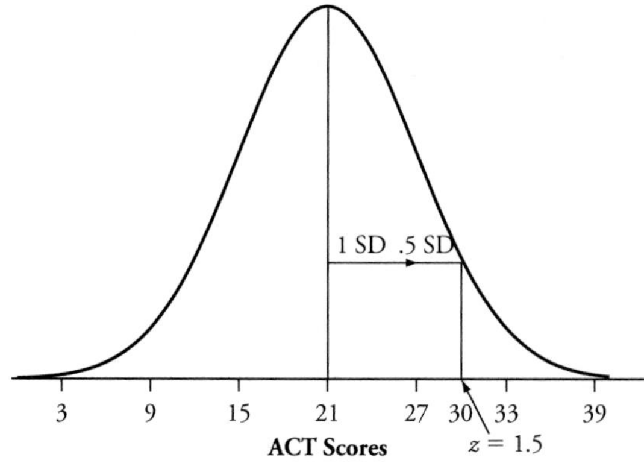

f. Which applicant has the higher z-score for his or her admissions test score?

g. Explain which applicant performed better on his or her admissions test compared to his or her peers.

h. Calculate the z-score for applicant Peter, who scored 1380 on the SAT, and for applicant Kelly, who scored 15 on the ACT.

Peter: Kelly:

i. Does Peter or Kelly have the higher z-score?

j. Under what conditions does a z-score turn out to be negative?

Activity 9-6: Marriage Ages

8-17, **9-6,** 16-19, 17-22, 23-1, 23-12, 26-4, 29-17, 29-18

Recall from Activity 8-17 that a student investigated whether people tend to marry spouses of similar ages and whether husbands tend to be older than their wives. He gathered data on the ages of a sample of 24 couples, taken from marriage licenses filed in Cumberland County, Pennsylvania, in June and July of 1993 (`MARRIAGEAGES`):

Couple	Husband's Age	Wife's Age	Difference	Couple	Husband's Age	Wife's Age	Difference
1	25	22	3	8	54	47	7
2	25	32	–7	9	31	30	1
3	51	50	1	10	54	44	10
4	25	25	0	11	23	23	0
5	38	33	5	12	34	39	–5
6	30	27	3	13	25	24	1
7	60	45	15	14	23	22	1

(continued)

Couple	Husband's Age	Wife's Age	Difference	Couple	Husband's Age	Wife's Age	Difference
15	19	16	3	20	62	60	2
16	71	73	–2	21	29	26	3
17	26	27	–1	22	31	23	8
18	31	36	–5	23	29	28	1
19	26	24	2	24	35	36	–1

a. Using your calculator, compute the median and mean age for each group (husbands/wives). Which spouse tends to be older and by how many years on average?

b. Using your calculator, compute the IQR and standard deviation of ages for each group. Does one group of spouses have more variability in their ages than the other group?

c. Comment on how the age distributions compare, citing numerical summaries for support.

d. Notice that the table also reports the difference in ages, subtracting the wife's age from the husband's age, for each couple. Calculate the mean and median of these age differences for the two groups. Do you notice anything about how these compare to the means and medians of the husbands and wives individually?

e. Calculate the IQR and standard deviation of these age differences. Compare these values to the IQR and standard deviation of the ages themselves in part d.

f. Determine how many and what proportion of the age differences fall within one standard deviation of the mean. Is this close to the percentage that the empirical rule predicts?

g. What do these calculations reveal about whether people tend to marry spouses of similar ages? Explain.

h. Based on the content, explain why the age differences have much less variability than the individual ages.

Solution

a. With 24 people in each group, the median ages are the average of the 12th and 13th ordered values. For husbands, the median is (30 + 31)/2 or 30.5 years. For wives, the median is (28 + 30)/2 or 29 years. For husbands, the mean age is 35.7 years and for wives, the mean age is 33.8 years. Husbands tend to be a little less than two years older than their wives.

b. The lower quartile is the median of the bottom 12 ordered values, so the average of the 6th and 7th values. For husbands, the lower quartile is (25 + 25)/2 or 25 years and the upper quartile is (51 + 38)/2 or 44.5 years. The IQR is, therefore, 44.5 − 25 or 19.5 years. You can see this by examining the sorted ages for husbands:

19 23 23 25 25 25 ↓ 25 26 26 29 29 30 ↓ 31 31 31 34 35 38 ↓ 51 54 54 60 62 71

For wives, the lower quartile is (24 + 24)/2 or 24 years and the upper quartile is (39 + 44)/2 or 41.5 years, so the IQR is 41.5 − 24 or 17.5 years. The standard deviations are 14.6 and 13.6 years for husbands and wives, respectively. These calculations indicate that the middle 50% of husbands' ages cover a slightly greater distance than the wives' ages by 2 years and that the husbands' ages typically lie slightly farther from the mean, by approximately 1 year on average.

c. The age distributions are quite similar for husbands and wives. Both are skewed to the right, centered around the low 30s or so, with considerable variability from the upper teens through low 70s. The husbands are a bit older on average, and their ages are a bit more spread out than the wives' ages.

d. The ordered differences in couples' ages are

−7, −5, −5, −2, −1, −1, 0, 0, 1, 1, 1, 1, 1, 2, 2, 3, 3, 3, 3, 5, 7, 8, 10, 15

The median is the average of the 12th and 13th ordered values: (1 + 1)/2 or 1 year.

The mean is the sum of these differences divided by 24, which turns out to be 45/24 or 1.9 years.

Notice that the mean of the age differences is equal to the difference in mean ages between husbands and wives: 1.9 = 35.7 − 33.8. But this property does not quite hold for the median.

e. The quartiles are −0.5 and 3, so the IQR is 3.5 years. The standard deviation of these age differences is 4.8 years.

f. To be within one standard deviation of the mean is to be within 1.9 ± 4.8 years, which means between −2.9 and 6.7 years. Seventeen of the age differences fall within this interval, which is a proportion of 17/24 or .708, or 70.8%. This percentage is quite close to 68%, which is what the empirical rule predicts. Because the distribution of the age differences does look fairly symmetric and mound-shaped, this outcome is not surprising.

g. The mean and median indicate that, on average, people marry someone within a couple years of their own age. More importantly, the measures of spread are fairly small for the differences, much smaller than for individual ages. This result suggests that there is not much variability in the differences, which suggests that people do tend to marry people of similar ages.

h. The differences have less variability because even though people get married from their teens to seventies (and beyond), they tend to marry people within a few years of their own age.

Wrap-Up...........

In this topic, you furthered your study of the fundamental concept of variability. You continued to see that even if two datasets have similar centers, the spread of the values might differ substantially. You also studied some common misconceptions about the meaning of variability; for example, variability is not the same as bumpiness or variety as you saw in Activity 4-3.

You also learned to calculate the range, interquartile range, and standard deviation as measures of the variability (spread) of a distribution, and you studied these measures' properties. You observed that range is not a very useful measure, because it considers only the extreme values in a dataset, and you learned that the standard deviation is not resistant to outliers.

You also explored the empirical rule and *z*-scores as applications of the standard deviation. These tools are particularly useful for enabling you to compare the proverbial apples and oranges, such as scores on SAT and ACT exams or heights of men and women, by converting measurements onto a common scale calculated in number of standard deviations away from the mean. The *z*-score serves as a "ruler" for measuring distances.

Some useful definitions to remember and habits to develop from this topic include

- Variability is a property of a distribution; standard deviation and interquartile range are two ways to measure variability for a particular dataset. (Similarly, as you saw in the previous topic, center, another property of distributions, can be measured by the mean and median.)
- In describing the variability in a dataset, focus on the bulk of the data and not on a few extreme values. Also remember that bumpiness and variety are not the same as variability.
- **Standard deviation** can be loosely interpreted as the typical deviation of an observation from the mean. Standard deviation is more cumbersome to calculate by hand but is more widely used than the **mean absolute deviation.**
- The **interquartile range** is the difference between the **upper quartile** and **lower quartile,** so the IQR is the width of the middle 50% of the ordered values in a dataset and is reported as a single number.
- The mean and standard deviation provide a useful summary for symmetric distributions. The median and IQR are always appropriate measures, but especially with skewed distributions.
- The **empirical rule** applies to mound-shaped distributions of data and says that approximately 68% of the observations fall within one standard deviation of the mean, approximately 95% fall within two standard deviations of the mean, and virtually all fall within three standard deviations of the mean. Apply the empirical rule only to mound-shaped data.
- The process of **standardization,** or calculating a ***z*-score,** converts data to a common scale so you can compare observations from apparently disparate distributions. You will use the *z*-score as a "ruler" to measure distances.

In the next topic, you will discover a new graphical display called a boxplot and the five-number summary on which it is based. You will learn a formal mechanism for identifying outliers, and you will learn some additional methods for analyzing data and making comparisons using your calculator.

Homework Activities

Activity 9-7: February Temperatures

2-5, 8-19, 9-7

Reconsider the data that you analyzed in Activity 2-5 concerning daily high temperatures in February 2006 for three locations. The dotplots are shown again here.

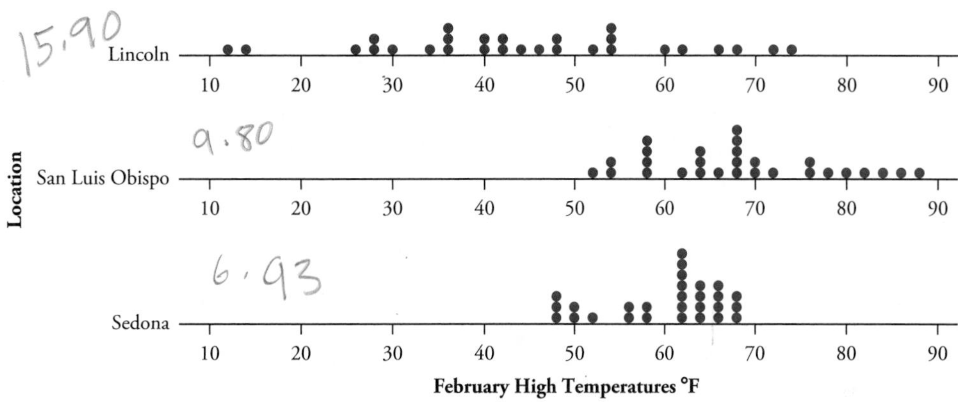

a. Before performing any calculations, predict which location will have the largest standard deviation and which will have the smallest.
b. Use your calculator (FEBTEMPS) to compute these standard deviations. Were your predictions correct?

Activity 9-8: Social Acquaintances

9-8, 9-9, 10-13, 10-14, 19-9, 19-10, 20-12

Consider the data collected in the Preliminaries section at the beginning of this topic on yourself and your classmates based on the "social acquaintance" exercise described by Malcolm Gladwell in *The Tipping Point.*

a. Produce a stemplot of these data, and comment on its features.
b. Calculate the median, quartiles, and IQR.
c. Using your calculator, compute the mean and standard deviation.
d. Determine how many and what proportion of the students' results fall within one standard deviation of the mean.
e. Is the proportion in part d close to what the empirical rule predicts? Is this surprising? Explain.

Gladwell writes that he has used this exercise with hundreds of people and in every group has found a few scores less than 20 and a few greater than 100. He marvels at the large variability in results and argues that, "Sprinkled among every walk of life are a handful of people with a truly extraordinary knack of making friends and acquaintances." He calls these people *connectors* and contends that they are the driving force behind social trends and fads. In particular, Gladwell claims that Paul Revere's midnight ride was so successful because Revere was a connector who knew the right people to contact in each village that he rode through.

f. Are your class results consistent with Gladwell's findings demonstrating considerable variability? Explain.

Activity 9-9: Social Acquaintances

9-8, 9-9, 10-13, 10-14, 19-9, 19-10, 20-12

Reconsider the previous activity. The file ACQUAINTANCESCP, which includes the list AQCNT, contains data from 99 undergraduate students at Cal Poly who engaged in this exercise during Winter 2006. Compare your class results with those from Cal Poly. Write a paragraph or two summarizing your comparison, including graphical displays and numerical summaries as appropriate.

Activity 9-10: Hypothetical Quiz Scores

The following histograms show the distributions of 16 quiz scores for each of four hypothetical students (named A, B, C, and D):

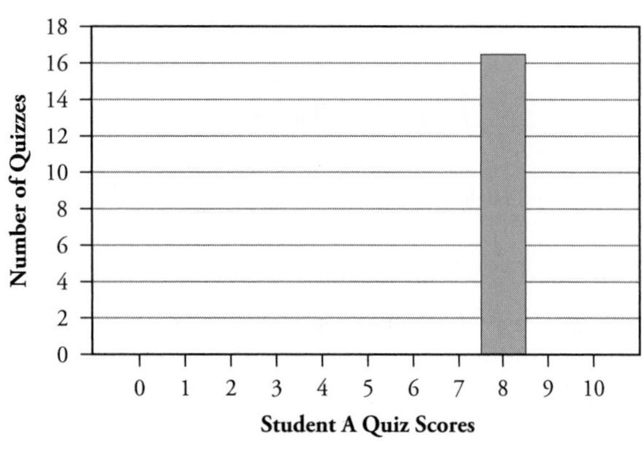

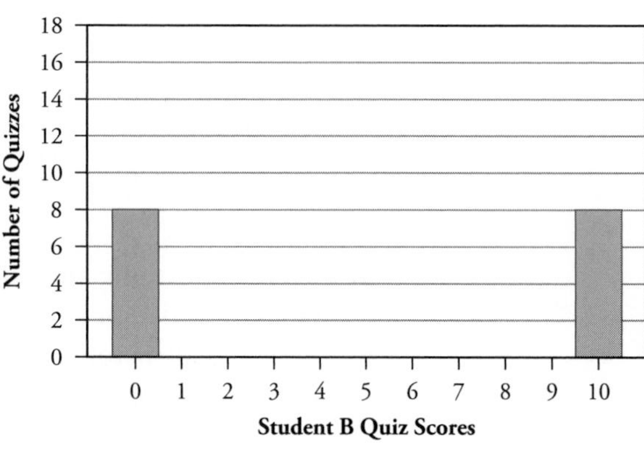

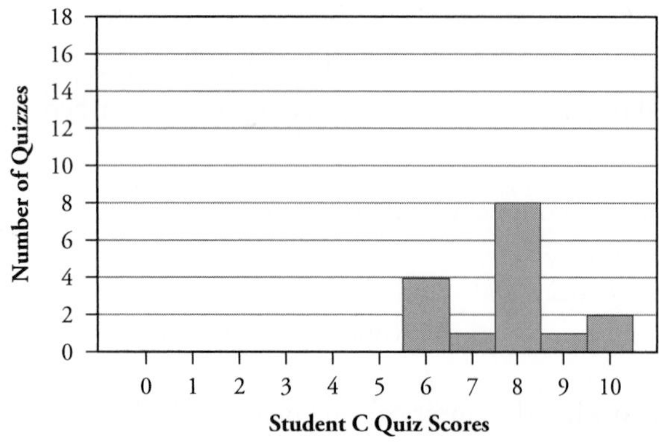

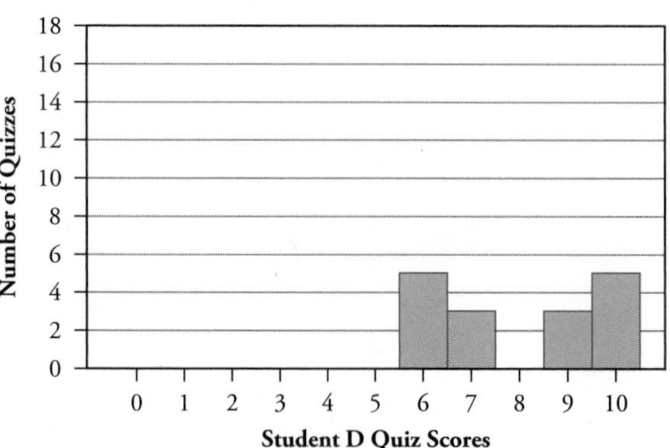

a. Without performing any calculations, arrange these students in order from smallest standard deviation to largest standard deviation. Explain the reasoning behind your choices.

b. Determine the value of the smallest standard deviation among these four students' quiz scores. Show your work or explain your answer.

c. Determine the value of the largest standard deviation among these four students' quiz scores. Show your work in calculating this value by hand.

Activity 9-11: Baby Weights

Three-month-old Benjamin Chance weighed 13.9 pounds. The national average for the weight of a three-month-old baby is 12.5 pounds, with a standard deviation of 1.5 pounds.

a. Determine the z-score for Benjamin's weight at age three months, and write a sentence interpreting the score.

b. For a six-month-old, the national average weight is 17.25 pounds, with a standard deviation of 2.0 pounds. Determine what Benjamin's weight would have been at the age of six months in order for him to have the same z-score at six months that he had at three months.

Activity 9-12: Student and Faculty Ages

a. Which group would you expect to have more variability in terms of age: the students at your school or the teachers at your school? Explain.

b. Make an educated guess as to the standard deviation of the ages of students at your school. Explain the reasoning behind your guess.

Activity 9-13: Baseball Lineups

9-1, 9-2, **9-13,** 9-17

This activity leads to the investigation of more properties of the measures of spread and center. Reconsider the players' ages from Activity 9-1. Suppose that every player in the Yankee lineup was two years older than reported.

a. How would you expect this change to affect the mean and median of those ages? Explain.

b. Calculate the new mean and median. (Feel free to use your calculator.) How did these values change?

c. How would you expect this change (adding two years to each player's age) to affect the IQR and standard deviation of those ages? Explain. *Hint:* Focus on how spread out the ages are.

d. Calculate the new IQR and standard deviation. (Feel free to use your calculator.) How did these values change?

Now suppose that the age of every player in the Tiger lineup was doubled.

e. Predict how the mean, median, IQR, and standard deviation will change.

f. Perform these calculations and report on your findings.

Activity 9-14: Pregnancy Durations

9-14, 12-6

Suppose the duration of human pregnancies follow a mound-shaped distribution with a mean of 266 days and a standard deviation of 16 days.

a. Approximately what percentage of human pregnancies last between 250 and 282 days?

b. Between what two values will the durations of 95% of all human pregnancy fall? Convert these values to months (with decimals), assuming the typical month has 30 days.

Suppose also that the durations of horse pregnancies follow a mound-shaped distribution with a mean of 336 days and a standard deviation of 3 days.

c. Is a horse or a human more likely to have a pregnancy that lasts within ± 6 days of its mean? Explain briefly.

Activity 9-15: Sampling Words

4-1, 4-2, 4-3, 4-4, 4-7, 4-8, 8-9, **9-15,** 14-6

Return to the `Sampling Words` applet that you used to explore properties of random sampling in Activity 4-3 (page 60).

a. Take **500** random samples of **5** words per sample, and look at the distribution of sample mean word lengths. Report the mean and standard deviation of these sample means.

b. Now take **500** random samples of **20** words per sample, and again look at the distribution of sample mean word lengths. Report the mean and standard deviation of these sample means.

c. How did the standard deviation of the sample means change as you quadrupled the sample size? *Hint:* Be more specific than saying "the standard deviation increased" or "the standard deviation decreased."

d. For the samples based on a sample size of 20, do you expect the empirical rule to hold fairly closely? Explain.

Activity 9-16: Tennis Simulations

7-22, 8-21, **9-16,** 22-18

Reconsider the data from Activities 7-22 and 8-21 on simulations of scoring systems for tennis.

a. Based solely on the frequency tables on pages 141 and 142 (before doing any calculations), indicate which scoring system appears to have the greatest variability in game lengths and which appears to have the least variability in game lengths.

b. For each of the three scoring systems, calculate by hand the quartiles and IQR.

c. Use your calculator (`TENNISSIM`) to compute the standard deviation of game lengths for each scoring system.

d. Do the IQRs and standard deviations agree in identifying the systems with the greatest and least variability? Do these values agree with your predictions in part a?

Activity 9-17: Baseball Lineups

9-1, 9-2, 9-13, **9-17**

Reconsider Activity 9-1, in which you compared the ages of players in the starting lineups of two baseball teams. Another variable on which you could compare those lineups is salary. The following table lists the 2006 salary (in millions of dollars, to the nearest 0.1 million) for those 20 players:

	Yankees		Tigers	
Position	Name	Salary	Name	Salary
C	Posada	12.0	I. Rodriguez	10.6
1B	Wilson	3.3	Casey	8.5
2B	Cano	0.4	Perez	2.5
3B	A. Rodriguez	25.6	Inge	3.0
SS	Jeter	20.6	Guillen	5.0

(*continued*)

	Yankees		Tigers	
Position	Name	Salary	Name	Salary
LF	Cabrera	0.3	Monroe	2.8
CF	Damon	13.0	Granderson	0.3
RF	Abreu	13.6	Gomez	0.3
DH	Giambi	20.4	Young	8.0
P	Wang	0.4	Robertson	0.4

Compare these two lineups with regard to salary. Include numerical summaries in your explanation. Address aspects of spread as well as center.

Activity 9-18: Population Growth

7-4, 8-15, **9-18,** 10-10

Reconsider the data from Activity 7-4 (page 126) on population growth in the 1990s for the 50 states.

a. Which region (eastern or western) appears to have more variability in population growth percentages?
b. Calculate the IQR of those percentage changes for the eastern states and for the western states. Comment on how the IQRs compare.
c. How would you expect the standard deviation of the percentage population change in the western states to change if Nevada were removed from the analysis? Explain.

Activity 9-19: Memorizing Letters

5-5, 7-15, 8-13, **9-19,** 10-9, 22-13, 23-3

Reconsider the data collected in Topic 5 for a memory experiment. Compare the performance scores of the two groups with regard to variability in scores. Use numerical summaries to support your comparison.

Activity 9-20: Monthly Temperatures

9-20, 26-12, 27-12

The following table reports the average monthly temperatures for San Francisco, California, and for Raleigh, North Carolina:

	Jan	Feb	Mar	Apr	May	Jun	Jul	Aug	Sep	Oct	Nov	Dec
Raleigh	39	42	50	59	67	74	78	77	71	60	51	43
San Francisco	49	52	53	56	58	62	63	64	65	61	55	49

a. Create dotplots (on the same scale and axis) to compare these temperature distributions for the two cities. (Ignore the month information in your dotplots.)
b. Calculate the median value of these temperatures for each city. Mark the medians on the respective dotplots. Are these medians fairly close?
c. Because the centers of these distributions are very close, can you conclude that there is not much difference between these two cities with regard to monthly temperatures? Explain.

d. Which city appears to have more variability in its monthly temperatures?
e. Calculate the range of monthly temperatures for these two cities, and comment on how the ranges compare.
f. Repeat part e with interquartile range.
g. Repeat part e with mean absolute deviation.
h. Repeat part e with standard deviation.

Activity 9-21: Nicotine Lozenge

1-16, 2-18, 5-6, **9-21,** 19-11, 20-15, 20-19, 21-6, 22-8

In Activity 5-6, you reviewed a study on the effectiveness of nicotine lozenges for smokers who want to quit smoking (Shiffman et al., 2002). The journal article in which the results of that study were published lists the means and standard deviations (SD) for the two treatment groups on a variety of baseline variables, such as subjects' age, weight, age of initiation to smoking, and number of cigarettes smoked per day. Some of this summary information is presented in the following table:

	Nicotine Lozenge		Placebo	
Variable	Mean	SD	Mean	SD
Age	41.11	12.06	40.48	11.94
Weight (in kg)	75.6	17.2	74.6	15.4
Age of Initiation	17.3	4.5	17.1	3.9
Cigarettes per Day	17.7	8.2	17.2	9.4

a. Notice that the mean age of smoking initiation is similar to the mean number of cigarettes smoked per day. Which of these variables has more variability? Explain how you can tell, and provide an interpretation of this comparison.
b. Why do you think the researchers provided this information in the article? What do you think they wanted readers to garner from this information?
c. Do you suspect that the empirical rule holds for any of these variables? If not, explain. If so, identify one variable for which you think the empirical rule holds and explain why.

Activity 9-22: Hypothetical Exam Scores

7-12, 8-22, 8-23, **9-22,** 10-22, 27-8

For each of the following properties, construct a dataset of ten hypothetical exam scores that satisfies the property. Assume the exam scores are integers between 0 and 100, inclusive, repeats allowed. You may use your calculator to perform the calculations.

a. Less than half of the exam scores fall within one standard deviation of the mean.
b. The standard deviation is positive, but the interquartile range equals 0.
c. All of the exam scores fall within one standard deviation of the mean.
d. The standard deviation is as large as possible.
e. The interquartile range is as large as possible, but the standard deviation is not as large as possible.
f. The mean absolute deviation is exactly half as large as the range.

Activity 9-23: More Measures

Consider the following two numerical measures:

- **Midrange:** the average of the minimum and maximum values
- **Midhinge:** the average of the lower and upper quartiles

a. For each of these, indicate whether it is a measure of center or a measure of spread. Explain how you can tell.

b. If a constant value is added to all of the data values in a dataset, does that change either (or both) of these measures? Explain. *Hints:* Feel free to try this with a small example. Also, you might want to reconsider your answer to part a after answering this question.

c. Which of these measures (or neither or both) is resistant to outliers? Explain, based on how the measures are defined.

d. Calculate the midrange and midhinge of the baseball players' ages, for both teams, from Activity 9-1 (page 165).

Activity 9-24: Hypothetical ATM Withdrawals

9-24, 19-21, 22-5

Suppose a bank wants to monitor the ATM withdrawals that its customers make from machines at three locations. Suppose they also sample 50 withdrawals from each location and tally the data in the following table. All missing entries are zeros. The raw (untallied) data are stored in the grouped file HYPOATM (with the lists **MACH1, MACH2,** and **MACH3,** respectively).

Cash Amount	20	30	40	50	60	70	80	90	100	110	120
Machine 1 Tally			25						25		
Machine 2 Tally	2	8	1	9	2	6	2	9	1	8	2
Machine 3 Tally	9					32					9

a. Produce visual displays of the distributions of cash amounts at each machine. Is each distribution perfectly symmetric?

b. Use your calculator to compute the mean and standard deviation of the cash withdrawal amounts at each machine. Are the mean and standard deviation identical for each machine?

c. Are the distributions of withdrawal amounts themselves identical for the three machines? Do the mean and standard deviation provide a complete summary of a distribution of data?

Activity 9-25: Guessing Standard Deviations

Notice that each of the following hypothetical distributions is roughly symmetric and mound-shaped.

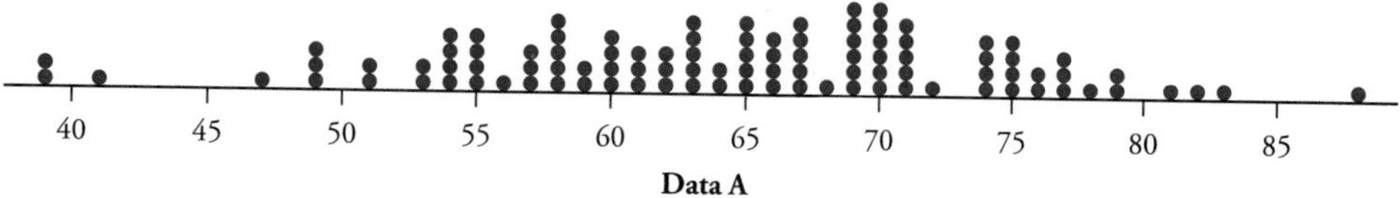

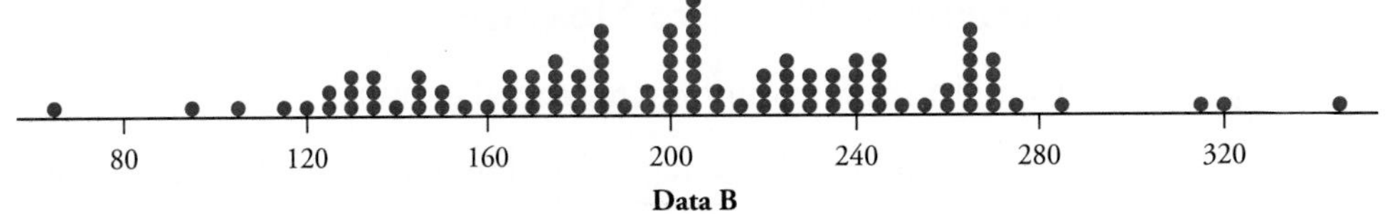

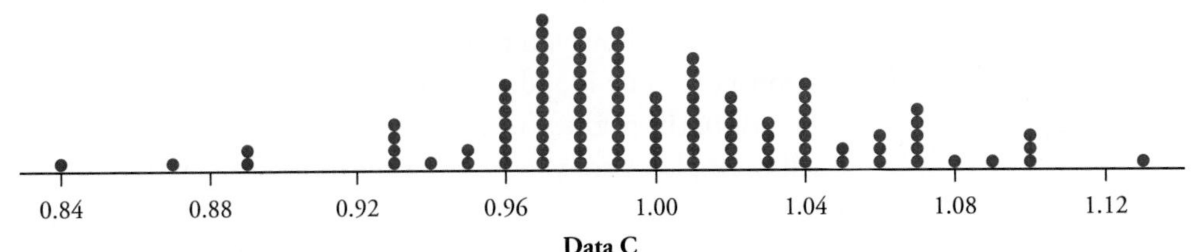

a. Use the empirical rule to make an educated guess about the mean and standard deviation of each distribution.

b. Use the data stored in the grouped file HYPOSTDDEV, which contains the lists **DATAA, DATAB, DATAC,** and **DATAD,** respectively, to calculate the actual means and standard deviations. Comment on the accuracy of your guesses.

TOPIC

10 More Summary Measures and Graphs

Now that you have learned the basics of summarizing and analyzing data, you will apply these techniques to an interesting variety of questions in this topic. For example, do sparrow deaths from a severe winter storm provide evidence for Darwin's theory of natural selection? Do steel roller coasters tend to go faster than wooden ones? Which brands of ice cream contain the most calories per serving? In this topic, you will investigate these questions and learn some new numerical and graphical summaries.

Overview

In the previous two topics, you studied numerical measures of the center and then of the spread of a distribution. In this topic, you will combine center and spread, as you work with the five-number summary, which conveys information about both. This summary will lead you to another visual display called a boxplot, which provides a useful way for comparing distributions, as well as a method for identifying outliers. You will also gain more experience using your calculator to analyze data and make comparisons.

Preliminaries

1. Would you expect larger or smaller sparrows to be more likely to survive a severe winter storm, or do you think that a sparrow's size is not related to its ability to survive?

2. How many calories would you expect to find in a one-half cup serving of *really good* ice cream?

3. Guess how much it cost for a family of four to attend a major-league baseball (MLB) game in 2006.

4. Name a city/team where you expect the cost of attending a major-league baseball game would be particularly expensive.

5. Name a city/team where you expect the cost of attending a major-league baseball game would be relatively inexpensive.

6. Name your favorite MLB team. If you do not have a favorite team, name the one closest to you or name a team that you have heard of.

In-Class Activities

Activity 10-1: Natural Selection

10-1, 10-6, 10-7, 12-20, 22-21, 23-3

A landmark study on the topic of natural selection was conducted by Hermon Bumpus in 1898. Bumpus gathered extensive data on house sparrows that were brought to the Anatomical Laboratory of Brown University in Providence, Rhode Island, following a particularly severe winter storm. Some of the sparrows were revived, but some sparrows perished. Bumpus analyzed his data to investigate whether those that survived tended to have distinctive physical characteristics related to their fitness.

The following sorted data are the total length measurements (in millimeters, from the tip of the sparrow's beak to the tip of its tail) for the 24 adult males that died and the 35 adult males that survived:

Sparrow Died						Sparrow Survived								
156	158	160	160	160	161	153	154	154	155	156	156	157	157	158
161	161	161	161	162	162	158	158	158	158	159	159	159	159	159
162	162	162	162	163	163	160	160	160	160	160	160	160	160	160
164	165	165	165	166	166	161	161	161	161	162	163	165	166	

a. Identify the observational units and explanatory and response variables in this study. Also classify the variables as categorical (also binary?) or quantitative.

Observational units:

Explanatory variable: Type:

Response variable: Type:

b. Is this study an observational study or an experiment? Explain your reasoning.

c. Determine the median and quartiles of the length measurements for the sparrows that *survived.* Also report the minimum and maximum values of these length measurements.

Median: 159

Lower quartile: 158 Upper quartile: 160

Minimum: 153 Maximum: 166

The median, quartiles, and extremes (minimum and maximum values) of a distribution constitute its **five-number summary (FNS).** The FNS provides a quick and convenient description of where the four quarters of the data in a distribution fall.

d. The following table reports the five-number summary for the lengths of the sparrows that died. Fill in the FNS for the lengths of the sparrows that survived:

	Minimum	Lower Quartile	Median	Upper Quartile	Maximum
Died	156 mm	161 mm	162 mm	163.5 mm	166 mm
Survived					

e. Compare the FNS for the two groups of sparrows. Is there evidence that one group of sparrows tended to be longer than the other? Explain.

The five-number summary forms the basis for a graph called a **boxplot.** Boxplots are especially useful for comparing distributions of a quantitative variable across two or more groups.

To construct a boxplot, you draw a *box* between the quartiles, indicating where the middle 50% of the data fall. You then extend horizontal lines called *whiskers* from the center of the sides of the box to the minimum and to the maximum values. You mark the median with a vertical line inside the box.

f. The following graph includes a boxplot of the distribution of lengths for the sparrows that died. Draw a boxplot of the distribution of lengths for the sparrows that survived on the same scale.

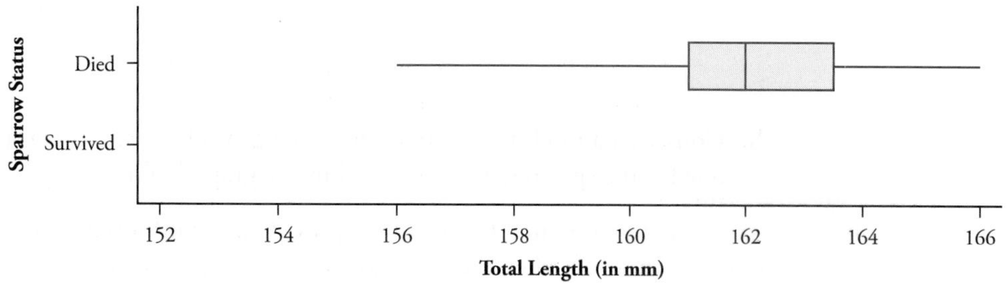

10

Modified boxplots convey additional information by treating outliers differently. On these graphs, you mark outliers using a special symbol and then extend the boxplot's whiskers only to the most extreme non-outlier value. This modification requires an explicit rule for identifying outliers. We will consider any observation falling more than 1.5 times the interquartile range away from the nearer quartile to be an outlier.

With the sparrows that perished, the IQR is 163.5 − 161 or 2.5 mm. Thus, 1.5 × IQR or 1.5(2.5) or 3.75, so any sparrow with a length greater than 3.75 mm away from its nearer quartile is considered an outlier. To look for such observations, you add 3.75 to the upper quartile, obtaining 163.5 + 3.75 or 167.25 mm. Because no deceased sparrow was longer than 167.25 mm, there are no outliers on the high end, and the whisker will still extend to the greatest length. On the low end, you subtract 3.75 from the lower quartile (161), obtaining 157.25. Any sparrows shorter than this will be labeled as outliers for the "died" distribution. The sparrow with a length of 156 mm is, therefore, an outlier, and the whisker will stop at 158, the lowest non-outlier.

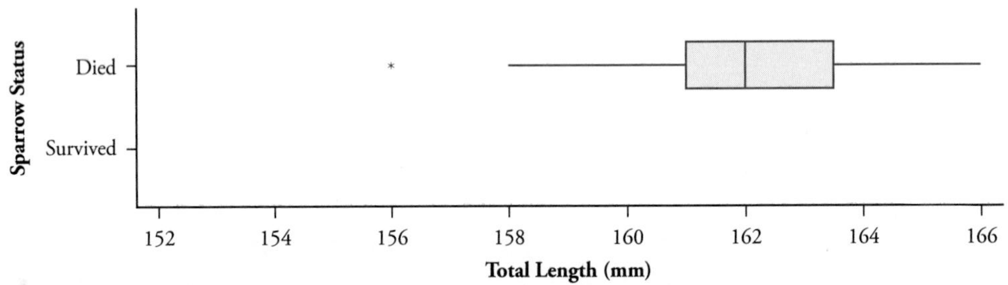

g. Use this rule to check for outliers among the sparrows that survived.

Q_L	Q_U	IQR = Q_U−Q_L	1.5 × IQR	Q_L − 1.5 × IQR	Q_U + 1.5 × IQR

Outliers:

h. Construct a modified boxplot for the lengths of the surviving sparrows next to the one for the sparrows who perished in the graph before part g.

Note: Because modified boxplots provide more information than unmodified ones, from now on when the text asks for a *boxplot,* you will create a *modified boxplot.*

i. Comment on what the five-number summaries and boxplots reveal about whether there appears to be a difference in lengths between the sparrows that survived and the sparrows that died in the winter storm.

j. Does the design of this study allow you to conclude that being shorter caused sparrows to be more likely to survive the storm? Explain.

k. Can you conclude from this study that shorter sparrows were more likely to survive the storm? Explain your answer, and also explain how this question differs from the question in part j.

Activity 10-2: Roller Coasters

2-17, **10-2**

Recall from Activity 2-17 that the Roller Coaster DataBase is an online resource that contains data on hundreds of roller coasters around the world. The following boxplots display the distributions of top speeds (in miles per hour) for coasters in the U.S., classified by whether the coaster is wooden or steel.

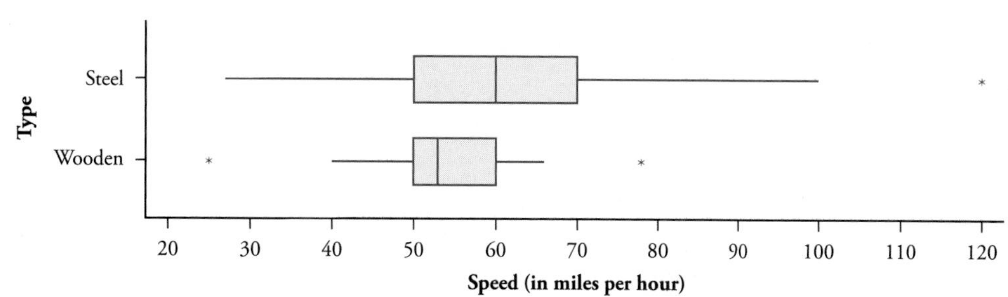

a. What proportion of the steel coasters have a top speed of 60 miles per hour (mph) or greater? Explain how you can tell from the boxplots.

b. What proportion of the wooden coasters have a top speed of 60 mph or greater? Explain how you can tell from the boxplots.

c. Which type of coaster (steel or wooden) has a higher proportion of coasters with a top speed greater than 50 miles per hour, or are the two types of coasters the same, or is it impossible to determine from the boxplot? Explain how you can tell (or why you cannot tell) from the boxplots.

d. Which type of coaster (steel or wooden) has a higher proportion of coasters with a top speed greater than 45 miles per hour, or are the two types of coasters the same, or is it impossible to determine from the boxplot? Explain how you can tell (or why you cannot tell) from the boxplots.

e. Which type of coaster (steel or wooden) has more variability in its speeds, or is the variability in the two types of coasters the same, or is it impossible to determine from the boxplot? Explain how you can tell (or why you cannot tell) from the boxplots.

f. Which type of coaster (steel or wooden) is more prevalent, or are there equal numbers of both types, or is it impossible to tell from the boxplot? Explain how you can tell (or why you cannot tell) from the boxplots.

Watch Out

- Boxplots can be tricky to read and interpret. The only information they provide is how the data divide into four pieces, each containing 25% of the data. For example, you know that half of the *steel* coasters travel faster than 60 mph, because 60 is the median speed, which you know because the middle line of the boxplot is located at 60 mph. Similarly, you know that only 25% of the *wooden* coasters travel faster than 60 mph, because the upper quartile (high end of the box) is at 60 mph. But for values that are not equal to one of the values in the five-number summary, such as 45, you can only say that *less than* 25% of the coasters have a top speed of 45 mph or less; you cannot determine which type (steel or wooden) has a greater percentage (closer to 25%) of coasters with a top speed greater than 45 mph. You also do not know how many coasters are in each quarter because you don't know the sample sizes involved and the boxplots only tell you about percentages.

- Remember that various software packages and calculators compute quartiles differently, so they might also construct boxplots slightly differently.

Activity 10-3: Ice Cream Calories

Food products are required by law to provide nutritional information on their labels, and many companies post such data on their websites. The grouped file ICECREAMCALORIES (with lists BJCAL, CSCAL, and DYCAL, respectively) contains data on calorie amounts per serving for various flavors of Ben & Jerry's, Cold Stone Creamery, and Dreyer's ice cream. Some of these values, for the first five flavors listed alphabetically, are detailed in the following table.

Ben & Jerry's		Cold Stone Creamery		Dreyer's	
Flavor	Calories per Serving	Flavor	Calories per Serving	Flavor	Calories per Serving
Black & Tan	230	Amaretto	390	Almond Praline	150
Brownie Batter	310	Banana	370	Andees Cool Mint	170
Butter Pecan	280	Black Cherry	390	Butter Pecan	170
Cherry Garcia	250	Bubble Gum	390	Cherry Chocolate Chip	160
Chocolate	260	Butter Pecan	390	Cherry Vanilla	140

a. Use the **1-Var Stat** command in your calculator to compute the five-number summary of the calorie amounts for all flavors in each brand. Report these in the following table:

	Minimum	Lower Quartile	Median	Upper Quartile	Maximum
Ben & Jerry's	110	220	265	290	360
Cold Stone	130	360	390	400	440
Dreyer's	90	110	120	150	190

b. Use your calculator to construct (modified) boxplots of the distribution of calorie amounts for the three brands on the same axis. To do this, press 2ND [STAT PLOTS] and then select either **Plot1, Plot2,** or **Plot3.** Turn on the plot, and arrow down and over to select the modified boxplot. For **Xlist,** enter one of the lists mentioned at the beginning of this activity. Set the **Freq** to **1. Mark**

can be set to one of the three symbols shown. *TI tip*: Once you have set up your plot, press the ZOOM key and then the 9 key to format the Graph screen automatically.

Discuss how the three brands' calorie amounts compare, based on these boxplots.

- BJcal has most variability
- CScal has highest median
- Dycal has lowest median

c. Can you speculate about a possible problem with making comparisons of calorie amounts among the three brands based on these data? *Hint:* You were not given one relevant piece of information in the description at the beginning of the activity.

Serving size

Five-number summaries and boxplots can be useful for making comparisons. There is one important problem with the previous analysis, though. Ben & Jerry's and Dreyer's both consider a serving to be ½ cup of ice cream, so their calorie amounts are directly comparable. But Cold Stone Creamery considers a serving to be 170 grams.

d. How could you adjust for this discrepancy (in principle, anyway)?

Convert

One difficulty with converting these calorie amounts to a common serving size is that ½ cup is a measure of volume and 170 grams is a measure of weight. The conversion is, therefore, not as simple as, say, converting inches to centimeters (both measures of length). The website `Gourmetsleuth.com` has a "gram conversion calculator" that applies to individual food items, however. `Gourmetsleuth.com` suggests that the conversion rate for ice cream is roughly 146 grams per cup.

e. How would you use this conversion information to convert Cold Stone's calorie amounts to a "per half cup" amount, comparable to Ben & Jerry's and Dreyer's?

f. Use your calculator to do this conversion. *Hint:* First, divide Cold Stone's calorie amounts by 170 to convert them to a "per gram" basis, and then multiply this amount by 146 to convert the calorie counts to a "per cup" basis, and finally, divide by 2 to convert the calorie counts to a "per half cup" basis and then store the data in a new named list.

g. Use your calculator to recalculate the five-number summary for Cold Stone's calorie amounts using the "per half cup" scale. How has the FNS changed?

h. Use your calculator to reproduce the boxplots for comparing the three distributions of calorie amounts. For this, you will select and set up all three stat plots (remember to turn on the plots) and use the [ZOOM] key mentioned in part b to set the screen. Comment on what the boxplots reveal about how the three ice cream brands compare.

Activity 10-4: Fan Cost Index

The data in the following table were compiled by the Team Marketing Report in an effort to measure the cost of attending major-league baseball games in 2006. The variables recorded were average price of adult tickets, average price of children's tickets, cost for parking, price of program, price for a medium baseball team cap, price for a small beer, how many ounces in a small beer, price for a small soda, how many ounces in a small soda, and price for a medium hot dog (all in U.S. dollars).

Team	Adult	Child	Park	Program	Cap	Beer	Oz	Soda	Oz	Hot Dog
Arizona	19.68	19.68	10	1	12	4.00	14	3.25	14	3
Atlanta	17.07	17.07	12	0	12	5.25	12	3.50	20	4.25
Baltimore	22.53	22.53	8	5	12	4.25	18	2.00	16	2.5
Boston	46.46	46.46	23	5	15	6.00	12	2.75	14	4
Chi Cubs	34.30	34.30	17	5	12	5.00	16	2.50	15	2.75

(continued)

10

Team	Adult	Child	Park	Program	Cap	Beer	Oz	Soda	Oz	Hot Dog
Chi White Sox	26.19	26.19	18	4	13	5.75	16	2.75	14	3
Cincinnati	17.90	17.90	12	4	15	6.00	20	2.50	16	3.25
Cleveland	21.54	21.54	12	1	15	4.25	14	2.25	12	2.5
Colorado	14.72	14.72	8	5	14	5.50	16	3.00	16	3.25
Detroit	18.48	18.48	15	5	15	5.00	16	3.00	20	3
Florida	16.70	15.61	8	5	15	5.75	20	3.50	12	4
Houston	26.66	26.23	10	4	11	7.00	16	4.00	21	4
Kansas City	13.71	13.71	6	5	12	3.75	12	2.00	14	2.5
LA Angels	18.97	18.70	8	3	6.99	4.50	14	2.50	14	3
LA Dodgers	20.09	20.09	10	5	12	8.00	20	4.25	12	4.5
Milwaukee	18.11	18.11	7	0	12	4.75	16	2.00	12	2.75
Minnesota	17.26	17.26	6	2	15	6.00	24	3.75	20	3.25
NY Mets	25.28	25.28	12	4	18	6.50	21	4.75	32	4.5
NY Yankees	28.27	28.27	12	7.75	15	6.00	16	3.50	16	3
Oakland	22.10	22.10	14	5	12	5.50	14	2.25	12	3.5
Philadelphia	26.73	26.73	10	5	15	5.00	21	3.25	20	3.5
Pittsburgh	17.08	17.08	10	5	12	4.25	21	2.25	20	2.25
San Diego	20.83	20.83	4	5	22	5.00	16	3.75	22	3.5
San Francisco	24.53	24.53	25	5	15	5.75	16	3.00	16	3.75
Seattle	24.01	24.01	17	4	16	5.00	12	2.50	16	3.25
St. Louis	29.78	28.57	10	2.5	14	7.75	24	4.50	14	3.5
Tampa Bay	17.09	13.85	0	0	15	5.00	16	3.75	16	3.25
Texas	15.81	15.81	8	5	10	6.00	22	2.75	16	2.5
Toronto	23.40	19.61	18.24	4.55	13.65	5.01	14	2.94	24	3.17
Washington	20.88	20.88	12	5	12	5.00	12	3.50	22	4

a. The fan cost index (FCI) is defined as the cost of two adult tickets, two children's tickets, parking, two programs, two baseball team caps, two small beers, four small sodas, and four hot dogs. Use your calculator to compute the FCI for each baseball team (*TI hint:* Remember that you can perform the algebra on a list and then save the results in a new named list). The data are stored in the grouped file FANCOST06 (with lists **ADLT, BEER, BEROZ, CAP, CHLD, HDOG, PRGM, PRK, SDA,** and **SDAOZ,** respectively). Report which team has the highest and which the lowest FCI, along with those values. *Hint:* To make sure your formula is entered properly, the FCI for Arizona is $147.72.

Highest FCI team: Value:

Lowest FCI team: Value:

b. Use your calculator to create a dotplot of the distribution of the FCI values. Comment on the dotplot's features.

c. Report the FCI value for the team that you identified in the Preliminaries section as your favorite team. Also comment on where this team's FCI falls relative to the other teams.

d. If you were to attend a game, who would you go with? Would your group typically be a family of four? Do you think the FCI accurately measures how much it would cost *your group* to attend a game?

e. Create a new variable, called the Me Cost Index (MCI), by changing the number of tickets and what is purchased. That is, follow what you did previously but change the coefficients to better reflect what your group would actually spend at a game. Report your new equation and which team has the highest and which has the lowest MCI, along with those values. Also comment on where your team falls in the distribution of MCI values.

Highest MCI team: Value:

Lowest MCI team: Value:

Your team's MCI value and relative position:

f. Which ballpark(s) charges the most for a small soda? Which charges the least? What are those values?

Highest team: Value:

Lowest team: Value:

g. Explain what is misleading about comparing the costs of a small soda or small beer in the different ballparks.

h. Create a new variable for comparing soda prices by converting to a "price per ounce" variable. Which ballpark has the most expensive soda per ounce, and which has the least expensive? What are those values?

Highest team: Value:

Lowest team: Value:

Activity 10-5: Digital Cameras

10-5, 26-18, 27-21

The July 2006 issue of *Consumer Reports* provided information and ratings of 78 brands of digital cameras. Some of the variables included were price, overall quality rating, and type of camera (compact, subcompact, advanced compact, and super-zoom). The grouped file DIGITALCAMERAS contains the lists **PRICE, SCORE, PRCCM** (price compact), **PRCSC** (price subcompact), **PRCAC** (price advanced compact), **PRCSZ** (price super-zoom), **SCRCM** (score compact), **SCRSC** (score subcompact), **SCRAC** (score advanced compact), and **SRCSZ** (score super-zoom), respectively; the first few lines of the data are shown here:

Brand	Model	Type	Price (in $)	Rating Score
Canon	Power Shot A620	Compact	325	76
Fujifilm	FinePix F10	Compact	315	76
Kodak	EasyShare Z700	Compact	245	76
Olympus	C-5500 Sport Zoom	Compact	280	76
Sony	Cyber-shot DSC-W45	Compact	220	76

a. Use your calculator to compute five-number summaries and produce comparative boxplots of the camera *prices,* based on the camera type (the TI-83/TI-84 Plus can only do three plots at a time. You will need to make four different 3-plot setups to do a complete comparison of the four camera types). Is there much overlap in prices among the four types of cameras? Do some types of cameras tend to cost more than others?

b. Arrange the camera types in order with regard to center, as measured by the median: Which camera type tends to cost the most, followed by second-most, second-least, and least?

Most: Second: Third: Least:

c. Arrange the camera types in order with regard to spread as measured by the IQR: Which camera type has the most variability in its prices, followed by second-most, second-least, and least variability in its prices?

Most: Second: Third: Least:

d. Now use your calculator to compute five-number summaries and to produce comparative boxplots of the camera *ratings*, based on the camera type. Write a paragraph summarizing what these boxplots reveal about the distributions of rating scores across the four types of cameras.

e. Comment on whether the types of cameras with the highest prices also tend to have the highest ratings.

Solution

a. The table below reports the five-number summaries, as reported by the software package Minitab. The boxplots of camera prices follow the table.

	Minimum	Lower Quartile	Median	Upper Quartile	Maximum
Advanced Compact	\$280	\$400	\$560	\$745	\$840
Compact	\$140	\$210	\$290	\$327.5	\$480
Subcompact	\$185	\$273.75	\$300	\$330	\$450
Super-zoom	\$250	\$311.25	\$357.5	\$487.5	\$720

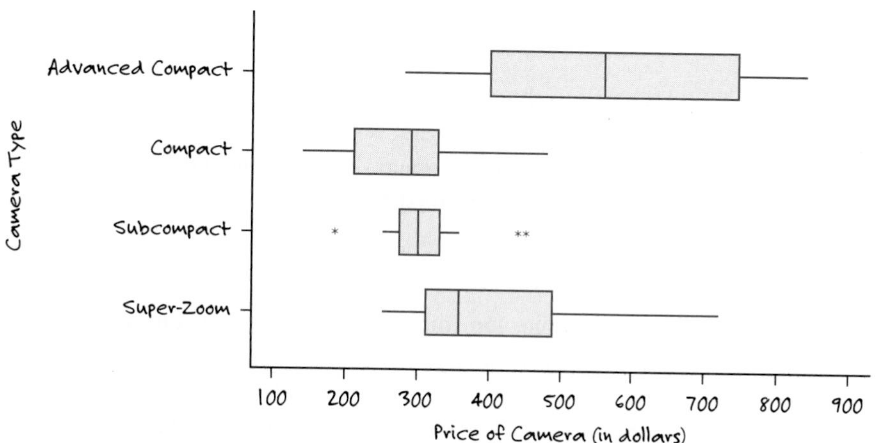

There is considerable overlap in prices among these four groups, but some camera types (e.g., advanced compact) do tend to cost more than other types.

b. Advanced compact cameras tend to cost the most, followed by super-zoom cameras. Compact and subcompact camera prices are similar, but the subcompact cameras cost a bit more on average than the compact cameras.

c. Advanced compact cameras have the most spread in terms of prices, again followed by super-zoom cameras. But compact cameras have more spread in prices than do subcompact cameras, which have the least variability in prices.

d. The boxplots of camera ratings are shown following the table. The five-number summaries, as reported by the software package Minitab, are as shown here:

	Minimum	Lower Quartile	Median	Upper Quartile	Maximum
Advanced Compact	63	69	70	73	78
Compact	62	65	71	73.5	76
Subcompact	53	61.75	65.5	69.25	75
Super-zoom	66	72.25	75.5	79	81

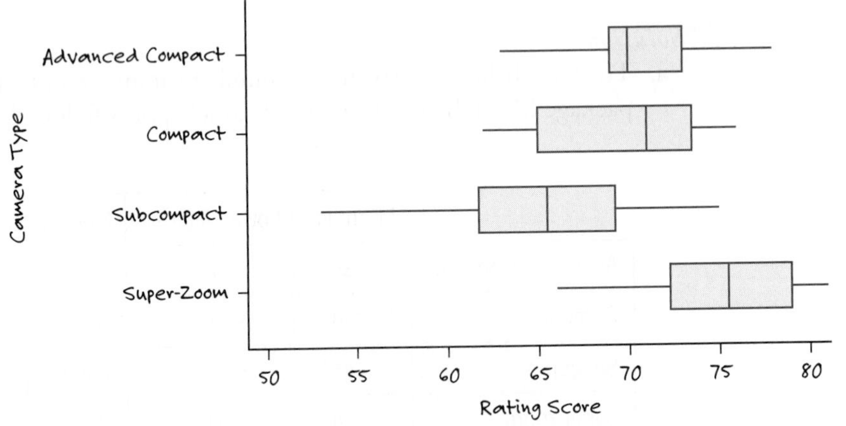

The super-zoom cameras tend to have the highest ratings, and the subcompact ones tend to rate the lowest. The subcompact cameras also have the most variability in ratings.

e. Even though the advanced compact cameras have the highest median price by far, their median rating is surpassed by both super-zoom and compact cameras. In fact, compact cameras have the second-highest median rating despite having the second-lowest median price.

Wrap-Up...........

In this topic, you studied the five-number summary as a convenient way to summarize a distribution, and you learned about new visual displays: the boxplot and modified boxplot. These displays are particularly useful for comparing distributions between two or more groups. For example, you found that sparrows who survived a winter storm tended to be longer than sparrows who perished.

You also used your calculator extensively to compare distributions using these statistical analysis tools, and along the way, you saw the utility of creating new variables and converting measurements to a common scale. Using a common scale was particularly important for drawing meaningful comparisons of calories per serving size across various brands of ice cream. The ability to understand and interpret your calculator's output is also paramount.

Some useful definitions to remember and habits to develop from this topic include

- The **five-number summary** consists of the minimum, lower quartile, median, upper quartile, and maximum values in a dataset. It provides a simple but useful numerical summary of a large distribution.
- A **boxplot** is a visual representation of the five-number summary. Boxplots are especially useful for comparing groups, but they do not provide as much information about distribution shape as a stemplot, dotplot, or histogram.
- **Modified boxplots** provide additional information by identifying outliers. Any observation more than 1.5 times the IQR (1.5 × IQR) away from its nearest quartile is considered an outlier.
- Variables might need to be transformed to a common scale in order to draw meaningful comparisons across groups.

This topic completes the unit on summarizing data. In the next unit, you will study the concept of randomness in more detail, learning how randomness in a data collection process can help you to draw confident conclusions from the resulting data.

Homework Activities

Activity 10-6: Natural Selection

10-1, **10-6,** 10-7, 12-20, 22-21, 23-3

Reconsider Activity 10-1, where you analyzed some of the famous Bumpus data on natural selection (Bumpus, 1898). Another variable that Bumpus recorded for the sparrows was their weight. Five-number summaries of sparrow weights (in grams) are provided here:

	Minimum	Lower Quartile	Median	Upper Quartile	Maximum
Sparrow Died	24.6	25.2	26.0	26.95	31.0
Sparrow Survived	23.2	24.3	25.7	26.50	28.0

a. Do the data suggest that the sparrows that survived differed from those that died with respect to weight? Refer to the five-number summaries to support your answer.

b. For each group of sparrows, determine how much (and how little) a sparrow would have to weigh in order to be considered an outlier (by the 1.5 × IQR rule).

c. Based on your answers to part b, along with the five-number summaries, what can you say about whether there are any outliers among the sparrows that survived? Repeat for the sparrows that died.

d. Explain why the five-number summaries do not provide enough information for you to draw complete (modified) boxplots of these weight distributions.

Activity 10-7: Natural Selection

10-1, 10-6, **10-7,** 12-20, 22-21, 23-3

Reconsider the Bumpus study (1898) from Activity 10-1 and the previous activity. In addition to total length and weight, Bumpus also recorded these data on the sparrows:

- Alar extent
- Length of head and beak

10

- Humerus bone length
- Femur bone length
- Tibiotarsus bone length
- Skull width
- Keel of sternum

These data are stored in the grouped file BUMPUS (with the respective died lists: **DALAR, DLBH, DLNG, DHUM, DFMR, DTBTR, DSKLW, DKLST,** and **DWGHT**; and the respective survived lists: **SALAR, SLBH, SLNG, SHUM, SFMR, STBTR, SSKLW, SKLST,** and **SWGHT**).

a. Use your calculator to produce boxplots comparing the distributions of surviving and perished sparrows for all of these variables.
b. For which variables do the distributions vary considerably? For which variables do they not vary much?
c. Write a paragraph summarizing what your analyses reveal about physical characteristics that distinguish those sparrows that survived from those sparrows that died.

Activity 10-8: Welfare Reform

In 1996, President Clinton and the Republican-controlled Congress passed sweeping welfare reform legislation (Wolf, 2008). One goal of this reform was to reduce the number of families receiving welfare assistance from the federal government. The following table (and the grouped file WELFAREREFORM with lists **AUG96, AG96N, AG96S, AG96M, AG96W, DEC05, DC05N, DC05S, DC05M,** and **DC05W,** respectively, for north, south, midwest, and west), contains data on the number of families in each state who received welfare assistance when the law was passed in August 1996 and the number of families receiving assistance in December 2005:

State	Aug 96 Families	Dec 05 Families	State	Aug 96 Families	Dec 05 Families
Alabama	41,032	20,316	Indiana	51,437	48,213
Alaska	12,159	3,590	Iowa	31,579	17,215
Arizona	62,404	41,943	Kansas	23,790	17,400
Arkansas	22,069	8,283	Kentucky	71,264	33,691
California	880,378	453,819	Louisiana	67,467	13,888
Colorado	34,486	15,303	Maine	20,007	9,516
Connecticut	57,326	18,685	Maryland	70,665	22,530
Delaware	10,585	5,744	Massachusetts	84,700	47,950
Florida	200,922	57,361	Michigan	169,997	81,882
Georgia	123,329	35,621	Minnesota	57,741	27,589
Hawaii	21,894	7,243	Mississippi	46,428	14,636
Idaho	8,607	1,870	Missouri	80,123	39,715
Illinois	220,297	38,129	Montana	10,114	3,947

(continued)

State	Aug 96 Families	Dec 05 Families	State	Aug 96 Families	Dec 05 Families
Nebraska	14,435	10,016	Rhode Island	20,670	10,063
Nevada	13,712	5,691	S Carolina	44,060	16,234
New Hamp	9,100	6,150	S Dakota	5,829	2,876
New Jersey	101,704	42,198	Tennessee	97,187	69,361
New Mexico	33,353	17,773	Texas	243,504	77,693
New York	418,338	139,220	Utah	14,221	8,151
N Carolina	110,060	31,746	Vermont	8,765	4,479
N Dakota	4,773	2,789	Virginia	61,905	9,615
Ohio	204,240	81,425	Washington	97,492	55,910
Oklahoma	35,986	11,104	W Virginia	37,044	11,275
Oregon	29,917	20,194	Wisconsin	51,924	17,970
Pennsylvania	186,342	97,469	Wyoming	4,312	294

a. Identify the observational units.

b. Use your calculator to create a new variable: *reduction in number of families receiving welfare assistance. Hint:* A positive value should indicate that the number of families receiving assistance declined over this period.

c. Now use your calculator to create another new variable: *percentage reduction.* Identify the three states with the greatest percentage reductions and the three states with the lowest reductions. *Hint*: Divide the previous variable (*percentage reduction*) by the number of families receiving assistance to begin with, and multiply that amount by 100 to convert to a percentage.

d. Create a dotplot of the *percentage reduction* variable, and comment on its key features.

e. Create comparative boxplots of the percentage reductions across these four regions. Comment on what these boxplots reveal about whether some regions of the country tended to see a greater percentage reduction in the number of welfare recipients than did other regions of the country.

Activity 10-9: Memorizing Letters

5-5, 7-15, 8-13, 9-19, **10-9,** 22-3, 23-3

Reconsider the class data from the memory experiment that you conducted in Activity 5-5 (page 81). Produce boxplots for comparing the distributions of scores between the two groups. Comment on what the boxplots reveal.

Activity 10-10: Population Growth

7-4, 8-15, 9-18, **10-10**

Reconsider the data from Activity 7-4 (page 126) concerning population growth in the 1990s across the 50 states. Produce boxplots for comparing the distributions of percentage changes in population between eastern and western states. Comment on what the boxplots reveal.

Activity 10-11: Sporting Examples

2-6, 3-4, 8-14, **10-11,** 22-26

Reconsider the data from Activity 2-6 (page 23) concerning total points earned in an introductory statistics course, comparing one section that used sports examples exclusively with a more traditional section using varied examples (SPORTSEXAMPLES with the lists RGLR [regular] and SPRTS [sports]). Produce boxplots for comparing the distributions of total points between the two sections. Comment on what the boxplots reveal.

Activity 10-12: Backpack Weights

2-13, **10-12,** 19-6, 20-17

Recall that Activity 2-13 described a study in which a group of undergraduate students at Cal Poly (Mintz, Mintz, Moore, and Schuh, 2002) examined how much weight students carry in their backpacks. They weighed the backpacks of 100 students and compared the distributions of backpack weights between men and women. The data are stored in the grouped file BACKPACK (**BDYWM, BPKWM, BDYWW, BPKWW, BDYWT, BPKWT, WGHT,** respectively, for body and backpack weights for men and women).

a. Use your calculator to produce comparative boxplots of the backpack weights carried by male and female students. Comment on what these boxplots reveal.
b. The *Ratio* attribute gives the ratio of backpack weight to body weight. Produce boxplots of these ratios, comparing male and female students. Comment on what these boxplots reveal.

Activity 10-13: Social Acquaintances

9-8, 9-9, **10-13,** 10-14, 19-9, 19-10, 20-12

Reconsider the class data collected in Topic 9 (Activities 9-8 and 9-9) concerning the number of social acquaintances you and your classmates have.

a. Calculate the five-number summary of this distribution.
b. Determine whether there are any outliers.
c. Construct a (modified) boxplot of the data.
d. Write a paragraph summarizing key features of the distribution. Refer to your calculations and boxplot for support.

Activity 10-14: Social Acquaintances

9-8, 9-9, 10-13, **10-14,** 19-9, 19-10, 20-12

The 1.5 × IQR rule is the most common one used for identifying outliers, but you can apply other rules as well. Here are four more rules:

- An outlier is more than 2 × IQR from its nearest quartile.
- An outlier is more than 3 × IQR from its nearest quartile.
- An outlier is more than two standard deviations from the mean.
- An outlier is more than three standard deviations from the mean.

Reconsider the class data collected in Topic 9 concerning the number of social acquaintances for you and your classmates.

a. Using each of these rules, determine whether the data contain any outliers.
b. Write a paragraph summarizing what these analyses reveal.

Activity 10-15: Diabetes Diagnoses

7-5, **10-15**

Reconsider the data from Activity 7-5 on ages at which a sample of 548 people with diabetes were first diagnosed with the disease. The frequency table is reproduced here:

Age	1	2	3	4	5	6	7	8	9	10	11	12	13	15	16	17	18
Tally	1	3	7	5	5	2	4	5	1	5	1	2	1	2	3	1	4
Age	19	20	21	24	25	26	27	28	29	30	31	32	33	34	35	36	37
Tally	3	1	3	2	1	4	4	2	1	6	5	3	8	5	15	4	7
Age	38	39	40	41	42	43	44	45	46	47	48	49	50	51	52	53	54
Tally	10	9	15	9	8	9	5	12	8	7	11	12	25	9	9	7	8
Age	55	56	57	58	59	60	61	62	63	64	65	66	67	68	69	70	71
Tally	22	13	8	11	16	19	5	16	9	10	10	9	6	12	7	13	4
Age	72	73	74	75	76	77	78	79	80	81	82	83	84	85	87	88	
Tally	4	9	4	8	7	5	6	3	2	3	2	1	2	1	1	1	

a. Use this frequency table to determine the five-number summary of these ages.
b. Check for outliers, according to the 1.5 × IQR criterion.
c. Produce a well-labeled, modified boxplot of the distribution of ages.
d. Does the boxplot display the shape of the distribution as well as the histogram from Activity 7-5 (page 128) did? Explain.

Activity 10-16: Hazardousness of Sports

The following table contains estimates of the number of sports-related injuries treated in U.S. hospital emergency departments in 1997, along with an estimate of the number of participants in the indicated sports taken from *Injury Facts* (National Safety Council, 1999). The data are stored in the grouped file SPORTSHAZARDS with lists **INJRS** and **PRTCP**, respectively.

Sport	Injuries	Participants	Sport	Injuries	Participants
Basketball	644,921	33,300,000	Ice hockey	77,491	1,900,000
Bicycle riding	544,561	45,100,000	Fishing	72,598	44,700,000
Football	334,420	20,100,000	Volleyball	67,340	17,800,000
Baseball, softball	326,714	30,400,000	Skateboarding	48,186	6,300,000
Roller skating	153,023	37,500,000	Golf	39,473	26,200,000
Soccer	148,913	13,700,000	Snowboarding	37,638	2,800,000
Weight lifting	86,024	47,900,000	Ice skating	25,379	7,900,000
Swimming	83,772	59,500,000	Bowling	23,317	44,800,000

(*continued*)

Sport	Injuries	Participants	Sport	Injuries	Participants
Tennis	22,294	11,100,000	Billiards, pool	3,685	36,000,000
Water skiing	10,657	6,500,000	Archery	3,213	4,800,000
Racquetball	10,438	4,500,000			

a. If you use the number of injuries as a measure of the hazardousness of a sport, which sport is more hazardous, bicycle riding or football? Ice hockey or soccer? Swimming or skateboarding?

b. Use your calculator to compute each sport's injury rate per thousand participants. *Hint:* Divide the number of injuries by the number of participants, and then multiply that amount by 1000 to produce an injury rate per thousand participants.

c. In terms of the injury rate per thousand participants, which sport is more hazardous, bicycle riding or football? Soccer or ice hockey? Swimming or skateboarding?

d. How do your answers to part a and part c compare? How do they compare to your intuitive perceptions of how hazardous these particular sports are?

e. List the three most and three least hazardous sports according to injury rate per thousand participants.

f. Identify some other factors related to the hazardousness of a sport. In other words, what information might you use to produce a better measure of a sport's hazardousness that is not already taken into account by the number or rate of injuries?

Activity 10-17: Gender of Physicians

Female physicians are more common in some medical specialties than in others. For each of 38 medical specialties, the following table lists the numbers of male and female physicians who identified themselves as practicing that specialty as of December 31, 2002, taken from the *World Almanac and Book of Facts 2006* (the data are stored in the grouped file PHYSICIANS with lists MEN and WOMEN):

Specialty	No. Men	No. Women	Specialty	No. Men	No. Women
Aerospace medicine	461	32	Family practice	63,194	23,317
Allergy & immunology	9,120	981	Forensic pathology	399	181
Anesthesiology	28,756	7,855	Gastroenterology	10,220	1,109
Cardiovascular disease	20,088	1,882	General practice	11,297	2,249
Child psychiatry	3,759	2,758	General preventive medicine	1,203	626
Colon/rectal surgery	1,084	113	General surgery	32,678	4,525
Dermatology	6,506	3,482	Internal medicine	101,633	41,658
Diagnostic radiology	18,086	4,698	Medical genetics	230	203
Emergency medicine	20,429	5,098	Neurological surgery	4,770	238

(continued)

Specialty	No. Men	No. Women	Specialty	No. Men	No. Women
Neurology	10,088	2,895	Physical med./rehab.	4,619	2,302
Nuclear medicine	1,200	255	Plastic surgery	5,822	713
Obstetrics/gynecology	25,606	15,432	Psychiatry	27,803	12,292
Occupational medicine	2,346	494	Public health	1,231	600
Ophthalmology	15,670	2,914	Pulmonary diseases	8,148	1,290
Orthopedic surgery	22,329	882	Radiation oncology	3,215	968
Otolaryngology	8,839	987	Radiology	7,574	1,218
Pathology–anat./clin.	12,575	5,604	Thoracic surgery	4,904	152
Pediatric cardiology	1,241	459	Transplantation surgery	75	8
Pediatrics	33,020	33,351	Urological surgery	10,048	383

a. Suppose that you use the number of women practicing a specialty as a measure of female participation in that specialty. Identify the three specialties with the most female physicians and the three specialties with the fewest female physicians; also record those values.

b. What aspect of the gender breakdown does the *number of women* variable not take into account?

c. Explain why *percentage of that specialty's practitioners who are women* is a useful variable for addressing the concern in part b.

d. Use your calculator to compute the percentage of each specialty's practitioners who are women. *Hint:* Divide the number of women in that specialty by the total number of practitioners. Then multiply that amount by 100 to convert the proportion into a percentage.

e. Identify the three specialties with the greatest percentage of female physicians and the three specialties with the lowest percentage of female physicians; also record those values.

f. Do your lists in part a and part e agree exactly? If not, explain (based on the data) why the lists differ.

g. Identify a specialty that relatively few women practice, yet a relatively greater percentage of women than is typical practice that specialty. Also record these values for that specialty, and explain why this occurs.

h. Produce visual displays of the distribution of the percentage of women in these medical specialties. Write a brief paragraph describing the distribution's key features.

Activity 10-18: Draft Lottery

10-18, 27-7, 29-9

In 1970, the United States Selective Service conducted a lottery to decide which young men would be drafted into the armed forces (Fienberg, 1971). Each of the 366 birthdays in a year was assigned a draft number. Young men born on days assigned low draft numbers were drafted. The grouped data file DRAFTLOTTERY contains lists of the draft numbers assigned to each birthday by month (JAN, FEB, MAR, APR, MAY, JUN, JUL, AUG, SEP, OCT, NOV, DEC, DRF70, DRF71, SEQ70, and SEQ71).

10

a. What is your birthday, and what draft number was assigned to it?

b. Use your calculator to compute the median of the draft numbers for each month. Report these, and comment on whether you observe any pattern that would appear to be suspicious coming from a truly random lottery.

c. Produce comparative boxplots for comparing the distributions of draft numbers across the twelve months. Again comment on whether you notice any pattern that strikes you as surprising coming from a truly random lottery.

Activity 10-19: Honda Prices

7-10, **10-19,** 12-21, 28-14, 28-15, 29-10, 29-11

The following three dotplots display data obtained for used Hondas for sale on the Internet. The three boxplots that follow also display this data. Your task is to match up each dotplot (a, b, c) with its corresponding boxplot (I, II, III). Also explain the reasoning behind your choices.

a.

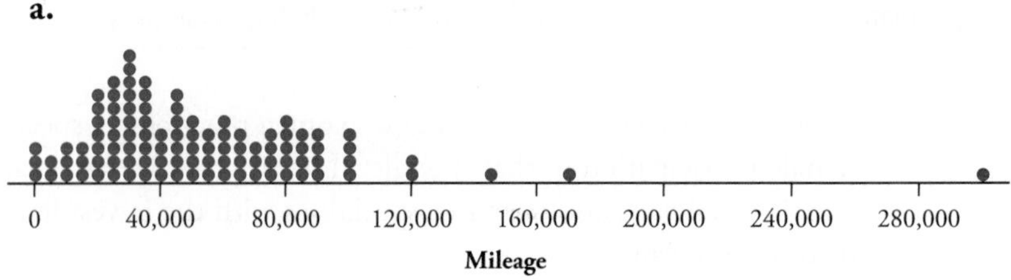

b.

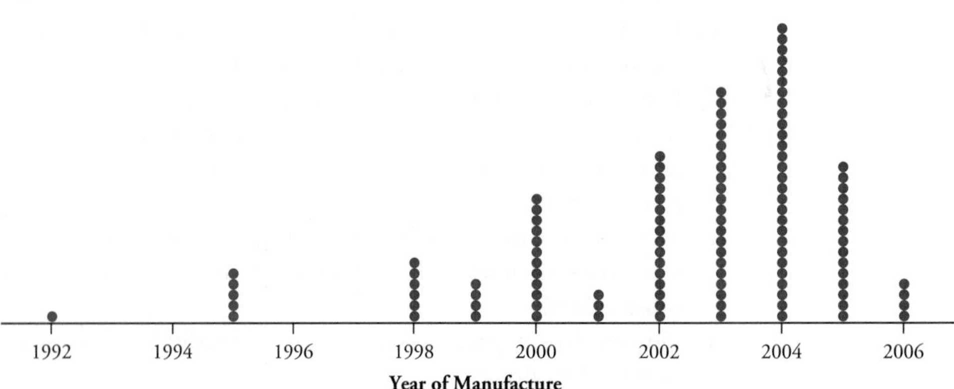

c.

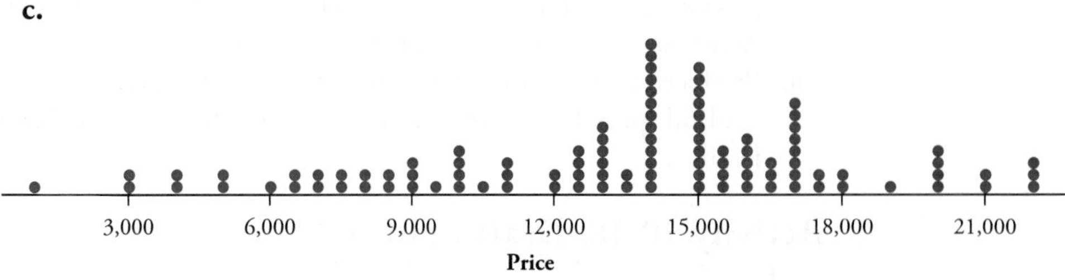

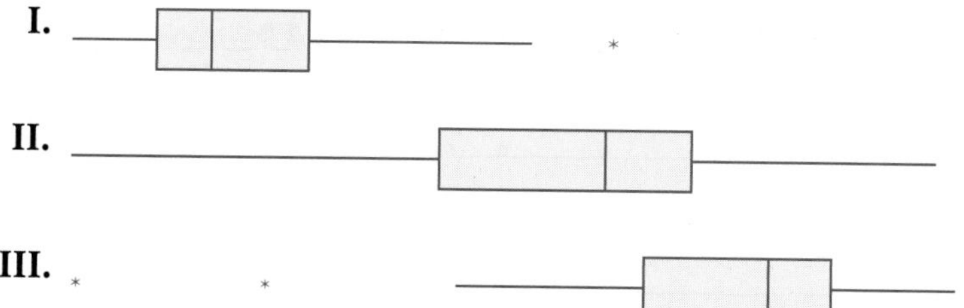

Activity 10-20: Planetary Measurements

8-12, **10-20,** 19-20, 27-13, 28-20, 28-21

Recall from Activity 8-12 the following diameters (in miles) of the nine planets:

Mercury	Venus	Earth	Mars	Jupiter	Saturn	Uranus	Neptune	Pluto
3,030	7,520	7,926	4,217	88,838	74,896	31,762	30,774	1,428

a. Calculate (by hand) the five-number summary of planet diameters.
b. Draw a boxplot (unmodified, by hand) of the distribution of planet diameters.
c. Would you classify this distribution as roughly symmetric, skewed to the left, or skewed to the right?

Activity 10-21: Rowers' Weights

7-2, 8-4, **10-21**

Refer to Activity 7-2 (page 123) and the data on the weights of rowers on the 2004 U.S. Olympic men's team.

a. Calculate the five-number summary for these weights.
b. Check the data for outliers using the 1.5 × IQR rule. Indicate how much, and how little, a rower would have to weigh in order to be considered an outlier. If any rowers are outliers, identify them by name and weight.
c. Draw (by hand) a boxplot of the distribution of rowers' weights.
d. Is the boxplot as informative as the dotplot in this situation? If so, explain your answer. If not, describe what the dotplot reveals that the boxplot does not.

Activity 10-22: Hypothetical Exam Scores

7-12, 8-22, 8-23, 9-22, **10-22,** 27-8

Consider the hypothetical exam scores presented in the next table for three classes of students. Dotplots of the distributions are shown following the table.

Class A	50	50	50	63	70	70	70	71	71	72	72	79	91	91	92
Class B	50	54	59	63	65	68	69	71	73	74	76	79	83	88	92
Class C	50	61	62	63	63	64	66	71	77	77	77	79	80	80	92

10

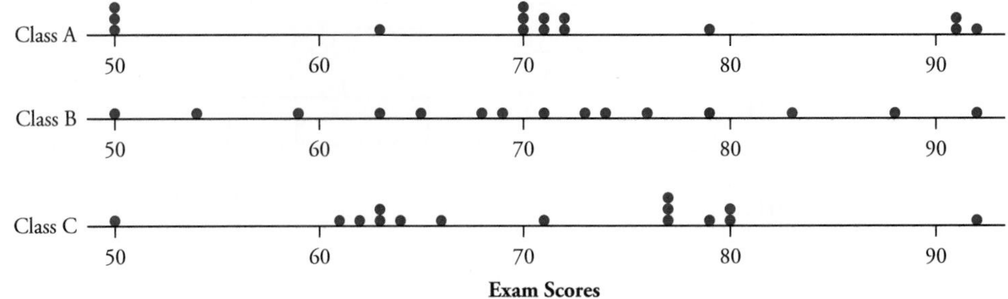

a. Do these dotplots reveal differences among the three distributions of exam scores? Explain briefly.

b. Calculate (by hand) five-number summaries of the three distributions. (Notice that the scores are already arranged in order.)

c. Create boxplots (unmodified, on the same scale as the dotplots) of these three distributions.

d. If you had not seen the actual data and had been shown only the boxplots, would you be able to detect the differences in the three distributions?

UNIT 3 • • •

Randomness in Data

TOPIC

11 • • • Probability

If four people drop their identical cell phones in a crowded elevator and then pick them up at random, how likely is it that at least one person will pick up the right phone? How many people will pick up the correct phone on average in the long run? In this topic, you will analyze this scenario in order to learn the basic ideas of probability. Other questions this topic addresses include: How likely is it that a family with two children will have one boy and one girl? What about a family with four children—are they more likely to have two children of each sex or a 3-2 gender breakdown? The study of probability will help you answer such questions, and you might even find yourself surprised by some of the answers. Applying probability methods to statistical issues will also help you answer questions such as how likely is it that random assignment will produce an equal split of genders between two treatment groups.

Overview

In the previous two units, you studied methods for analyzing data, from displaying them graphically to describing them verbally and numerically. In Topics 4 and 5, you learned how to collect data by taking a random sample from a population or by randomly assigning subjects to treatment groups. At first glance, you might think that introducing randomness into the process would make it more difficult to draw reliable conclusions. Instead, you will find that randomness actually produces predictable, long-run patterns that allow you to quantify how closely the sample is expected to come to the population result and how closely two experimental groups should resemble each other. This topic introduces you to probability, the mathematical study of randomness, and asks you to explore some of its properties.

Preliminaries

1. What does it mean to say there is a 30% chance of rain tomorrow?

2. If four people in a crowded elevator get their cell phones mixed up and decide that each of them will simply take one at random, would it be more likely for everyone to get the right phone or for nobody to get the right phone?

3. In the previous scenario, do you think it would be more likely for exactly one person to get the right phone or for exactly two people to get the right phone?

4. Suppose a couple has four children. Which gender breakdown do you think is more likely, a 2-2 or a 3-1 breakdown?

5. Which family is more likely to have an exact 50/50 gender breakdown, a family with four children or with ten children?

In-Class Activities

Activity 11-1: Random Babies

11-1, 11-2, 14-8, 16-17, 20-16, 20-22

Suppose that on one night at a certain hospital, four mothers (named Johnson, Miller, Smith, and Williams) give birth to baby boys. Each mother gives her child a first name alliterative to his last name: Jerry Johnson, Marvin Miller, Sam Smith, and Willy Williams. As a very sick joke, the hospital staff decides to return babies to their mothers completely at random. (We apologize for this absurd and objectionable context, but we wanted to choose a situation that you might find memorable. Please realize we are only kidding, and do not try this at home!)

Using this scenario, you will investigate these questions: How often will at least one mother get the right baby? How often will every mother get the right baby? What is the most likely number of correct "matches" of baby to mother? On average, how many mothers will get the right baby?

Because it is clearly not feasible to actually carry out this exercise over and over again, to investigate what would happen in the long run, you will use simulation.

Simulation is an artificial representation of a random process used to study the process's long-term properties.

To represent the process of distributing babies to mothers at random, you will shuffle and deal cards (representing the babies) to regions on a sheet of paper (representing the mothers).

a. Take four index cards and one sheet of scratch paper. Write one of the babies' first names on each index card, and divide the sheet of paper into four regions, writing one of the mothers' last names in each region. Shuffle the four index cards well, and then deal them out randomly, placing one card face down on each region of the sheet. Finally, turn over the cards to reveal which babies were randomly assigned to which mothers. Record how many mothers received their own baby.

b. Repeat the random "dealing" of babies to mothers a total of five times, recording in each case how many mothers received the correct baby:

Number of Repetitions	1	2	3	4	5
Number of Matches					

c. Combine your results sequentially with those of your classmates until you obtain a total of 100 repetitions (also called trials). Create a table that keeps track of how many of the five trials for each student had at least one match, the cumulative total number of trials, the cumulative number of trials where at least one mother gets the right baby, and the cumulative proportion of trials in which at least one mother gets the right baby. The table should begin as follows:

Student #	Column 1: Number of Trials with at Least One Match	Column 2: Cumulative Number of Trials	Column 3: Cumulative Number of Trials with at Least One Match	Column 4: Cumulative Proportion of Trials with at Least One Match
1		5	(student 1's count)	(column 3 ÷ column 2)
2		10	(student 1's + student 2's count)	
3		15		
4		20		
...				

d. Construct a graph of the cumulative proportion of trials with at least one match vs. the cumulative number of trials.

e. Does the proportion of trials that result in at least one mother getting the right baby fluctuate more at the beginning or at the end of this process?

f. Does the cumulative proportion (also known as the **relative frequency**) appear to be "settling down" and approaching one particular value? What do you think that value will be?

The **probability** of a random event occurring is the long-run proportion (or **relative frequency**) of times the event would occur if the random process were repeated over and over under identical conditions. You can *approximate* a probability by simulating the process many times. Simulation leads to an **empirical estimate** of the probability, which is the proportion of times that the event occurs in the simulated repetitions of the random process.

g. Now combine your results on number of matches with the rest of the class's results, obtaining a tally of how often each *number of matches* outcome occurred. Record the counts and proportions in the following table:

Number of Matches	0	1	2	3	4	Total
Count						
Proportion						1.00

h. In what proportion of these simulated cases did at least one mother get the correct baby?

i. Based on the class's simulation results, what is your empirical estimate of the probability of obtaining no matches?

j. Based on the class's simulation results, what is your empirical estimate of the probability of obtaining at least one match?

You can approximate these probabilities more accurately by performing more trials.

k. Use the `Random Babies` applet, shown here, to simulate this random process. Leave the number of trials at **1,** and click Randomize five times. Then turn off the Animate feature and ask for **995** more trials, for a total of **1000** trials. (Be forewarned that the applet provides a graphic illustration of where babies come from.)

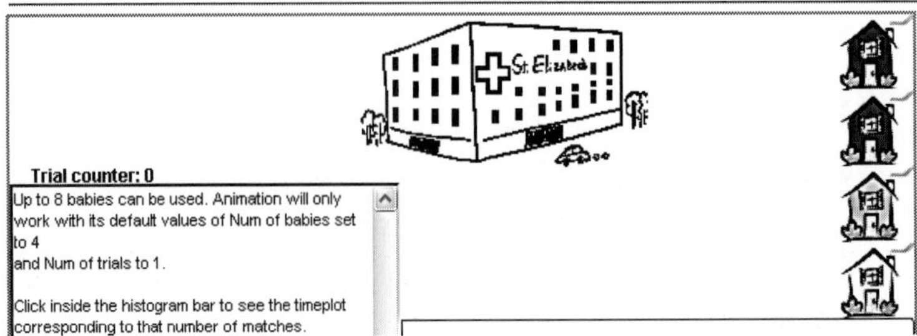

Record the counts and proportions in the following table:

Number of Matches	0	1	2	3	4	Total
Count						
Proportion						1.00

l. Are these simulation results reasonably consistent with the class results?

m. Report the new (likely more accurate) empirical estimate that at least one mother gets the correct baby.

n. Click the bar of the histogram corresponding to **0** matches to see a graph of how the relative frequency changes over time. Does this graph appear to be fluctuating less as more trials are performed, approaching a limiting value?

o. Explain why an outcome of exactly three matches is impossible.

p. Is it impossible to get four matches? Would you say it is rare? Or unlikely?

q. Would you consider a result of zero matches or of one match or of two matches to be unlikely? Explain.

Watch Out

This question of whether an outcome is unlikely or rare is an important one that we will return to often. To some extent deciding when an outcome is unlikely or rare is a matter of personal judgment, but in this text, we will generally not regard an outcome as *unlikely* unless its probability is .10 or less, and we will generally not regard an outcome as *rare* unless its probability is .05 or less.

Activity 11-2: Random Babies

11-1, **11-2,** 14-8, 16-17, 20-16, 20-22

Reconsider the "random babies" scenario. Now you will determine the relevant probabilities exactly.

> In situations where the outcomes of a random process are **equally likely,** you can calculate theoretical exact probabilities by listing all of the possible outcomes and determining the proportion that corresponds to the event of interest. The listing of all possible outcomes is called the **sample space.**

The sample space for the "random babies" scenario consists of all possible ways to distribute the four babies to the four mothers. Let 1234 mean that the first baby went to the first mother, the second baby to the second mother, the third baby to the third mother, and the fourth baby to the fourth mother. In this scenario, all four mothers get the correct baby. As another example, 1243 means that the first two mothers get the right baby, but the third and fourth mothers have their babies switched. All of the possibilities are listed here:

1234	1243	1324	1342	1423	1432
2134	2143	2314	2341	2413	2431
3124	3142	3214	3241	3412	3421
4123	4132	4213	4231	4312	4321

a. How many different arrangements are there for returning the four babies to the four mothers?

b. For each of these arrangements, indicate how many mothers get the correct baby. The first two have already been entered.

1234	4	1243	2	1324		1342		1423		1432	
2134		2143		2314		2341		2413		2431	
3124		3142		3214		3241		3412		3421	
4123		4132		4213		4231		4312		4321	

c. In how many arrangements is the number of "matches" equal to exactly

4: 3: 2: 1: 0:

d. Calculate the (exact) probabilities by dividing your answers to part c by your answer to part a.

4: 3: 2: 1: 0:

e. Comment on how closely the empirical probabilities from your class simulation analysis in the previous activity approximated these probabilities. Then comment on how closely the applet simulation analysis, using a greater number of repetitions or trials, approximated these probabilities.

An empirical estimate from a simulation generally gets closer to the actual probability of an event occurring as the number of repetitions increases.

f. For your class simulation results from the previous activity, calculate the average (mean) number of matches per repetition of the process. *Hint:* Multiply each number of matches outcome by the number of occurrences for that number, sum those products, and then divide by the total number of trials.

The listing of possible values of a numerical random variable, along with the probabilities for those values, is called a **probability distribution.** The long-run average value achieved by a numerical random process is called its **expected value.** To calculate this expected value from the probability distribution, multiply each outcome by its probability, and then add these values over all of the possible outcomes: *outcome1* × *probability of outcome1* + *outcome2* × *probability of outcome2* + …

g. Calculate the expected value for the number of matches from the probability distribution, and compare that value to the average number of matches from the simulated data found in part f.

h. What is the probability that the result of this random process equals this expected value? Based on this probability, would you say that you "expect" this outcome to occur most of the time? Explain.

Watch Out

The expected value might not literally be expected at all. An expected value is interpreted as the long-run average value of a numerical random process. But in the short run, or particularly with a single occurrence, the expected value might not be likely to occur. In the "random babies" scenario, the probability of obtaining exactly one match (the expected value) is only 1/3, and obtaining zero matches is actually a more probable outcome. A more dramatic example is rolling a single fair die: The expected value is 3.5, but of course it's impossible for a die to land on the value 3.5 in a single roll.

Activity 11-3: Family Births

11-3, 11-12

Suppose a couple has two children. Assume that each child is equally likely to be a boy or a girl, regardless of the outcome of previous births.

a. Suppose someone argues that the couple is guaranteed to have one boy and one girl, because that's what a .5 probability means: 50% should be boys and 50% should be girls, and 50% of two children is one child. Do you believe this argument? (Do you know of any two-child families with two boys or two girls?) How would you respond to this argument to help the person see their faulty reasoning?

Watch Out

Many people fall into the trap of believing that probabilities should also hold in the short run. Remember, probability is a long-term property: If you observe a very large number of births, then the proportion who are girls will be very close to one-half. But in a small sample of two or four or even ten births, it's quite possible that the proportion of girls born might not even be close to one-half of the births.

b. Now, suppose someone else claims that there are three possible outcomes for this family: two boys, two girls, or one of each. Equal likeliness therefore establishes that the probability of each of these outcomes is one-third. Do you believe this argument? If so, why? If not, how would you respond to it?

Now you will use simulation to approximate the probabilities associated with a couple having two children.

c. Use the Random Digits Table (Table I) found at the back of the book to simulate the children's genders for four families of two children each. Start at any line and let each even digit represent a girl and each odd digit represent a boy. Record the random digits and the corresponding genders in the table:

	Family 1		Family 2		Family 3		Family 4	
	Child 1	Child 2	Child 1	Child 2	Child 1	Child 2	Child 1	Child 2
Random Digit								
Gender								
Number of Girls								

d. Continue until you have simulated the gender breakdowns for a total of 20 two-child families. Record how many and what proportion you have of each type:

	Two Girls	Two Boys	One of Each
Tally (Count)			
Proportion			

e. Based on your (very small) simulation analysis, does it appear that the probability of each of these three outcomes is one-third? Explain.

f. How could you obtain better empirical estimates of these probabilities?

g. Combine your simulation results with those of your classmates:

	Two Girls	Two Boys	One of Each
Tally (Count)			
Proportion			

h. Based on these (more extensive) simulation results, does it appear that the probability of each of these outcomes is one-third? Explain.

Because each of the two children is equally likely to be a boy or girl, the correct way to list the sample space of equally likely outcomes is $\{B_1B_2, B_1G_2, G_1B_2, G_1G_2\}$, where the subscript indicates the first or second child born. It is these four outcomes that are equally likely to occur.

i. Use this sample space to determine the exact probabilities of a two-child couple having two girls, two boys, or one of each. Are these probabilities reasonably close to the empirical estimates from your class simulation?

Two girls: Two boys: One of each:

Activity 11-4: Jury Selection

11-4, 12-1, 15-12

About 20% of the adult residents in San Luis Obispo County (California) are age 65 or older.

a. If one of these adults is chosen at random, what is the probability that he/she will be age 65 or older?

Suppose that a random sample of 12 adults is taken from this county, perhaps to form a jury.

b. Guess the probability that at least one-third (four or more people) of the jury are age 65 or older.

The following histogram displays the results of simulating 1000 such randomly selected juries. The observational units in this graph are the simulated juries, and the variable measured is the number of senior citizens on the jury.

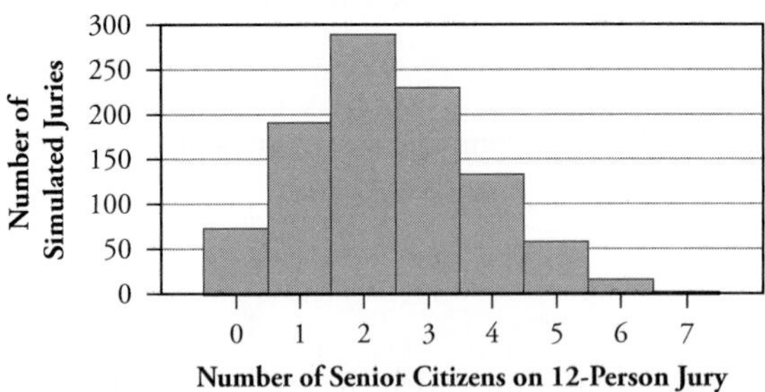

c. Use this histogram of simulation results to approximate the probability that senior citizens (age 65 or older) would comprise at least one-third of the members of a 12-person jury. How close to this probability was your guess in part b?

Now suppose that a random sample of 75 adult residents is taken from this county, perhaps to form a pool from which to choose a jury.

d. Guess the probability that at least one-third of this pool of 75 people are aged 65 or older.

The following histogram displays the results of simulating 1000 such randomly selected jury pools.

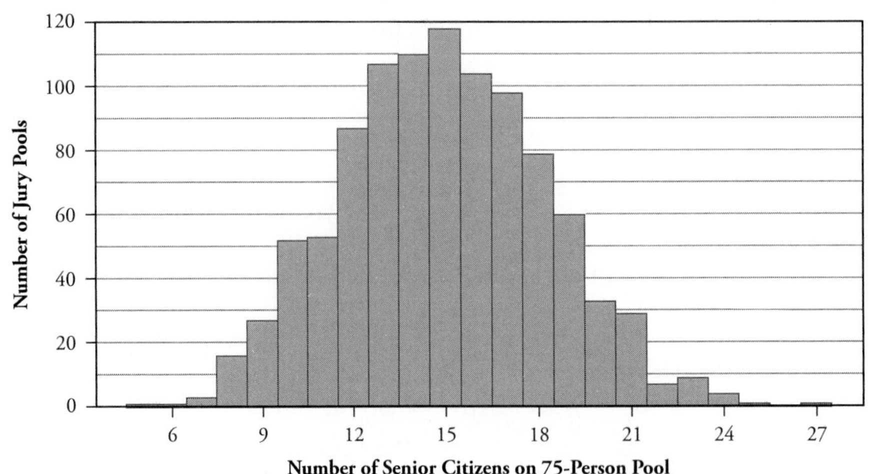

e. Use this histogram of simulation results to approximate the probability that senior citizens (age 65 or older) would comprise at least one-third (i.e., 25 or more) of the members in a 75-person jury pool. How close to this probability was your guess in part d?

f. For which sample size (12 or 75) is the sample (from a population with 20% senior citizens) more likely to contain at least one-third senior citizens? Explain why this makes sense.

g. For which sample size (12 or 75), is the sample more likely to contain between 15% and 25% senior citizens? Refer to the previous histograms, and explain why this answer makes sense.

h. Explain what the previous questions reveal about the effect of sample size on how closely the sample proportion of senior citizens comes to its population value.

This activity reveals that a larger sample size is more likely to produce a sample proportion close to its population value. Also note that a larger sample size produces a probability distribution that is quite symmetric and mound-shaped.

Watch Out

It is easy to confuse *sample size* and *number of samples.* In both of these simulations, the number of samples is 1000, but the sample sizes are 12 and 75. *Sample size* refers to the number of observational units in each sample, not to the number of samples generated in the simulation.

√ Activity 11-5: Treatment Groups

11-5, 11-13

Suppose that six subjects are to be randomly assigned to two treatment groups (call them the old treatment and the new treatment), with three subjects assigned to each group. Suppose that three subjects are female (Kelly, Fantasia, and Carrie) and three are male (Ruben, Taylor, and Bo). You saw in Topic 5 that randomization tends to balance out the gender breakdown between treatment groups. You will now use simulation to approximate the probabilities of the various possible outcomes and then use a sample space to calculate exact probabilities.

a. Describe how you could use a die or the Random Digits Table to randomly assign these six subjects to the two groups.

b. Conduct 40 repetitions of this random assignment. For each repetition, keep track of the number of women assigned to the new treatment group. (Notice that this one number is enough to tell you the complete gender breakdown of both groups.) Summarize the results in this table:

Number of Women in "New" Treatment Group	0 (0/3 split)	1 (1/2 split)	2 (2/1 split)	3 (3/0 split)
Count (Number of Randomizations)				
Proportion of Randomizations (Count/40)				

c. What is your empirical estimate of the probability that the gender breakdown is either 1/2 or 2/1?

d. How could you determine a more accurate empirical estimate of this probability?

e. Determine an empirical estimate of the expected value of the number of women assigned to the new treatment group.

The sample space of all possible outcomes of this random assignment process can be written as follows, where the three initials given correspond to the three subjects assigned to the new treatment group: {KFC, KFR, KFT, KFB, KCR, KCT, KCB, KRT, KRB, KTB, FCR, FCT, FCB, FRT, FRB, FTB, CRT, CRB, CTB, RBT}.

f. Explain why you can reasonably assume that these 20 outcomes are equally likely.

g. Use this sample space to determine the exact probabilities that you approximated in part b:

Number of Women in "New" Group	0	1	2	3
Probability				

h. Determine the probability that the gender breakdown is either 1/2 or 2/1 in both groups. How close to this probability was your empirical estimate in part c?

i. Report the probability that all the men are assigned to one group and all the women are assigned to the other group (an 0/3 or 3/0 split in the new treatment group). Would you consider this result to be not surprising, mildly surprising, or very surprising?

j. Determine the expected value of the number of women assigned to the new treatment group. How close to this value was your empirical estimate in part e?

k. Explain why this expected value makes sense.

Solution

a. You could assign each subject a number from 1–6. Then, as the die is rolled, you could assign the subjects corresponding to the first three (distinct) numbers rolled, to the new treatment group. The process would be similar using the Random Digits Table, where you could simply skip over the digits 7–9 and 0.

b. Answers will vary. (You should find *roughly* 2 randomizations with 0 women in the new group, 18 randomizations with 1 woman, 18 randomizations with 2 women, and 2 randomizations with 3 women.)

c. Answers will vary. Determine the total number of repetitions with 1 woman or 2 women in the new group, and divide that number by 40. (Your empirical estimate should be fairly close to .9.)

d. Performing many more than 40 repetitions of the random assignment should produce a more accurate estimate.

e. Answers will vary. Calculate this empirical estimate of the expected value by multiplying the number of women by the count of repetitions that produced that number, and then divide that value by 40. (Your empirical estimate should be fairly close to 1.5.)

f. Random assignment ensures that all possible ways to assign these subjects to the treatment groups are equally likely to occur.

g. The exact probabilities are 1/20 or .05 for 0 women in the new group, 9/20 or .45 for 1 woman in the new group, 9/20 or .45 for 2 women in the new group, and 1/20 or .05 for 3 women in the new group.

h. The probability of a 1/2 or 2/1 gender breakdown is 18/20 or .9.

i. The probability that the two genders are completely separated into two groups (all men in one group and all women in the other) is 2/20 or .1. This result is small enough to be perhaps mildly surprising, but it is not small enough to be very surprising. Although randomization *should* balance out the gender breakdown between the two groups *in the long run,* with such a small study group, you could, simply by chance, still end up with all males in one group and all females in the other group.

j. The expected value is 0(1/20) + 1(9/20) + 2(9/20) + 3(1/20) or 30/20 or 1.5 women.

k. This expected value makes sense, because with three women randomly assigned among two groups, you would expect half of them to be assigned to each group in the long run, and half of 3 is 1.5.

Wrap-Up. . . . .

This topic initiated your formal study of randomness by introducing you to the concept of probability. You learned that probability is a long-run property—namely, the long-run proportion, or relative frequency, of times that an event occurs when a random process is repeated many times. For example, the probability that all four mothers get the right baby means the long-run proportion of times that all four mothers get the right baby, if the random process of distributing babies to mothers were repeated indefinitely.

You also learned two ways to estimate or calculate probabilities. The first is to approximate probabilities by simulating the random process. The more times you simulate the random process, the closer the empirical probabilities will approximate the exact theoretical probabilities. The second way discussed to calculate probabilities, which is appropriate when you have equally likely outcomes, is to list all possible outcomes of the random process and count how many of the possible outcomes result in the event of interest. For example, the exact probability that all four mothers would get the correct baby is 1/24, or about 4.2%.

You also encountered the notion of expected value, calculated by multiplying each possible value by its probability and then adding those products. Expected value refers to the long-run average result of a numerical random process. For example, when distributing babies to their mothers at random, the expected value is 1, meaning that in the long run, the average number of correct matches will be 1. This expected value might not be likely or even possible. For example, when rolling a fair, six-sided die, the expected value is 3.5, even though the result cannot possibly be 3.5 dots showing on the roll of one die.

Some useful definitions to remember and habits to develop from this topic include

- The **probability** that a random process results in a particular event is the long-run proportion (**relative frequency**) of times that the event would occur if the random process were repeated a great many times.

- Probabilities can be approximated by **simulation,** which means to artificially re-create the random process a large number of times. The observed proportion of times that the event occurs is an **empirical estimate** of the probability. Generally speaking, as the process is simulated more times, empirical estimates come closer to the actual probabilities.
- In situations where the outcomes of a random process are **equally likely,** theoretical probabilities can be calculated exactly by listing all possible outcomes. This listing is called the **sample space** of the random process. Then the probability of an event is the number of outcomes in the sample space that comprise the event divided by the total number of outcomes in the sample space. Before using this method to determine theoretical probabilities, make sure you can reasonably assume equal likeliness.
- The **expected value** of a numerical random process refers to the long-run average result of that process. Both of these terms are important in this interpretation: *long-run* and *average.*
- Expected value does not literally mean "expected." The expected value might be quite unlikely and might not even be possible.
- Be careful when determining probabilities. Research shows that many people have faulty intuition regarding probabilities.
- Sample size plays an important role in probability calculations. With a larger sample size, the sample results are more likely to come close to the theoretical probabilities.

The next topic introduces you to the most important probability model in all of statistics: the normal distribution.

Homework Activities

Activity 11-6: Random Cell Phones

Suppose that three executives bump into each other in an elevator and drop their identical cellular phones as the doors are closing, leaving them with no alternative but to pick up a phone at random.

a. Describe in detail how you could conduct a simulation of this situation to produce empirical estimates of the probabilities involved.
b. List all the possible outcomes in the sample space for this situation. *Hint:* Compare it to the random babies scenario in Activity 11-1.
c. Use the sample space to determine the probability that

- Nobody gets the correct phone
- Exactly one person gets the correct phone
- Exactly two people get the correct phone
- All three people get the correct phone
- At least one person gets the correct phone

d. If this situation were to befall a group of five people, you could calculate the probability that nobody gets the correct phone as 11/30. Explain in your own words what this probability means about the likelihood of nobody getting the correct phone. Include the long-run interpretation of probability in your answer and relate your response to the situation's context.

Activity 11-7: Equally Likely Events

Indicate which of the following scenarios have outcomes that are equally likely. Provide a brief justification in each case.

a. Whether a fair die lands on 1, 2, 3, 4, 5, or 6
b. The sum of two fair dice landing on 2, 3, 4, 5, 6, 7, 8, 9, 10, 11, or 12
c. A coin landing heads or tails when flipped
d. A coin landing heads or tails when spun on its side
e. A tennis racquet landing with the label "up" or "down" when spun on its end
f. Your grade in this course being A, B, C, D, or F
g. Whether California experiences a catastrophic earthquake within the next year
h. Whether your waitress or waiter brings you the correct meal you ordered in a restaurant
i. Whether there is intelligent life on Mars
j. Whether a woman will be elected president in the next U.S. presidential election
k. Whether a woman will be elected president before the year 2100
l. Colors of Reese's Pieces candies: orange, yellow, and brown

Activity 11-8: Interpreting Probabilities

Explain in your own words what is meant by the following statements. Include the long-run interpretation of probability in your answer and relate your response to the situation's context.

a. There is a .3 probability of rain tomorrow.
b. Your probability of winning at a "daily number" lottery game is 1/1000.
c. The probability that a five-card poker hand contains "four of a kind" is .00024.
d. The probability of getting a red M&M candy is .2.
e. The National Climatic Data Center estimates a 70% probability of having a white Christmas in Minneapolis.
f. A researcher from Ohio State University estimates that entrepreneurs who open a restaurant business have a 60% probability of failing.

Activity 11-9: Racquet Spinning

11-9, 13-11, 17-3, 17-18, 18-3, 18-12, 18-13

Tennis players often spin a racquet on its end to observe whether it lands with the handle's label facing "up" or "down" as a random method for determining who serves first in a game. To estimate the probability that a particular racquet lands "up" and to see whether the probably of the racquet landing "up" is about 50%, author A spins his racquet 100 times. A graph of the relative frequency of "up" results after each spin is shown here.

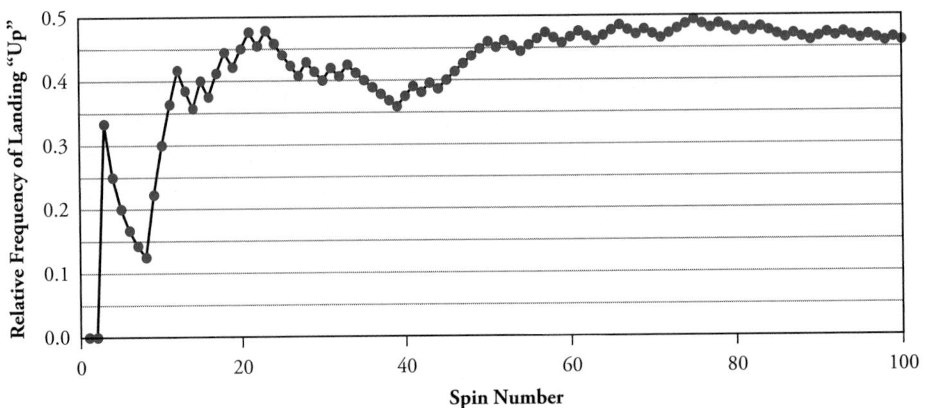

a. Explain in your own words what the phrase "probability of the racquet landing up" means and how this graph relates to that phrase.
b. Using the graph, indicate the empirical estimate of this probability after 10 spins, after 20 spins, after 40 spins, and after 100 spins.
c. Does this sample of spins appear to provide evidence that the racquet when spun is not equally likely to land "up" or "down"? Explain.
d. Spin your own racquet 100 times. How does your estimate of the probability of this racquet landing up compare?

Activity 11-10: Committee Assignments

11-10, 11-11

A college professor was assigned to a committee of six people, composed of four men and two women. The committee selected two officers to carry out the majority of its administrative work, and both of the women were selected. The professor wondered whether this constituted evidence of subtle discrimination, so she considered how unlikely such an event would be if the two officers had been chosen at random from the six committee members.

a. Suppose the six committee members are named Alice, Bonnie, Carlos, Danny, Evan, and Frank. Describe how you might use a fair, six-sided die to simulate the random selection of two people to be officers for this committee. *Hint:* Keep in mind the officers have to be different people.
b. Use a die to carry out this simulation for a total of 50 repetitions. In how many of these 50 repetitions did the two women end up being the officers? What is your empirical estimate of that probability?
c. Calculate the mean number of men in these 50 simulated pairs of officers.

Activity 11-11: Committee Assignments

11-10, **11-11**

Reconsider the previous activity.

a. To pursue a theoretical analysis, begin by listing all possible pairs of officers that could be chosen from among these six people.
b. How many pairs are possible? How many of the pairs consist of two women?
c. What is the theoretical probability of obtaining two women officers if you randomly choose two people from among these committee members? Is this outcome impossible? Rare? Uncommon? Likely?
d. If the process of randomly selecting two people from among these six people were repeated over and over, in the long run what percentage of the time would two men be selected as officers? Would such an outcome be a surprising result using random selection?
e. Repeat part d for the outcome of one man and one woman.
f. If the two officers were chosen at random, what would be the most likely gender breakdown, two men, two women, or one of each?
g. Use the sample space in part a to calculate the theoretical expected number of men among the two officers. How close did your simulated estimate come to the theoretical expected value?

Activity 11-12: Family Births

11-3, **11-12**

Reconsider Activity 11-3, but suppose now that a family has *four* children rather than two.

a. Conduct a simulation to produce approximate probabilities for the various possible gender breakdowns. Describe how you conducted the simulation, and also report the approximate probabilities.
b. Based on your simulation results, which is more likely—a 2/2 split of boys and girls or a 3/1 split in either direction?
c. Write out the sample space and calculate the exact probabilities for the various possible gender breakdowns.
d. Re-answer part b using exact probabilities.

Activity 11-13: Treatment Groups

11-5, **11-13**

Suppose that four subjects, two men and two women, will be randomly assigned to two groups of two people each.

a. Conduct a simulation to produce approximate probabilities for the various possible gender breakdowns for the two groups. Describe how you conducted the simulation, and also report the approximate probabilities.
b. Write out the sample space and calculate the exact probabilities for the various possible gender breakdowns.

Activity 11-14: Die Rolling

a. Verify the claim made in the Watch Out at the end of Activity 11-2 that the expected value of rolling a single fair six-sided die is 3.5. Show the details of your calculation. *Hint:* First, list all the possible values and their probabilities.
b. Provide an interpretation of what the value 3.5 means in this context.

Activity 11-15: Simulating the World Series

Suppose two baseball teams, the Domestic Shorthairs and the Cache Cows, are to play each other in a best-of-three series and that the Shorthairs have a .7 probability of winning any one game, regardless of the outcomes of preceding games. The winner of the best-of-three series is the first team to win two games.

a. Do you suspect that the probability of the Shorthairs' winning the series will be greater than .7, less than .7, or exactly .7? Explain your reasoning.
b. Follow these steps to use simulation to estimate the probability that the Shorthairs will win the series: Use the Random Digits Table at the back of the book. (Report the line number that you use.) Start at any line and let 1–7 represent a win for the Shorthairs and 8, 9, and 0 represent a win for the Cows. Record the digits until one team has won two games; this team is the winner of your first simulated series. Repeat this process for a total of 50 simulated series. In each case, record the team who won and also how many games the series entailed.
c. What proportion of the 50 simulated series did the Shorthairs win? Is this more or less than .7?
d. Would you expect the Shorthairs to have a higher, lower, or the same probability of winning a best-of-seven series (meaning the winner of the series is the first team to win four games)? Explain your reasoning.

e. Conduct a simulation of 50 best-of-seven series to estimate the probability that the Shorthairs win such a series.
f. Which length of series gives the greater advantage to the stronger team (the Shorthairs, in this case)? Explain in your own words why this result makes sense.

Activity 11-16: Dating Game Show

Consider the following scenario for a reality game show: Two women (Allyson and Elsa) and two men (Bart and Dwayne) spend an hour together, and then each one selects the person of the opposite sex he or she is more interested in. Any pair for which both people pick each other is treated to an extravagant date together. Suppose that every person chooses randomly, so all possible outcomes are equally likely. (For example, one possible outcome is that Allyson chooses Bart, Bart chooses Allyson, Elsa chooses Dwayne, and Dwayne chooses Allyson. With this outcome, Allyson and Bart have chosen each other, so they get to go on a date, but Elsa and Dwayne did not choose each other, so they do not go on a date.)

a. List all possible outcomes in the sample space of this random process. *Hint:* You might want to do a few repetitions of a simulation to help you understand this random process. If you use any kind of shorthand notation, define it clearly.
b. How many possible outcomes are there?
c. Determine the probability that all four people will go on a date.
d. Determine the probability that at least one couple will go on a date.
e. Determine the expected value of the number of people who will go on a date.

Activity 11-17: Dice-Generated Ice Cream Prices

11-17, 11-18

Suppose the owners of an ice cream shop hold a promotional special: Customers roll two dice, and the price of a small ice cream cone (in cents) is the greater number followed by the lower number. (For example, if you roll a 5 and a 3 in either order, the price of an ice cream cone is 53 cents.)

a. What is the probability that the price of an ice cream cone turns out to be 32 cents? *Hint:* There are 36 possible and equally likely outcomes from rolling a pair of dice. Count how many of those 36 outcomes produce a price of 32 cents.
b. What is the probability of the price being 33 cents?
c. What is the probability of the price being 34 cents?
d. What is the probability of the price being less than 40 cents?
e. What is the probability of the price being greater than 50 cents?
f. Determine the expected value of the price.
g. Write a sentence interpreting what this expected value means.

Activity 11-18: Dice-Generated Ice Cream Prices

11-17, **11-18**

Reconsider the previous activity.

a. Download the ICPRICE program to your calculator and use it to conduct a simulation of **999** repetitions of generating the ice cream price (*TI tip:* You might need either to delete or to archive your other programs, lists, and variables, so you do not run out of memory). Create and then comment on a histogram of the resulting prices. The results are stored in a list named ICPRC. *TI hint:* You can also use the **randInt** command to simulate rolling a die. Press [MATH] and select **randInt** from the PRB menu. For example, to simulate 50 repetitions, you would enter **randInt(1,6,50)** at the prompt and then press the [ENTER] key.

b. Calculate the average of these 999 prices. Is the average price close to the expected value? Should it be? Explain.

Activity 11-19: Hospital Births

According to *The World Almanac and Book of Facts 1999,* approximately 25% of all babies born in Texas are Hispanic. Consider a city in Texas with two hospitals, and suppose that hospital A has 10 births per day, whereas hospital B has 50 births per day.

a. Download the SIMSAMP program into your calculator and use it to simulate the births for 365 days at each of these two hospitals. The number of samples is **365** and the true population proportion is **.25.** The sample sizes are **10** and **50,** respectively. The results are stored in a list named **PROP**. You will need to run this simulation twice: once for hospital A and then again for hospital B (before rerunning the SIMSAMP program, copy the **PROP** list to another named list). Calculate the proportion of Hispanics born per day at each hospital, and create histograms for displaying these distributions.
b. Which hospital has more days on which more than 40% of the babies born are Hispanic?
c. Which hospital has more days on which between 15% and 35% of the babies born are Hispanic?
d. Which hospital has more days on which less than 40% of the babies born are Hispanic?
e. Explain how you could have answered parts b–d without the benefit of the simulation, based on what you learned in this topic about randomness and sample size.

Activity 11-20: Collecting Prizes

Suppose a brand of breakfast cereals runs a promotion whereby they put prizes in each box of cereal, and you set out to buy as many boxes as necessary in order to obtain a complete set of the four prizes offered.

a. Make an educated guess of the expected value of the number of boxes you will have to buy.
b. Describe how you could use the Random Digits Table to simulate this process. Mention any assumptions you are making about this random process.
c. Use the Random Digits Table to simulate this random process once. Record the line number that you use in the table, and also show how the digits translate into prizes. How many boxes did you have to buy to obtain a complete set of prizes?
d. Now simulate this random process for a total of 25 repetitions. Create and comment on a dotplot or histogram of your resulting values. (Label the horizontal axis of the dotplot or histogram appropriately. Think about what the observational units and variable are in this situation.)
e. Use your simulation results to approximate the expected value of the number of boxes you have to buy in order to obtain a complete set of four prizes.
f. Use your simulation results to approximate the probability that you can obtain a complete set of prizes by buying 10 or fewer boxes.
g. How could you obtain a more accurate approximation of this probability and expected value?

Activity 11-21: Runs and "Hot" Streaks

Reconsider the setting of Activity 11-9, involving the spinning of a tennis racquet. Notice from the graph shown in that activity that among the first ten spins, there occurred a string (or "run") of five consecutive "down" results.

a. In ten flips of a fair coin, would you be surprised to observe a string of five or more consecutive heads or tails? Guess how often such a string would appear in, say, 1000 repetitions of ten coin flips.
b. Download the STREAK program into your calculator and use it to simulate **100** repetitions of **10** flips of a fair coin. The results are stored in the named list **PSTRK**, which contains the length of the longest streak (run) of either heads or tails.
c. Create a histogram of the distribution of the longest streaks among these 100 simulations.
d. In how many of these 100 simulations of ten coin flips did a streak of five or more occur? Would you characterize this event as very surprising?
e. What is the most common length of the longest streak?
f. Calculate and record the mean and median of the distribution of these 100 longest streaks.
g. Would it be very surprising to flip a coin ten times and find that the longest streak is one? Explain based both on your simulation results and on the ways in which such an event could occur.

Activity 11-22: Solitaire

11-22, 11-23, 15-4, 15-14, 21-20, 27-18

To simulate wins and losses at the card game solitaire, you must first make an assumption about (or an estimate of) the probability of winning a single game. Enamored of the solitaire game on his new computer, author A sets out to estimate this probability and wins 25 games while losing 192 games.

a. What proportion of games played did he win?

For ease of simulation, let us simplify this a bit by assuming that the probability of this author winning a game of solitaire is 1/9.

b. Describe how you might use the Random Digits Table to simulate repetitive plays of solitaire games with this success probability. *Hint:* Use only one digit per simulated game.
c. Use the Random Digits Table to simulate the playing of solitaire until the first win is achieved. Record the number of games the author needed to play (including the win) as your first observation. Then repeat this process a total of 25 times.
d. Create a dotplot of the distribution of the number of solitaire games that the author will need to play before his first win in 25 repetitions of the game.
e. Calculate the mean and median of these values.
f. Based on your simulation, about how many games of solitaire can this author expect to play before winning for the first time?
g. Based on your simulation, about how many games must this author play to have a 50% chance of winning at least once?

Activity 11-23: Solitaire

11-22, **11-23,** 15-4, 15-14, 21-20, 27-18

Refer to the simulation study in the previous activity. Anxious to outperform author A, author B plays 444 games of solitaire and wins 74.

a. What proportion of games played did she win?
b. Assuming the winning probability to be 1/6, simulate the number of games she needs to play before she wins a game. (You may use your calculator, the Random Digits Table, or a physical device such as a die to do the simulation.) Repeat this process (playing until the first win) a total of 25 times.

c. Create a dotplot of the distribution of the number of games she needs to play to obtain the first victory in these 25 repetitions of the game.
d. Calculate the mean and median of these values.
e. Based on your simulation, about how many games of solitaire can this author expect to play before winning for the first time?
f. Based on your simulation, about how many games must this author play to have a 50% chance of winning at least once?
g. Describe how your findings differ when the probability of winning is 1/6 as opposed to 1/9.

Activity 11-24: AIDS Testing

The ELISA test for AIDS was used in the screening of blood donations in the late 1980s. As with most medical diagnostic tests, the ELISA test was not infallible (Gastwirth, 1987). If a person actually carried the AIDS virus, experts estimated this test gave a positive result 97.7% of the time. (This number is called the *sensitivity* of the test.) If a person did not carry the AIDS virus, ELISA gave a negative result 92.6% of the time (this number is called the *specificity* of the test). Estimates at the time were that 0.5% of the American public carried the AIDS virus (the *base rate* of the disease). Determine the (conditional) probability that a person actually carries the AIDS virus given that he/she tests positive on the ELISA test.

a. First, without performing any calculations, guess the (conditional) probability that a person who tests positive carries the AIDS virus.

Imagine a hypothetical population of 1,000,000 people for whom these percentages hold exactly. (The population size is large so the calculations all work out to be integers.) For parts b–e, record your answers in the appropriate cells of a 2 × 2 table like the one shown following part d.

b. Assuming that 0.5% of the population of 1,000,000 people carries the AIDS virus, how many such carriers are in the population? How many noncarriers are in the population?
c. Consider just the *carriers*. If 97.7% of them test positive, how many people test positive? How many carriers does that leave who test negative?
d. Now consider only the *noncarriers*. If 92.6% of them test negative, how many people test negative? How many noncarriers does that leave who test positive?

	Positive Test	Negative Test	Total
Carries AIDS Virus	(c)	(c)	(b)
Does Not Carry AIDS Virus	(d)	(d)	(b)
Total	(e)	(e)	1,000,000

e. Determine the total number of positive and the total number of negative test results.
f. Of those who test positive, what proportion actually carry the disease? How does this compare to your prediction in part a?
g. Explain why this probability turns out to be small compared to the sensitivity and specificity of the test (and to most people's guesses). Refer to calculations in the table in your answer.

TOPIC

12 • • • Normal Distributions

Do most healthy adults have a healthy body temperature close to 98.6 degrees? What proportion of newborn babies are considered to be of "low birth weight"? If you observe a footprint at the scene of a crime and predict the gender of the criminal based on the length of the footprint, how often will your prediction be incorrect? These are some of the questions that can be addressed with the most important probability distribution in all of statistics—the normal distribution.

Overview

You began your formal study of randomness and probability in the previous topic. Toward the end of that topic, you learned that mound-shaped distributions often model the outcomes of different variables, such as counts and proportions. Such a pattern arises so frequently that it has been extensively studied mathematically. In this topic, you will investigate the mathematical models known as normal distributions, which describe this symmetric pattern of variation very accurately. You will learn how to use normal distributions to calculate probabilities in a variety of contexts.

Preliminaries

1. How much do you think a typical baby weighs at birth?

2. Approximately what proportion of babies would you guess weigh less than 6 pounds? Greater than 10 pounds?

3. Let's say the distribution of men's foot lengths has mean 25 centimeters (cm) and standard deviation 4 cm, whereas the distribution of women's foot lengths has mean 19 cm and standard deviation 3 cm. Suppose you have to predict the gender of someone with a 21 cm foot length; would you predict this person was a man or a woman? Is your answer guaranteed to be correct? What if the person's foot length were 23 cm? At what point would you consider the foot length long enough that you could predict the person was a man?

4. Using the cut-off value you just specified, guess the probability of a male footprint being mistakenly classified as a female footprint.

5. Using this cut-off value again, guess the probability of a female footprint being mistakenly classified as a male footprint.

In Class Activities

Activity 12-1: Body Temperatures and Jury Selection

11-4, **12-1,** 12-19, 15-3, 15-12, 15-18, 15-19, 19-3, 19-7, 20-11, 22-10, 23-3

The first histogram displays the body temperatures (in degrees Fahrenheit) of 130 healthy adults (Shoemaker, 1996). The second histogram displays the number of senior citizens in simulated samples of jury pools composed of 75 adult residents of San Luis Obispo county, taken from Activity 11-4.

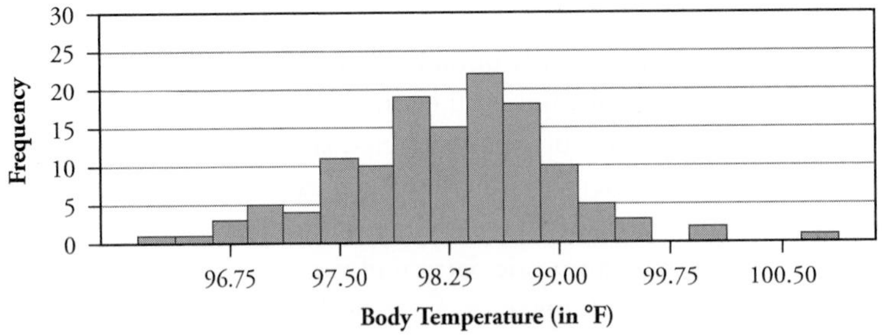

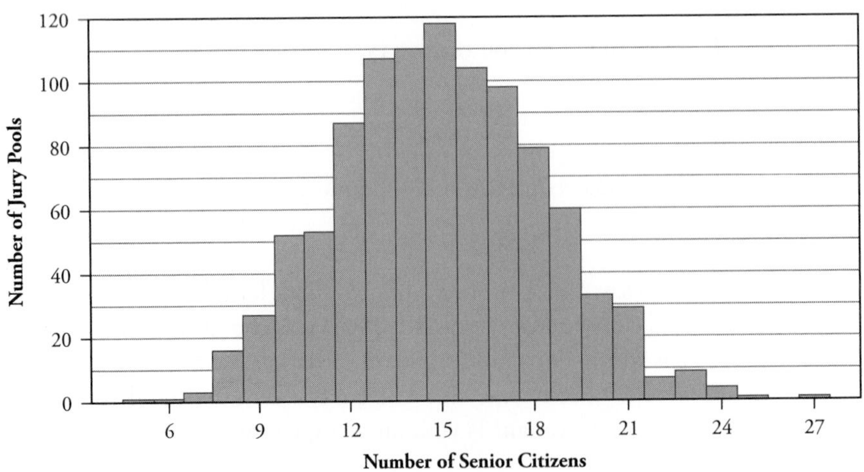

a. What similarities do you notice regarding the shapes of these distributions?

b. Draw a smooth curve (*smooth* meaning having no breaks or jagged edges) that seems to approximate the general shape apparent in the two histograms.

Data that display the general shape seen in the these examples occur frequently. The theoretical mathematical **models** used to approximate such distributions are called **normal distributions.** Every normal distribution shares three distinguishing characteristics: all are symmetric, have a single peak at their center, and follow a bell-shaped curve. Two things, however, distinguish one normal distribution from another: its *mean* and *standard deviation.* The mean of a normal distribution, represented by μ, determines where its center is; the peak of a normal curve occurs at its mean, which is also its point of symmetry. The standard deviation of a normal distribution, represented by σ, indicates the spread of the distribution. (*Note:* We reserve the symbols $\bar{x}$ and s to refer to the mean and standard deviation computed from sample data rather than a mathematical model.) The distance between the mean μ and the points where the curvature changes is equal to the standard deviation σ. The drawing shown here illustrates this relationship.

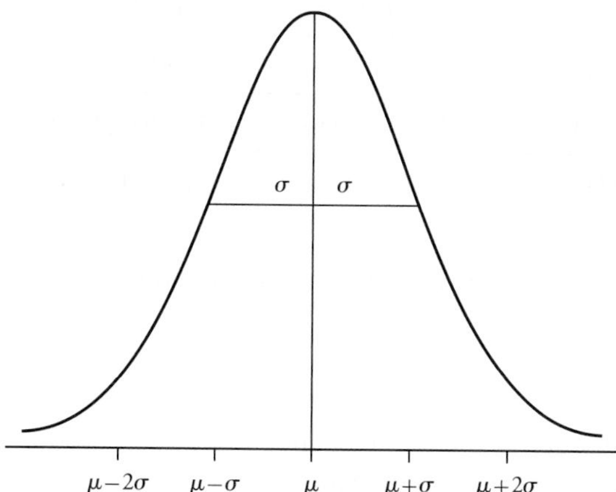

c. The following drawing contains three normal curves; think of them as approximating the distribution of exam scores for three different classes. One curve (call it A) has a mean of 70 and a standard deviation of 5; another curve (call it B) has a mean of 70 and a standard deviation of 10; the third curve (call it C) has a mean of 50 and a standard deviation of 10. Identify the curves by labeling each curve with its appropriate letter.

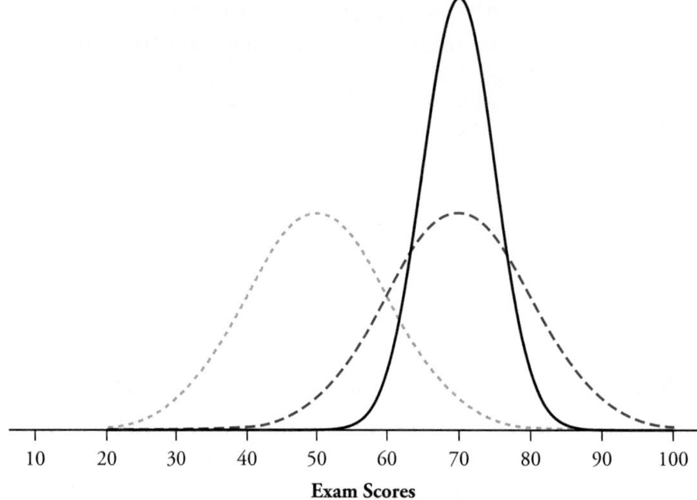

As a consequence of sharing a common shape, the body temperature and simulated jury pool distributions share other features such as the empirical rule, which was discussed in Topic 9. The following table reports the proportions of observations falling within one, two, and three standard deviations of the mean for each dataset.

	Body Temperature Dataset	Number of Senior Citizens Dataset	Empirical Rule
Mean, SD	98.249, 0.733	14.921, 3.336	
Within One SD of Mean	90/130 (69.2%)	703/1000 (70.3%)	68%
Within Two SDs of Mean	123/130 (94.6%)	957/1000 (95.7%)	95%
Within Three SDs of Mean	129/130 (99.2%)	998/1000 (99.8%)	99.7%

d. Are these proportions quite close to each other and to what the empirical rule would predict?

Recall also from Topic 9 the idea of standardization to produce a *z*-score, which indicates a value's relative position in a dataset:

$$z\text{-score} = \frac{x - \mu}{\sigma} = \frac{\text{observation} - \text{mean}}{\text{standard deviation}}$$

e. Find the *z*-score for the value 97.5 in the distribution of body temperatures (mean 98.249; SD 0.733).

f. Find the *z*-score for the value 11.5 in the distribution of number of senior citizens in jury pools (mean 14.921; SD 3.336).

g. What is true about these two *z*-scores?

h. How would you interpret these *z*-scores?

Examining the data reveals that 19 of the 130 body temperatures (14.6%) are less than 97.5 degrees and 153 of the 1000 samples of jury pools (15.3%) have less than 11.5 senior citizens.

The closeness of these percentages indicates that to find the (approximate) proportion falling in a given region for normal distributions, all you need to determine is the *z*-score. Values with the same *z*-score will have the same percentage lying below them for any normal distribution. Thus, instead of finding percentages for all normal distributions, you need only the percentages corresponding to these *z*-scores.

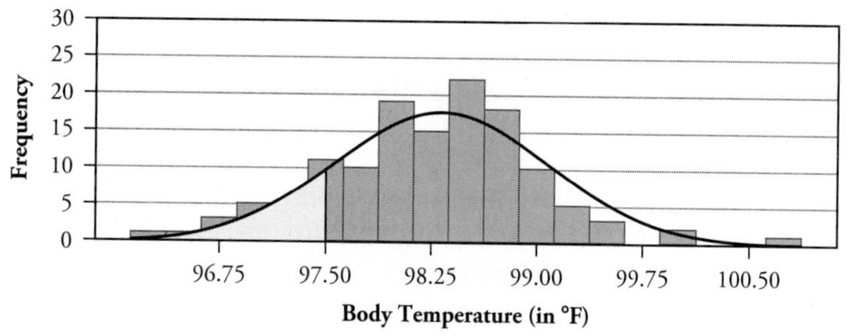

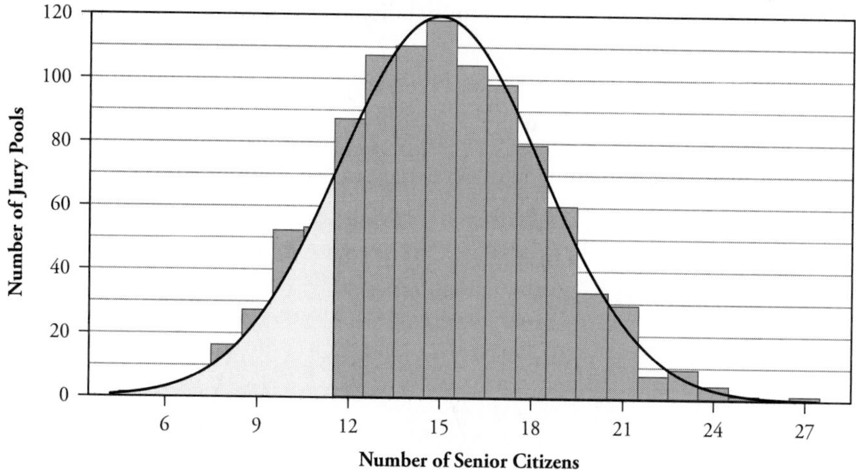

The probability of a randomly selected observation falling in a certain interval is equivalent to the proportion of the population's observations falling in that interval. Because the total area under the curve of a normal distribution is 1, this probability can be calculated by finding the area under the normal curve for that interval.

To find the area under a normal curve, you can use either your calculator or tables. Table II in the back of the book (the Standard Normal Probabilities Table) reports the area to the left of a given *z*-score under the normal curve. Enter the table at the row corresponding to the first two digits in the *z*-score. Then move to the column corresponding to the hundredths digit.

i. Use Table II to look up the area to the left of −1.03 under the normal curve. *Hint:* Find −1.0 along the left and the .03 along the top.

j. Is this value reasonably close to the proportions of adult body temperatures and senior citizens less than endpoints with *z*-scores of −1.03?

It is customary to use the symbol Z to denote observations from the **standard normal distribution,** which has mean 0 and standard deviation 1. The notation $\Pr(a < Z < b)$ denotes the probability lying between the values a and b, calculated as the area under the standard normal curve in that region. The notation $\Pr(Z < c)$ denotes the area to the left of a particular value c, whereas $\Pr(Z > d)$ refers to the area to the right of a particular value d.

Activity 12-2: Birth Weights

12-2, 14-9, 15-15

Birth weights of babies in the U.S. can be modeled by a normal distribution with mean 3300 grams (about 7.3 pounds) and standard deviation 570 grams (about 1.3 pounds). Babies weighing less than 2500 grams (about 5.5 pounds) are considered to be of low birth weight.

a. A graph of this normal distribution appears next. Shade in the region whose area corresponds to the probability that a baby will have a low birth weight.

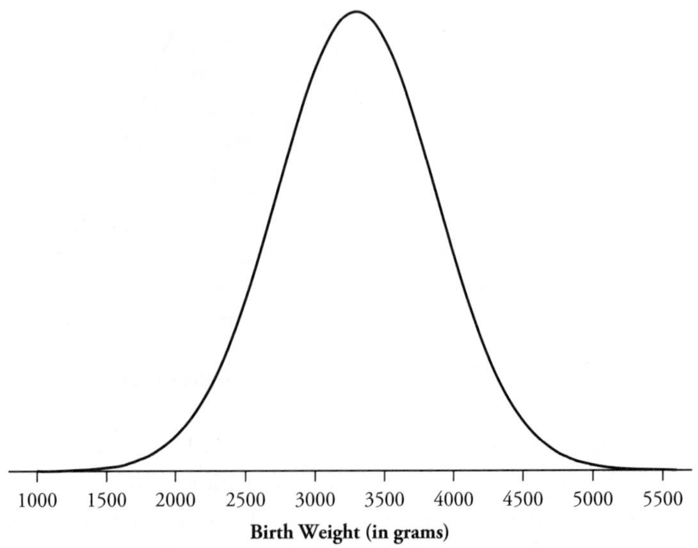

b. Based on this shaded region (remembering that the total area under the normal curve is 1), make an educated guess as to the proportion of babies born having a low birth weight.

c. Calculate the z-score for a birth weight of 2500 grams.

d. Look up this z-score in Table II to determine the proportion of babies born having a low birth weight. In other words, find $\Pr(Z < z)$, where z represents the z-score calculated in part c.

e. Confirm your answer to part d by using the `Normal Probability Calculator` applet. Enter the mean and standard deviation and press Scale to Fit; in the second region, check the first box and enter **2500** in the X column; and make sure the direction button is set to $<$. Then when you press Enter, the applet calculates the z-score and probability, and it also produces a shaded sketch. You can enter **birth weight** in the Variable field and press Scale to Fit to label the horizontal axis.

Normal Probability Calculator

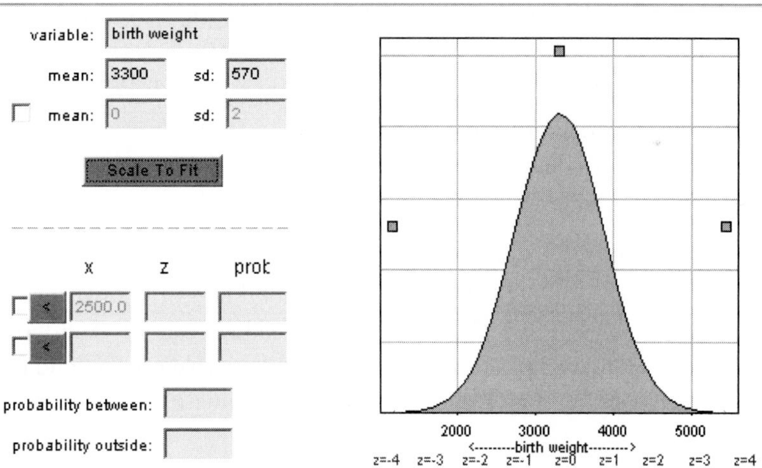

f. What proportion of babies does the normal distribution predict as weighing greater than 10 pounds (4536 grams) at birth? *Hint:* Always start with a sketch of the normal curve and shade in the area you are looking for. Then use the applet or your calculator, or compute the z-score and find the proportion below this z-score using Table II. If you use the table, recognize that what it reports is not the probability you are looking for exactly, but it allows you to determine the desired probability with one more step.

TI hint: To use your calculator to compute the probability (the area under the curve), you will use the **normalcdf** command. The syntax for this function is **normalcdf** (*lowerbound, upperbound, mean, standard deviation*), where *lowerbound* and *upperbound* are the boundaries of the area you are about to compute. The **normalcdf** function can be found by pressing the [2ND] [DISTR] keys and then pressing [2] to select it. Your Home screen should look like the following once you have entered the appropriate values:

```
normalcdf(4536,1
00000,3300,570)
```

Notice that we chose 100,000 for the upperbound. Entering a large number is necessary to capture the area to the right of 4536 (remember that the normal curve goes to $\pm\infty$ at the tails, respectively). Although we used 100,000 in our calculation, any large number works just as well.

g. Describe two different ways that you could have used Table II to answer part f. *Hint:* One way makes use of the normal curve's *symmetry.*

h. Determine the probability that a randomly selected baby weighs between 3000 and 4000 grams at birth. *Hint:* Start with a sketch. If you use the table, subtract the proportions obtained by Table II for the two relevant z-scores. If you use the applet, check the second box and enter both X values.

i. Data from the *National Vital Statistics System* indicate that in 2004 there were 4,112,052 births in the U.S. A total of 331,772 babies were of low birth weight, whereas 2,697,819 babies weighed between 3000 and 4000 grams. Calculate the observed proportions in each of these two groups, and comment on how well the calculations in parts d and h approximate these values.

j. How little would a baby have to weigh to be among the lightest 2.5% of all newborns? (This number is called the 2.5th **percentile** of the birth weight distribution.) *Hint:* Start with a sketch. You will then need to read Table II "in reverse," looking up the probability in the middle of the table and reading backward to find the relevant z-score. Then you will have to unconvert the z-score back to the birth weight scale. Use the `Normal Probability Calculator` applet to check your answer: enter the probability **.025,** in the Prob box and press Enter.

k. Now use your calculator to check your answer. To do this, press 2ND [DISTR] and then press 3 to select the **invNorm** command. The syntax for this function is **invNorm**(*area, mean, standard deviation*). For this activity, enter the values shown here on your Home screen:

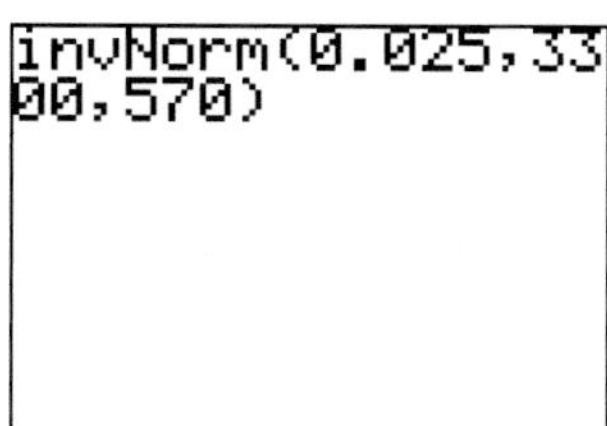

l. How much would a baby have to weigh to be among the heaviest 10% of all newborns?

Watch Out

- It is *always* helpful to draw a sketch of the relevant normal curve, shade in the region of interest, and check whether the probability calculated seems reasonable in light of the sketch.
- When you are using the Standard Normal Probabilities Table to find the probability to the *right* of the *z*-score, remember to subtract the probability reported by the table from 1. Do not subtract the *z*-score from 1, but subtract the probability to the *left* (as reported in the table) from 1.
- When you are using the table to find the probability between two values, remember to subtract the table's probabilities from each other. Again, do not subtract the *z*-scores; subtract the probabilities.
- When you are using the applet, make sure that the inequality symbol (as in part f) is set in the correct direction for the question at hand.
- When using your calculator, make sure you set the bounds correctly (as in the TI hint in part f).
- Be careful with phrases such as "at least" and "at most." Weighing at least 3000 pounds means to weigh 3000 or more pounds, so that indicates the area to the right of 3000. Weighing at most 2500 pounds is to weigh 2500 or less pounds, so that indicates the area to the left of 3000.
- But with a normal distribution, you do not need to worry about the difference between the phrases "at least" and "more than." The probability of any one specific value is zero because the area above one specific value is zero.

Activity 12-3: Blood Pressure and Pulse Rate Measurements

The following dotplots pertain to a random sample of 63 subjects from the 2003–2004 National Health and Nutrition Examination Survey (NHANES) study. The variables displayed are systolic blood pressure, diastolic blood pressure, and pulse rate:

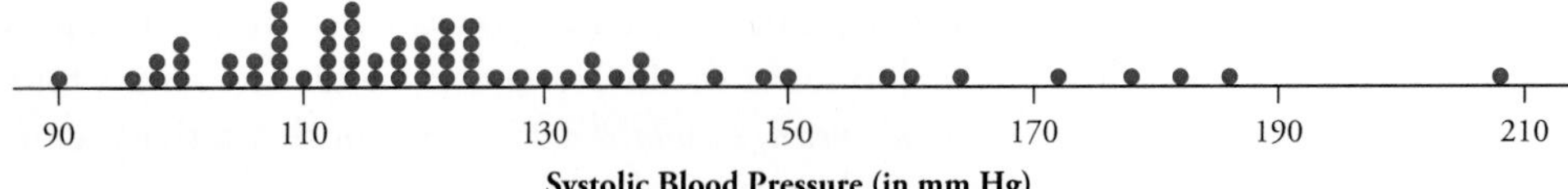

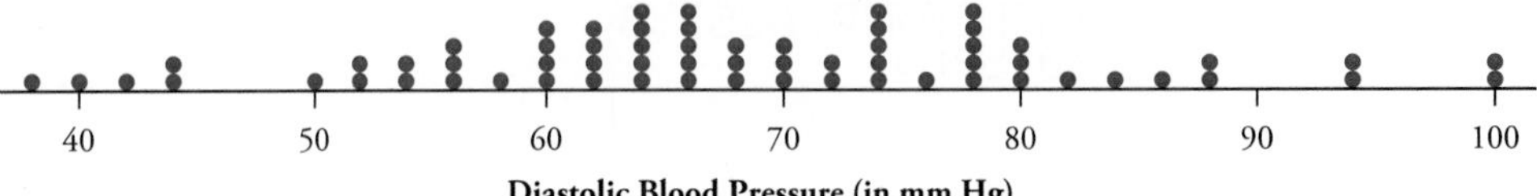

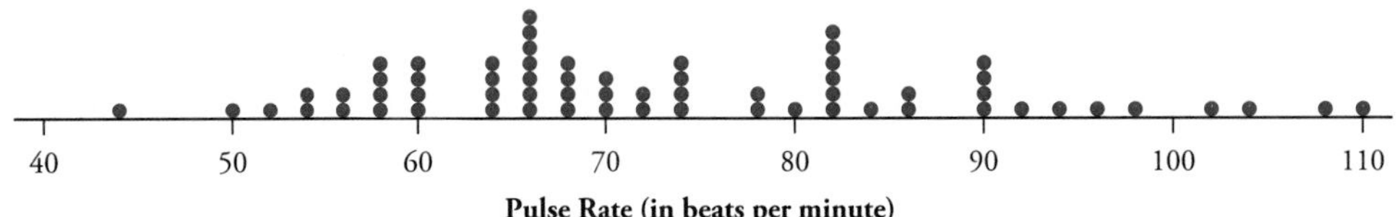

a. Which of these dotplots could *most* plausibly have come from a population that follows a normal distribution? Explain.

b. Which of these dotplots could *least* plausibly have come from a population that follows a normal distribution? Explain.

An important step in many statistical analyses is to judge whether sample data could have plausibly come from a population that follows a normal distribution. You can learn a lot from examining a dotplot or a histogram, but another graphical display has been developed especially for this purpose. A **normal probability plot** graphs the observed data against what would be expected from a theoretical normal distribution. If the data were perfectly normal, this plot would display a straight line. With real data, you look to see whether the normal probability plot roughly follows a straight line (an easier task than judging the fit of a curve to a histogram or dotplot). Skewed data reveal a curved pattern in a normal probability plot.

c. Use your calculator to produce a normal probability plot for these three variables. (The data are stored in the grouped file BLOODPRESSURES with lists **DIABP, SYSBP,** and **PLRT,** respectively.) Press [2ND] [**STAT PLOT**] and select one of the plots from the **STAT PLOTS** menu to create a normal probability plot. Arrow down to **Type** and over to the last icon to select it, enter the name of the **Data List,** and for the **Data Axis,** select the **X** variable. Make sure the other plots are turned off, and then press [ZOOM] followed by the [9] key. Comment on what these plots reveal, and whether they confirm your answers to parts a and b.

Activity 12-4: Criminal Footprints

Men tend to have longer feet than women. So, if you find a really long footprint at the scene of a crime, then in the absence of any other evidence, you would probably conclude that the criminal was a man. And conversely, if you find a really short footprint at the scene of a crime, then (again in the absence of any other information) you would probably conclude that the criminal was a woman. Where should the cut-off value be for concluding that the criminal is a man or a woman? And what is the probability that you will make a mistake?

Suppose that men's foot lengths are normally distributed with mean 25 centimeters and standard deviation 4 centimeters, and women's foot lengths are normally distributed with mean 19 centimeters and standard deviation 3 centimeters.

a. Sketch these two normal curves on the same axis, and label them (the curves and the axis) clearly.

A reasonable starting point to deciding on a cut-off value is to split the difference: conclude a footprint belongs to a man if it is longer than 22 centimeters (the midpoint of the means 19 and 25).

b. Using this rule, what is the probability that you will mistakenly identify a man's footprint as having come from a woman?

c. Using this rule again, what is the probability that you will mistakenly identify a woman's footprint as having come from a man?

d. Change the cut-off value (from 22 centimeters to some new value) so that the error probability in part b is reduced to .08.

e. Determine the probability of mistakenly identifying a woman's footprint as having come from a man using the new cut-off value in part d.

f. Comment on how changing the cut-off value affected the two kinds of error probabilities. Also explain why it makes sense that the probabilities changed as they did.

Solution

a.

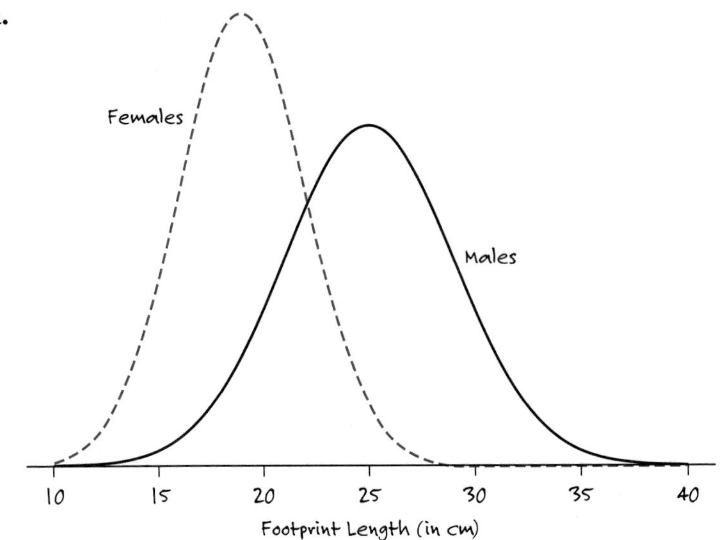

b. Because you are dealing with a male footprint, the z-score is $(22 - 25)/4$ or -0.75. Using Table II (or your calculator), the probability of the footprint being less than 22 centimeters is .2266. So, roughly 22.66% of men have a footprint smaller than 22 centimeters and would be misclassified as female.

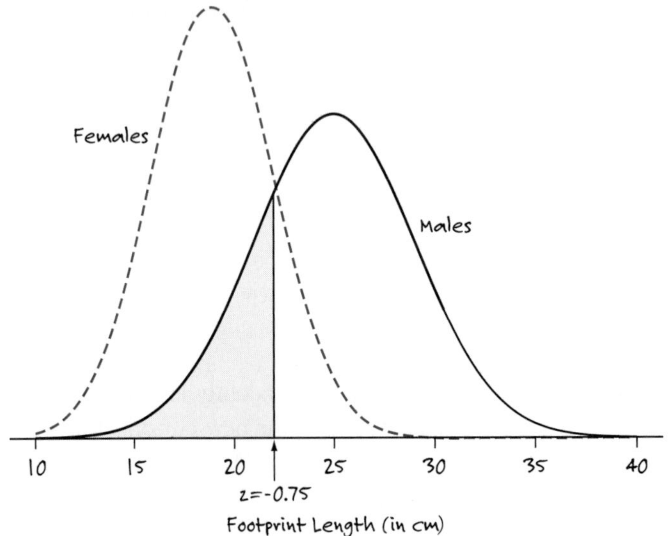

c. Now you are dealing with a female footprint, so the z-score is $(22 - 19)/3$ or 1.00. Using Table II (or your calculator), the probability of the footprint being longer than 22 centimeters is 1 − .8413 or .1587. This indicates that about 16% of females would be mistakenly identified as male.

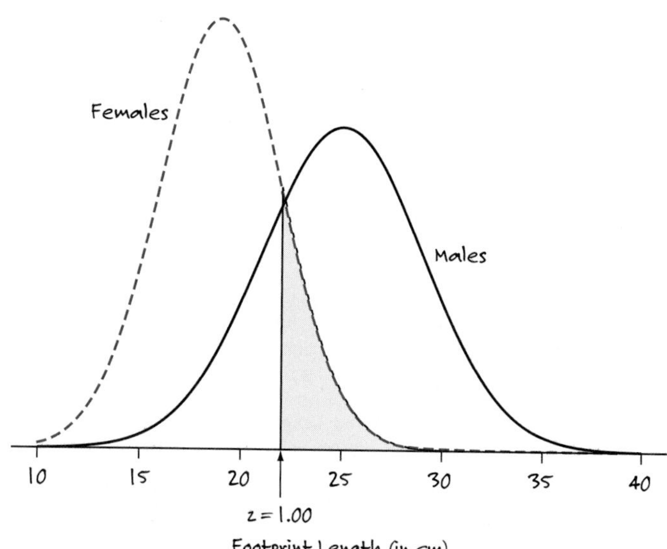

d. Using Table II or your calculator, the z-score to produce a probability of .08 is −1.41. To find the corresponding male foot length, you need to solve $-1.41 = (x - 25)/4$, which gives you $x = 25 - 1.41(4) = 25 - 5.64 = 19.36$ centimeters. You can also think of this as subtracting 1.41 standard deviations of 4 from the mean of 25. Notice that this new cut-off value (19.36 centimeters) is much smaller than before (22 centimeters) in order to reduce the probability of classifying a male footprint as having come from a woman.

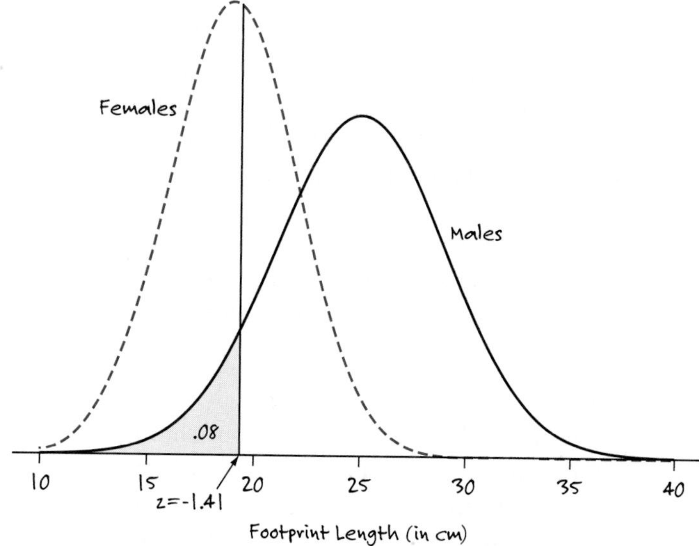

e. Using this new cut-off value of 19.36 centimeters, the z-score for a female footprint is $(19.36 - 19)/3$ or 0.12. The probability of a female footprint being longer than 19.36 centimeters is 1 − .5478 or .4522 (using Table II or your calculator).

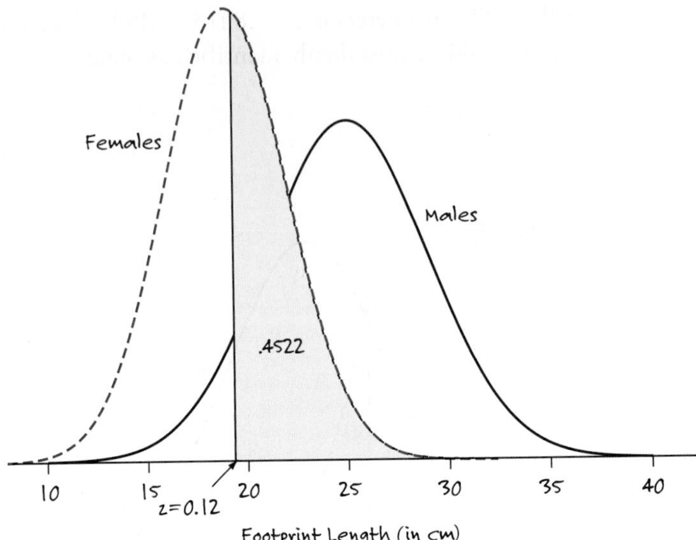

f. The probability of misclassifying a female footprint as a male footprint is much greater than before (.4522 as opposed to .1587). In order to reduce the probability of one type of error (misclassifying a man's footprint as having come from a woman) from .2266 to .08, the probability of making the other kind of error increases substantially. This exercise reveals that there is a trade-off between the probabilities of making the two kinds of errors that can occur: you can reduce one error probability, but only by increasing the other error probability.

Watch Out

- When working with two normal curves such as this, it's easy to get confused about which curve to use for a given question. Read the questions carefully.
- As we've said before, it really is *always* helpful to draw a sketch and check whether your answer makes sense.
- Try to recognize right away whether you have been given a value for the variable and asked for a probability (as in parts b, c, and e in this activity), or whether you have been given a probability and asked for the value of the variable (as in part d). In the latter case, you need to use the table "in reverse."
- Avoid sloppy notation. For example, do not say that $z = .12 = .5478$. Instead, say that the z-score is .12 and the probability (or area) to its left is .5478. You could also say that $\Pr(Z < .12) = .5478$.
- Remember the normal distribution is a mathematical model, so it is an idealization that never describes real data perfectly.

Wrap-Up...........

This topic introduced you to the most important mathematical model in all of statistics—the normal distribution. You have seen that data from many quantitative variables follow a familiar bell-shaped curve. This normal (bell-shaped) model can, therefore, be used to approximate the behavior of many real-world phenomena, such as birth weights and foot lengths. In addition to histograms and other common graphical displays, you can use normal probability plots to assess whether sample data can reasonably be modeled with a normal distribution, which is particularly helpful with smaller sample sizes.

You have also seen that the process of standardization, or calculating a z-score, allows you to use a table of standard normal probabilities to perform calculations related to normal distributions. You practiced using this table, and also using your calculator, both to calculate probabilities and to determine percentiles. For example, instead of finding the proportion of newborn babies who weigh less than 8 pounds, you can determine the weight such that 90% of newborn babies weigh less than that.

Some useful definitions to remember and habits to develop from this topic include

- The **normal distribution** is a useful model for summarizing the behavior of many quantitative variables.
- A **normal probability plot** is a useful tool for judging whether sample data could plausibly have come from a normally distributed population. With sample data, a normal probability plot that is roughly linear suggests the data can reasonably be modeled using a normal distribution.
- You can calculate probabilities from normal distributions by determining the area under the normal curve over the interval of interest. These areas can be interpreted either as the probability that a randomly selected value falls in the interval or as the proportion of values in the distribution that fall in the interval.
- To calculate probabilities from normal distributions, you can standardize (use the z-score) the values of interest and use a table of standard normal probabilities.
- The Standard Normal Probabilities Table only gives the area to the left of a z-score. To find the area to the right of a z-score, subtract the table's area from 1. To find the area in between two z-scores, subtract the smaller table area from the larger table area.
- You calculate **percentiles** from a normal probability table by reading the table "in reverse."

The next two topics will reveal how normal distributions describe the pattern of variation that arises when you repeatedly take samples from a population. In Topic 13, you will explore how a sample *proportion* varies from sample to sample, and in Topic 14, studying the variation of a sample *mean* will occupy your attention. These topics point toward the key role of normal distributions in the most important theoretical result in all of statistics—the Central Limit Theorem.

Homework Activities

Activity 12-5: Normal Curves

For each of the following normal curves, identify (as accurately as you can from the graph) the mean μ and standard deviation σ of the distribution.

a.

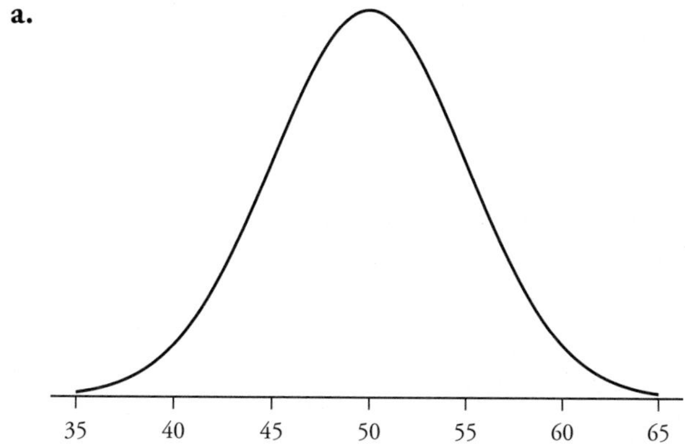

b.

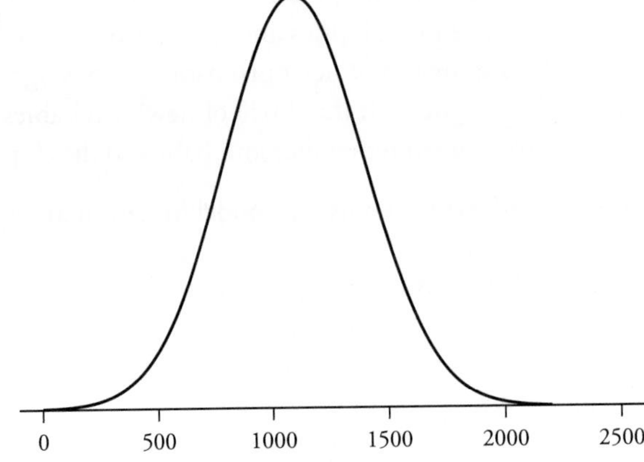

c.

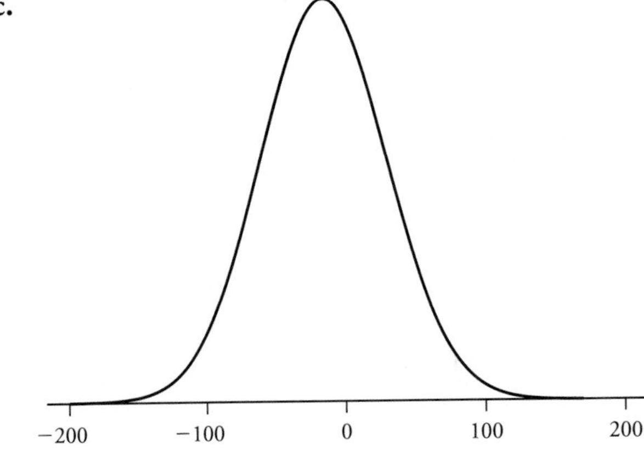

d.

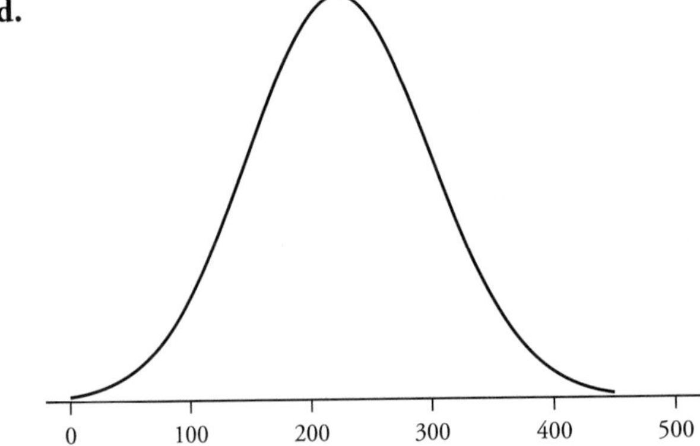

Activity 12-6: Pregnancy Durations

9-14, **12-6**

Data from the *National Vital Statistics System* reveal that the distribution of the duration of human pregnancies (i.e., the number of days between conception and birth) is approximately normal with mean $\mu = 270$ and standard deviation $\sigma = 17$ days. Use this normal model to determine the proportion of all pregnancies that come to term (have a duration) in

a. Less than 244 days (approximately 8 months)
b. More than 275 days (approximately 9 months)

c. More than 300 days
d. Between 260 and 280 days
e. Data from the *National Vital Statistics System* reveal that of 4,112,052 births in the U.S. in 2004, the number of pregnancies that resulted in a preterm delivery, defined as 36 or fewer weeks since conception, was 508,356. Compare this observed proportion to the prediction made using the normal model.

Activity 12-7: Professors' Grades

12-7, 12-8

Suppose you are deciding whether to take Professor Fisher's class or Professor Savage's class next semester. You happen to know that each professor gives As to those students scoring greater than 90 on the final exam and Fs to those students scoring less than 60 on the final exam. You also happen to know that the distribution of scores on Professor Fisher's final is approximately normal with mean 74 and standard deviation 7 and that the distribution of scores on Professor Savage's final is approximately normal with mean 78 and standard deviation 18.

a. Draw a sketch of both teachers' grade distributions using the same scale.
b. Which professor gives the higher proportion of As? Show the appropriate calculations to support your answer.
c. Which professor gives the higher proportion of Fs? Show the appropriate calculations to support your answer.
d. Suppose that Professor DeGroot has a policy of giving As to the top 10% of student scores on his final, regardless of the actual scores. If the distribution of scores on his final exam turns out to be normal with mean 69 and standard deviation 9, how high does your score have to be to earn an A?

Activity 12-8: Professors' Grades

12-7, **12-8**

Suppose that Professors Wells and Zeddes have final exam scores that are approximately normally distributed with mean 75. The standard deviation of Wells's scores is 10 and the standard deviation of Zeddes's scores is 5.

a. With which professor is a score of 90 on the final exam more impressive? Support your answer with appropriate probability calculations and with a well-labeled sketch.
b. With which professor is a score of 60 more discouraging? Again support your answer with the appropriate probability calculations and with a well-labeled sketch.

Activity 12-9: IQ Scores

12-9, 14-13, 15-20

Suppose the IQ scores of students at a certain college follow a normal distribution with mean 115 and standard deviation 12.

a. Draw a sketch of this distribution. Be sure to label the horizontal axis.
b. Shade in the area corresponding to the proportion of students with an IQ less than 100. Based on this shaded region, make an educated guess as to this proportion of students.

c. Use the normal model to determine the proportion of students with an IQ score less than 100.
d. Find the proportion of these undergraduates having IQs greater than 130.
e. Find the proportion of these undergraduates having IQs between 110 and 130.
f. With his IQ of 75, Forrest Gump would have a higher IQ than what percentage of these undergraduates?
g. Determine how high a student's IQ must be to be in the top 1% of all IQs at this college.

Activity 12-10: Candy Bar Weights

12-10, 14-10, 15-10

Suppose the wrapper of a certain candy bar lists its weight as 2.13 ounces. Naturally, the weights of individual bars vary somewhat. Suppose also that the weights of these candy bars vary according to a normal distribution with mean $\mu = 2.2$ ounces and standard deviation $\sigma = 0.04$ ounces.

a. What proportion of candy bars weighs less than the advertised weight?
b. What proportion of candy bars weighs more than 2.25 ounces?
c. What proportion of candy bars weighs between 2.2 and 2.3 ounces?
d. If the manufacturer wants to adjust the production process so only 1 candy bar in 1000 weighs less than the advertised weight, what will the mean of the actual weights be (assuming the standard deviation of the weights remains 0.04 ounces)?
e. If the manufacturer wants to adjust the production process so the mean remains at 2.2 ounces but only 1 candy bar in 1000 weighs less than the advertised weight, how small does the standard deviation of the weights need to be?
f. If the manufacturer wants to adjust the production process so the mean is reduced to 2.15 ounces but only 1 candy bar in 1000 weighs less than the advertised weight, how small does the standard deviation of the weights need to be?

Activity 12-11: SATs and ACTs

9-5, 12-11

Refer to the information presented in Activity 9-5 about SAT and ACT tests. Recall that among the college's applicants who take the SAT, scores have a mean of 1500 and a standard deviation of 240. Further recall that among the college's applicants who take the ACT, scores have a mean of 21 and a standard deviation of 6. Consider again applicant Bobby, who scored 1740 on the SAT, and applicant Kathy, who scored 30 on the ACT.

a. Assuming that SAT scores of the college's applicants are normally distributed, what proportion of applicants scored higher than Bobby on the SATs?
b. Assuming that ACT scores of the college's applicants are normally distributed, what proportion of applicants scored higher than Kathy on the ACTs?
c. Which applicant seems to be stronger in terms of standardized test performance? Explain.

Activity 12-12: Heights

Heights of American men aged 20 to 29 approximately follow a normal distribution with mean 70 inches and standard deviation 3 inches (reported in the *Statistical Abstract of the United Sates 1998*). The same shape and spread hold for heights of women in this age group, but with mean 65 inches.

a. Use this normal model to determine the probability that an American man in this age group will be shorter than 66 inches tall.
b. According to this normal model, what proportion of American men aged 20–29 are 6 feet or taller?
c. How tall must a man be to be among the tallest 10% of men in this age group?
d. Answer parts a–c for American women in this age group.
e. Sample data from the National Center for Health Statistics (reported in the *Statistical Abstract of the United States 1998*) indicate that 11.7% of a sample of men and 74.0% of a sample of women are shorter than 56 inches. Are these results generally consistent with your calculations?
f. Sample data from the National Center for Health Statistics indicate that 29.9% of a sample of men and 0.5% of a sample of women are 6 feet or taller. Are these results generally consistent with your calculations?

Activity 12-13: Weights

Sample data from the National Center for Health Statistics (reported in the *Statistical Abstract of the United States 1998*) reveal that weights of American men aged 20–29 have a mean of approximately 175 pounds and a standard deviation of approximately 35 pounds. For women, the mean weight is approximately 140 pounds and the standard deviation is approximately 30 pounds.

a. If these distributions are roughly normal, what percentage of men does the model predict to weigh less than 150 pounds? Less than 200 pounds? Less than 250 pounds?
b. Answer part a for women.
c. Sample data from the National Center for Health Statistics reveal that the observed percentages in these ranges are 29.0%, 82.1%, and 96.2% for men, compared to 70.4%, 92.5%, and 99.0% for women. How well does the normal model predict these percentages?

Activity 12-14: Dog Heights

Suppose that heights of male German Shepherds follow a normal distribution with mean 25 inches and standard deviation 2.5 inches. Suppose further that heights of male Shelties follow a normal distribution with mean 15 inches and standard deviation 1.5 inches.

a. What percentage of male German Shepherds have a height between 22.5 and 27.5 inches?
b. Between what two values do roughly 95% of male Sheltie's heights fall? *Hint:* Use the empirical rule.
c. How tall must a male Sheltie be to be among the tallest 10% of Shelties?
d. Which is more unusual: a male German Shepherd having a height of 28 inches or a male Sheltie having a height of 18 inches? Explain your answer, and justify with appropriate calculations.

Activity 12-15: Baby Weights

When he was three months old, Benjamin Chance weighed 13.9 pounds. The national average for the weight of a three-month-old baby is 12.5 pounds, with a standard deviation of 1.5 pounds.

a. Determine the *z*-score for Ben's weight, and write a sentence interpreting it.
b. Determine what proportion of three-month-olds weigh more than Ben. Be sure to state any assumptions you make in order to do this calculation.

c. For a six-month-old, the national average weight is 17.25 pounds, with a standard deviation of 2.0 pounds. Determine what Ben's weight would have to be at the age of six months, in order for him to be among the middle 68% of six-month-old babies.

Activity 12-16: Coin Ages

12-16, 14-1, 14-2, 19-15

A person with too much time on his hands collected 1000 pennies that came into his possession in 1999 and calculated the age (as of 1999) of each penny. The distribution of penny ages has mean 12.264 years and standard deviation 9.613 years. Knowing these summary statistics but without seeing the distribution, can you comment on whether the normal distribution is likely to provide a reasonable model for the ages of these pennies? Explain.

Activity 12-17: Empirical Rule

a. Use the Standard Normal Probabilities Table (Table II) or your calculator to find the proportion of the normal curve that falls within *one* standard deviation of the mean, in other words between *z*-scores of −1 and +1.
b. Repeat part a for *two* standard deviations from the mean.
c. Repeat part a for *three* standard deviations from the mean. These calculations provide the theoretical basis for the empirical rule.
d. Determine the *z*-scores such that the area under a normal curve between this *z*-score and its negative is 0.5000. Then subtract these two *z*-scores to find the interquartile range (IQR) of a normal distribution.
e. Recall that the outlier rule in Topic 10 identified an outlier as any observation falling more than 1.5 × IQR away from its nearer quartile. Use Table II or your calculator and your answer to part d to determine the *z*-scores for outliers and then the probability that an observation from a normal distribution will be classified as an outlier.

Activity 12-18: Critical Values

12-18, 16-2, 16-20, 19-13

a. Use the Standard Normal Probabilities Table (Table II) or your calculator to find the *z*-score such that the area to the score's right under a normal curve equals .10 (or a value as close as possible to this value from the table). In other words, find the value z^* such that $\Pr(Z > z^*)$ or .10.
b. Repeat part a for an area of .05.
c. Repeat part a for an area of .025.
d. Repeat part a for an area of .01.
e. Repeat part a for an area of .005.

These values are called the *critical values* of the normal distribution.

Activity 12-19: Body Temperatures

12-1, **12-19,** 15-3, 15-18, 15-19, 19-3, 19-17, 20-11, 22-10, 23-3

Reconsider the data on body temperatures from Activity 12-1. The data is stored in a grouped file BODYTEMPS that contains the lists **BDTMP** (body temperatures), **BTMPF** (female body temperatures), and **BTMPM** (male body temperatures).

a. Use your calculator to produce a histogram and normal probability plot of these data. Comment on whether a normal model seems to be appropriate for these data.
b. Produce separate histograms and normal probability plots for men and women. For which gender does a normal model seem to be more appropriate? Explain.

Activity 12-20: Natural Selection

10-1, 10-6, 10-7, **12-20,** 22-21, 23-3

Recall from Activity 10-1 the sparrow data collected by the scientist Bumpus to investigate Darwin's theory of natural selection (BUMPUS).

a. Produce and comment on normal probability plots of the total sparrow length measurements, separated by sparrows that survived the storm and those that did not survive the storm. Comment on whether a normal model seems to be appropriate for these data.
b. Examine normal probability plots for all of the other quantitative response variables in the data file, again separated by sparrows that survived the storm and those that did not. Identify which variables are well modeled by a normal distribution and which are not.

Activity 12-21: Honda Prices

7-10, 10-19, **12-21,** 28-14, 28-15, 29-10, 29-11

Reconsider the data on year, mileage, and price for a sample of used Honda Civics for sale on the Internet (HONDAPRICES). Create and examine normal probability plots for each of these variables, and summarize what the plots reveal. The list names are **YEAR, PRICE,** and **MILES,** respectively.

Activity 12-22: Your Choice

6-30, **12-22,** 19-22, 21-28, 22-27, 26-23, 27-22

a. Identify at least three variables that you have studied in this course (either data collected in class or supplied in the text) that have a roughly normal distribution.
b. List at least three variables that you have studied in this course that clearly do not have a normal distribution.

TOPIC

13 • • •

Sampling Distributions: Proportions

Kissing couples lean their heads either to the right or to the left. Do they lean their heads to the right more often than to the left? After all, most people are right-handed, and for most people, the right eye is dominant. Scientists have suggested that late-stage human embryos even turn their heads to the right in the womb. Perhaps this "right" tendency is reflected in how couples kiss as well. But by studying a sample, even a random sample of kissing couples, you expect that the sample proportion who lean right might not exactly reflect the population proportion. How much do you expect a sample proportion to differ from the population proportion? For example, if 65% of kissing couples in a sample lean their heads to the right, would that convince you that more than half of the population lean their heads to the right? You will explore these issues in this topic as you begin to study the important topic of sampling distributions.

Overview

You have been studying the ideas of probability and normal distributions. These ideas arise in the practice of statistics because (as you studied in Topics 4 and 5) proper data collection strategies involve the deliberate introduction of randomness into the process. Therefore, drawing meaningful conclusions from sample data requires an understanding of the properties of randomness. In this topic, you will study how sample proportions vary from sample to sample. You will see, however, that these sample results vary not in a haphazard way, but in a predictable manner, related to the normal distribution, that enables you to draw conclusions about the underlying population proportion. This topic will also mark your introduction to the two essential concepts of statistical inference: confidence and significance.

Preliminaries

1. Which color of Reese's Pieces candies do you think is the most common: orange, brown, or yellow?

2. Guess the proportion of all Reese's Pieces candies that are that color.

3. If each student in this class takes a random sample of 25 Reese's Pieces candies, would you expect every student to obtain the same number of orange candies?

4. If someone asks you to call the outcome of a coin flip, would you call "heads" or "tails"?

5. Record the number of heads and tails responses from yourself and your classmates to the previous question.

In-Class Activities

Recall from Topic 4 that a population consists of the entire group of observational units of interest to a researcher, whereas a sample refers to the (often small) part of the population that the investigator actually studies. Also remember that a parameter is a numerical characteristic of a population, whereas a statistic is a numerical characteristic of a sample.

In certain contexts, a population can also refer to a *process* (such as flipping a coin or manufacturing a candy bar) that, in principle, can be repeated indefinitely. Using this interpretation of population, a sample is a specific collection of process outcomes.

Throughout this topic, we will be careful to use different symbols to denote parameters and statistics. For example, we use the following symbols to denote proportions, means, and standard deviations (note that we consistently use Greek letters for parameters):

	(Population) Parameter		(Sample) Statistic	
Proportion	π	"pi"	$\hat{p}$	"p-hat"
Mean	μ	"mu"	$\bar{x}$	"x-bar"
Standard Deviation	σ	"sigma"	s	

Activity 13-1: Candy Colors

1-13, 2-19, **13-1,** 13-2, 15-7, 15-8, 16-4, 16-22, 24-15, 24-16

Consider the population of the Reese's Pieces candies manufactured by Hershey. Suppose you want to learn about the color distribution of these candies but you can afford to take a sample of only 25 candies.

a. Take a random sample of 25 candies and record the count and proportion of each color in your sample.

	Orange	Yellow	Brown
Count			
Proportion (Count/25)			

b. Is the proportion of orange candies among the 25 candies you selected a parameter or a statistic? What symbol is used to denote the proportion?

c. Is the proportion of orange candies manufactured by the Hershey's process a parameter or a statistic? What symbol represents the proportion?

d. Do you *know* the value of the proportion of orange candies manufactured by Hershey?

e. Do you know the value of the proportion of orange candies among the 25 candies that you selected?

These simple questions point out the important fact that although you typically know (or can easily calculate) the value of a sample statistic, only in very rare cases do you know the value of a population parameter. Indeed, a primary goal of sampling is to *estimate* the value of the parameter based on the statistic.

f. Do you suspect that every student in the class obtained the same proportion of orange candies in his or her sample?

g. Use the following axis to construct a dotplot of the sample proportions of orange candies obtained by students in the class.

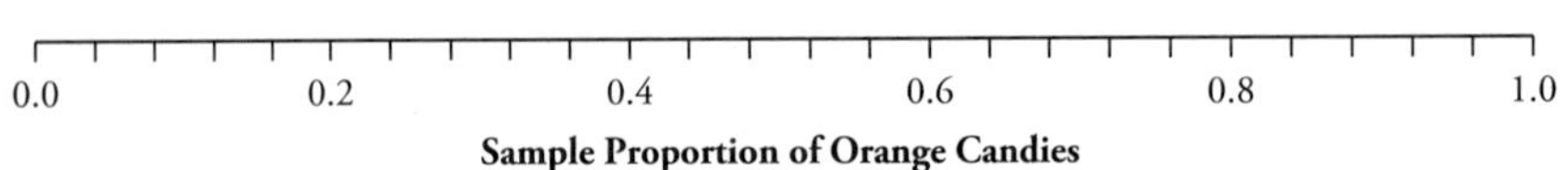

h. Identify the observational units in this graph and the variable being measured from unit to unit.

i. Comment on the shape, center, and spread of these sample proportions.

These results should remind you of the important statistical property known as **sampling variability:** values of sample statistics vary from sample to sample.

j. Based on what you learned in Topic 4 about *unbiased* statistics, and having the benefit of seeing the sample results from the entire class, guess the value of the population proportion of orange candies.

k. Assuming that each student has access only to his/her sample, would most estimates be reasonably close to the true parameter value? Would some estimates be way off? Explain.

l. Remembering what you learned about the effect of sample size in Topic 11, in what way would the dotplot look different if each student took a sample of 10 candies instead of 25? Explain.

m. In what way would the dotplot look different if each student took a sample of 75 candies instead of 25? Explain.

Activity 13-2: Candy Colors

1-13, 2-19, 13-1, **13-2,** 15-7, 15-8, 16-4, 16-22, 24-15, 24-16

Your class results should suggest that even though sample values vary depending on which sample you happen to pick, there seems to be a *pattern* to this variation. To investigate this pattern more thoroughly, however, you need more samples. Because it is time-consuming (and possibly fattening) to literally sample candies, you will use technology to *simulate* the sampling process.

To perform these simulations, you need to assume that you know the actual value of the parameter. Let us suppose that 45% of the population of Reese's Pieces candies is orange ($\pi = .45$).

a. Open the `Reese's Pieces` applet. Set the sample size to **25,** the number of samples to **1,** and the population proportion π to **.45.** Then click Draw Samples.

Reese's Pieces Samples

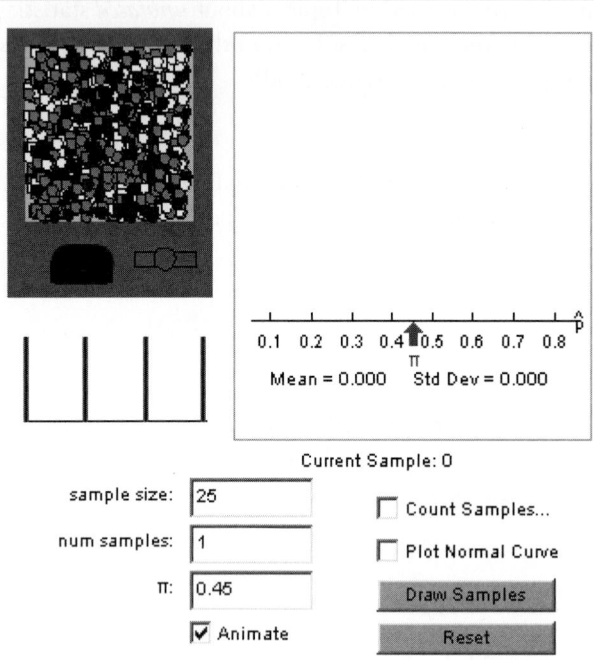

Report the sample proportion of orange candies in the selected sample, using the appropriate symbol to represent the proportion.

b. Click Draw Samples four more times, recording the value of the sample proportion obtained each time. Notice that the applet plots the values of the resulting sample proportions on the dotplot. Did you get the same sample proportion each time?

c. Now uncheck the Animate box and enter **495** in the Num Samples field, for a total of **500** samples. Click Draw Samples. Reproduce a rough sketch of the dotplot (conveying the overall behavior), labeling the axis appropriately.

d. Do you notice any pattern in the variation of the resulting 500 sample proportions? Explain.

e. Select Plot Normal Curve, and comment on how well the curve appears to model the simulated sample proportions.

The distribution of the sample proportions from sample to sample is called the **sampling distribution** of the sample proportion. Even though the sample proportion of orange candies varies from sample to sample, that variation has a recognizable long-term pattern. These simulated sample proportions approximate the theoretical sampling distribution derived from all possible samples.

13

f. Record the mean and standard deviation of these sample proportions, as reported by the applet:

Mean of $\hat{p}$ values: Standard deviation of $\hat{p}$ values:

g. Roughly speaking, are more sample proportions *close* to the population proportion (which, you will recall, is .45) than are *far* from the population proportion?

h. Let us quantify the previous question. Select Count Samples… and then use the drop-down menu to choose the Count Between option. Enter the appropriate values to count how many of the 500 sample proportions are within ± .10 of .45 (i.e., between .35 and .55). *Note:* Because .10 should be close to the standard deviation you calculated in part d, you are looking for roughly one standard deviation on each side of π. Then count how many of the sample proportions are within ± .20 of .45 and within ± .30 of .45. Record the results in this table:

	Number of 500 Sample Proportions	Percentage of 500 Sample Proportions
Within ± .10 of .45		
Within ± .20 of .45		
Within ± .30 of .45		

i. Forget for a moment that you have designated the population proportion of orange candies to be .45. Suppose each of these 500 imaginary students (the different samples) estimated the population proportion of orange candies by creating an interval extending a distance of .20 on either side of her or his sample proportion. What percentage of the 500 students would capture the actual population proportion of .45 within this interval?

j. Still forgetting that you actually know the population proportion of orange candies to be .45, suppose that you are one of those 500 imaginary students. Would you have any way of knowing definitively whether your sample proportion was within .20 of the population proportion? Could you be reasonably "confident" that your sample proportion was within .20 of the population proportion? Explain.

Although you cannot use a sample proportion to determine a population proportion exactly, you can be reasonably **confident** that the population proportion is within a certain distance of the sample proportion. This distance depends primarily on how confident you want to be and on the size of the sample. You will study this notion extensively when you encounter confidence intervals in Topic 16.

k. Use the applet to simulate drawing **500** samples of **75** candies each (so the samples are three times larger than the ones you gathered in class and simulated earlier). Report the mean and standard deviation of the simulated sample proportions:

Mean of $\hat{p}$ values: Standard deviation of $\hat{p}$ values:

l. How have the shape, center, and spread of the sampling distribution changed from when the sample size was only 25 candies?

m. Use the applet to count how many of these 500 sample proportions are within $\pm$ .10 of .45. Record this number and the percentage here.

n. How do the percentages of sample proportions falling within $\pm$ .10 of .45 compare between the sample sizes of 25 and 75?

o. In general, is a sample proportion more likely to be closer to the population proportion with a larger sample size or with a smaller sample size?

Because these sample proportions follow approximately a normal distribution, the *empirical rule* establishes that approximately 95% of the sample proportions fall within two standard deviations of the mean of these sample proportions.

p. In part k, you found the mean and standard deviation of these sample proportions for a sample size of 75. Double the standard deviation. Then subtract this value from .45 and also add this value to .45. Record the results.

q. Use the applet to count how many of the sample proportions for samples of size 75 fall within the interval you calculated in part p. What percentage of the 500 sample proportions does this represent? Is this percentage close to the 95% predicted by the empirical rule?

r. If each of the 500 imaginary students subtracted this value (twice the standard deviation) from her or his sample proportion and also added this value to her or his sample proportion, what percentage of the students' intervals would contain the actual population proportion of .45?

This activity reveals that if you want to be about *95% confident* of capturing the population proportion within a certain distance of your sample proportion, that distance should be approximately twice the standard deviation of the sampling distribution of sample proportions.

In fact, there is a theoretical result that tells you what to expect for the shape, center, and spread of the sampling distribution of the sample proportion without having to use simulations.

Central Limit Theorem (CLT) for a Sample Proportion
Suppose a simple random sample of size n is to be taken from a large population in which the true proportion possessing the attribute of interest is π. The sampling distribution of the sample proportion $\hat{p}$ is approximately normal with mean equal to π and standard deviation equal to

$$\sqrt{\frac{\pi\,(1-\pi)}{n}}$$

This normal approximation becomes more and more accurate as the sample size n increases, and it is generally considered to be valid as long as $n\pi \geq 10$ and $n(1-\pi) \geq 10$.

The following graph illustrates this result.

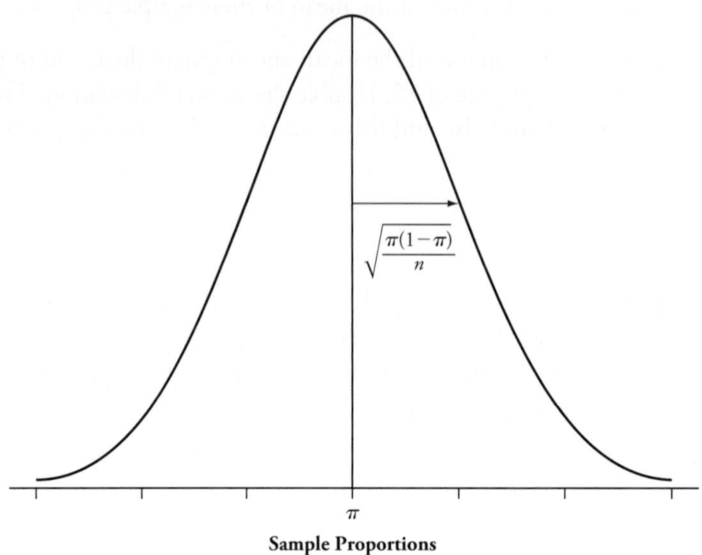

s. If you continue to assume that the population proportion of orange candies is $\pi = .45$, what does the Central Limit Theorem say about the mean and standard deviation of the sampling distribution of sample proportions when the sample consists of $n = 25$ candies? Do these values come close to your simulated results in part f?

Theoretical mean of $\hat{p}$ values:

Theoretical standard deviation of $\hat{p}$ values:

t. Repeat the previous question for a sample size of $n = 75$. Compare your theoretical answers to your simulated results in part k.

Theoretical mean of $\hat{p}$ values:

Theoretical standard deviation of $\hat{p}$ values:

u. Now suppose that only 10% of Reese's Pieces are orange. Use the applet to draw **500** random samples of size n = **25** in this case. Then use the applet to plot a normal curve on the resulting sample proportions of orange candies. Does the normal model summarize this distribution well? Explain why this is not a contradiction of the Central Limit Theorem.

Watch Out

- The concept of sampling distribution is one of the most difficult statistical concepts to firmly grasp because of the different "levels" involved. For example, here the original observational units are the candies, and the variable is the color (a categorical variable). But at the next level, the observational units are the samples, and the variable is the proportion of orange candies in the sample (a quantitative variable). Try to keep these different levels clear in your mind.

- It's essential to distinguish clearly between parameters and statistics. A parameter is a fixed numerical value describing a population. Typically, you do not know the value of a parameter in real life, but you might perform calculations assuming a particular parameter value. On the other hand, a statistic is a number describing a sample, which varies from sample to sample if you were to repeatedly take samples from the population.
- Notice that the Central Limit Theorem (CLT) specifies *three* things about the distribution of a sample proportion: shape, center (as measured by the mean), and spread (as measured by the standard deviation). It's easy to focus on one of these aspects of a distribution and ignore the other two. As with other normal distributions, drawing a sketch can help you to visualize the CLT.
- Ensure that these conditions hold before you apply the CLT: the sample needs to have been chosen randomly, and the sample size condition requires that $n\pi \geq 10$ and $n(1 - \pi) \geq 10$. In part u, the sample size is not large enough, considering the fairly small value of the population proportion (.10), and so the sampling distribution is not well modeled by a normal distribution. (Technically, it's the normal shape that depends on this condition. The results about the mean and standard deviation hold regardless of this condition.)
- As long as the population size is much larger than the sample size (say, 20 times larger), the *population* size itself does not affect the behavior of the sampling distribution. This sounds counterintuitive to most people, because it means that a random sample of size 1000 from one state will have the same sampling variability as a random sample of size 1000 from the entire country. But think about it: If chef Julia prepares soup in a regular-sized pot and chef Emeril prepares soup in a restaurant-sized vat, you can still learn the same amount of information about either soup from one spoonful. You don't need a larger spoonful to decide if you like the taste of Emeril's soup.
- As we've said before, try not to confuse the sample size with the number of samples. The sample size is the important number that affects the behavior of the sampling distribution. In practice, you typically get only one sample. We have asked you to simulate a large number of samples only to give you a sense for how sample statistics vary under repeated sampling; we have tried to ask for enough samples (typically 500 or 1000) to give you a sense of what would happen in the long run. In fact, now that you know the Central Limit Theorem's description of how sample proportions vary under repeated sampling, you no longer need to simulate taking many samples from the population.

Activity 13-3: Kissing Couples

13-3, 13-4, 16-6, 17-12, 24-4, 24-14

As reported in the journal *Nature* (2003), German bio-psychologist Onur Güntürkün conjectured that the human tendency to turn to the right manifests itself in other ways as well, so he studied kissing couples to see whether they tended to lean their heads to the right while kissing. He and his researchers observed couples from age 13 to 70 in public places such as airports, train stations, beaches, and parks in the United States, Germany, and Turkey. They were careful not to include couples who were holding objects such as luggage that might have affected which direction couples leaned. They observed a total of 124 kissing couples.

Suppose for now that one-half of all kissing couples lean to the right.

a. Is this number a parameter or a statistic? Explain. Also indicate what symbol denotes the number.

The following histogram displays the results of 1000 simulated samples of 124 couples, assuming that one-half of all kissing couples lean their heads to the right.

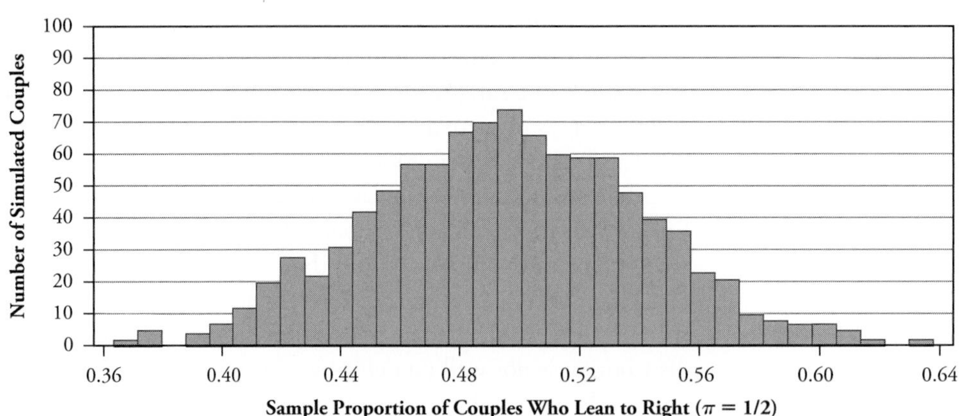

b. Check whether the CLT applies in this case. If it does, describe what the CLT says about the theoretical sampling distribution of the sample proportion.

Shape:

Center:

Spread:

c. Does the histogram of simulation results appear to be consistent with what the CLT predicts? Explain.

In the sample of 124 couples observed by the researcher, 80 couples leaned their heads to the right and 44 leaned their heads to the left.

d. Calculate the sample proportion of couples who leaned their heads to the right. Also use the appropriate symbol to denote this proportion.

e. Judging from the histogram of simulation results, would it be surprising to observe such a sample proportion (.645) of couples leaning their heads to the right when one-half of all kissing couples lean their heads to the right? Explain.

f. Based on the theoretical mean and standard deviation of the sampling distribution (from the CLT), calculate the z-score for the observed sample proportion (.645).

g. Do the sample data cast doubt on the hypothesis that one-half of all kissing couples lean to the right? Explain, based on the histogram and z-score.

This activity has introduced you to the concept of **statistical significance.** You determine the statistical significance of a sample statistic by exploring the sampling distribution of the statistic, investigating how often an observed sample result occurs simply by random chance. Roughly speaking, a sample result is said to be **statistically significant** if it is unlikely to occur due to random sampling variability alone.

In this example, it would be highly unlikely for the observed sample proportion ($\hat{p}$ = .645 in a sample of size n = 124) to arise if one-half of the population leaned their heads to the right (π = .5). Therefore, the sample data cast strong doubt on the claim that π = .5. You will study this reasoning process in detail when you encounter tests of significance in Topic 17.

Activity 13-4: Kissing Couples

13-3, **13-4,** 16-6, 17-12, 24-4, 24-14

Reconsider the study described in Activity 13-3 concerning the direction that kissing couples lean their heads. The following two histograms display simulated sample proportions of kissing couples who lean their heads to the right, again based on a sample size of 124. The first histogram assumes that two-thirds of the population lean their heads to the right when kissing ($\pi \approx$ .667), and the second histogram assumes that three-fourths of the population lean their heads to the right when kissing (π = .75).

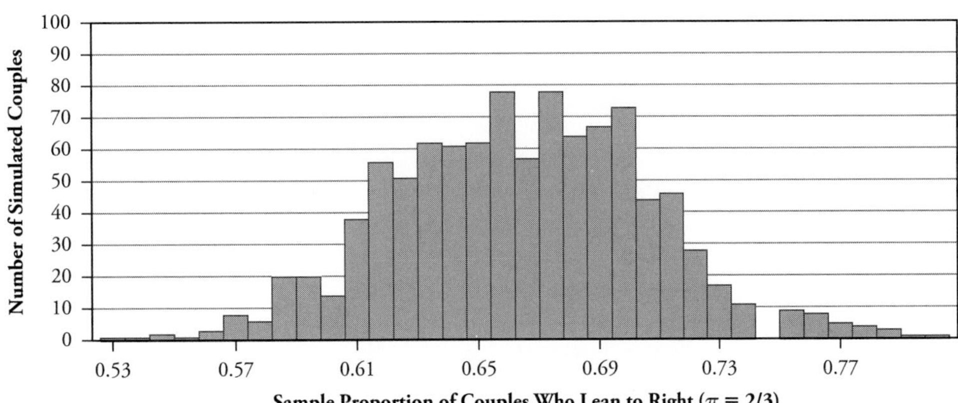

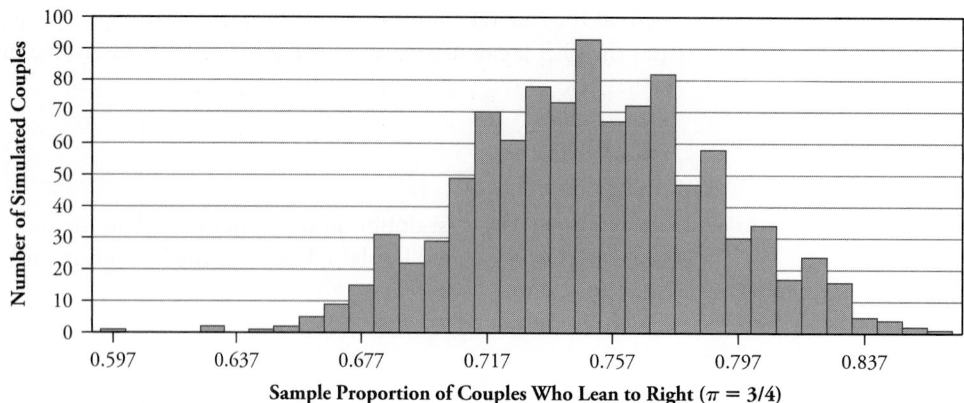

a. Based on the first histogram, comment on how surprising the sample result (80 of 124 couples in the sample leaning their heads to the right when kissing) would be if the population proportion were equal to 2/3.

b. Based on the theoretical mean and standard deviation of the sampling distribution (from the CLT, again assuming that 2/3 of all couples lean their heads to the right when kissing), calculate the z-score for the observed sample proportion ($\hat{p} = .645$).

c. Based on your answers to parts a and b, do the sample data provide strong evidence to doubt that 2/3 of all kissing couples lean their heads to the right? Explain your reasoning.

d. Repeat parts a–c for the second histogram, which assumes that 3/4 of all kissing couples lean their heads to the right.

e. Estimate the standard deviation of the sampling distribution of $\hat{p}$, using a reasonable estimate for the population proportion π based on the sample data. Then double this standard deviation and form an interval of that length centered at the sample proportion. Comment on how this interval relates to your analysis in parts a–d.

Solution

a. Recall that the observed sample proportion of kissing couples who lean their head to the right is $\hat{p} = 80/124 = .645$. This value is not at all uncommon in the first histogram.

b. The CLT says that the sample proportion in this case would vary approximately normally with mean equal to .667 and standard deviation equal to

$$\sqrt{\frac{(.667)(.333)}{124}} \approx .042$$

The z-score for the observed sample proportion of 0.645 is, therefore,

$$z = \frac{.645 - .667}{.042} \approx -0.52$$

so the observed sample proportion .645 lies only about half of a standard deviation from the population proportion when $\pi = .667$.

c. The observed sample proportion is barely one-half of a standard deviation away from what you would expect if the population proportion were equal to 2/3, not a surprising result at all. Therefore, the sample data provide no reason to doubt that the population proportion of kissing couples who lean their head to the right equals 2/3.

d. The value .645 is pretty far along the lower tail of the second histogram. This indicates that the observed sample proportion would rarely occur if the population proportion were equal to 3/4. Further evidence of this result is provided by the rather large negative z-score:

$$\frac{.645 - .750}{\sqrt{\frac{(.750)(.250)}{124}}} = \frac{.645 - .750}{.039} \approx -2.69$$

Therefore, the sample data provide fairly strong evidence that the population proportion of kissing couples who lean their head to the right is not 3/4 (because it would be rather surprising to find a sample proportion so far from this population proportion by chance alone).

e. A reasonable estimate of the population proportion π is the sample proportion .645. An estimate of the standard deviation of $\hat{p}$ would then be

$$\sqrt{\frac{(.645)(.355)}{124}} \approx .043$$

Doubling this standard deviation gives .086. The interval is, therefore, $.645 \pm .086$, which runs from .559 to .731. Notice that 1/2 and 3/4 are not in this interval, but 2/3 is. The interval is consistent with the earlier analysis of the plausibility of the values 1/2, 2/3, and 3/4 for the population proportion of kissing couples who lean their head to the right.

Wrap-Up

In this topic, you explored the obvious (but crucial) concept of sampling variability, and you learned that the distribution of sample proportions displays a predictable long-term pattern. For example, even though your sample proportion of orange Reese's Pieces was probably different from your neighbor's, when you took hundreds of random samples of Reese's Pieces and graphed the resulting sample proportions of orange candies, you discovered a rather obvious bell-shaped pattern. This distribution is known as the sampling distribution of the statistic, the statistic being, in this case, the sample proportion of orange candies.

You investigated properties of this sampling distribution, such as how the normal distribution often provides a reasonable model for the behavior of a sample proportion as it varies from sample to sample. You also discovered that larger sample sizes produce less variation among sample proportions than do smaller sample sizes.

In addition, you began to explore how sampling distributions relate to the important idea of statistical *confidence*: you can have a certain amount of confidence that the unknown value of a population parameter is within a certain distance of the observed value of a sample statistic. For example, you were able to conclude with high confidence that the population proportion of kissing couples who lean their head to the left is within $\pm$ .086 of the observed sample proportion of .645.

You also encountered the issue of statistical *significance*: You can evaluate a conjecture for the value of the population parameter by determining how often an observed sample statistic result occurs by sampling variability or chance alone when the parameter has that value. For example, you found that the observed proportion of kissing couples who leaned their head to the right would have been incredibly unlikely if half of the population leaned to the right.

Some useful definitions to remember and habits to develop from this topic include

- A parameter corresponds to the population and a statistic to a sample. Parameters and statistics are both *numbers,* but the value of a parameter is not usually known because it is often unfeasible to gather data about the entire population. However, you can still describe the parameter in words (e.g., the proportion of right-leaning kissing couples in the world).
- This topic dealt only with *categorical* variables. Therefore, the statistic of interest in this topic has been a sample *proportion*. This statistic is denoted by $\hat{p}$, and its population parameter counterpart is denoted by π.
- If you take random samples from a population over and over again, the value of the sample proportion varies from sample to sample. This phenomenon is known as **sampling variability.** When you graph these varying sample proportions, the observational units of that graph are the *samples,* and the variable is the *sample proportion.*
- Even though the value of a sample proportion varies from sample to sample, the distribution of sample proportions, known as its **sampling distribution,** follows a predictable long-term pattern.
- Simulation is a valuable tool for studying a sampling distribution, where you can actually simulate many random samples from a population where you have specified the parameter value(s).
- Perhaps the most important theoretical result in all of statistics is the **Central Limit Theorem:** Whenever the sample size is large enough, the sampling distribution of sample proportions will have a shape that is approximately normal,

with its mean equal to the value of the population proportion (π) and with its standard deviation equal to

$$\sqrt{\frac{\pi(1-\pi)}{n}}$$

- Notice that the sample size n is in the denominator of this standard deviation expression, indicating that larger samples produce less variability than smaller samples.
- The empirical rule establishes that in about 95% of all samples, the sample proportion falls within two standard deviations of the population proportion. This means that when you only observe one sample (which is almost always the case), you can be **confident** that the unknown population value falls within two standard deviations of your sample proportion $\hat{p}$.
- This sampling distribution result also enables you to judge whether an observed sample proportion is unlikely to occur for a conjectured value of the population proportion. This issue is known as **statistical significance.**

In the next topic, you will study the sampling distribution not of a sample *proportion,* but of a sample *mean.* In the process, you will discover many similarities between the two types of sampling distributions. You will also continue to explore the connection between a sampling distribution and the fundamental concepts of statistical confidence and statistical significance.

Homework Activities

Activity 13-5: Miscellany

Identify each of the following as a parameter or a statistic, and also indicate the symbol used to denote the parameter or statistic. In some cases, you might have to form (and specify) your own conclusion as to what the population of interest is.

a. The proportion of students in your class who prefer to hear "good news" before bad
b. The proportion of all students at your school who prefer to hear "good news" before bad
c. The mean number of states visited by the students in your class
d. The mean number of states visited by all students in your school
e. The standard deviation of the lengths of words in the Gettysburg Address
f. The standard deviation of the lengths of words in a random sample that you select from the Gettysburg Address
g. The proportion of American voters who voted for the Republican candidate in the most recent presidential election
h. The proportion of people at the next party you attend who voted in the most recent presidential election
i. The proportion of American voters interviewed by CNN who voted for the Democratic candidate in the most recent presidential election
j. The mean amount spent on Christmas presents by American adults last year
k. The proportion of American adults who told a Gallup pollster that they prefer to abolish rather than retain the penny
l. The mean number of cats in the households of your school's faculty members
m. The mean number of cats among all American households
n. The proportion of "heads" in 100 coin flips
o. The mean weight of 20 bags of potato chips

Activity 13-6: Generation M

3-8, 4-14, **13-6,** 16-1, 16-3, 16-7, 18-1, 21-11, 21-12

Recall from Activity 3-8 the survey of a nationally representative sample of 2032 teenagers, asking about their use of various types of media. Identify each of the following as a parameter or a statistic, and provide an appropriate symbol to represent the value.

a. The proportion of all American teenagers who have a television in their bedrooms
b. The proportion of these 2032 teenagers who have a television in their bedrooms
c. The proportion of boys in this sample who have a television in their bedrooms
d. The average number of hours of recreational computer use per week by the teenagers in this study
e. The average number of hours of recreational computer use per week among all American teenagers

Activity 13-7: Presidential Approval

13-7, 13-8

Let π represent the population proportion of adult Americans who respond favorably when asked about the President's job performance. Suppose you repeatedly take simple random samples (SRSs) of 1000 adult Americans and determine the sample proportion who respond favorably.

a. Use the Central Limit Theorem to calculate the standard deviation of the sampling distribution of these sample proportions for each of the following values of π: 0, .2, .4, .5, .6, .8, 1.
b. Which value(s) of π produces the most variability in sample proportions?
c. Which value(s) of π produce the least variability in sample proportions?
d. Explain why your answers to parts b and c make sense.
e. Without doing more calculations, describe whether, and if so how, your answers to parts a–c might change using a sample size of 500 rather than 1000 people.

Activity 13-8: Presidential Approval

13-7, **13-8**

Suppose that 40% of adult Americans respond favorably when asked about the President's job performance, and suppose you plan to take an SRS of n adult Americans.

a. Use the Central Limit Theorem to calculate the standard deviation of the sampling distribution of the sample proportion of adult Americans who respond favorably, for each of the following values of n: 100, 200, 400, 800, 1600.
b. Comment on how the standard deviation changes as the sample size increases.
c. By how many times does the sample size have to increase in order to cut this standard deviation in half?
d. Would your answer to part c change if the population proportion who responded favorably were equal to some other value than .4? Explain.

Activity 13-9: Pet Ownership

13-9, 13-14, 13-15, 18-2, 20-21

Suppose you want to estimate the proportion of all households in your hometown that have a pet cat. (Let us call this proportion π.) Suppose further that you take an SRS of 200 households, whereas your polling competitor takes an SRS of only 50 households.

a. Can you be certain that the sample proportion of cat households in your sample will be closer to π than your competitor's sample proportion?

b. Do you have a better chance than your competitor of obtaining a sample proportion of cat households that fall within $\pm$ .05 of the actual value π? Explain.

Suppose now that the actual proportion of households with a pet cat (among the entire population of households in your hometown) is $\pi = .25$.

c. Use the expression for the standard deviation of the sampling distribution of sample proportions specified by the Central Limit Theorem result to determine the standard deviation when the sample size is $n = 200$. Then perform the same calculation for a sample size of $n = 50$. Which sample size produces the smaller standard deviation? How many times smaller is it than the previous standard deviation?

d. Use the SIMSAMP program (download it to your calculator if you have not done so already) to simulate the random selection of **500** samples, with each sample containing **200** households and $\pi = \mathbf{.25}$. The results are stored in a list named PROP. (*Hint:* You can use the `Reese's Pieces` applet if you disregard the candy context and consider orange candies to represent households with cats.) How many of these 500 samples produce a sample proportion within $\pm$ .05 of the population proportion?

TI note: This program might take quite awhile to run and use a significant amount of memory. Please make sure you have set aside enough time to run this program and deleted or archived all unnecessary data and programs.

e. Again use your calculator to simulate the random selection of **500** samples, this time with each sample composed of only **50** households. How many of these 500 samples produce a sample proportion within $\pm$ .05 of the population proportion?

f. Write a paragraph commenting on the similarities and differences between the two distributions.

Activity 13-10: Calling "Heads" or "Tails"

13-10, 17-14, 17-15, 24-19

Consider the responses to the question of whether students would call "heads" or "tails" if asked to predict the result of a coin flip.

a. What proportion of students said that they would call "heads"? Is this proportion a parameter or a statistic? Indicate the appropriate symbol to represent the value.

Now suppose that 50% of the population of students would call "heads".

b. Use the SIMSAMP program (see the note in part d of the previous activity), or possibly the `Reese's Pieces` applet, to simulate **999** random samples of size n, where n is the number of students in your class who responded to the question. Produce graphs of the distribution of the 999 sample proportions, and describe the shape, center, and spread of this distribution.

c. Comment on where the sample proportion for your class falls in the distribution. Would the sample result you obtained in class be surprising if, in fact, 50% of the population of students would call "heads"? Explain.

In his book *Statistics You Can't Trust,* Steve Campbell (1998) claims that people call "heads" 70% of the time.

d. Assuming that Campbell is correct, repeat parts b and c. Is your class sample surprising if Campbell is right? Explain.

Activity 13-11: Racquet Spinning

11-9, **13-11,** 17-3, 17-18, 18-3, 18-12, 18-13

Reconsider the situation described in Activity 11-9. A tennis racquet is spun 100 times, resulting in 46 "up" and 54 "down" results. This activity asks you to assess more formally whether these sample data provide much reason to doubt that the spinning process would produce 50% for each result in the long run.

a. Is .50 (the proportion form of 50%) a parameter or a statistic? Explain.
b. Is .46 a parameter or a statistic? Explain.
c. Use the SIMSAMP program (see note in Activity 13-9, part d), or possibly the `Reese's Pieces` applet, to simulate **999** repetitions of this process (spinning a tennis racquet 100 times), assuming the results are 50/50 in the long run ($\pi = .5$). Also, use your calculator or applet to calculate the sample proportion of "up" outcomes for each of the 999 repetitions.
d. Produce a dotplot or histogram of the 999 sample proportions. Comment on the shape of this distribution, and also calculate its mean and standard deviation.
e. Compare your answers to part d with those predicted by the Central Limit Theorem.
f. In how many and what proportion of your 999 samples was the proportion of "up" results either 46% or less or 54% or greater?
g. Does your answer to part f suggest that the observed sample result (46 "up" results in 100 spins) is very unlikely to occur by chance alone if, in fact, the results are 50/50 in the long run? Explain.

Activity 13-12: Halloween Practices

A 1999 Gallup survey of a random sample of 1005 adult Americans found that 69% planned to give out Halloween treats from the door of their home.

a. Is .69 a parameter or a statistic? Explain.
b. Does this finding necessarily prove that 69% of all adult Americans planned to give out treats? Explain.
c. If the population proportion planning to give out treats was really .7, would the sample result have fallen within two standard deviations of .7 in the sampling distribution? Support your answer with the appropriate calculation. *Hint:* To fall "within two standard deviations" means that the difference between the observed statistic $\hat{p} = .69$ and the parameter value .7 is less than two times the standard deviation of $\hat{p}$.
d. Repeat part c if the population proportion were really 0.6.
e. Working only with multiples of .01 (e.g., .60, .61, .62, ...), determine and list all potential values of the population proportion π for which the observed sample proportion .69 falls within two standard deviations of π in the sampling distribution. Show calculations to support your list. *Hint:* Because the standard deviation of $\hat{p}$ is similar for all of these π values, you might want to calculate it once and then use that approximation throughout.
f. Summarize your findings in part e in terms of which numbers are plausible values for the proportion of the population who planned to give out Halloween treats from the door of their home in 1999, based on this observed sample result.

Activity 13-13: Distinguishing Between Colas

13-13, 17-24, 18-9

In an experiment to determine whether people can distinguish between two brands of cola, subjects are presented with three cups of cola. Two cups contain one brand of cola, and the third cup contains the other brand. Subjects taste from all three cups and then identify the one cola that differs from the other two. Suppose an experiment consists of 30 trials.

a. If a subject cannot distinguish between the colas and, therefore, guesses each time, what proportion does the subject get correct in the long run?
b. Describe how you could use a fair six-sided die to simulate this experiment over and over for a person who simply guesses for each of the 30 trials.
c. Use the SIMSAMP program (see note in Activity 13-9, part d), or possibly the `Reese's Pieces` applet, to simulate **999** repetitions of this exercise for a subject who is simply guessing.
d. Produce a histogram of the distribution of the 999 sample proportions of correct answers in this experiment. Comment on its shape. Is the shape consistent with the shape predicted by the Central Limit Theorem?
e. Calculate the mean and standard deviation of your 999 simulated sample proportions. Compare their values to those predicted by the CLT.
f. In how many and what proportion of your 999 simulated repetitions was the guessing subject correct 40% or more of the time?
g. If a subject were correct 40% of the time in this experiment, would you be fairly convinced that the subject did better than simply guessing would allow? Explain, relating your answer to the simulation results.
h. If a subject were correct 60% of the time in this experiment, would you be fairly convinced that the subject did better than simply guessing? Explain, relating your answer to the simulation results.
i. Copy the named list **PROP** to a new named list **PROP1.** Now use the SIMSAMP program to simulate **999** repetitions of this experiment for a subject who is able to distinguish between the colas correctly 2/3 of the time in the long run.
j. Produce dotplots (using the DOTPLOT program) of both sets (**PROP** and **PROP1**) of simulated sample proportions on the same scale. Comment on similarities and differences in the distributions. *Hint:* As you learned in Topic 7, comment on shape, center, and spread, as well as on any unusual features of interest in the distribution. Also comment on the amount of overlap between the two distributions.
k. In how many and what proportion of these 999 new repetitions is the subject correct at least 60% of the time?

Activity 13-14: Pet Ownership

13-9, **13-14,** 13-15, 18-2, 20-21

The *Statistical Abstract of the United States 2006* reports on a survey that asked a national sample of 80,000 American households about pet ownership. Suppose (for now) that one-third of all American households own a pet cat.

a. Is this number (one-third) a parameter or a statistic? Explain. What symbol is used to represent the number?
b. What does the Central Limit Theorem say about how the sample proportion of households who own a pet cat would vary under repeated random sampling?

13

Comment on shape, center, and spread of the distribution. Also draw a well-labeled sketch of this sampling distribution.

c. Still supposing that one-third of all American households own a pet cat, between what two values do you expect 95% of all sample proportions to fall?

d. Explain why the interval in part c turns out to be so narrow.

e. This survey found that 31.6% of the households sampled owned a pet cat. Is this number a parameter or a statistic? Explain. What symbol is used to represent the value?

f. Indicate this sample proportion (.316) on your sketch of the sampling distribution in part b, and calculate its *z*-score.

g. Do the sample data provide evidence that the population proportion of households who own a pet cat is not one-third? Explain.

Activity 13-15: Pet Ownership

13-9, 13-14, **13-15**, 18-2, 20-21

Reconsider the previous activity. Suppose that 5% of all American households own a pet bird.

a. What does the Central Limit Theorem say about how the sample proportion of households who own a pet bird would vary under repeated random sampling? Comment on shape, center, and spread of the distribution. Also draw a well-labeled sketch of this sampling distribution.

b. How does the standard deviation of this sampling distribution compare to the standard deviation of the sampling distribution for pet cats from the previous activity? Explain why this change happens, despite the two samples having the same sample size.

c. This survey found that 4.6% of the households sampled own a pet bird. Does this result provide evidence that the population proportion who own a pet bird is not 5%? Explain.

Activity 13-16: Volunteerism

13-16, 15-16, 21-17

The Bureau of Labor Statistics reported in December 2005 that 28.2% of a random sample of 60,000 American adults had participated in a volunteer activity between September 2004 and September 2005.

a. Is 28.2% a parameter or a statistic? Convert it to a proportion and indicate the symbol used to denote the value.

b. Suppose that 25% of the population had participated in a volunteer activity during this time period. What does the CLT say about how the sample proportion would vary, based on a sample size of 60,000 individuals? Draw a well-labeled sketch of this sampling distribution.

c. Calculate the *z*-score for the observed sample proportion of .282, based on the CLT.

d. Is this *z*-score extreme enough to cast substantial doubt on the assertion that 25% of the population participated in a volunteer activity? Explain your reasoning.

Suppose a small polling organization takes a random sample of 500 people and finds that 28.2% claim they participated in a volunteer activity in the past year.

e. Re-answer parts b–d for this scenario (still assuming that 25% of the population participated in volunteer activity).

f. Explain why it makes sense for your answers to differ as much as they do.

Activity 13-17: Pursuit of Happiness

2-16, 3-25, **13-17**, 25-1, 25-2, 25-4

Recall from Activity 2-16 that the Pew Research Center conducted a survey in February 2006 in which American adults were asked, "Generally, how would you say things are these days in your life—would you say that you are very happy, pretty happy, or not too happy?" Of the 3014 respondents, 84% answered either "very happy" or "pretty happy."

a. Suppose that 80% of the population actually felt very happy or pretty happy. Calculate the z-score for the observed sample proportion, based on the CLT, assuming this .8 value for the population proportion.
b. Is this z-score extreme enough to cast doubt on the assertion that 80% of the population felt happy? Explain.
c. Repeat parts a and b for each of the following assertions about the proportion of the population who felt very happy: 82%, 83%, 84%, 85%, 86%, 87%, and 88%. Which of these values appear to be plausible for the population proportion, based on the observed sample proportion and the CLT?

Activity 13-18: Cursive Writing

13-18, 16-8

An article in the October 11, 2006, issue of the *Washington Post* claimed that 15% of high school students used cursive writing on the essay portion of the SAT exam in the academic year 2005–2006 (Pressler). Suppose you take a random sample from these exams and see what proportion of the sample used cursive writing for the essay.

a. Report an expression for the standard deviation of the sample proportion, in terms of sample size n.
b. Use the empirical rule to determine how large the sample size would have to be for 95% of the sample proportions to fall between .10 and .20.
c. Use the empirical rule to determine how large the sample size would have to be for 95% of the sample proportions to fall between .13 and .17.

13

TOPIC

14 • • • Sampling Distributions: Means

Most of us look up to tall people, literally. But do tall people, by virtue of their height, also have an advantage in society and in business? Does their height lead to a perception of high competence or strong leadership? One recent investigation of this issue took a random sample of chief executive officers (CEOs) to decide if their average height is taller than the national average. But, as you saw in the previous topic, the sample result would probably not mirror the population exactly and, in fact, would vary from sample to sample. In this topic, you will examine how a sample mean varies, which enables you to address larger (taller?) issues such as the CEO height question.

Overview

In the previous topic, you studied how a sample proportion, which summarizes the distribution of a categorical variable, varies from sample to sample. In this topic, you will explore how the sample mean, which summarizes the center of the distribution of a quantitative variable, varies from sample to sample. The sampling distribution of the sample mean is a bit more complex than the sampling distribution of the sample proportion, because the shape of the underlying population comes into play—but a variety of similarities between the two sampling distributions emerge. As in the previous topic, you will again find that these statistics do not vary haphazardly, but according to a predictable, long-term pattern, and you will see that sample size affects the amount of variation that occurs. You will also discover more connections between a sampling distribution and the fundamental concepts of confidence and significance.

Preliminaries

1. How long do you think a typical U.S. penny has been in circulation?

2. If you look at different samples of pennies and their mean age, would you expect to see more variability across sample means of penny ages if the sample size were 5 or if the sample size were 25?

3. How much do you think a typical American spends on Christmas gifts in one year?

4. Guess the standard deviation of those Christmas gift expenditures.

In-Class Activities

Note: In Activity 14-2, part j, you will be asked to study the result from a simulation of 500 random samples of size $n = 30$ from a population representing hypothetical amounts of change (in cents) given out by a store. In order to do this, you need to start the simulation now so the results will be ready by the time you get to Activity 14-2, part j. The name of the program is CHANGE, and it will take approximately 10 minutes to run. If you haven't already, download it to your calculator and then begin the simulation. The means of the results are stored in the list PCHNG.

Activity 14-1: Coin Ages

12-16, **14-1,** 14-2, 19-15

The following histogram displays the distribution of ages (in years) for a population of 1000 pennies in circulation, collected by one of the authors in 1999. Some summary data for this distribution of ages are listed here:

Number of Pennies	Mean	Standard Deviation	Min	Lower Quartile	Median	Upper Quartile	Max
1000	12.264	9.613	0	4	11	19	59

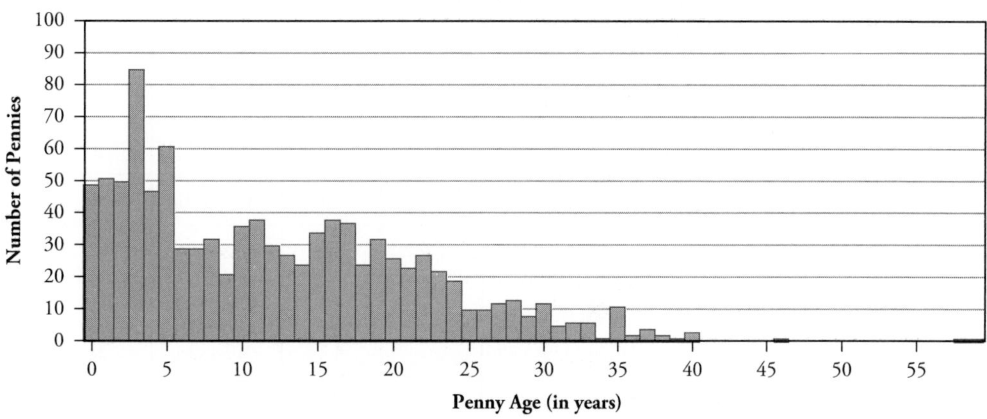

a. Identify the observational units and variable of interest here. Is this variable quantitative or categorical?

Observational units: Variable:

Quantitative or categorical?

b. If you regard these 1000 pennies as a population from which you can select samples, are the values in the previous table parameters or statistics? What symbols represent the mean and standard deviation?

c. Does this population of coin ages roughly follow a normal distribution? If not, what shape does it have?

Rather than ask you to sample actual pennies from a container with all 1000 of these pennies, you will use the Random Digits Table to draw random samples of pennies from this population. To do this, we will assign a three-digit label to each of the 1000 pennies. The following table reports the number of pennies of each age and their three-digit numbers.

Age (in yrs)	Count	ID #s	Age (in yrs)	Count	ID #s	Age (in yrs)	Count	ID #s
0	49	001–049	15	34	610–643	30	12	945–956
1	51	050–100	16	38	644–681	31	5	957–961
2	50	101–150	17	37	682–718	32	6	962–967
3	85	151–235	18	24	719–742	33	6	968–973
4	47	236–282	19	32	743–774	34	1	974
5	61	283–343	20	26	775–800	35	11	975–985
6	29	344–372	21	23	801–823	36	2	986–987
7	29	373–401	22	27	824–850	37	4	988–991
8	32	402–433	23	22	851–872	38	2	992–993
9	21	434–454	24	19	873–891	39	1	994
10	36	455–490	25	10	892–901	40	3	995–997
11	38	491–528	26	10	902–911	46	1	998
12	30	529–558	27	12	912–923	58	1	999
13	27	559–585	28	13	924–936	59	1	000
14	24	586–609	29	8	937–944	Total	1000	

Notice that each age has a number of ID labels assigned to it equal to the number of pennies of that age in the population. Thus, for example, the age of 10 years has 36 ID labels because 36 of the 1000 pennies are 10 years old, whereas an age of 30 years has one-third as many ID labels because only 12 of the 1000 pennies are 30 years old.

d. Use the Random Digits Table to draw a random sample of five penny ages from this population. (Do this by selecting five three-digit numbers from the table and then finding the age assigned to each number in the table shown here. If you happen to get the same three-digit number twice, ignore the repeat and choose another number.) Record the penny ages, and draw a dotplot of your sample distribution on the axis shown here:

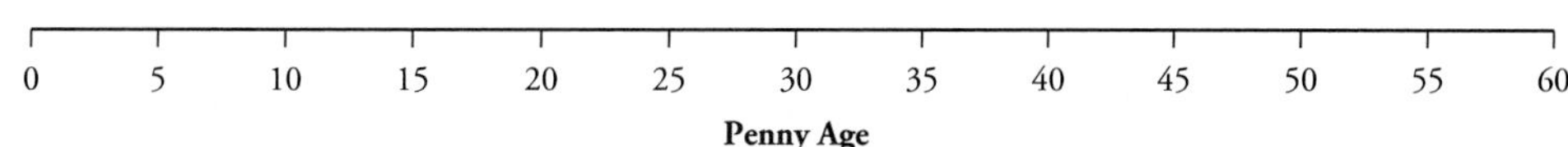

e. Calculate the sample mean of your five penny ages.

f. Take four more random samples of five pennies each. Calculate the sample mean age for each sample, and record the results in the table:

	1	2	3	4	5
Sample Mean ($\bar{x}$)					

g. Did you get the same value for the sample mean all five times? What phenomenon that you studied earlier does this result again reveal? How is this variable fundamentally different from the variable in the Candy Colors activity in the previous topic?

You are again encountering the notion of **sampling variability.** Because age is a quantitative and not a categorical variable, you are observing sampling variability as it pertains to sample *means* instead of sample *proportions*. As was the case with the ample proportion, the sample mean varies from sample to sample not in a haphazard manner, but according to a predictable long-term pattern known as a **sampling distribution.**

h. Use your calculator to compute the mean and standard deviation of your five sample means.

Mean of $\bar{x}$ values: SD of $\bar{x}$ values:

i. Is this mean (of the $\bar{x}$ values) reasonably close to the population mean (μ = 12.264 years)? Is the standard deviation greater than, less than, or about equal to the population standard deviation (σ = 9.613 years)?

As was the case with the sample proportion, the sample mean is an *unbiased* estimator of the population mean. In other words, the center of the sampling distribution of sample means is the population mean. Also evident is that variability in the sampling distribution of the sample mean is smaller than variability in the original population (as long as the sample size is greater than one).

Now consider taking a random sample of 25 pennies from this same population. By taking five samples of five pennies each, you have essentially done this already. Consider all of your observations as a random sample of size 25. (Notice that you are ignoring the small possibility that a coin could be repeated in your sample of 25.) The sample mean of this sample of 25 pennies is exactly the mean of your five sample means recorded in part h.

j. Pool your sample mean from the sample of size 25 with those of your classmates. Produce a dotplot of these sample means. Label the axis clearly. What are the observational units in this plot?

k. Does this distribution appear to be centered at the population mean (μ = 12.264)?

l. Do the values appear to be more or less spread out than either the population distribution or the distribution of your five sample means of size 5?

m. Does this distribution appear to be closer to a normal (bell) shape than the distribution of ages in the original population? (Recall the histogram of the population distribution shown at the start of this activity.)

Watch Out

As mentioned in the previous topic, the idea of a sampling distribution is one of the most complex that you will encounter. Notice again that different levels are involved. The original observational units were pennies, with the variable being age. But once you took several samples, the observational units became the *samples,* with the variable being the *average* (mean) penny age of each sample.

This topic can be especially confusing because we are using the word *mean* in three different ways:

- The *population* mean, denoted by μ, is equal to 12.264 years for this population of 1000 pennies. Notice that this number (a parameter) is fixed; it does not vary from sample to sample.
- The *sample* mean, denoted by $\bar{x}$, varies from sample to sample. But there is a predictable, long-term pattern to this variation.
- The *mean* of the sample means is also known as the *mean of the sampling distribution.* You will learn more about this in the next activity.

Activity 14-2: Coin Ages

12-16, 14-1, **14-2,** 19-15

To study the sampling distribution of sample means more thoroughly requires simulating many more samples, so you now turn to technology to select many random samples.

Consider again the population of 1000 penny ages described in Activity 14-1. In particular, recall that the distribution is skewed to the right, with a mean of μ = 12.264 years and a standard deviation of σ = 9.613 years.

a. Use the `Sampling Pennies` applet to take a random sample of size n = **1** penny each, by clicking the Draw Samples button. Keep clicking Draw Samples until you have taken five samples. Then turn off the Animate feature and ask for **495** more samples (for a total of **500,** still with a sample size of **1** penny per sample). Notice that the applet creates a dotplot of the sample means, but with a sample size of 1 the sample mean is equal to the age of the penny itself.

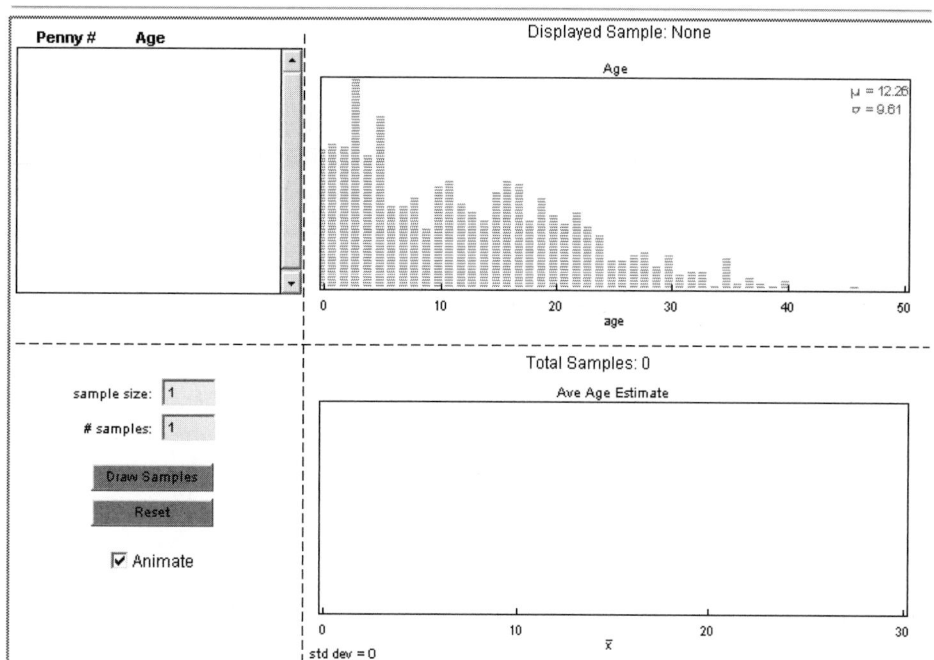

Reproduce a rough sketch of the dotplot you would generate. Does the dotplot resemble the distribution of ages in the population?

b. Report the mean and standard deviation of these 500 mean ages. Record them in the first empty row of the table shown following part h. Are the mean and standard deviation reasonably close to their population counterparts ($\mu = 12.264$ and $\sigma = 9.613$)? Describe the shape of this distribution.

c. Now use the `Sampling Pennies` applet to take **500** random samples of $n =$ **5** pennies each and to calculate the sample mean age for each of the 500 samples. Examine the dotplot of these 500 sample means and reproduce a rough sketch of the dotplot here. Comment on how this distribution differs from the population and from the distribution in part a.

d. Report the mean and standard deviation of these 500 sample means. Record them and the general shape of the distribution in the second row of the table following part h. Are the mean and standard deviation reasonably close to their population counterparts? If not, how do they differ?

e. Now use the applet to take **500** random samples of $n =$ **25** pennies each and to calculate the sample mean age for each of the 500 samples. Examine the dotplot of these 500 sample means, and sketch its shape here. Comment on how this distribution differs from the population and from the distributions in parts a and c.

f. Report the mean and standard deviation of these 500 sample means. Record them and the general shape of the distribution in the third row of the table following part h. Are these values reasonably close to their population counterparts? If not, how do they differ?

g. Now use the applet to take **500** random samples of n = **50** pennies each and to calculate the sample mean age for each of the 500 samples. Examine the dotplot of these 500 sample means, and sketch its shape here. Comment on how this distribution differs from the population and from the distributions in parts a, c, and e.

h. Report the mean and standard deviation of these 500 sample means. Record them and the general shape of the distribution in the following table. Are these values reasonably close to their population counterparts? If not, how do they differ?

	Population Mean: μ = 12.264 years	Population SD: σ = 9.613 years	Population Shape: Skewed to the Right
Sample Size	Mean of Sample Means	SD of Sample Means	Shape of Sample Means
1			
5			
25			
50			

Each of these distributions approximates the sampling distribution of the sample mean from this population for that sample size. These simulations reveal that when the sample size is fairly large, the distribution of sample means follows approximately a normal distribution. The distribution of a sample of individual penny ages (a sample size of 1) resembles that of the population, but the sampling distribution of the sample mean becomes more and more normal-shaped as the sample size increases. The theoretical means of these distributions are all equal to the population mean (and the means of your simulated sample means should be very close to the population mean), but the variability in sample means decreases as the sample size increases.

Consider a different population, as shown in the following histogram, representing hypothetical amounts of change (in cents) given out by a store:

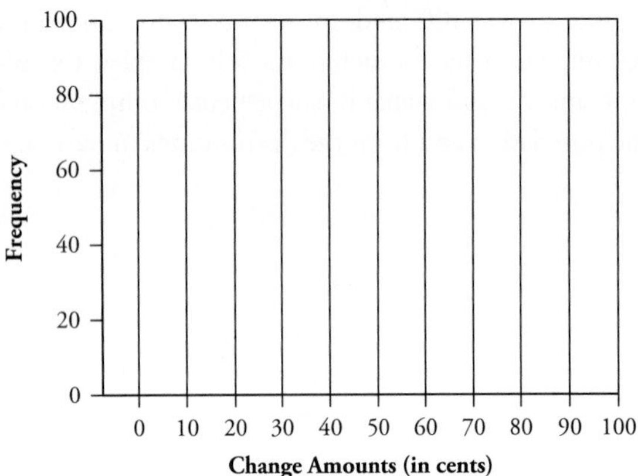

i. Does this population follow a normal distribution? How would you describe its shape?

j. Examine a dotplot or histogram of the results (PCHNG) from the simulation started at the beginning of this topic of 500 random samples of size $n = 30$ from this population. Comment on the shape of the distribution. Also report the mean and standard deviation of these sample means.

k. Use your calculator to produce a normal probability plot (as you studied in Activity 12-3) of the sample means. Does the distribution appear to be approximately normal?

Even with these two nonnormal populations (penny ages and change amounts), the distribution of sample means from these populations follows roughly a normal distribution when you use a relatively large sample size. A theoretical result affirms your simulation findings.

Central Limit Theorem (CLT) for a Sample Mean

Suppose a simple random sample of size n is taken from a large population in which the variable of interest has mean μ and standard deviation σ. Then, provided that n is large (at least 30 as a general guideline), the sampling distribution of the sample mean $\bar{x}$ is approximately *normal* with mean equal to μ and standard deviation equal to $\sigma/\sqrt{n}$. This normal approximation holds with large sample sizes *regardless* of the shape of the population distribution. The accuracy of the approximation increases as the sample size increases. For populations that are themselves normally distributed, the sampling distribution of the sample mean is normal for all sample sizes.

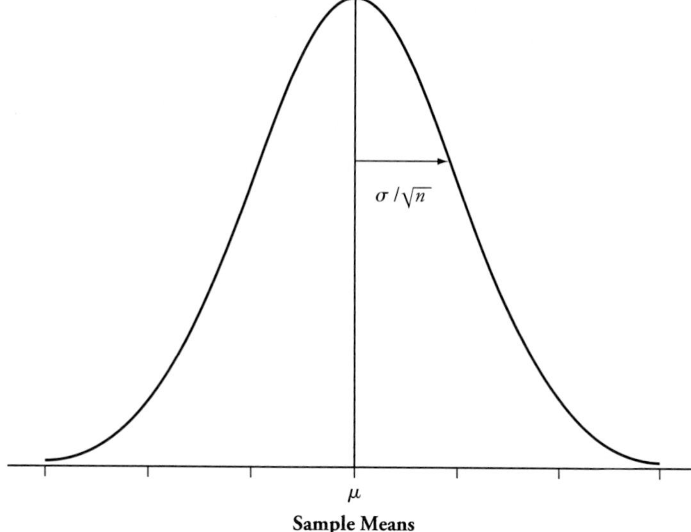

l. For a sample size of $n = 50$ from the penny age population, calculate $\sigma/\sqrt{n}$. Then confirm that its value is reasonably close to the standard deviation of your 500 simulated sample means reported in part h.

m. Repeat part l for a sample size of $n = 30$ from the (uniform) change amount population, for which the population standard deviation is $\sigma = 28.866$. Then confirm that its value is reasonably close to the standard deviation of your 500 simulated sample means reported in part j.

Watch Out

As with the CLT for a population proportion, the CLT for a sample mean specifies three things: the shape, center, and spread of the sampling distribution of a sample mean. Also notice several similarities between the two versions of the CLT:

- The shape of the sampling distribution is (approximately) normal.
- The mean of the sampling distribution is equal to the population parameter of interest.
- The standard deviation of the sampling distribution decreases by a factor of $1/\sqrt{n}$ as the sample size n increases.

Always make sure the CLT conditions are met before you try to apply it:

- The sample must be selected randomly.
- The sample size must be large (at least 30), or else the population itself must follow a normal distribution.

Do not confuse the two versions of the Central Limit Theorem; the key difference between them is the type of variable involved (as you first experienced in Topic 1):

- The CLT for a sample proportion (Topic 13) applies when the underlying variable is categorical.
- The CLT for a sample mean (this topic) applies when the underlying variable is quantitative.

Activity 14-3: Christmas Shopping

14-3, 14-7, 15-11, 19-1

In a survey conducted by the Gallup organization during the period November 18–21, 1999, they interviewed 922 American adults who expected to buy Christmas gifts that year. These persons were asked how much money they personally expected to spend on Christmas gifts, and the sample mean of the responses was $857.

a. Is $857 a parameter or a statistic? Explain, and indicate what symbol represents this value.

b. For this study, identify the

i. Population of interest:

ii. Sample selected:

iii. Parameter of interest:

iv. Statistic calculated:

Let μ denote the mean dollar amount the population of American adults expected to spend on Christmas gifts, and let σ represent the standard deviation of those amounts.

c. Does μ necessarily equal $857? Is it *possible* that the sample mean could have turned out to be $857 even if μ were equal to $850? What if μ were equal to $800? What about $1000? Explain.

Assume for now that μ is $850 and that σ is $250.

d. What does the Central Limit Theorem say about how the sample mean $\bar{x}$ would vary if samples of size $n = 922$ were taken over and over again? Does your answer depend on the shape of the distribution of expected expenditures in the population? Explain.

e. Draw a sketch to represent this distribution. (Please be sure to label and scale the horizontal axis.)

f. Would a sample mean of \$857 be a surprising result if μ were \$850 and σ were \$250? Explain, basing your answer on parts d and e.

g. Repeat parts d–f using \$800 as the population mean of interest.

These questions are a reminder of the concept of **statistical significance,** which asks whether a sample result is unlikely to have occurred by sampling variability or chance alone.

Continue to assume the population standard deviation is $\sigma = \$250$, but assume you do not know the value of the population mean μ.

h. What does the CLT say about the standard deviation of the sampling distribution of the sample mean?

i. Begin to apply the empirical rule to this sampling distribution by doubling the standard deviation calculated in part h. Then subtract that amount from the sample mean to get a lower bound, and add that amount to the sample mean to get an upper bound. Report the *interval* of values that you have constructed.

This question should bring to mind the concept of **statistical confidence,** which creates an interval of parameter values that could have plausibly generated the sample statistic actually obtained. Roughly speaking, a parameter value is considered plausible if it is within two standard deviations of the sample statistic.

Activity 14-4: Looking Up to CEOs

14-4, 20-18

Recall from Activity 1-4 that a study mentioned in the book *Blink* concerned a psychologist who conjectured that taller people tend to be more respected (Gladwell, 2005). This psychologist hypothesized that chief executive officers (CEOs) of American companies tend to be taller than the average height of 69 inches for an adult American male. She investigated this hypothesis by recording the heights for a random sample of 100 male American CEOs.

a. Is 69 a parameter or a statistic? Explain.

b. What term applies to the number 100, and what symbol is used to represent this number?

c. Suppose that the population mean height of male CEOs is the same 69 inches as for adult American males in general. What additional information do you need to know to specify the sampling distribution of the sample mean CEO height in this study?

Suppose that the standard deviation of the heights of adult American males is 3 inches.

d. Assuming that heights of male CEOs have the same distribution as the heights of American males in general, describe the sampling distribution of the sample mean height in this study. Also draw and label a sketch of this sampling distribution.

e. Suppose that the psychologist will only be persuaded that the average height of CEOs is taller than 69 inches if the sample mean CEO height exceeds 69 inches by at least two standard deviations (we are now talking about the standard deviation of the sample *mean*). Determine how large the sample mean would have to be.

f. Repeat parts d and e if the sample size were 30 instead of 100, and comment on how this changes your answers.

Solution

a. The mean height among all adult American males, 69 inches, is a parameter because it describes the entire population. It is denoted by the symbol μ.

b. The sample size is 100 and is denoted by the symbol n.

c. You need to know the population standard deviation of the heights of adult American males, denoted by σ. Because the sample size is fairly large (100 is larger than 30), you do not need to know whether the population distribution of heights is normal because the Central Limit Theorem tells you the shape of the sampling distribution of the sample mean will be approximately normal with this sample size, regardless of the shape of the original population distribution.

d. The CLT establishes that the sampling distribution of the sample mean height is approximately normal with mean 69 inches and standard deviation $3/\sqrt{100} = 0.3$ inches.

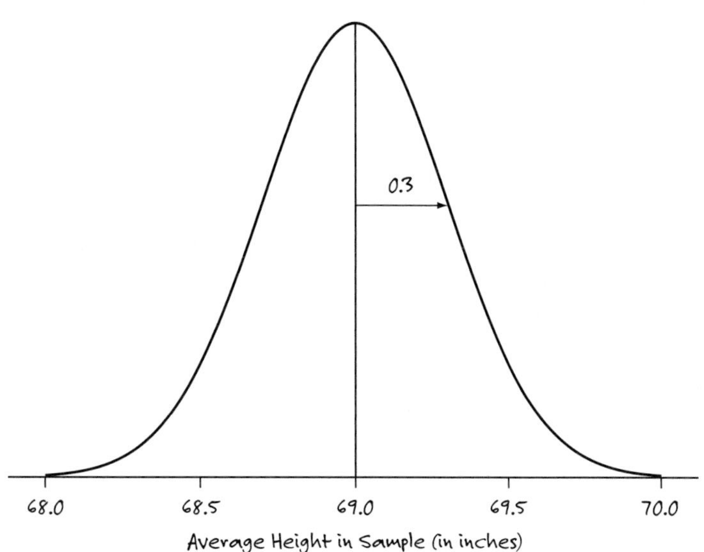

e. Doubling the standard deviation of the sample mean gives you 0.6 inches, so the sample mean height among the CEOs would have to be at least 69.6 inches to persuade the psychologist that, on average, CEOs are indeed taller than the average adult male.

f. If the sample size were 30, the normal approximation should still be valid, and the standard deviation of the sampling distribution would increase to $3/\sqrt{30} =$ 0.548 inches. Doubling this standard deviation gives you 1.096 inches, so the sample mean height would have to be at least 70.096 inches to be persuasive. The smaller sample size produces more sampling variability, hence a larger standard deviation. As a result, the cut-off value needed for a persuasive sample mean height is larger.

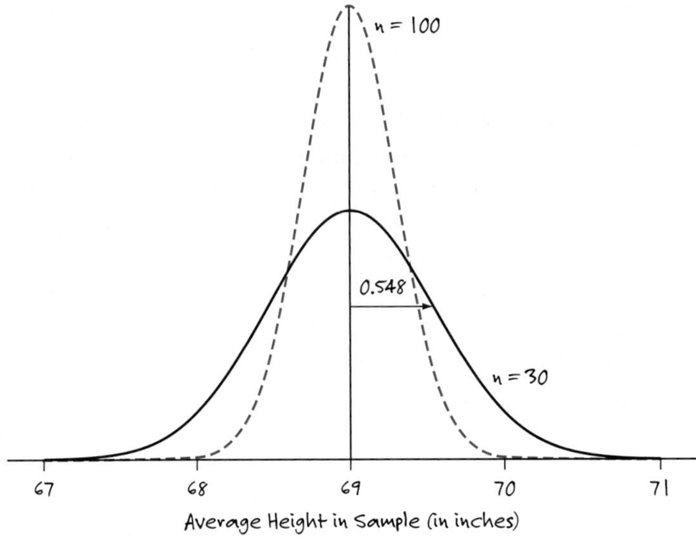

Watch Out

- A common error when working with the Central Limit Theorem is to forget to incorporate the sample size *n* into the standard deviation of the sample mean. It's tempting (but wrong) to say that the standard deviation is 3 inches, period. That value is the standard deviation of individual heights in the population. As you saw with the pennies, the sample mean height varies much less than individual heights.
- Don't worry about the shape of the population distribution when the sample size is large. As you saw with the penny ages (quite skewed to the right) and the change amounts (symmetric but uniform rather than bell-shaped), the sample mean still varies normally when the sample size is fairly large. So, the previous calculations hold even if the distribution of heights in the population of adult American males is not normal.

Wrap-Up

This topic continued your study of the fundamental concept of a sampling distribution. You discovered that just as a sample proportion varies from sample to sample according to a normal distribution, so too (under the right conditions) does a sample mean. Moreover, you learned that for large sample sizes this result is true regardless of the shape of the population from which the samples are drawn. For example, the distribution of the average (sample mean) age on large samples of pennies turns out to be close to normal even though the population of penny ages is quite skewed. However, if the sample size is small, then you need evidence that the population distribution is normal before you can apply the Central Limit Theorem.

You also saw that the sampling distribution of the sample mean is centered at the population mean and that sampling variability decreases as sample size increases. You again learned that the ideas of confidence and significance are closely related to a sampling distribution. For example, knowing how the sample mean varies enables you to determine how close a sample result regarding the average amount spent on Christmas presents is likely to come to the population average.

Some useful definitions to remember and habits to develop from this topic include

- The sampling distributions that you have studied in this topic pertain to *quantitative* variables.

- The **sampling distribution** of a sample mean is the long-term pattern of variation in sample means that results from taking many random samples from a population.
- Knowing how sample means vary under repeated sampling enables you to draw inferences about the population mean from a sample mean.
- When the population being sampled has a normal distribution, then the sample means vary *normally* (i.e., with the shape of a bell curve) regardless of sample size.
- Even when the population being sampled does not have a normal distribution, the sample means vary approximately normally whenever the sample size is large. The convention is to regard samples of 30 or more observational units as large enough for this approximation to be valid.
- The sampling distribution of a sample mean $\bar{x}$ is centered around the population mean μ.
- As with a sample proportion, larger samples produce less variability in sample means than smaller samples do. Specifically, the standard deviation of a sample mean is $\sigma/\sqrt{n}$.
- Keep in mind three different distributions:
    - The *population* distribution. This distribution is almost never observed, and the goal in a statistical study is usually to learn about this population distribution from the sample distribution.
    - The *sample* distribution. This distribution consists of the sample data that you actually observe and analyze. With random sampling, the sample distribution should roughly resemble the population distribution.
    - The *sampling* distribution. This distribution describes how a sample statistic (such as a proportion $\hat{p}$ or a mean $\bar{x}$) varies if random samples are repeatedly taken from the population. You used simulation to study sampling distributions and now know the theoretical result—the **Central Limit Theorem**—that describes them. These sampling distributions often behave very differently than the population or sample distribution.
- Make sure the technical conditions are satisfied before you apply the Central Limit Theorem. For a sample mean, the population must follow a normal distribution or the sample size must be large (at least 30 observations). Furthermore, all the statements regarding the sampling distribution assume you are working with simple *random* samples.

The next topic asks you to work with the Central Limit Theorem more formally. You will use what you learned about normal distribution calculations in Topic 12 to perform probability calculations based on these results.

Homework Activities

Activity 14-5: Heart Rates

The following histogram displays the distribution of heart rates for 1934 subjects evaluated in the 2003–2004 NHANES study. The data are stored in the file HEARTRATE, which includes the lists HRTRT and FREQH, as well as the program HEART. Download this grouped file to your calculator.

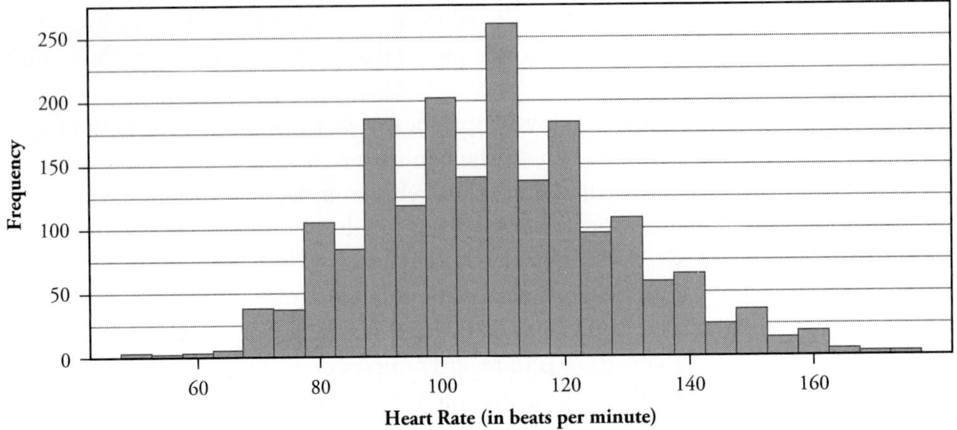

a. Does this distribution appear to be approximately normal?

b. Produce a normal probability plot of these heart rates. Comment on what this plot reveals about the normality of the distribution. You will need to use the list HRTRT as the Xlist and FREQH as the frequency list.

Regard these 1934 measurements as a population from which you can select samples.

c. Use the HEART program to take **999** random samples of size n = **3** from this population. Calculate the sample mean heart rate for each sample. Examine graphical displays and numerical summaries of the distribution of the 999 sample means. Describe this distribution.

d. Does the distribution of sample mean heart rates appear to be normal, despite the small sample size? Explain why this makes sense.

e. Repeat parts c and d for **999** random samples of size n = **10.** Comment on how the results differ with this larger sample size.

Activity 14-6: Sampling Words

4-1, 4-2, 4-3, 4-4, 4-7, 4-8, 8-9, 9-15, **14-6**

As you first did in Activity 4-1, consider the 268 words in the Gettysburg Address to be a population from which to take random samples. Open the applet `Sampling Words` for this purpose.

a. Is the population of word lengths approximately normal? Explain.

b. Take **500** random samples of size n = **1** from this population, and examine the distribution of sample means. Is it approximately normal? Is it closer to normal than the original population? Does it vary more or less than the original population?

c. Repeat part b for **500** random samples of size n = **5.**

d. Repeat part b for **500** random samples of size n = **10.**

Activity 14-7: Christmas Shopping

14-3, **14-7,** 15-11, 19-1

Reconsider Activity 14-3, in which you examined the dollar amount that individuals expected to spend on Christmas presents in 1999.

a. Suppose the population mean amount were really μ = \$850. Would a sample mean as large or larger than the mean actually obtained (\$857) be more likely if the population standard deviation were σ = \$250 or if it were σ = \$1250? Explain.

b. Suppose the population mean were really $\mu = \$800$. Would a sample mean as large or larger than the mean actually obtained (\$857) be more likely if the population standard deviation were $\sigma = \$250$ or if it were $\sigma = \$1250$? Explain.

c. Answer parts d–i of Activity 14-3 under the assumption that $\sigma = \$1250$ rather than $\sigma = \$250$.

d. Comment on how your answers differ when $\sigma = \$1250$ as opposed to $\sigma = \$250$.

Activity 14-8: Random Babies

11-1, 11-2, **14-8,** 16-17, 20-16, 20-22

In Activity 11-2, the (exact) probability distribution of the number of matches in the "random babies" activity is given:

Number of Matches	0	1	2	3	4
Probability	3/8	1/3	1/4	0	1/24

Recall also that the mean of this population is $\mu = 1$ match, and the standard deviation is $\sigma = 1$ match.

a. Is the population distribution normal? Is it close to normal? Explain.

b. Does the Central Limit Theorem say that if you repeatedly take samples of size 5 from this distribution, the resulting sample means will closely follow a normal distribution?

c. Does the Central Limit Theorem say that if you repeatedly take samples of size 100 from this distribution, the resulting sample means will closely follow a normal distribution?

d. What does the CLT say about the mean and standard deviation of the sampling distribution in part c?

e. Draw a sketch of the (approximate) sampling distribution of the sample mean number of matches when the sample size is 100, labeling the horizontal axis clearly.

f. If you repeatedly take samples of size 100 from this distribution and calculate the sample mean number of matches each time, approximately what percentage of these sample means would fall between 0.8 and 1.2? Explain.

g. Continuing the scenario presented in part f, if for each of the samples you construct an interval by subtracting 0.2 from the sample mean to obtain a lower bound and adding 0.2 to the sample mean to obtain an upper bound, approximately what percentage of those intervals would capture the population mean of $\mu = 1$? Explain.

Activity 14-9: Birth Weights

12-2, **14-9,** 15-15

Recall Activity 12-2, where you assumed that birth weights of babies could be modeled as a normal distribution with mean $\mu = 3300$ grams and standard deviation $\sigma = 570$ grams. One of the following histograms displays the sample mean birth weights in 1000 samples of $n = 5$ babies, and the other displays the sample mean birth weights in 1000 samples of $n = 10$ babies:

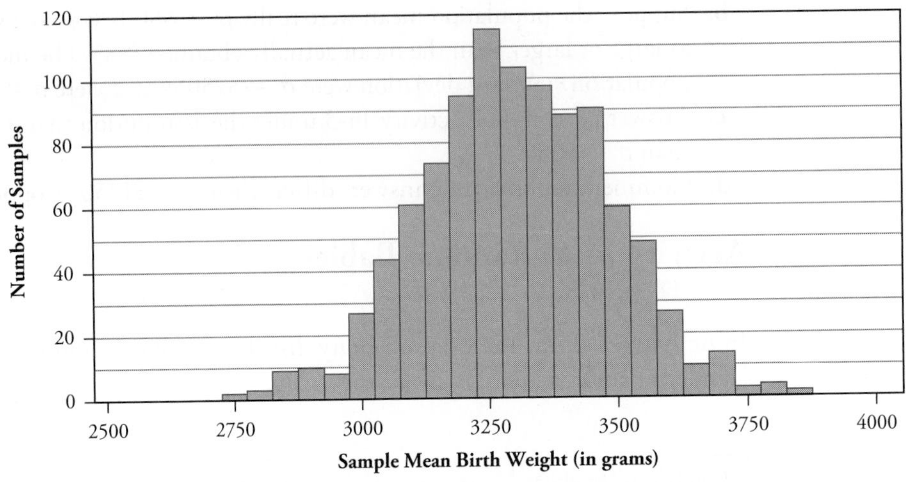

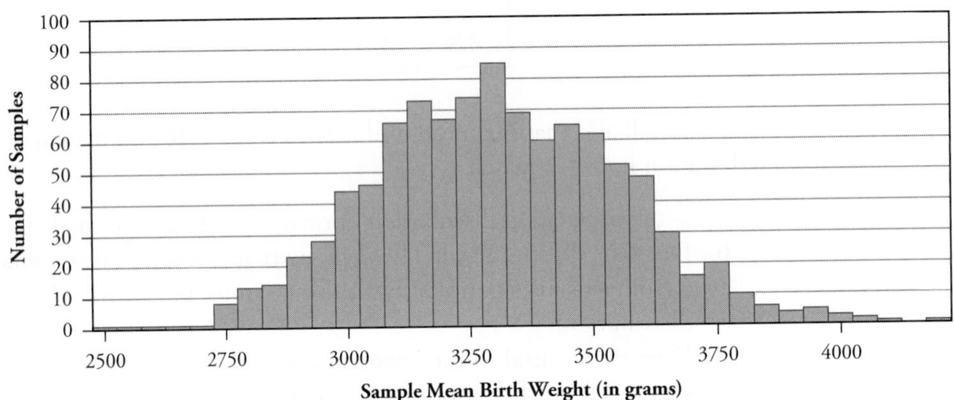

a. Which histogram goes with which sample size? Explain.

b. Judging from these histograms, which sample size is more likely to produce a sample mean birth weight less than 2500 grams?

c. Judging from these histograms, which sample size is more likely to produce a sample mean birth weight less than 3000 grams?

d. Judging from these histograms, which sample size is more likely to produce a sample mean birth weight greater than 3500 grams?

e. Judging from these histograms, which sample size is more likely to produce a sample mean birth weight between 3000 grams and 3500 grams?

f. Write a paragraph summarizing what this activity reveals about the effect of sample size on the sampling distribution of a sample mean.

Activity 14-10: Candy Bar Weights

12-10, **14-10,** 15-10

Recall Activity 12-10, which asked you to assume that the actual weight of a certain candy bar, whose advertised weight is 2.13 ounces, varies according to a normal distribution with mean $\mu = 2.2$ ounces and standard deviation $\sigma = 0.04$ ounces.

a. What does the CLT say about the distribution of sample mean weights if samples of size $n = 5$ are taken repeatedly?

b. Draw a sketch of this sampling distribution, labeling the horizontal axis clearly.

Suppose you are skeptical about the manufacturer's claim that the mean weight is $\mu = 2.2$ ounces, so you take a random sample of $n = 5$ candy bars and weigh them. You find a sample mean weight of 2.15 ounces.

c. Is it possible to get a sample mean weight this low even if the manufacturer's claim that $\mu = 2.2$ ounces is valid? Explain, referring to the graph you sketched in part b.
d. Is it very *unlikely* to get a sample mean weight this low even if the manufacturer's claim that $\mu = 2.2$ ounces is valid? Explain.
e. Would finding the sample mean weight to be 2.15 ounces provide strong evidence to doubt the manufacturer's claim that $\mu = 2.2$ ounces? Explain, referring to the sampling distribution.
f. Would finding the sample mean weight to be 2.18 ounces provide strong evidence to doubt the manufacturer's claim that $\mu = 2.2$ ounces? Explain, again referring to the sampling distribution.
g. What values for the sample mean weight would provide fairly strong evidence against the manufacturer's claim that $\mu = 2.2$ ounces? Explain, once again referring to the sampling distribution. *Hint:* Reconsider the empirical rule.

14

Activity 14-11: Cars' Fuel Efficiency

The highway miles per gallon rating of the 1999 Volkswagen Passat was 31 mpg (*Consumer Reports,* 1999). The fuel efficiency that a driver obtains on an individual tank of gasoline naturally varies from tankful to tankful. Suppose the mpg calculations per tank of gas vary normally with a mean of $\mu = 31$ mpg and a standard deviation of $\sigma = 3$ mpg.

a. Would it be surprising to obtain 30.4 mpg on one tank of gas? Explain.
b. Would it be surprising for a sample of 30 tanks of gas to produce a sample mean of 30.4 mpg or less? Explain, referring to the CLT and to a sketch that you draw of the sampling distribution.
c. Would it be surprising for a sample of 60 tanks of gas to produce a sample mean of 30.4 mpg or less? Explain, again referring to the CLT and to your sketch of the sampling distribution.
d. Would it be surprising for a sample of 150 tanks of gas to produce a sample mean of 30.4 mpg or less? Explain, again referring to the CLT and to your sketch of the sampling distribution.
e. Do any of your responses depend on knowing the shape of the population distribution? Explain.

Activity 14-12: Got a Tip?

1-10, 5-28, **14-12,** 15-17, 22-4, 22-9, 25-3

Suppose a waitress keeps track of her tips, as a percentage of the bill, for a random sample of 50 tables.

a. Identify the observational units in this study.
b. Identify the variable in this study. Classify its type as categorical or quantitative.

Suppose that, unknown to the waitress, her population mean tip percentage is 15%, with a standard deviation of 4%.

c. Describe and sketch the sampling distribution of her sample mean tip percentage.
d. Between what two values is there a 68% chance that her sample mean tip percentage will fall?

Activity 14-13: IQ Scores

12-9, **14-13**, 15-20

Suppose that the IQ scores of your hometown's residents follow a normal distribution with mean 105 and standard deviation 12. Which is more likely: that a randomly selected resident will have an IQ greater than 120, or that the average IQ in a random sample of 10 residents will be greater than 120? Explain briefly, without performing any calculations and using language that a person who has not studied statistics can understand.

TOPIC

15

Central Limit Theorem

On October 5, 2005, a tour boat named the *Ethan Allen* capsized on Lake George in New York with 47 passengers aboard. In the inquiries that followed, it was suggested that the tour operators should have realized that so many passengers were likely to exceed the weight capacity of the boat. In this topic, you will see how to put together what you have learned in this unit to calculate this probability and other similar questions.

Overview

In previous topics, you used hands-on and technology simulations to discover that although the value of a sample proportion or a sample mean varies from sample to sample, that variation has a predictable long-term pattern. The Central Limit Theorems (CLT), introduced in Topics 13 and 14, specify that for large sample sizes this pattern follows a normal distribution. In this topic, you will apply what you learned in Topic 12 about calculating normal probabilities to examine the implications and applications of this theorem in detail. You will continue to focus on how the CLT lays the foundation for widely used techniques of statistical inference.

Preliminaries

1. Guess the percentage of adult Americans who smoke regularly.

2. Guess which of the 50 states has the highest percentage of smokers and which has the lowest percentage of smokers.

3. If 45% of all Reese's Pieces candies are orange, are you more likely to find less than 40% orange candies in a sample of 75 candies or in a sample of 175 candies?

4. If 45% of all Reese's Pieces candies are orange, are you more likely to find between 35% and 55% orange candies in a sample of 75 candies or in a sample of 175 candies?

In-Class Activities

In the previous two topics, you discovered that sample statistics (such as proportions and means) vary from sample to sample according to a predictable long-term pattern. In many situations, this pattern follows a normal distribution centered at the population parameter. Thus, you can use your knowledge of normal distributions to calculate the probability that a sample statistic will fall within an interval of values of interest.

The Central Limit Theorems for a sample proportion and for a sample mean state conditions under which a sample statistic follows a normal distribution very closely. As we noted before, these two results are very similar in specifying that

- The shape of the sampling distribution is normal or approximately normal.
- The center (mean) of the sampling distribution equals the value of the population parameter.
- The variability of the sampling distribution decreases as the sample size n increases, by a factor of $1/\sqrt{n}$.

In fact, the proportion of "successes" from a binary variable can be thought of as the mean of that variable if the "successes" are coded as "1" and the "failures" are coded as "0." For example, if females are coded as 1 ("success") and males are coded as 0 ("failure"), then the proportion of females in a dataset is equal to the average of the zeros and ones. Thus, the two Central Limit Theorems for categorical and quantitative variables can be considered the same. Accordingly, we will refer to either version as the Central Limit Theorem, or CLT for short.

The following table summarizes the results of the CLT:

		Categorical Variable	Quantitative Variable
Sampling Distribution	Sample Statistic	$\hat{p}$ (sample proportion)	$\bar{x}$ (sample mean)
	Center (Mean)	π (population proportion)	μ (population mean)
	Standard Deviation	$\sqrt{\pi(1-\pi)/n}$	$\sigma/\sqrt{n}$
	Shape for Large Sample Size (n)	Approximately normal	Normal or approximately normal
	When Is n Large?	When $n\pi \geq 10$ and $n(1-\pi) \geq 10$	When population is normal or $n \geq 30$

In the following activities, you will apply the Central Limit Theorem to various contexts. As this table indicates, the first question you will need to ask is whether the underlying variable of interest is quantitative or categorical; this answer will tell you which version of the result to use.

Activity 15-1: Smoking Rates

15-1, 15-2, 15-9

The Centers for Disease Control and Prevention reported that 20.9% of American adults smoked regularly in 2004. Treat this as the parameter value for the current population of American adults.

a. What symbol represents the proportion .209?

b. Suppose you plan to take a random sample of 100 American adults. Will the sample proportion who smoke equal exactly .209? (Is this result even possible with a sample of exactly 100 people?) Explain.

c. What does the CLT say about how this sample proportion who smoke would vary from sample to sample? (Comment on shape, center, and spread of this sampling distribution.)

d. On the following graph of this sampling distribution, shade the area corresponding to a sample proportion exceeding .25. Based on this shaded area, guess the probability of obtaining a sample proportion who smoke that exceeds .25.

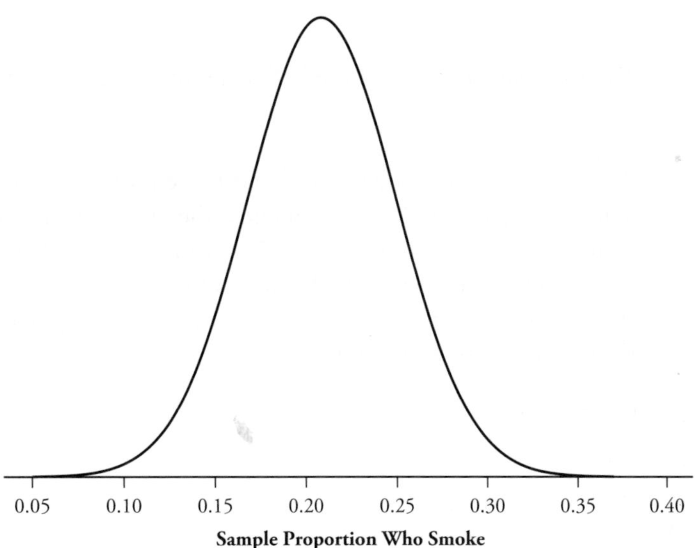

Prediction:

e. Using the CLT result, find the *z*-score corresponding to a sample proportion who smoke equal to .25.

f. Use this z-score and Table II or your calculator to compute the probability that the sample proportion who smoke from this population exceeds .25. Is this value reasonably close to your prediction in part d?

g. Suppose you instead plan to take a random sample of 400 American adults. How do you expect the probability of the sample proportion who smoke exceeding .25 to differ from when the sample size was 100? Explain your reasoning, including a well-labeled sketch of the sampling distribution when $n = 400$.

h. Calculate the probability asked for in part g. Was your conjecture correct? Explain.

i. Did the *population* size of the U.S. enter into your calculations?

j. The Centers for Disease Control and Prevention also reported that 20.9% of adults in the state of Virginia smoked, the same percentage as for the entire country. Suppose you select a random sample of only Virginia residents, using the same sample sizes as in parts b and g. How do the previous calculations change if you want to find the probability of the sample proportion of Virginia smokers exceeding .25 for each sample size?

Watch Out

- The variable here is whether a person smokes regularly, so the relevant parameter and statistic are a proportion, not a mean. But it's quite common to hear and read about *percentages,* such as the 20.9% smoking rate reported here. Be sure to convert this percentage (20.9%) to a proportion (.209) before applying the CLT and especially in calculating the standard deviation.
- As long as the population is large relative to the size of the *sample,* the actual size of the *population* does not affect CLT calculations. This is hard for many people to believe. But think about sampling Reese's Pieces: the population size of all such candies in the world is enormous, yet that population had no impact on

- the pattern of variation in your samples. Also recall that you encountered this phenomenon when sampling words in the Gettysburg Address in Activity 4-4.
- The *sample size,* however, does have a large impact on CLT calculations.

Activity 15-2: Smoking Rates

15-1, **15-2,** 15-9

The Centers for Disease Control and Prevention further reported that 10.5% of Utah residents smoked regularly in 2004. Treat .105 as the current parameter value for Utah.

a. Suppose you plan to take a random sample of 100 Utah residents and find the proportion who smoke in the sample. Sketch (and label) the sampling distribution of the sample proportion in this case.

b. Use the CLT to calculate the probability that a sample proportion from this population exceeds .25.

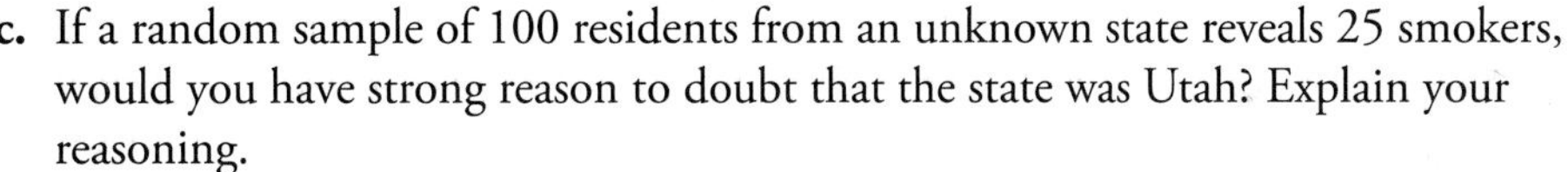

c. If a random sample of 100 residents from an unknown state reveals 25 smokers, would you have strong reason to doubt that the state was Utah? Explain your reasoning.

This activity should reinforce the idea of **statistical significance,** which you encountered in Topics 13 and 14. A statistically significant result is one that is unlikely to occur by random variation alone. Previously, you used simulations to investigate this issue; now the Central Limit Theorem enables you to perform probability calculations to assess this "unlikeliness."

Activity 15-3: Body Temperatures

12-1, 12-19, **15-3,** 15-18, 15-19, 19-3, 19-7, 20-11, 22-10, 23-3

Consider taking a simple random sample of 130 healthy adults and measuring their body temperature. Suppose the mean body temperature is 98.6 degrees Fahrenheit with a standard deviation of 0.7 degrees in the population of all healthy adults.

a. Are these numbers (98.6 degrees and 0.7 degrees) parameters or statistics? Explain. Also indicate the symbols used to represent them.

b. Does the CLT apply in this situation? Explain.

c. What does the CLT say about how the sample mean body temperature would vary if random samples are taken repeatedly from this population? Draw a well-labeled sketch of this sampling distribution.

d. Determine the probability that the sample mean body temperature will fall between 98.5 and 98.7 degrees (in other words, within ± 0.1 degrees of the population mean).

e. Now suppose the *population* mean body temperature is actually 98.3 degrees (and the population standard deviation is still 0.7 degrees). Determine the probability that the *sample* mean body temperature falls between 98.2 and 98.4 degrees (again within ± 0.1 degrees of the population mean).

f. How do your answers to parts d and e compare?

g. Now return to the more realistic assumption that you do not know the value of the population mean body temperature μ. What is the probability that a random sample of size 130 results in a sample mean body temperature within ± 0.1 degrees of the actual population mean μ? Justify your answer. *Hint:* Include a z-score calculation in your justification. Also suppose you still know the population standard deviation to be 0.7 degrees.

This activity should confirm and solidify the notion of **statistical confidence** that you encountered in Topics 13 and 14. Even when you do not know the value of a population parameter, you can be confident that the sample statistic will fall within a certain distance of that unknown parameter value. The Central Limit Theorem allows you to determine the probability that a sample statistic will fall within a certain distance of the population parameter.

Activity 15-4: Solitaire

11-22, 11-23, **15-4,** 15-14, 21-20, 27-18

In Activity 11-22, you modeled author A's games of solitaire as having a probability of success $\pi = 1/9$. Suppose he plays $n = 10$ games of solitaire.

a. Use the Central Limit Theorem to approximate the probability that author A wins 10% or less of those games. *Hint:* Find the mean and standard deviation of the sampling distribution for the sample proportion and standardize (find the *z*-score for) the value .10.

The exact probability for each outcome of number of games won can be determined as follows:

Number of Wins	0	1	2	3	4	5	6	7	8	9	10
Proportion Wins	0	.1	.2	.3	.4	.5	.6	.7	.8	.9	1.0
Probability	.308	.385	.217	.072	.016	.004	.000	.000	.000	.000	.000

(The last four probabilities listed are not exactly zero, but they are less than .001.)

b. Use these probabilities to calculate his (exact) probability of winning 10% or less of the ten games played.

c. Are the probabilities in parts a and b close?

d. Explain why the CLT provides such a poor approximation for the calculated probability in this situation. Are the technical conditions concerning n and π needed for the validity of the CLT satisfied in this case? Explain.

Watch Out

Always remember to check the technical conditions on which the validity of the Central Limit Theorem rests. If these conditions are not satisfied, calculations from the CLT can be erroneous and misleading. For sample proportions, the convention is that $n\pi \geq 10$ and $n(1 - \pi) \geq 10$. For sample means, the result holds exactly when the population itself is normally distributed and approximately for large sample sizes, $n \geq 30$.

15

Activity 15-5: Capsized Tour Boat

On October 5, 2005, a tour boat named the *Ethan Allen* capsized on Lake George in New York with 47 passengers aboard. The maximum weight capacity of the boat was estimated to be 7500 pounds.

Data from the Centers for Disease Control and Prevention indicate that weights of American adults in 2005 had a mean of 167 pounds and a standard deviation of 35 pounds. Use this information to estimate the probability that the total weight in a random sample of 47 American adults exceeds 7500 pounds in 2005. Explain your reasoning and justify your calculations throughout. *Hint:* You may first want to re-express the question in terms of a sample mean or a sample proportion, whichever is appropriate considering the underlying variable in this study.

Solution

First, weight is a quantitative variable, so the relevant statistic is the sample mean weight of the 47 passengers. Because the question is phrased in terms of the *total* weight in a sample of 47 adults, you must rephrase it in terms of the sample *mean* weight. If total weight exceeds 7500 pounds, then the sample mean weight must exceed 7500/47 or 159.574 pounds. So, you want to find the probability that $\bar{x} > 159.574$ (with $n = 47$ and $\sigma = 35$).

The CLT applies because the sample size ($n = 47$) is fairly large, greater than 30. The sampling distribution of $\bar{x}$ is, therefore, approximately normal with mean 167 pounds and standard deviation equal to $\sigma/\sqrt{n} = 35/\sqrt{47} = 5.105$ pounds. A sketch of this sampling distribution is shown here:

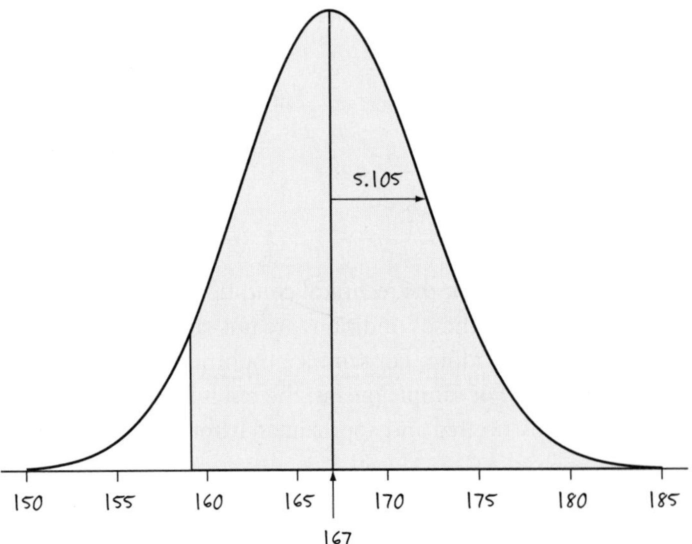

Now you can use the `Normal Probability Calculator` applet or your calculator or the Standard Normal Probabilities Table to find the probability of interest. The z-score corresponding to a sample mean weight of 159.574 pounds is $(159.574 - 167)/5.105 = -1.45$. The probability of the weight being less than 159.574 pounds is found from the table to be .0736, so the probability of exceeding this weight is $1 - .0736 = .9264$. It's not surprising the boat capsized with 47 passengers!

Watch Out

- The first step in any CLT problem is to identify which version of the result to use. Determining this involves asking whether the question is about a sample proportion or a sample mean. Often the question itself will use the word *proportion* or *mean*. That is not the case with this question, however, so you have to identify the underlying variable of interest and ask whether that variable is categorical or quantitative. In this case, weight is a quantitative variable, so the CLT for a sample mean applies.
- Ensure the conditions required for the CLT to be valid are satisfied. In this case, the fairly large sample size assures you that the sampling distribution of $\bar{x}$ will be approximately normal whether the distribution of weights is normal or not.

Wrap-Up...........

This topic has given you practice using the result of the Central Limit Theorem to calculate the probability that a sample statistic will fall within a certain interval of values (assuming the population parameter is known). This topic also reinforced the fundamental idea that sampling distributions have less variability with larger sample sizes, which in turn affects the likelihood of certain sample results occurring.

In practice, of course, you want to go in the other direction: to make inferences about an unknown population parameter based on the observed value of a sample statistic. Even though these two approaches seem to pull in opposite directions, the Central Limit Theorem actually provides the justification for much of statistical inference.

This topic also explored ideas related to statistical confidence and statistical significance. These are the two most important concepts in statistical inference; understanding them is crucial to being able to produce and interpret statistical reasoning.

Some useful definitions to remember and habits to develop from this topic include

- The first step in applying the **Central Limit Theorem** is to recognize whether you are dealing with a sample *proportion* (for example, the proportion who smoke in a sample) or a sample *mean* (for example, the average body temperature in a sample). A proportion applies to a categorical variable (whether a person is a smoker) and a mean applies to a quantitative variable (a person's body temperature).
- Make sure the conditions are satisfied before applying the CLT. With a proportion, check that $n\pi \geq 10$ and $n(1 - \pi) \geq 10$. With a mean, check that $n \geq 30$ or that you can reasonably assume the population itself follows a normal distribution. Also remember the CLT applies *only* if the sample is randomly chosen from the population.
- A larger sample produces less variability (with either a sample mean or a sample proportion) than a smaller sample does. Therefore, the sampling distribution gets taller and skinnier as sample size increases. The sample statistic is, therefore, more likely to be close to the value of the population parameter with a larger sample.

- The concept of **statistical confidence** relates to how close you expect a sample statistic to come to its corresponding population value.
- The concept of **statistical significance** concerns how unlikely an observed sample statistic is to have occurred, assuming some conjectured value for the population parameter.

These concepts of confidence and significance are the focus of the next unit of the course. You will continue to explore what these concepts mean and how to apply them to answer research questions. You will also learn how to construct confidence intervals and perform tests of significance. As you progress through the next few topics, you can return to this unit to remember the basic reasoning behind the calculations you are performing.

Homework Activities

Activity 15-6: Miscellany

For each of the following, indicate whether the question pertains to a sample *mean* or a sample *proportion*.

a. How likely is it that the average height in a random sample of 100 chief executive officers exceeds 69 inches?
b. How surprising would it be for more than two-thirds of the CEOs in a random sample of 100 to be taller than 69 inches?
c. What is the probability that a random sample of 25 students at your school has a grade point average greater than 3.0?
d. What is the probability that a random sample of 25 students at your school produced a sample mean sleep time last night of less than seven hours?
e. What is the probability that a random sample of 25 students at your school finds that less than 90% of students know their mothers' birthdays?

Activity 15-7: Candy Colors

1-13, 2-19, 13-1, 13-2, **15-7**, 15-8, 16-4, 16-22, 24-15, 24-16

As you did in Activity 13-2, consider a simple random sample of size $n = 75$ from the population of Reese's Pieces. Continue to assume that 45% of this population is orange ($\pi = .45$).

a. According to the Central Limit Theorem, how will the sample proportion of orange candies vary from sample to sample? Describe not only the shape of the distribution but also its mean and standard deviation.
b. Sketch this sampling distribution, and shade the area under this curve corresponding to the probability that a sample proportion of orange candies will be less than 0.4. By examining this area, guess the value of the probability.
c. Use Table II or your calculator to compute the probability that a sample proportion of orange candies will be less than .4.
d. Calculate the probability that a sample proportion of orange candies falls between .35 and .55 (within ± .10 of .45).
e. Compare this probability with part m in Activity 13-2, in which you found the percentage of 500 simulated sample proportions that fell within ± .10 of .45. Are the probabilities reasonably close?

Activity 15-8: Candy Colors

1-13, 2-19, 13-1, 13-2, 15-7, **15-8,** 16-4, 16-22, 24-15, 24-16

Reconsider the previous activity. Now suppose you take simple random samples of size $n = 175$ from the population of Reese's Pieces candies, still assuming that 45% of this population is orange ($\pi = .45$).

a. According to the Central Limit Theorem, how will the sample proportion of orange candies vary from sample to sample? Describe not only the shape of the distribution but also its mean and standard deviation. What has changed about this sampling distribution from when the sample size was 75?

b. Sketch this sampling distribution, and shade the area corresponding to the probability that a sample proportion of orange candies will be less than .4 (with a sample size of 175). By looking at this area, guess the value of the probability.

c. Calculate the probability that (with a sample size of 175) a sample proportion of orange candies will be less than .4.

d. How does this probability differ from the probability when the sample size is 75? Explain why this result makes sense.

e. Calculate the probability that a sample proportion of orange candies (using a sample size of 175) falls between .35 and .55 (within $\pm$ .10 of .45).

f. How does the probability in part e differ from when the sample size is 75? Explain why this result makes sense.

15

Activity 15-9: Smoking Rates

15-1, 15-2, **15-9**

The proportion who smoke among adult residents of Kentucky in 2004 was .276, the highest of any state in the U.S. Treat this as the population proportion, and suppose you take a random sample of 400 Kentucky residents.

a. Draw and label a sketch of the sampling distribution of the sample proportion of Kentucky smokers.

b. Determine the probability of obtaining a sample proportion of Kentucky smokers more than .025 away from the population proportion (in other words, either greater than 0.301 or less than .251). *Hint:* Either calculate both probabilities ($\hat{p} \geq 0.301$ and $\hat{p} \leq .251$) separately and add the values, or calculate the middle probability($.251 \leq \hat{p} \leq .301$) and subtract from 1, or calculate one probability and explain why it is valid to double it.

c. Repeat part b with a distance of .05 instead of .025.

d. If you are presented with a random sample of 400 residents from an unknown state, and you find that 25% of the sample are smokers, would you have reason to believe that the state is not Kentucky? Explain, based on your calculations in parts b and/or c.

e. Repeat part d if you find that 22.5% of the sample are smokers.

Activity 15-10: Candy Bar Weights

12-10, 14-10, **15-10**

Activity 12-10 asked you to assume that the actual weight of a certain candy bar, whose advertised weight is 2.13 ounces, varies according to a normal distribution with mean $\mu = 2.20$ ounces and standard deviation $\sigma = 0.04$ ounces.

a. What is the probability that an individual candy bar weighs between 2.18 and 2.22 ounces?

b. Suppose you plan to take a sample of five candy bars and calculate the sample mean weight. Does the CLT apply in this instance? Explain.
c. What does the CLT say about how these sample means will vary from sample to sample? Draw a sketch of the sampling distribution.
d. Shade in the region on your sketch corresponding to the probability that the sample mean weight of these five candy bars falls between 2.18 and 2.22 ounces. Use your sketch to guess the value of this probability. Is this value greater or less than your answer to part a?
e. Use the Standard Normal Probabilities table or your calculator to compute the probability that the sample mean weight of these five candy bars falls between 2.18 and 2.22 ounces. Comment on the accuracy of your guess in part d.
f. How do you expect this probability to change if the sample size were 40 instead of 5? Explain your reasoning.
g. Calculate this probability, and comment on your conjecture.
h. Which of your previous calculations (parts a, d, and f) remain approximately correct even if the candy bar weights themselves have a skewed, nonnormal distribution?

Activity 15-11: Christmas Shopping

14-3, 14-7, **15-11**, 19-1

Recall Activity 14-3 and the population of American households that purchase Christmas presents. Consider the variable *amount expected to be spent on Christmas presents as reported in late November*. Suppose this population has mean $\mu = \$850$ and standard deviation $\sigma = \$250$.

a. If a random sample of 5 households is selected, is it valid to use the Central Limit Theorem to describe the sampling distribution of the sample mean? Explain.
b. If a random sample of 500 households is selected, does the Central Limit Theorem tell you the sampling distribution of the sample mean? Draw and label a sketch of the sampling distribution.
c. Using a random sample of 500 households, determine the probability that the sample mean falls within $\pm$ \$18.39 of the population mean \$850 (in other words, between \$831.61 and \$868.39).
d. Repeat part c for the probability that the sample mean falls within $\pm$ \$21.91 of \$850.
e. Repeat part c for the probability that the sample mean falls within $\pm$ \$28.80 of \$850.
f. Try to find a value k such that there is a.8 probability of the sample mean falling within $\pm$ k of \$850. Fill in the appropriate area on a sketch of the sampling distribution.
g. If the population mean were actually $\mu = \$1000$, what is the probability that a random sample of 500 households would produce a sample mean within $\pm$ \$18.39 of \$1000? Does this probability look familiar? Explain.

Activity 15-12: Jury Selection

11-4, 12-1, **15-12**

Reconsider Activity 11-4, which asked about the number of senior citizens (age 65 or older) selected for a jury (sample size 12) or a jury pool (sample size 75) from a county in which 20% of the adults are senior citizens.

a. Does the Central Limit Theorem apply to either, both, or neither of these sample sizes?

b. For the sample size(s) that you answered in part a, sketch the sampling distribution of the sample proportion of senior citizens.
c. For the sample size(s) that you answered in part a, calculate the probability that the sample contains at least one-third senior citizens.
d. Compare this probability to your empirical estimate in Activity 11-4 on pages 223 and 224.

Activity 15-13: Non-English Speakers

In the state of California in 1990, 31.5% of the residents spoke a language other than English at home. Suppose you take a simple random sample of 100 California residents and determine the sample proportion who speak a language other than English at home.

a. Draw and label a sketch of this sampling distribution.
b. Determine the probability that more than half of the residents sampled speak a language other than English at home.
c. Determine the probability that fewer than one-quarter of the residents sampled speak a language other than English at home.
d. Determine the probability that between one-fifth and one-half of the residents sampled speak a language other than English at home.
e. In the state of Ohio in 1990, 5.4% of the residents spoke a language other than English at home. Suppose you select a simple random sample of 100 Ohio residents. Draw a sketch of the sampling distribution for Ohio on the same scale as you used for California in part a.
f. Without doing the calculations, indicate how you expect the answers to parts b, c, and d to change for Ohio as opposed to California.

15

Activity 15-14: Solitaire

11-22, 11-23, 15-4, **15-14,** 21-20, 27-18

Recall from Activity 11-22 that author A has a 1/9 probability of winning a game of solitaire.

a. How many games does author A have to play before you can use the CLT to reasonably approximate the sampling distribution of the sample proportion of wins for author A? Explain.
b. Repeat part a for author B, who has a 1/6 probability of winning a game of solitaire.
c. Suppose that author B beats author A 80% of the time at a different card game. How many games do they have to play for the CLT to reasonably approximate the sampling distribution of the sample proportion of wins for author B? Explain.

Activity 15-15: Birth Weights

12-2, 14-9, **15-15**

In Activity 12-2, you learned that birth weights of newborn babies in the U.S. can be modeled by a normal curve with mean 3300 grams and standard deviation 570 grams. Also recall that babies weighing less than 2500 grams are considered to have low birth weight.

a. Calculate the probability that a single, randomly selected newborn baby is of low birth weight.
b. Determine the probability that the average birth weight in a random sample of size $n = 2$ babies is less than 2500 grams.

c. Is this probability less than, greater than, or the same as the probability in part a? Explain why this result makes sense.
d. Repeat parts b and c for the average birth weight in a random sample of size $n = 4$.
e. Determine the probability that a single, randomly selected newborn baby weighs between 3000 and 3600 grams.
f. Do you expect this probability to be greater than, less than, or the same as the average birth weight in a sample of $n = 20$ babies? Explain.
g. Calculate the probability that the average birth weight in a sample of 20 babies is between 3000 and 3600 grams, and comment on whether the calculation confirms your conjecture in part f.

Activity 15-16: Volunteerism

13-16, **15-16,** 21-17

Recall from Activity 13-16 that the Bureau of Labor Statistics reported in December of 2005 that 28.2% of a random sample of 60,000 American adults had participated in a volunteer activity between September 2004 and September 2005.

a. Use the CLT to calculate the probability of obtaining such a large sample proportion if the proportion of all Americans who participated in volunteering equaled .25.
b. Repeat part a, supposing that the study had been conducted by a small polling organization that had taken a random sample of 500 people and found that 28.2% claimed to have participated in a volunteer activity.
c. Which of these two scenarios provides stronger evidence against the claim that 25% of the population served as volunteers? Explain your reasoning.

Activity 15-17: Got a Tip?

1-10, 5-28, 14-12, **15-17,** 22-4, 22-9, 25-13

In Activity 14-12, you learned about a waitress who keeps track of her tips, as a percentage of the bill, for a random sample of 50 tables. Suppose that, unknown to her, the population mean tip percentage is 15% with a standard deviation of 4%.

a. Determine the probability that her sample mean tip percentage exceeds 16.4%. (Treat the sample as if it were a random one.)
b. If her sample mean tip percentage turns out to be 16.4%, does that provide strong evidence that the population mean tip percentage is actually greater than 15%? Explain your reasoning.
c. If her sample mean tip percentage turns out to be 14.4%, does that provide strong evidence that the population mean tip percentage is actually less than 15%? Explain your reasoning.

Activity 15-18: Body Temperatures

12-1, 12-19, 15-3, **15-18,** 15-19, 19-3, 19-7, 20-11, 22-10, 23-3

Reconsider Activity 15-3. Describe how your answers to parts b–g would change if the sample size were 40 rather than 130. Also explain why it makes sense for your answers to change in this way.

Activity 15-19: Body Temperatures

12-1, 12-19, 15-3, 15-18, **15-19,** 19-3, 19-7, 20-11, 22-10, 23-3

Reconsider again Activity 15-3. Recall that the data stored in the file BODYTEMPS are the body temperature measurements for a sample of 130 healthy adults (Shoemaker, 1996).

a. Create and comment on graphical displays of the distribution of body temperatures.
b. Calculate and report the sample mean and standard deviation.
c. Suppose the population mean were 98.6 degrees with a standard deviation of 0.7 degrees. Use the CLT to determine the probability of obtaining a sample mean as low as the one in this sample, given these assumptions.
d. Is the probability in part c low enough to provide compelling evidence that the population mean body temperature is not 98.6 degrees? Explain your reasoning.

Activity 15-20: IQ Scores

12-9, 14-13, **15-20**

Reconsider Activity 14-13, where you supposed that the IQs of your hometown's residents follow a normal distribution with mean 105 and standard deviation 12.

a. Calculate the probability that a randomly selected resident has an IQ greater than 110.
b. Calculate the probability that the average IQ in a random sample of 10 residents exceeds 110.
c. Calculate the probability that the average IQ in a random sample of 40 residents exceeds 110.
d. Would any of the calculations in parts a–c be valid (approximately), even if the distribution of IQs in the population were skewed rather than normal? Explain.

UNIT

4 • • •

Inference from Data: Principles

TOPIC

16 Confidence Intervals: Proportions

The generation of children being raised in the 2000s has been dubbed Generation M, because of the extent to which various forms of media (hence the *M*) permeate their lives. The Kaiser Family Foundation commissioned an extensive survey to investigate this phenomenon. In this topic, you will learn more about how to use sample results to estimate population values. For example, what proportion of all American teens have a television in their rooms? Is this proportion higher for boys or for girls? Is the proportion higher or lower for other forms of media, such as CD players and video game players and computers? How can we answer these questions without asking all American teens?

Overview

In the last unit, you explored how sample statistics vary from sample to sample. You studied this phenomenon empirically through simulations and theoretically with the Central Limit Theorem. You learned that this variation has a predictable long-term pattern. This pattern enables you to make probability statements about sample statistics, provided you know the value of the population parameter. These probability statements allow you to turn the tables and address the much more common goals of estimating and making decisions about an unknown population parameter based on an observed sample statistic. These are the goals of statistical inference.

There are two major techniques for classical statistical inference: confidence intervals and tests of significance. Confidence intervals seek to estimate a population parameter with an interval of values calculated from an observed sample statistic. Tests of significance assess the extent to which sample data refute a particular hypothesis concerning the population parameter. This topic extends your study of the concept of statistical confidence, begun in Topic 13, by introducing you to confidence intervals for estimating a population proportion.

Preliminaries

1. Guess the proportion of American youth from ages 8–18 who have a television in their bedrooms.

2. Mark on the following number line an interval that you believe with 80% confidence to contain the actual proportion of American youth from ages 8–18 who have a television in their bedrooms. (In other words, if you were to create a large number of these intervals, 80% of them should succeed in *capturing*—containing within the interval—the actual population proportion.)

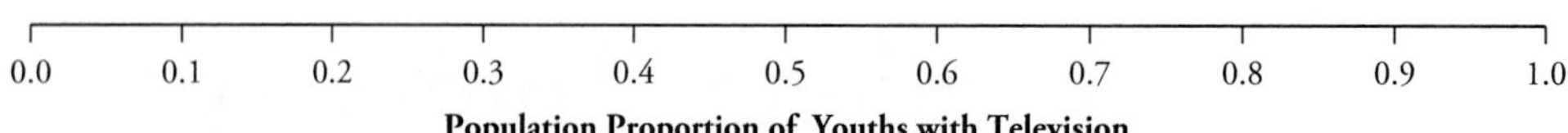

3. Mark on the following number line an interval that you believe with 99% confidence to contain this population proportion.

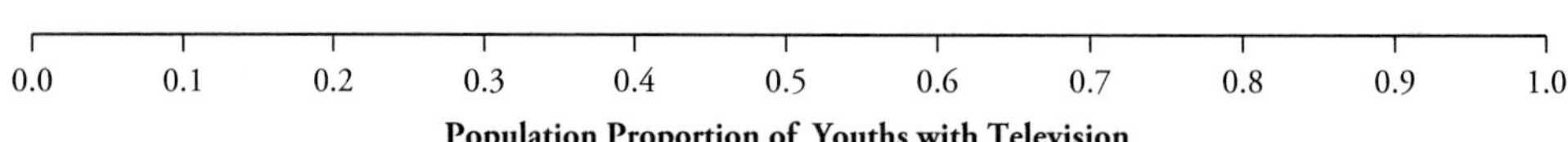

4. Which of these two intervals is wider, the 80% interval or the 99% interval? Explain why this makes sense.

In-Class Activities

Activity 16-1: Generation M

3-8, 4-14, 13-6, **16-1,** 16-3, 16-7, 18-1, 21-11, 21-12

Recall from Activity 3-8, the Kaiser Family Foundation commissioned an extensive survey in 2004 that investigated the degree to which American youth from ages 8–18 have access to various forms of media. They distributed written questionnaires to a random sample of 2032 youths in this age range. One of the many questions asked youths whether they have a television in their bedrooms, to which 68% answered yes.

a. What are the observational units in this study?

b. Identify the variable, and classify it as quantitative or categorical.

Because the variable of interest is categorical and binary, the relevant parameter of interest is a *proportion*.

c. Is .68 a parameter or a statistic? What symbol would you use to represent it?

d. Describe in words the relevant population parameter of interest in this study. What symbol would you use to represent it?

e. Does the Kaiser survey allow the researchers to determine the exact value of the parameter? Explain.

f. Is the parameter value more likely to be close to the Kaiser survey's sample proportion than to be far from it? Explain.

These questions deal with the issue of statistical confidence that you explored in Topic 13. Although a sample statistic from a simple random sample provides a reasonable estimate of a population parameter, you certainly do not expect the sample statistic to equal the (unknown) population parameter exactly (sampling variability). It is likely, however, that the unknown parameter value is "in the ballpark" of the sample statistic. The purpose of confidence intervals is to use the sample statistic to construct an interval of values that you can be reasonably confident contains the actual, though unknown, parameter.

Through simulations you discovered earlier that extending two standard deviations on either side of the sample proportion $\hat{p}$ produced an *interval* (a set of plausible values) that succeeded in capturing the value of the population proportion, π, for about 95% of all samples (Activity 13-2). The Central Limit Theorem revealed this standard deviation of $\hat{p}$ to be $\sqrt{\pi(1-\pi)/n}$.

However, you cannot use this formula in practice because you do not know the value of π. (Indeed, the whole point of establishing the interval is to estimate the value of π based on the sample proportion $\hat{p}$.)

g. What value (in general) seems like a reasonable replacement to use as an estimate for π in this standard deviation expression?

The estimated standard deviation

$$\sqrt{\frac{\hat{p}(1-\hat{p})}{n}}$$

of the sample statistic $\hat{p}$ is called the **standard error** of $\hat{p}$.

h. Calculate the standard error of $\hat{p}$ for the Kaiser survey about the proportion of American children who have a television in their bedrooms.

i. Now, go two standard errors on either side of $\hat{p}$ by doubling this standard error, subtracting the result from $\hat{p}$, and adding it to $\hat{p}$. This technique forms a reasonable interval estimate (set of plausible values) of π, the population proportion of all American 8–18-year-olds who have a television in their bedrooms.

j. Do you know for sure whether the actual value of π is contained in this interval?

All you can say for certain is that if you were to construct many intervals this way using different random samples from the same population, then approximately 95% of the resulting intervals would contain π. Note that it is not technically appropriate to say that this interval (.66, .70) has a .95 probability of containing π (see Activity 16-4 for more discussion on this point). This has led statisticians to coin a new term, **confidence,** to describe the level of certainty in the interval estimate, and so your conclusion is that you are "95% confident" that the interval you calculated contains π.

The other complication is that so far we have limited you to the value 2 for the number of standard errors and the resulting approximate 95% confidence interval. Returning to the empirical rule, you know that going just one standard error (standard deviation) on either side of $\hat{p}$ corresponds to 68%, whereas going three standard errors corresponds to 99.7%. By using a different multiple of the standard error to vary the distance that you go from $\hat{p}$, you can vary the **confidence level,** which is the measure of how confident you are that the interval does, in fact, contain the true parameter value. You find this multiplier, called the **critical value,** using the normal distribution, and you will denote it by z^* (see Activity 16-2). You specify the confidence level by deciding the level of confidence necessary for the given situation; common values are 90%, 95%, and 99%. The confidence level, in turn, determines the critical value.

Confidence interval for a population proportion π:

$$\hat{p} \pm z^* \sqrt{\frac{\hat{p}\,(1-\hat{p})}{n}}$$

where $\hat{p}$ denotes the sample proportion, n is the sample size, and z^* represents the critical value from the standard normal distribution for the confidence level desired.

The "±" in this expression means that you subtract the term following it from the term preceding it to get the lower endpoint of the interval, and then you add the terms to get the upper endpoint of the interval. This expression is often denoted as

$$\left(\hat{p} - z^* \sqrt{\frac{\hat{p}\,(1-\hat{p})}{n}},\ \hat{p} + z^* \sqrt{\frac{\hat{p}\,(1-\hat{p})}{n}}\right)$$

For this confidence interval procedure to be valid, the sample size needs to be large relative to the value of the population proportion. This assures you that the sampling distribution of $\hat{p}$ is approximately normal. To check this condition, you will see whether $n\hat{p} \geq 10$ and if $n(1-\hat{p}) \geq 10$, which says there must be at least ten "successes" and ten "failures" in the sample. If this condition is not met, then the stated confidence level might be inaccurate, and the interval could be very misleading.

Furthermore, if you do not have a random sample from the population of interest, then the confidence interval might not be estimating the desired parameter. Without a random sample, you might still be able to argue that the sample is representative of the population of interest, but you should only do so with caution.

Therefore, before interpreting your confidence interval, make sure these technical conditions are met.

Technical conditions:

- The sample is a simple random sample from the population of interest. Check the data collection protocol.
- The sample size is large relative to the value of the population proportion. Check whether or not $n\hat{p} \geq 10$ and $n(1 - \hat{p}) \geq 10$.

Activity 16-2: Critical Values

12-18, **16-2,** 16-20, 19-13

Before calculating a confidence interval, you need to be able to determine the critical value, z^*. As an example, suppose you want to find the value of z^* for a 98% confidence interval. In other words, find the value z^* such that the area under the standard normal curve between $-z^*$ and z^* is equal to .98 by following these steps:

a. Draw a sketch of the standard normal curve and shade the area corresponding to the middle 98% of the distribution. Also indicate roughly where $-z^*$ and z^* fall in your sketch.

b. Based on your sketch, what is the total area under the standard normal curve to the left of the value z^*?

c. Use the Standard Normal Probabilities Table (Table II) or your calculator and the **invNorm** command (found by pressing [2ND] [DISTR]) to find the value (z^*) that has this area (your answer to part b) to its left under the standard normal curve.

This value z^*, which you should have found to be approximately 2.33, is called the **upper .010 critical value** of the standard normal distribution because the area to its right under the standard normal curve is .010. This critical value is used for 98% confidence intervals.

d. Repeat parts a–c to find the critical value corresponding to a 95% confidence interval.

(*Note:* 1.96 is the more exact value that we previously approximated as being 2.) Critical values for any confidence level can be found in this manner. Some commonly used confidence levels and critical values are listed here:

Confidence Level	80%	90%	95%	99%	99.9%
Critical Value z^*	1.282	1.645	1.960	2.576	3.291

Activity 16-3: Generation M

3-8, 4-14, 13-6, 16-1, **16-3,** 16-7, 18-1, 21-11, 21-12

a. Recall that the survey of 2032 children in Activity 16-1 found that 68% had a television in their bedrooms. Use this sample to construct a 95% confidence interval for π, the proportion of all American 8–18-year-olds who have a television in their bedrooms.

b. Write a sentence interpreting what this interval reveals.

c. Can you be *certain* that this interval contains the actual value of π?

d. Calculate the *width* of this interval. *Hint:* The width is the difference between the upper and lower endpoints of the interval.

e. Determine the *half-width* of this interval. *Hint:* The half-width is the width divided by 2.

The half-width of a confidence interval is often called the survey's **margin-of-error.**

f. Explain how you could have determined the half-width of the interval without first calculating the width. *Hint:* What part of the confidence interval formula corresponds to this margin-of-error?

g. Determine the midpoint of this interval. *Hint:* You can find the interval's midpoint by calculating the average of its endpoints.

h. Does this midpoint value look familiar? Explain why this value makes sense, based on how the confidence interval is constructed.

i. What technical conditions underlie the validity of this procedure? Check whether the conditions seem to be satisfied here.

j. How would you expect a 99% confidence interval to differ from the 95% confidence interval? Provide an intuitive explanation for your answer without doing the calculation and not based on the formula.

k. Determine a 99% confidence interval for the population proportion of American 8–18-year-olds who have a television in their bedrooms.

l. How does the 99% interval compare with the 95% interval? Compare the midpoints as well as the margins-of-error.

As the confidence level increases, the margin-of-error and width of the confidence interval also increase.

m. Based on these intervals, does it seem plausible that .5 is the proportion of all American youths who have a television set in their bedrooms? Does .75 seem plausible? What about two-thirds? Explain.

.50: .75: Two-thirds:

Explanation:

The survey also provided sample results broken down separately for boys and girls. There were 1036 girls and 996 boys in the sample. Among the boys, 72% had a television in their bedroom, as compared to 64% of the girls.

n. Determine a 95% confidence interval for the proportion of American boys in this age group who have a television in their bedrooms. Then do the same for the analogous proportion of American girls.

Boys: Girls:

o. How do these intervals compare? Do the intervals seem to indicate that the proportion of boys who have a television in their bedrooms is different than that proportion for girls? Explain.

p. Report the margins-of-error for these intervals. Are these greater than or less than (or the same as) the margin-of-error based on the entire sample (from part e)? Explain why this makes sense intuitively.

A larger sample size produces a narrower confidence interval whenever other factors remain the same.

Watch Out

A confidence interval is just that—an *interval*—so it includes all values between its endpoints. For example, the 95% confidence interval (.660, .700) means you are 95% certain that the population proportion of 8–18-year-olds who have a TV in their bedrooms is somewhere between .660 and .700. Do not mistakenly think that only the endpoints matter or that only the margin-of-error matters. The midpoint and actual values within the interval matter.

q. Now suppose you want to do a follow-up survey for which the margin-of-error (with 95% confidence) will equal .01. Use the confidence interval formula to determine what sample size you will need to achieve this margin-of-error. *Hint:* Based on this study, assume that the sample proportion of American youths with a television in their bedrooms will be about .68. Set the margin-of-error expression equal to .01, and solve algebraically for the sample size n.

Because you cannot sample a fraction of a person, the convention is to round up to the next integer when performing these sample size calculations.

r. How would you expect the necessary sample size to change if you want the confidence level to be 99% instead of 95%, with a .01 margin-of-error? Explain your reasoning.

s. Perform the sample size calculation in part r. Do you need to change your answer to part r?

Note: When you do not have a prior estimate of the population proportion to use in the sample size calculation, using .5 is conservative in the sense that it leads to a larger value for the sample size than would any other value.

Activity 16-4: Candy Colors

1-13, 2-19, 13-1, 13-2, 15-7, 15-8, **16-4,** 16-22, 24-14, 24-15

Assume for the moment that 45% of all Reese's Pieces are orange, i.e., that the population proportion of orange candies is $\pi = .45$. To examine more closely what the interpretation of "confidence" means in this statistical context, you will use an applet to simulate confidence intervals for 200 random samples of 75 candies each.

a. Open the `Simulating Confidence Intervals` applet and make sure that it is set for Proportions with the Wald method (the name for the method you just learned). Set the population proportion value to $\pi =$ **.45,** the sample size to $n =$ **75,** the number of intervals to **1,** and the confidence level to **95%.** Click Sample. The horizontal line that appears represents the confidence interval resulting from the sample.

Simulating Confidence Intervals

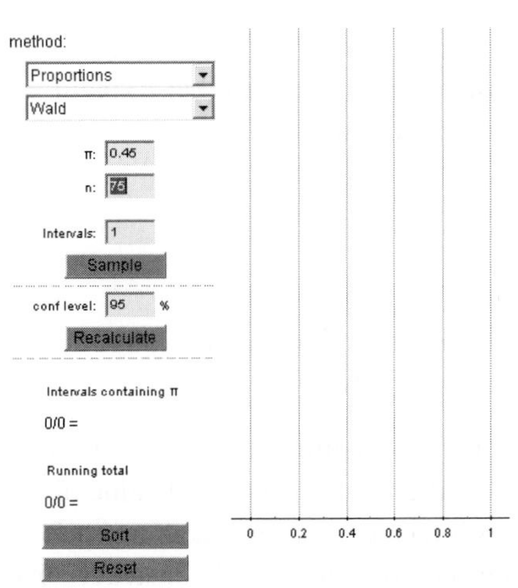

Click on the line, and report the endpoints of the interval. Does this interval succeed in capturing the true value of the population proportion (which you have set to be $\pi = .45$)?

b. Click Sample four more times. Does the interval change each time? Do all of the intervals succeed in capturing the true value of π?

c. Now click Reset and then change the number of intervals to 200 and click Sample. How many and what proportion of these 200 intervals succeed in capturing the true value of π? (Notice that the intervals that succeed are colored green and those that fail to capture π are colored red.)

d. Click Sort. What do you notice about the intervals that fail to capture the true value of π? *Hint:* What is true about those intervals' sample proportions, or midpoints?

e. If you had taken a *single* sample of 75 candies (in a real situation, not in this applet environment), would you have any *definitive* way of knowing whether your 95% confidence interval contained the actual value of π? Explain.

f. Keep clicking Sample until you have generated 1000 samples and intervals. Report the Running Total of intervals that succeed in capturing the true value of π and the percentage of intervals this represents.

g. Is this percentage close to 95%? Should it be? Explain what this simulation reveals about the phrase "95% confident."

This simulation illustrates the following points about the proper use and interpretation of confidence intervals:

- Interpret a 95% (for instance) confidence interval by saying that you are 95% *confident* that the interval contains the actual value of the population proportion.
- More specifically, this interpretation means that if you repeatedly take simple random samples from the population and construct 95% confidence intervals for each sample, then in the long run approximately 95% of these confidence intervals will succeed in capturing the actual population proportion.

h. Predict how the resulting confidence intervals will change if the sample size were increased to 150.

i. Change the sample size to **150** and generate 1000 intervals (200 intervals at a time). How many and what proportion of the intervals succeed in capturing the actual value of π? Is this reasonably close to the percentage from part f? What is different about these intervals compared to those generated with a sample size of 75 candies?

j. Before you change the confidence level to 90%, predict *two* things that will change about the intervals. *Hint:* One change concerns their length and the other concerns their success rate.

k. Change the confidence level to 90% and click Recalculate. How many and what proportion of the intervals succeed in capturing the actual value of π? Is this reasonably close to the percentages from parts f and i? What is different about these intervals?

Watch Out

- It is incorrect to say that π has a .95 probability of falling within the 95% confidence interval. It is also technically incorrect to say that the probability is .95 that a particular calculated 95% confidence interval contains the actual value of π. The technicality here is that π is not random; it is some fixed (but unknown) value. What is random, changing from sample to sample, is the sample proportion and thus the interval based on it. Thus the probability statement applies to what values an interval will take prior to the sample being collected (i.e., to the *method*), not whether a particular interval contains the fixed parameter value once it has been calculated.
- Confidence intervals estimate the value of a population *parameter;* they do not estimate the value of a sample statistic or of an individual observation.

Activity 16-5: Elvis Presley and Alf Landon

3-1, 3-6, **16-5**

Recall from Activity 3-1 that in 1936 the *Literary Digest* received 2.4 million responses to their survey about the upcoming presidential election, with 57% of the sample indicating support for Alf Landon over Franklin Roosevelt.

a. Use your calculator's **1-PropZInt** command and this sample result to form a 99.9% confidence interval for π, the actual proportion of all adult Americans who preferred Landon over Roosevelt. Press the STAT key, and from the **TESTS** menu, arrow down to select **1-PropZInt.** Enter **57%** of **2.4** million for ***x*** (multiply **2.4** million by **.57**), **2.4** million for ***n*,** and **.999** for **C-Level.** Arrow down to select **Calculate** and press ENTER.

b. Explain why this interval turns out to be so very narrow, even with such a large confidence level.

16

c. In the actual election, 37% of the population voted for Landon. Explain why the confidence interval in part a did such a poor job of predicting the election result.

This activity serves as a reminder of the critical importance of collecting data using sound strategies. Inference procedures such as confidence intervals are invalid, and can produce very misleading results, when applied to data obtained through a biased sampling plan.

√ Activity 16-6: Kissing Couples

13-3, 13-4, **16-6,** 17-12, 24-4, 24-13

Recall from Activity 13-3 the study in which a sample of 124 kissing couples were observed, with 80 of them leaning their heads to the right.

a. Identify the observational units and variable in this study.

b. Describe the sample data numerically and graphically.

c. Determine and interpret a 95% confidence interval for the population proportion of all kissing couples who lean to the right. Also provide an interpretation of what is meant by "95% confidence" in this context.

d. Also determine a 90% and a 99% confidence interval for this population proportion. Comment on how these intervals compare.

e. Based on these intervals, comment on whether it seems plausible to believe that 50% of all kissing couples lean to the right. Also comment on the plausibility of 2/3, and of 3/4, as the value of the population proportion of kissing couples who lean to the right.

f. Comment on whether the technical conditions necessary for these confidence intervals to be valid are satisfied here.

16

Solution

a. The observational units are the kissing couples. The variable is which direction they lean their heads while kissing.

b. The sample consists of the 124 kissing couples observed by the researchers in various public places. The statistic is the sample proportion of couples who lean to the right when kissing: $\hat{p} = 80/124 = .645$. A bar graph is shown here.

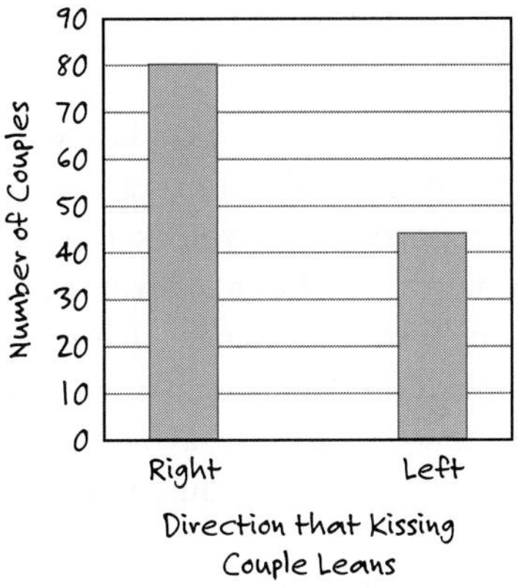

c. A 95% confidence interval for the population proportion of couples who lean to the right is

$$.645 \pm 1.96\sqrt{\frac{(.645)(1-.645)}{124}}$$

which is .645 ± .084, or (.561, .729). You are 95% confident that the population proportion of kissing couples who lean to the right is somewhere between .561 and .729. This "95% confidence" means that if you were to take many random samples and generate a 95% confidence interval (CI) from each, then in the long run, 95% of the resulting intervals would succeed in capturing the actual value of the population proportion, in this case the proportion of all kissing couples who lean their heads to the right.

d. A 90% CI is

$$.645 \pm 1.645\sqrt{\frac{(.645)(1-.645)}{124}}$$

which is .645 ± .071, or (.574, .716).

A 99% CI is

$$.645 \pm 2.576\sqrt{\frac{(.645)(1 - .645)}{124}}$$

which is .645 ± .111, or (.534, .756).

The higher confidence level produces a wider confidence interval. All of these intervals have the same midpoint: the sample proportion .645.

e. Because none of these intervals includes the value .5, it does not appear to be plausible that 50% of all kissing couples lean to the right. In fact, all of the intervals lie entirely above .5, so the data suggest that more than half of all kissing couples lean to the right. The value 2/3 is quite plausible for this population proportion, because .667 falls within all three confidence intervals. The value 3/4 is not very plausible, because only the 99% CI includes the value .75; the 90% CI and 95% CI do not include .75 as a plausible value.

f. The sample size condition is clearly met, as $n\hat{p} = 80$ is greater than 10, and $n(1 - \hat{p}) = 44$ is also greater than 10. But the other condition is that the sample be randomly drawn from the population of all kissing couples. In this study, the couples selected for the sample were those who happened to be observed in public places while the researchers were watching. Technically, this is not a random sample, and so you should be cautious about generalizing the results of the confidence intervals to a larger population.

Watch Out

Don't forget about the necessity for random sampling. When you don't have a true random sample, which might often be the case, proceed cautiously and ask yourself what population you might be willing to believe the sample results represent. For example, in this kissing study, you would probably generalize the results only to the population of couples who kiss in public places in the countries where the researchers gathered their data.

Wrap-Up...........

This topic provided your first formal exposure to statistical inference by introducing you to confidence intervals, a widely used technique. You learned how to construct a confidence interval for a population proportion, and you examined how to interpret both the resulting interval and also what the **confidence level** means. For example, using a 95% confidence level gives you "95% confidence" that the interval contains the actual value of the (unknown) population parameter. As you saw by exploring the `Simulating Confidence Intervals` applet, saying you are 95% confident means that 95% of all intervals generated by the procedure in the long run will succeed in capturing the unknown population value.

You also investigated the effects of sample size and confidence level on the interval and its **margin-of-error.** The ideal confidence interval is very narrow with a high confidence level. But there's a trade-off: using a higher confidence level produces a wider interval if all else remains the same. One solution is to use a larger sample because a larger sample produces a narrower interval (for the same level of confidence). This sample size issue explains why the margin-of-error was greater for estimating the proportion of 8–18-year-old boys with a TV than for estimating the proportion of 8–18-year-olds (boys and girls combined) with a TV. You also learned how to plan ahead by determining, prior to collecting data, the sample size needed to achieve a certain margin-of-error for a given confidence level.

Some useful definitions to remember and habits to develop from this topic include

- The purpose of a **confidence interval** is to estimate the value of a population parameter with an interval of values that you believe with high confidence to include the actual parameter value. The confidence level indicates how confident you are in the interval.
- The general form of all confidence intervals in this course is estimate ± *margin-of-error,* where *margin-of-error* = (*critical value*) × (*standard error of estimate*). The estimate is a sample statistic, calculated from sample data. The **standard error** is an estimate of that statistic's standard deviation (how much the statistic varies from sample to sample), also calculated from sample data.
- A specific confidence interval procedure for estimating a population proportion is

$$\hat{p} \pm z^* \sqrt{\frac{\hat{p}\,(1-\hat{p})}{n}}$$

- The margin-of-error is affected by several factors, primarily
  - A higher confidence level produces a greater margin-of-error (a wider interval).
  - A larger sample size produces a smaller margin-of-error (a narrower interval).
- Common confidence levels are 90%, 95%, and 99%. The phrase "95% confidence" means that if you were to take a large number of random samples and use the same confidence interval procedure on each sample, then in the long run 95% of those intervals would succeed in capturing the actual parameter value. Note that this is not the same as saying there is a 95% probability that the parameter is inside the calculated interval.
- Always check the technical conditions before applying this procedure. The sample is considered large enough for this procedure to be valid as long as $n\hat{p} \geq 10$ and $n(1-\hat{p}) \geq 10$. If this condition is not met, then the normal approximation of the sampling distribution is not valid and the reported confidence level might not be accurate.
- Always consider how the sample was selected to determine the population to which the interval applies. If the sample was randomly selected from the population of interest, then the interval applies to that population.
- Remember that you can plan ahead to determine the sample size necessary to achieve a desired margin-of-error for a given confidence level.

In the next topic, you will continue to study inference procedures for a population proportion. You will investigate the second major type of statistical inference procedure: a test of significance.

Homework Activities

Activity 16-7: Generation M

3-8, 4-14, 13-6, 16-1, 16-3, **16-7,** 18-1, 21-11, 21-12

Recall the study from Activity 16-1 about media in the lives of American youths aged 8–18. You were told in Activity 16-1 that of the 2032 youths sampled, 68% reported having a television in their bedrooms. It turns out that 86% of the youths sampled had a CD/tape player, 49% had a video game player, and 31% had a computer in their bedrooms.

a. For each of these devices, determine the margin-of-error for a 95% confidence interval.

b. Does a survey's margin-of-error depend on more than its sample size? Explain how you can tell.
c. Which device produces the greatest margin-of-error, and which produces the least?
d. Based on these calculations, make a conjecture about the value for the sample proportion that produces the greatest margin-of-error.

Activity 16-8: Cursive Writing

13-18, **16-8**

Recall from Activity 13-18 that an article in the October 11, 2006, issue of the *Washington Post* claimed that 15% of high school students used cursive writing on the essay portion of the SAT exam in the academic year 2005–2006. Suppose you want to design a study to estimate this proportion for the following year.

a. How many essays would you need to sample in order to estimate this proportion to within ± .01 with 99% confidence? (Use the 2005–2006 proportion as a reasonable starting guess for how the following year's proportion will turn out.)
b. Describe the two changes you could make to this goal (estimating this proportion to within ± .01 with 99% confidence) if you could not afford to sample so many essays.

Activity 16-9: Penny Activities

16-9, 16-10, 16-11, 17-9, 18-14, 18-15

a. Flip a penny in the air 50 times and keep track of how many heads and tails result. Use your sample results to construct a 95% confidence interval for the probability (long-term relative frequency) that a *flipped* penny lands heads.
b. Instead of flipping a penny in the air, consider spinning it on its side and seeing whether it lands on heads or tails. Do this 50 times, and use your sample results to construct a 95% confidence interval for the probability that a *spun* penny lands heads.
c. Now consider balancing a penny on its side and then striking the table nearby so the penny tilts to land heads or tails. Do this 50 times, and use your sample results to construct a 95% confidence interval for the probability that a *tilted* penny lands heads.
d. Based on these results, which activities (flipping, spinning, or tilting) could plausibly be 50/50 for landing heads or tails? Explain.
e. Which activity appears to result in the highest probability of landing heads?

Activity 16-10: Penny Activities

16-9, **16-10,** 16-11, 17-9, 18-14, 18-15

Refer to the previous activity.

a. Determine the sample size necessary to estimate the probability that a *flipped* penny lands heads to within a margin-of-error of ± .02 with confidence level 95%. *Hint:* Use your estimate of this probability from part a in the previous activity.
b. How would you expect this sample size to change if you increase the confidence level to 99%? Explain.
c. Perform the sample size calculation requested in part b, and see whether your prediction was right.
d. How would you expect this sample size to change if you also adjusted the margin-of-error to ± .01 with 99% confidence? Explain.
e. Perform the sample size calculation requested in part d, and see whether your prediction was right.

Activity 16-11: Penny Activities

16-9, 16-10, **16-11,** 17-9, 18-14, 18-15

Refer to the previous two activities.

a. Determine the sample size necessary to estimate the probability that a *spun* penny lands heads to within a margin-of-error of ± .02 with a confidence level of 95%. Use your estimate from part b of Activity 16-9.
b. Determine the sample size necessary to estimate the probability that a *tilted* penny lands heads to within a margin-of-error of ± .02 with a confidence level of 95%.
c. With which penny activity (flipping, spinning, or tilting) is the necessary sample size largest? Why do you think this is the case? Use your estimate from part b of Activity 16-9.

Activity 16-12: Credit Card Usage

1-9, **16-12,** 17-11, 19-8, 20-10

Recall from Activity 1-9 that the Nellie Mae organization conducts an extensive annual study of credit card usage by college students. For their 2004 study, they analyzed credit bureau data for a random sample of 1413 undergraduate students between the ages of 18 and 24. They found that 76% of the students sampled held a credit card.

a. Determine this result's margin-of-error with 95% confidence.
b. Determine and interpret a 95% confidence interval for the proportion of all undergraduate college students in the U.S. who held a credit card in 2004.
c. Comment on whether the technical conditions required for the validity of this procedure are satisfied here.

Activity 16-13: Responding to Katrina

2-12, 4-10, **16-13**

On September 8–11, 2005, less than two weeks after the destruction caused by Hurricane Katrina, a CNN/*USA Today*/Gallup poll asked, "Just your best guess, do you think one reason the federal government was slow in rescuing these people was because many of them were black, or was that not a reason?" Of the 848 non-Hispanic white adults who were interviewed, 12% said yes. Of the 262 black adults interviewed, 60% said yes.

a. Determine the margin-of-error and a 95% confidence interval for the population proportion of white adults who answered yes.
b. Repeat for the black adults.
c. Compare these intervals (not just their widths).
d. Which group has the greater margin-of-error? Explain why.
e. What additional information (not given above) would you like to know in order to determine if these intervals are valid?

Activity 16-14: *West Wing* Debate

16-14, 18-6

On Sunday, November 6, 2005, the popular television drama *The West Wing* held a live debate between two fictional candidates for President. Immediately afterward, an MSNBC/Zogby poll found that 54% favored Democratic Congressman Matt Santos, played by Jimmy Smits, whereas 38% favored Republican Senator Arnold Vinick, played by Alan Alda. The poll was conducted online with a sample of 1208 respondents; the Zogby company screened the online respondents to try to ensure they were representative of the population of adult Americans.

a. Produce a 95% confidence interval for the population proportion who favored Santos.
b. Does this interval suggest that more than half of the population favored Santos? Explain.

Activity 16-15: Magazine Advertisements

16-15, 17-20, 21-21

The September 13, 1999, issue of *Sports Illustrated* contained 116 pages, and 54 of those pages contained an advertisement. The September 14, 1999, issue of *Soap Opera Digest* consisted of 130 pages, including 28 pages with advertisements.

a. What are the observational units in this study?
b. For each magazine, treat this issue's pages as a sample from the population of all the magazine's pages over all issues, and construct a 95% confidence interval for the population proportion of pages that contain ads.
c. Write a sentence or two interpreting these intervals. (Be sure to relate your interpretation to the context.)
d. Clearly explain what is meant by the phrase "95% confidence" in this context.
e. Does each interval contain the sample proportion of pages with ads?
f. Explain why the previous question is silly and did not require you to even look at the intervals.
g. Find a recent issue of another magazine and repeat this analysis to produce a confidence interval for the proportion of its pages that contain ads.

Activity 16-16: Phone Book Gender

4-17, **16-16,** 18-11

Suppose you want to estimate the proportion of women among the residents of San Luis Obispo County, California. A random sample of columns from the phone book reveals 36 listings with both male and female names, 77 listings with a male name only, 14 listings with a female name only, 34 listings with initials only, and 5 listings with a pair of initials.

a. How many first names are supplied in these listings altogether? *Hint:* Ignore the listings with initials, and remember to count both male and female names for the couples.
b. How many and what proportion of those names are female?
c. Use these sample data to form a 90% confidence interval for the proportion of women in San Luis Obispo County.
d. Do you have any concerns about the sampling method that might render this interval invalid? Explain.
e. Suggest a more reasonable, but still practical, sampling method for estimating the proportion of women in the county.

Activity 16-17: Random Babies

11-1, 11-2, 14-8, **16-17,** 20-16, 20-22

Reconsider the simulation data collected by the class in Activity 11-1 on page 216.

a. What proportion of the simulated repetitions resulted in no mothers getting the correct baby?

b. Use this simulated sample information to form a 95% confidence interval for the long-term proportion of times that no mother would get the right baby.
c. Write a sentence or two interpreting this interval. (Be sure to relate your interpretation to the context.)

This is a rare instance in which you can actually calculate the population parameter and check whether the confidence interval succeeds in capturing it. The parameter here is the theoretical probability of zero matches, which you calculated in Activity 11-2 on page 220.

d. Report the value of this population parameter.
e. Does the 95% confidence interval succeed in capturing the population parameter?
f. If 1000 different statistics classes carried out this simulation and calculated a 95% confidence interval as you did in part b, approximately how many of their intervals would you expect to succeed in capturing the parameter value? Explain.
g. Use the simulated sample data to form an 80% confidence interval for the parameter. Does this interval succeed in capturing its value? Re-answer part f for this confidence level.

Activity 16-18: Charitable Contributions

16-18, 18-7

The 2004 General Social Survey found that 78.9% of the adult Americans interviewed claimed to have given a financial contribution to charity in the previous twelve months.

a. Describe the parameter value of interest in this situation.
b. If the survey had involved 250 households, determine a 99% confidence interval for this parameter.
c. Repeat part b, supposing the survey had involved 500 households.
d. Repeat part b, supposing the survey had involved 1000 households.
e. Repeat part b, supposing the survey had involved 2000 households.
f. Compare the margins-of-error for these four intervals. Describe how they are related. (Be as specific as possible.)
g. Does doubling the sample size cut the margin-of-error in half? If not, by what factor does the sample size have to be increased in order to cut the margin-of-error in half?
h. The actual survey involved 1334 households. Determine and interpret a 99% confidence interval for the parameter.

Activity 16-19: Marriage Ages

8-17, 9-6, **16-19,** 17-22, 23-1, 23-12, 26-4, 29-17, 29-18

Reconsider Activity 9-6, in which you analyzed the ages of couples who applied for marriage licenses. Now consider the issue of estimating the proportion of all marriages in this county for which the bride is younger than the groom. These 24 couples are actually a subsample from a larger sample of 100 couples who applied for marriage licenses in Cumberland County, Pennsylvania, in 1993. Marriage ages for this larger sample appear in the table that follows:

Husband's Age	Wife's Age	Husband's Age	Wife's Age	Husband's Age	Wife's Age	Husband's Age	Wife's Age	Husband's Age	Wife's Age
22	21	40	46	23	22	31	33	24	25
38	42	26	25	51	47	23	21	25	24
31	35	29	27	38	33	25	25	46	37
42	24	32	39	30	27	27	25	24	23
23	21	36	35	36	27	24	24	18	20
55	53	68	52	50	55	62	60	26	27
24	23	19	16	24	21	35	22	25	22
41	40	52	39	27	34	26	27	29	24
26	24	24	22	22	20	24	23	34	39
24	23	22	23	29	28	37	36	26	18
19	19	29	30	36	34	22	20	51	50
42	38	54	44	22	26	24	27	21	20
34	32	35	36	32	32	27	21	23	23
31	36	22	21	51	39	23	22	26	24
45	38	44	44	28	24	31	30	20	22
33	27	33	37	66	53	32	37	25	32
54	47	21	20	20	21	23	21	32	31
20	18	31	23	29	26	41	34	48	43
43	39	21	22	25	20	71	73	54	47
24	23	35	42	54	51	26	33	60	45

a. For how many of these 100 marriages can you determine which partner was younger? (In other words, eliminate the cases in which both the bride and groom listed the same age on their marriage license because you cannot tell which partner is younger in those cases.)

b. In how many of *these* marriages is the bride younger than the groom?

c. Find a 90% confidence interval for the proportion of all marriages (excluding those where the couple have the same yearly age) in this county for which the bride is younger than the groom.

d. Repeat part c with 95% confidence.

e. Repeat part c with 99% confidence.

f. Do any of these intervals include the value .5?

g. Comment on whether the sample data suggest that the bride is younger than the groom in more than half of the marriages in this county.

Activity 16-20: Critical Values

12-18, 16-2, **16-20**, 19-13

Determine the critical value z^* corresponding to

a. 85% confidence

b. 97.5% confidence

c. 51.6% confidence

d. Which of these three z^* values is largest? Which is smallest? Explain why this makes sense.

Activity 16-21: Wrongful Conclusions

8-6, **16-21,** 17-8, 28-25

Suppose that Andrew and Becky both study a random sample that has a sample proportion of $\hat{p} = .4$. Each uses the sample data to produce a confidence interval for the population proportion, with Andrew obtaining the interval (.346, .474) and Becky obtaining the interval (.286, .514).

a. One of these intervals has to be incorrect. Identify which one, and explain why.

Suppose that Andrew and Becky decide to gather new random samples, with one using a sample size of 100 and the other using a sample size of 200. Andrew's confidence interval turns out to be (.558, .682), and Becky's is (.611, .779).

b. Report the sample proportion $\hat{p}$ obtained by each researcher. (Assume that both intervals were calculated correctly.)

c. Report the margin-of-error for each interval.

d. Explain why the information provided does not enable you to tell which sample size was used by which researcher.

e. Suppose that Andrew had the sample size of 100 and Becky 200. Determine the confidence level used by each. *Hint:* First determine the critical value z^* used by each.

Suppose that Andrew and Becky decide to study another issue with new samples, one with a sample size of 250 and one with a sample size of 1000. Both decide to form 90% confidence intervals from their samples. Andrew obtains the interval (.533, .635), and Becky gets (.550, .602).

f. Report the sample proportion and the margin-of-error for each interval.

g. Which sample size goes with which researcher? Explain.

h. Which researcher is more likely to obtain an interval that succeeds in capturing the population proportion? Explain.

Activity 16-22: Candy Colors

1-13, 2-19, 13-1, 13-2, 15-7, 15-8, 16-4, **16-22,** 24-15, 24-16

Assume for now that 15% of all Reese's Pieces are orange, i.e., that the population proportion of orange candies is $\pi = .15$.

a. Open the `Simulating Confidence Intervals` applet and make sure it is set for Proportions with the Wald method. Set the population proportion value to $\pi =$ **.15,** the sample size to $n =$ **10,** the number of intervals to **200,** and the confidence level to **95%.** Click Sample, and keep clicking until you have generated 1000 intervals. What percentage of these intervals succeed in capturing the true value of the population proportion (which you have set to be $\pi = .15$)? Is this percentage close to 95%?

b. Explain why it is not surprising that the success rate was not very close to 95% in this case. *Hint:* Check the technical conditions for the validity of the confidence interval procedure.

Activity 16-23: Penny Thoughts

2-1, 3-10, **16-23**

Recall from Activity 3-10 that a Harris Poll in 2003 asked a national sample of 2316 adults whether they favored or opposed abolishing the penny, with 59% saying that they oppose it.

a. Clearly define the population parameter of interest.

b. Determine and interpret a 95% confidence interval for the population parameter of interest.

c. Comment on whether the technical conditions are satisfied.

TOPIC

17

Tests of Significance: Proportions

How often does the winning team in a baseball game score more runs in one inning than the losing team scores in the entire game? Does this happen three-quarters of the time, as someone once claimed in a letter to the "Ask Marilyn" columnist? In this topic, you will learn to assess the evidence that sample data provide about such a claim. Another example that you will investigate concerns whether people who need to make up an answer to a question on the spot tend to select certain answers more than others.

Overview

In the previous topic, you took what you learned about the concept of statistical confidence in Unit 3 and studied a procedure known as confidence intervals. In this topic, you will augment what you learned about the concept of statistical significance by studying a procedure known as a test of significance. Such a procedure assesses the degree to which sample data provide evidence against a particular conjecture about the value of the population parameter of interest. Your understanding of the reasoning process behind these tests will deepen, and you will study the formal structure of a test of significance, while learning a specific procedure for conducting a test concerning a population proportion.

Preliminaries

1. A campus legend tells the story of two friends who lied to their professor by blaming a flat tire for their having missed an exam. The professor sent them to separate rooms to take a make-up exam. The first question (worth 5 points) was easy, but the second question, worth 95 points, asked, "Which tire was it?" If you were caught in a situation where you suddenly had to choose a hypothetically flat tire, which of the four tires would you say went flat (left front, right front, left rear, or right rear)?

2. Record the response count for yourself and your classmates:

   Left front: Right front:

   Left rear: Right rear:

3. In what percentage of major-league baseball games would you predict that the winning team scores more runs in one inning than the losing team scores in the entire game?

4. Consider a cola taste test in which a subject is presented with three cups, two that contain the same brand of cola and one that contains a different brand. The subject is to identify which one of three cups contains the different brand. In a test consisting of 60 trials, how convinced would you be that the subject did better than simply guessing if he or she was correct 21 times?

5. How many would a person have to get right out of 60 trials for you to be reasonably convinced that he or she does better than simply guessing at which one is the different brand?

6. If a subject identifies the different brand correctly in 40% of a sample of trials, would you be more impressed if it were a sample of 200 trials or a sample of 20 trials?

● ● ● In Class Activities

Activity 17-1: Flat Tires

17-1, 17-2, 17-4, 17-10, 17-16, 18-18, 24-16

Consider the "which tire?" question asked in the Preliminaries section.

a. If there is nothing special about the right-front tire, what proportion of people will select that tire if a very large population is asked this question? In other words, what is the probability that a person selects the right-front tire, if there's nothing special about it that makes a person prone to select it more or less than the other tires?

b. Is this value a parameter or a statistic? Explain. What symbol represents this value?

It has been conjectured that people tend to pick the right-front tire more than you would expect by random chance.

c. What does this conjecture imply about the value of the population parameter?

In Fall 2005, one of the authors asked her students to answer this question, using a sample size of 73 students.

d. According to the Central Limit Theorem for a sample proportion (Topic 13), if π actually equals .25, what pattern will the sampling distribution of the sample proportion follow? *Hint:* Describe its shape, center, and spread, and also draw a well-labeled sketch of this distribution.

e. Are the conditions met here for the Central Limit Theorem to be valid? Explain.

Of the 73 students sampled, 34 of them selected the right-front tire.

f. Determine the sample proportion of students who chose the right-front tire, and indicate the appropriate symbol for this proportion. Is this sample proportion greater than one-fourth?

g. Shade the area under your curve corresponding to the probability of getting a sample proportion at least as large as the one observed. Calculate the *z*-score and probability (as in Topic 15), still assuming that the right-front tire would be chosen by one-fourth of the population.

h. Based on this probability, would you consider such a sample result (at least 34 of 73 choosing the right-front tire) surprising, if there's nothing special about the right-front tire? Would you consider it so surprising that you would conclude the right-front tire would be chosen by more than one-fourth of the population?

A sample result that is very unlikely to occur by random chance alone is said to be **statistically significant.** We now formalize this process of determining whether a sample result provides statistically significant evidence against a conjecture about the population parameter. The resulting procedure is called a **test of significance.**

The **null hypothesis** is denoted by $\mathbf{H_0}$. It states that the parameter of interest is equal to a specific, hypothesized value. In the context of a population proportion, $\mathbf{H_0}$ has the form

$$H_0: \pi = \pi_0$$

where π represents the population proportion of interest and π_0 is replaced by the specific conjectured value of interest. The null hypothesis is typically a statement of "no effect" or "no difference." A significance test is designed to assess the strength of evidence *against* the null hypothesis.

i. Clearly define π (in words) for the "flat tires" study:

Let π represent the . . .

j. The null hypothesis here is that the right-front tire is chosen no more (or less) often than any other tire. Translate this into a null hypothesis statement by referring to your answer to part a:

H_0: π =

The **alternative hypothesis** is denoted by $\mathbf{H_a}$. It states what the researchers suspect or hope to be true about the parameter of interest. It depends on the purpose of the study and must be specified *before* the sample data are examined. The alternative hypothesis can take one of three forms:

i. $\pi < \pi_0$ *or*
ii. $\pi > \pi_0$ *or*
iii. $\pi \neq \pi_0$

The first two forms are called **one-sided** alternatives, whereas the last form is a **two-sided** alternative.

k. Refer back to the researchers' conjecture and your answer to part c to complete the following alternative hypothesis statement about π:

H_a: π

The **test statistic** is a value computed by *standardizing* the observed sample statistic on the basis of the hypothesized parameter value. It is used to assess the evidence against the null hypothesis. In the context of a population proportion, the test statistic is denoted by z and calculated as follows:

$$z = \frac{\hat{p} - \pi_0}{\sqrt{\frac{\pi_0(1-\pi_0)}{n}}}$$

Note: During significance test calculations, you assume that the null hypothesis is true, so use the hypothesized value of the population proportion (π_0) in the denominator of this test statistic, rather than the sample statistic $\hat{p}$ as you did with a confidence interval.

l. Report the value of the test statistic. *Hint:* This is the z-score you calculated in part g.

The ***p*-value** is the probability, assuming the null hypothesis to be true, of obtaining a test statistic at least as extreme as the one actually observed. *Extreme* means "in the direction of the alternative hypothesis," so the p-value takes one of three forms (corresponding to the appropriate form of H_a):

i. $\Pr(Z \leq z)$ (area to left of z-score) *or*
ii. $\Pr(Z \geq z)$ (area to right of z-score) *or*
iii. $2 \times \Pr(Z \geq |z|)$ (area more extreme than the z-score in both directions)

m. Report the probability that you calculated in part g, which is the p-value of this test.

n. Does this probability suggest that it is very unlikely for 34 or more of 73 students to choose the right-front tire if one-fourth of the population would choose the right-front tire? For instance, would such a sample result occur less than 10% of the time in the long run? Less than 5%? Less than 1%?

Judge the strength of the evidence that the data provide against the null hypothesis by examining the p-value. The *smaller* the p-value, the stronger the evidence *against* H_0 (and thus the stronger the evidence in favor of H_a). For instance, typical evaluations are

- p-value $> .1$: *little* or *no* evidence against H_0
- $.05 < p$-value $\leq .10$: *some* evidence against H_0
- $.01 < p$-value $\leq .05$: *moderate* evidence against H_0
- $.001 < p$-value $\leq .01$: *strong* evidence against H_0
- p-value $\leq .001$: *very strong* evidence against H_0

The **significance level,** denoted by α (alpha), is an optional "cut-off" level for the p-value that the experimenter decides, in advance, to regard as decisive. Common values are $\alpha = .10$, $\alpha = .05$, and $\alpha = .01$. The smaller the significance level, the more evidence you require in order to be convinced that H_0 is not true.

If the p-value of the test is less than or equal to the significance level α, the **test decision** is to *reject* H_0; otherwise, the decision is to *fail to reject* H_0. Another common expression is to say that the data are statistically significant at the α level if the p-value is less than or equal to α. Thus, a result is statistically significant if it is unlikely to have occurred by chance or sampling variability alone (assuming that the null hypothesis is true).

This definition says that a result is statistically significant if it is unlikely to have occurred by chance or sampling variability alone (assuming that the null hypothesis is true). Note that failing to reject H_0 is not the same as affirming its truth; you are simply declaring that the evidence is not convincing enough to reject it.

As with all inference procedures, you also need to consider the conditions that must hold in order for this significance test to be valid. The technical conditions needed to establish the validity of this significance testing procedure are similar to those for the confidence interval procedure.

Technical conditions:

- The data are a simple random sample from the population of interest.
- The sample size is large relative to the hypothesized population proportion. You can check this by seeing whether $n\pi_0 \geq 10$ and $n(1 - \pi_0) \geq 10$.

o. Verify that this condition about the sample size is satisfied for the "flat tires" study.

p. Summarize the conclusion you would draw from this study from the context of the data.

Watch Out

- It might take some time for you to fully understand the reasoning process of a test of significance. Some people find it to be a bit convoluted to start by assuming the null hypothesis to be true and ask how unlikely the observed sample data would be, given that hypothesis. If the answer is that the observed sample data would be *very unlikely* if the hypothesis were true, then the sample data provide strong evidence against that null hypothesis.
- The *p*-value is *not* the probability that the null hypothesis is true. Rather, it is the probability of obtaining such an extreme sample result (or one even more extreme) if the null hypothesis is true.
- As always, be sure to relate your conclusions to the context. For example, do not think that "reject H_0" is a complete conclusion. In fact, it's even incomplete to say "reject H_0 and conclude that $\pi > .25$." You must instead express this conclusion in context by saying, "The sample data provide very strong evidence that more than one-fourth of the population would choose the right-front tire."
- Always include all steps of a significance test:

  1. Identify and define the parameter.
  2. State the null and alternative hypotheses, preferably in words as well as symbols.
  3. Check the technical conditions.

4. Calculate the test statistic.
5. Report the p-value.
6. Summarize your conclusion in context, including a test decision if a significance level α is provided.

Activity 17-2: Flat Tires

17-1, **17-2,** 17-4, 17-10, 17-16, 18-18, 24-16

Reconsider the scenario from the previous activity. In Fall 2006, the other author asked his students to answer this same question. Of the 74 students sampled, 24 selected the right-front tire.

a. Conduct a test of significance (of whether students tend to choose the right-front tire more than one-fourth of the time) based on these sample data. Report the null and alternative hypotheses (in symbols and in words), test statistic, and p-value. Summarize your conclusion, and indicate whether the sample result is statistically significant at the $\alpha = .05$ level. Also check whether the technical conditions required for this procedure hold.

Define parameter of interest:

H_0:

H_a:

Check technical conditions:

Test statistic $z =$

p-value $=$

Test decision at $\alpha = .05$ level:

Conclusion in context:

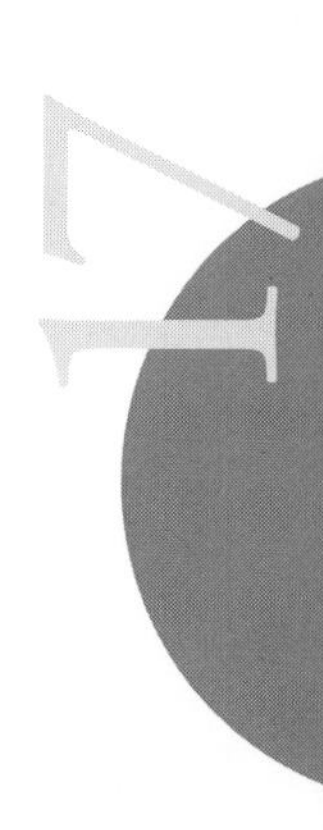

b. Confirm your calculation of the test statistic and p-value by using the `Test of Significance Calculator` applet. Make sure the procedure is set to One Proportion. Enter the value of π_0 for the null hypothesis and click the inequality symbol (>) to specify the correct form for the alternative hypothesis. Then enter the sample size n and sample number of successes (Count) and click Calculate. The applet will automatically calculate the value of the sample proportion $\hat{p}$.

Test of Significance Calculator

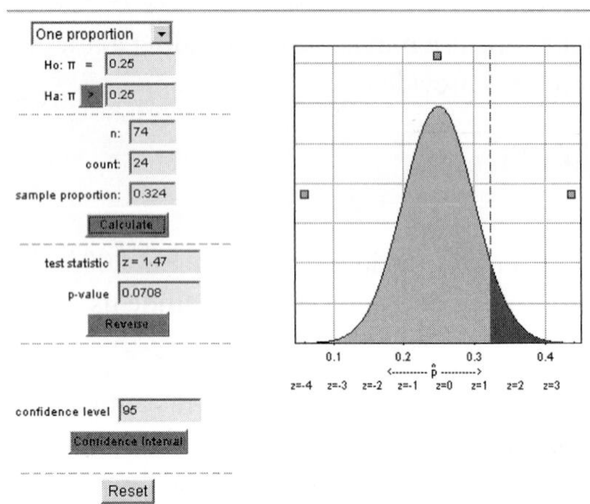

Activity 17-3: Racquet Spinning

11-9, 13-11, **17-3,** 17-18, 18-3, 18-12, 18-13

Recall from Activity 11-9 that a spun tennis racquet yielded 46 "up" results in a sample of 100 spins. You will test whether this sample result provides strong evidence that the racquet will *not* land "up" 50% of the time in the long run.

a. Describe the population parameter of interest.

b. Specify the null and alternative hypotheses, in symbols and in words. *Hint:* Is a direction specified for which side of 50% the proportion is suspected to fall on? If not, use a two-sided alternative.

c. Comment on whether the technical conditions for the validity of this test procedure are satisfied.

d. Calculate the test statistic.

e. Calculate the p-value. *Hint:* With a two-sided alternative, *extreme* is measured in both directions, so double the area under the standard normal curve to the right of the absolute value of the test statistic. It is very helpful to include a sketch of the standard normal curve with the p-value shaded.

f. Based on this p-value, would you reject the null hypothesis at the .05 significance level? Explain.

g. Summarize your conclusion in context.

Watch Out

Be careful never to "accept" a null hypothesis. Even when the p-value is not terribly small, the appropriate conclusion is that the sample data do not provide *enough* evidence to reject the null hypothesis in favor of the alternative. In other words, a test of significance only assesses the evidence *against* the null hypothesis, not in favor of it.

Activity 17-4: Flat Tires

17-1, 17-2, **17-4,** 17-10, 17-16, 18-18, 24-16

Reconsider the "flat tires" situation from Activities 17-1 and 17-2. Now suppose that 30% of a random sample select the right-front tire.

a. Consider the question of whether this sample result constitutes strong evidence that the right-front tire would be chosen more than one-quarter of the time in the long run. Do you need more information to answer this question? Explain.

b. Suppose that this sample result (30% answering "right-front") had come from a sample of $n = 50$ people. Use your calculator to conduct the appropriate test of significance by using the **1-PropZTest** command. The **1-PropZTest** can be found by pressing STAT and then selecting the TESTS menu. (*TI hint:* Your TI-83 Plus or TI-84 Plus calculator uses p_0 instead of π_0.) Record in the first row of the following table the value of the test statistic and the test's p-value. Also indicate (yes or no) whether the test is significant at each of the listed significance levels.

c. Repeat part b for the other sample sizes listed in the table.

Sample Size	# "right-front"	$\hat{p}$	z statistic	p-value	$\alpha = .10$?	$\alpha = .05$?	$\alpha = .01$?	$\alpha = .001$?
50	15	.30						
100	30	.30						
150	45	.30						
250	75	.30						
500	150	.30						
1000	300	.30						

d. Write a few sentences summarizing what your analysis reveals about whether a sample result of 30% is significantly greater than a hypothesized value of 25%.

Watch Out

You cannot conduct a significance test without knowing the sample size involved. Sample size plays a key role in tests of significance. The statistical significance of a sample result depends largely on the sample size involved. With large sample sizes, even small differences (such as 30% in a sample compared to a hypothesized population percentage of 25%) can be statistically significant, because they are unlikely to occur by chance.

Activity 17-5: Baseball "Big Bang"

17-5, 17-17

A reader wrote in to the "Ask Marilyn" column in *Parade* magazine to say that his grandfather told him that in three-quarters of all baseball games, the winning team scores more runs in one inning than the losing team scores in the entire game. (This phenomenon is known as a "big bang.") Marilyn responded that this proportion seemed too high to be believable. Let π be the proportion of all major-league baseball games in which a "big bang" occurs.

a. Restate the grandfather's assertion as the null hypothesis, in symbols and in words.

b. Given Marilyn's conjecture, state the alternative hypothesis, in symbols and in words.

To investigate this claim, we randomly selected one week of the 2006 major-league baseball season, which turned out to be July 31–August 6, 2006. Then we examined the 95 games played that week to determine which had a big bang and which did not.

c. Sketch and label the sampling distribution for the sample proportion of games containing a big bang, according to the Central Limit Theorem, assuming that the grandfather's null hypothesis is true. Also check whether the conditions hold for the CLT to apply.

Of the 95 games in our sample, 47 contained a big bang.

d. Calculate the sample proportion of games in which a big bang occurred. Use an appropriate symbol to denote it.

e. Is this sample proportion less than three-fourths and therefore consistent with Marilyn's (alternative) hypothesis? Shade the area under your sampling distribution curve corresponding to this sample result in the direction conjectured by Marilyn.

f. Calculate the test statistic and use the Standard Normal Probabilities Table (Table II) your calculator to find its p-value.

$z =$

p-value $=$

g. Based on this p-value, would you say that the sample data provide strong evidence to support Marilyn's contention that the proportion cited by the reader's grandfather is too high to be the actual value? Explain. Also indicate what test decision you would reach at the $\alpha = .01$ level.

In her response, Marilyn went on to conjecture that the actual proportion of "big bang" games is one-half.

h. Using a two-sided alternative, state the null and alternative hypotheses (in symbols and in words) for testing Marilyn's claim.

i. Use your calculator to determine the test statistic and p-value for this test.

j. What conclusion would you draw concerning Marilyn's conjecture?

k. Use the sample data to produce a 95% confidence interval to estimate the proportion of all major-league baseball games that contain a big bang. Interpret this interval, and comment on what it reveals about the grandfather's claim and Marilyn's response.

Solution

a. The null hypothesis is that the proportion of all major-league baseball games that contain a big bang is three-fourths. In symbols, the null hypothesis is H_0: $\pi = .75$.

b. The alternative hypothesis is that less than three-fourths of all major-league baseball games contain a big bang. In symbols, the alternative hypothesis is H_a: $\pi < .75$.

c. The CLT applies here because $95(.75) = 71.25$ is greater than 10, and $95(.25) = 23.75$ is also greater than 10. According to the CLT, the sample proportion would vary approximately normally with mean .75 and standard deviation

$$\sqrt{\frac{(.75)(.25)}{95}} \approx .0444$$

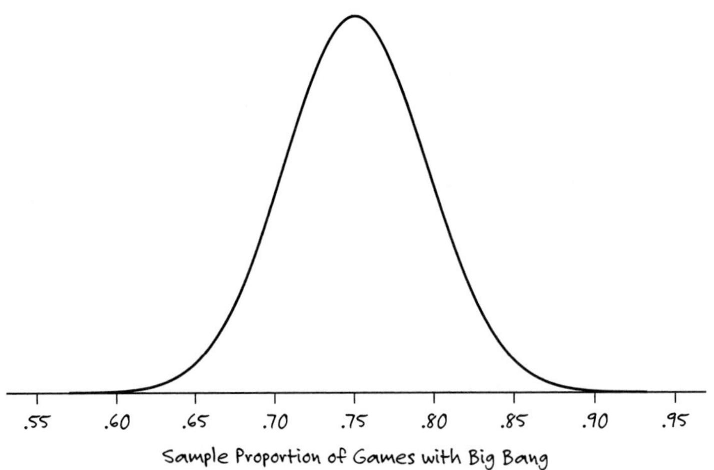

d. The sample proportion of games in which a big bang occurred is

$$\hat{p} = \frac{47}{95} \approx .495$$

e. Yes, this sample proportion is less than .75, as Marilyn conjectured.

f. The test statistic is

$$z = \frac{.495 - .75}{\sqrt{\frac{(.75)(.25)}{95}}} \approx \frac{.495 - .75}{.0444} \approx -5.74$$

This statistic says that the observed sample result is almost six standard deviations below what the grandfather conjectured. This z-score is way off the chart in Table II, indicating that the p-value is virtually zero.

g. Yes, this very small p-value indicates that the sample data provide extremely strong evidence against the grandfather's claim. There is extremely strong evidence that less than 75% of all major-league baseball games contain a big bang. The null (grandfather's) hypothesis would be rejected at the $\alpha = .01$ level.

h. The hypotheses for testing Marilyn's claim are H_0: $\pi = .5$ vs. H_a: $\pi \neq .5$.

i. The test statistic is

$$z = \frac{.495 - .5}{\sqrt{\frac{(.5)(.5)}{95}}} \approx \frac{.495 - .5}{.0513} \approx -0.10$$

The p-value is $2(.4602) = .9204$.

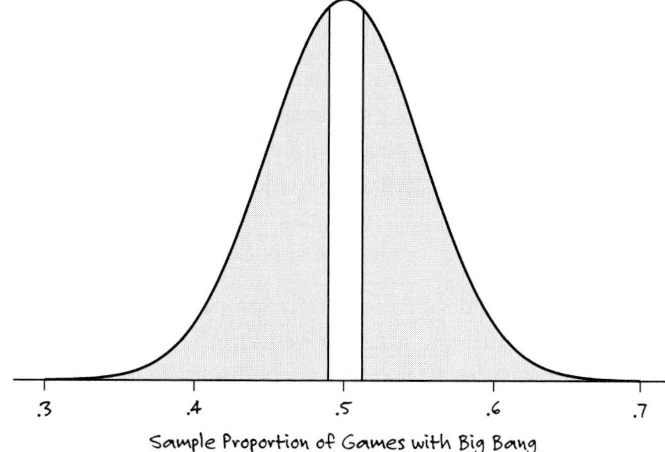

j. This p-value is not small at all, suggesting that the sample data are quite consistent with Marilyn's hypothesis that half of all games contain a big bang. The sample data provide no reason to doubt Marilyn's hypothesis.

k. A 95% confidence interval for π (the population proportion of games that contain a big bang) is given by

$$\hat{p} \pm z^* \sqrt{\frac{\hat{p}(1-\hat{p})}{n}}$$

which is

$$.495 \pm 1.96\sqrt{\frac{.495^*.505}{95}}$$

which is .495 ± .101, which is the interval from .394 to .596. Therefore, you are 95% confident that between 39.4% and 59.6% of all major-league baseball games contain a big bang. The grandfather's claim (75%) is not within this interval or even close to it, which explains why it was so soundly rejected. Marilyn's conjecture (50%) is well within this interval of plausible values, which is consistent with it not being rejected.

Watch Out

- Remember that the hypotheses are always statements about a parameter, not about a statistic. The whole point is to see what you can infer about an unknown parameter value based on a sample statistic.
- Remember to use the hypothesized value of the parameter (denoted by π_0) of the test statistic calculation under the square root in the denominator and also in checking the technical conditions. It is easy to mistakenly use the sample proportion $\hat{p}$ in those calculations.
- Try to carry many decimal places of accuracy in intermediate calculations. If you round too much in an early calculation, that error can get magnified in subsequent calculations.
- Again remember that you do not "accept" a null hypothesis, even one with a p-value as great as Marilyn's. The sample data are in very close agreement with Marilyn's hypothesis, but you still should not conclude that exactly 50% of all games contain a big bang.
- Even with an extremely small p-value, stop short of saying that the data *prove* that the null hypothesis is false. Even though you have overwhelming evidence against the grandfather's claim, you have not technically *proven* that his claim is wrong.

Wrap-Up...........

This topic has introduced you to the reasoning process and formal structure of a **test of significance.** The reasoning process asks how often the observed sample result, or one even more extreme, would occur purely by chance, given the hypothesized value of the population parameter. If such an extreme sample result turns out to be unlikely, then the sample data provide evidence against the hypothesized parameter value.

This basic structure of a test of significance involves the following steps:

1. Identify and define the population parameter of interest.
2. State the **null** and **alternative hypotheses** based on the study question.
3. Check whether the technical conditions required for the procedure to be valid are satisfied.
4. Calculate the **test statistic,** which measures the distance between the observed sample statistic and the hypothesized value of the parameter.
5. Calculate the ***p*-value,** which reports the probability of obtaining such an extreme test statistic value when the null hypothesis is true. It is often very helpful to include a sketch of the sampling distribution, shading in the appropriate *p*-value area.
6. Draw a conclusion about the study question (in context) based on the magnitude of this *p*-value.

You studied how to implement this process with a *z*-test about a population proportion. In later topics, you will see that the structure and reasoning is the same for many other situations (e.g., comparing two population means).

You also learned that the smaller the *p*-value, the stronger the evidence against the null hypothesis. In other words, a small *p*-value indicates that the observed result would be surprising if the null hypothesis were true, and so provides evidence that the null hypothesis is not true. For example, the sample baseball data provided very strong evidence against the grandfather's "big bang" claim (that $\pi = .75$) because the tiny *p*-value revealed that the sample data observed ($\hat{p} = .495$) would have been extremely unlikely to occur if the grandfather's claim had been true. Instead of thinking that we happened to observe an incredibly unlucky sample, we will conclude that the data provide overwhelming evidence against the grandfather's claim. On the other hand, the *p*-value associated with Marilyn's claim (that $\pi = .5$) was not small, and so the sample data were not inconsistent with her claim (are consistent with sampling variability), providing no reason to doubt her hypothesis.

You also discovered that the sample size in a study plays a large role in calculating the *p*-value and, therefore, in determining whether the sample result is statistically significant (i.e., unlikely to occur due to sampling variability alone). For example, simply learning that 30% of a sample chooses the right-front tire does not enable you to say whether this is (statistically) significantly greater than 25%. If 30% of a sample of 10 people answer "right front," then this result is not at all significant. But if the sample contains 1000 people, then 30% answering "right front" is significantly greater than 25%, because such an extreme result would almost never happen by random chance alone.

Some useful definitions to remember and habits to develop from this topic include

- State the null and alternative hypotheses based on the research question *before* examining the sample data.
- The alternative hypothesis can have one of three different forms, corresponding to $<$, $>$, and $\neq$. The first two of these are **one-sided** alternatives and the last is a **two-sided** alternative.

- Always check the technical conditions before applying the test procedure.
- The test statistic provides a measure of how far the observed sample result is from the hypothesized value. With this test for a population proportion, the test statistic is the z-score for the sample proportion under the assumption that the null hypothesis is true.
- The p-value reports the probability of obtaining such an extreme sample result or more extreme by chance alone when the null hypothesis is true.
- The direction of the alternative hypothesis indicates how to calculate the p-value from the standard normal table.
- The smaller the p-value, the stronger the evidence against the null hypothesis.
- When the p-value is less than (or equal to) a prespecified **significance level,** the **test decision** is to reject the null hypothesis. Otherwise, the test decision is to fail to reject the null hypothesis.
- The greater the sample size, the smaller the p-value (if all else remains the same), and, therefore, the stronger the evidence against the null hypothesis (as long as the observed sample result is in the direction of the alternative hypothesis).
- As always, remember to relate your conclusions to the study's context and research question!

Confidence intervals and significance tests are the most widely used techniques in statistical inference, so you will continue to study them in the next topic. You will discover how these two techniques are related to each other, and you will learn to avoid some common misinterpretations.

Homework Activities

Activity 17-6: Interpreting p-values

17-6, 17-7

Suppose you conduct a significance test and decide to reject the null hypothesis at the $\alpha = .05$ level.

a. If you were to use the $\alpha = .10$ level instead, would you reject the null hypothesis, fail to reject it, or do you not have enough information to know? Explain. *Hint:* What must be true about the p-value, knowing that you reject the null hypothesis at the $\alpha = .05$ level?

b. If you were to use the $\alpha = .01$ level instead, would you reject the null hypothesis, fail to reject it, or do you not have enough information to know? Explain.

Now suppose you conduct a different significance test on a different set of data, and your test decision is to fail to reject the null hypothesis at the $\alpha = .05$ level.

c. If you were to use the $\alpha = .03$ level instead, would you reject the null hypothesis, fail to reject it, or do you not have enough information to know? Explain.

d. If you were to use the $\alpha = .07$ level instead, would you reject the null hypothesis, fail to reject it, or do you not have enough information to know? Explain.

Activity 17-7: Interpreting p-values

17-6, **17-7**

a. Is it possible for a p-value to be greater than .5? If so, explain the circumstances under which this could happen. If not, explain why not.

b. Is it possible for a p-value to be greater than 1? If so, explain the circumstances under which this could happen. If not, explain why not.

Activity 17-8: Wrongful Conclusions

8-6, 16-21, **17-8,** 28-25

Describe what is incorrect about each of the following hypotheses.

a. $H_0: \hat{p} = .5$
b. $H_0: \pi = 1.2$
c. $H_0: \pi = .5, H_a: \pi \geq .5$
d. $H_0: \pi = .5, H_a: \pi \neq .6$
e. $H_0: \pi \neq .5, H_a: \pi = .5$

Activity 17-9: Penny Activities

16-9, 16-10, 16-11, **17-9,** 18-14, 18-15

Suppose we claim to have a special penny that lands "heads" when flipped more than half the time. Suppose that we also tell you that we tossed this penny multiple times and obtained 75% "heads". Would you be reasonably convinced that this was not a fair coin? If so, explain why. If not, describe what additional information you would ask for and explain why that information is necessary.

Activity 17-10: Flat Tires

17-1, 17-2, 17-4, **17-10,** 17-16, 18-18, 24-16

Conduct a significance test of whether *your* class data (collected in the Preliminaries section) provides strong evidence that more than one-fourth of the students at your school would select the right-front tire. Report all components of the significance test. Summarize your conclusion, and explain the reasoning process that follows from your test.

Activity 17-11: Credit Card Usage

1-9, 16-12, **17-11,** 19-8, 20-10

Reconsider the study described in Activity 16-12. The Nellie Mae organization found that in a random sample of 1413 undergraduate students taken in 2004, 76% of the students sampled had a credit card. Conduct a significance test of whether this sample result provides strong evidence that more than 75% of all undergraduate college students in the U.S. had a credit card in 2004. Report the hypotheses, test statistic, and *p*-value. Also check the technical conditions of the test and summarize your conclusion.

Activity 17-12: Kissing Couples

13-3, 13-4, 16-6, **17-12,** 24-4, 24-13

Recall from Activities 13-3, 13-4, and 16-6 the study in which a sample of 124 kissing couples were observed, with 80 of them leaning their heads to the right.

a. Use these sample data to test whether more than half of the population of kissing couples lean their heads to the right. Report the hypotheses, test statistic, and *p*-value. Summarize your conclusion at the $\alpha = 0.01$ significance level, and explain how your conclusion follows from your test.
b. Repeat part a, testing whether the population proportion of kissing couples who lean to the right is less than three-fourths.
c. Repeat part a, testing whether the population proportion of kissing couples who lean to the right differs from two-thirds.

Activity 17-13: Political Viewpoints

17-13, 24-10

The 2004 General Social Survey asked a random sample of 1309 American adults to report their political viewpoint as liberal, moderate, or conservative. The number who classified themselves as moderate was 497.

a. Can you determine whether more than one-third of the sample consider themselves to be political moderates? If so, answer the question. If not, explain why not.

b. Suppose that one-third of the population consider themselves to be political moderates. What then would be the probability that 497 or more in a random sample of 1309 people would call themselves political moderates? *Hint:* Answer this question by determining the *p*-value of the relevant test.

c. How would you expect the *p*-value to change if the sample had contained 327 people, of whom 124 called themselves political moderates? Explain. *Hint:* Note that the sample size and number of moderates have been reduced by a factor of four.

d. Determine the *p*-value based on the sample data in part c. Was your answer to part c correct?

Activity 17-14: Calling "Heads" or "Tails"

13-10, **17-14,** 17-15, 24-18

Refer to the data collected in Topic 13 and analyzed in Activity 13-10 about whether you would call "heads" or "tails" if asked to predict the result of a coin flip.

a. What proportion of the responses were "heads"?

b. Is this proportion a parameter or a statistic? Explain.

c. Write a sentence identifying the parameter (and population) of interest in this situation.

d. Specify the null and alternative hypotheses, in words and in symbols, for testing whether the sample result differs significantly from .5.

e. Sketch and label the sampling distribution for the sample proportion specified by the null hypothesis. Shade the region corresponding to the *p*-value. Also provide a check of the technical conditions.

f. Calculate the test statistic and *p*-value for this test.

g. Is the sample result statistically significantly different from .5 at the .10 level? At the .05 level? At the .01 level?

h. Write a few sentences summarizing and explaining your conclusion.

i. Describe specifically what would have changed in this analysis if you had worked with the proportion of "tails" responses rather than "heads" responses.

Activity 17-15: Calling "Heads" or "Tails"

13-10, 17-14, **17-15,** 24-18

Refer to the previous activity, where you examined data on the proportion of students who would respond "heads" if asked to predict a coin flip. In his book *Statistics You Can't Trust,* Steve Campbell claims that people call "heads" 70% of the time when asked to predict the result of a coin flip. Conduct a test of whether your sample data provide evidence against Campbell's hypothesis. Report the hypotheses, sketch the sampling distribution specified by the null hypothesis, check the technical conditions, and calculate the test statistic and *p*-value. Write a few sentences describing your conclusion.

Activity 17-16: Flat Tires

17-1, 17-2, 17-4, 17-10, **17-16,** 18-18, 24-16

Reconsider the hypothetical results presented in Activity 17-4 for the "flat tires" question. Determine the smallest sample size n for which a sample result of 30% answering "right-front" would be significant at the .10 level. *Hint:* You may either use trial and error with your calculator or work analytically with the normal table and the formula for the test statistic.

Activity 17-17: Baseball "Big Bang"

17-5, **17-17**

Consider again the "big bang" phenomenon described in Activity 17-5. Statistician Hal Stern examined all 968 baseball games played in the National League in 1986 and found that 419 of them contained a "big bang."

a. Perform the appropriate test to see whether this sample proportion differs significantly from .5 at the $\alpha = .02$ level. Report your hypotheses in symbols and in words, your well-labeled sketch of the sampling distribution under the null hypothesis, your check of the technical conditions, the test statistic, and the p-value, in addition to stating and explaining your conclusion.

b. If you redefine "big bang" to mean that the winning team scores *at least* as many (instead of more) runs in one inning as the losing team scores in the entire game, then 651 of those 968 games contained a big bang. Does this sample proportion differ significantly from the "Ask Marilyn" reader's grandfather's assertion of .75 at the $\alpha = .08$ level? Again report the details of your analysis.

Activity 17-18: Racquet Spinning

11-9, 13-11, 17-3, **17-18,** 18-3, 18-12, 18-13

Refer to Activity 17-3, where you performed a test of whether a spun tennis racquet would *not* land "up" 50% of the time in the long run.

a. Report the p-value, and indicate whether you would reject the null hypothesis at the .05 level.

b. Does the test result indicate that the racquet would definitely land up 50% of the time in the long run? Explain.

c. What is the smallest significance level at which you would reject the null hypothesis? *Hint:* Do not confine your consideration to common α levels.

d. Explain precisely how your analysis would change depending on whether you work with the proportion landing "up" or the proportion landing "down."

e. Use the sample data to find a 95% confidence interval for the long-run proportion of times that the racquet would land "up."

f. Does this interval include the value .5?

g. Explain the consistency between your answers to parts a and f.

Activity 17-19: Therapeutic Touch

5-21, **17-19**

In the "therapeutic touch" experiment described in Activity 5-21 on page 89, subjects identified which of their hands the experimenter had placed her hand over.

a. Identify the null and alternative hypotheses, in symbols and in words, for testing whether the subjects could distinguish more often than not over which hand the experimenter's hand was held. Also clearly identify the parameter of interest in words.

b. Combining the results of the 21 subjects, there were a total of 123 correct identifications in 280 repetitions of the experiment. Use these sample data to conduct the test of the hypotheses specified in part a. Report the test statistic and *p*-value.
c. Explain why it makes sense that the *p*-value is greater than .5 in this situation.
d. Is it fair to conclude that you should "accept the null hypothesis" in this situation? Explain.
e. What conclusion do you draw from this study about the effectiveness of therapeutic touch?

Activity 17-20: Magazine Advertisements

16-15, **17-20,** 21-21

Recall from Activity 16-15 that the September 13, 1999, issue of *Sports Illustrated* had 336 pages, and 54 of them contained an advertisement. Prior to collecting these data, subscriber Frank Chance conjectured that 30% of the magazine's pages contain ads.

a. Write a sentence identifying the parameter of interest here. Also indicate a symbol used to represent that parameter.
b. Express the subscriber's conjecture in symbols. Is this the null or the alternative hypothesis?
c. Treat the pages of this issue as a sample of the population of all *Sports Illustrated* pages. Conduct a significance test of whether the sample data provide strong evidence against the subscriber's conjecture.
d. Write a few sentences describing and explaining your conclusion.
e. Do you think that the technical conditions for the procedure in part c to be valid have been met? Explain.

17

Activity 17-21: Hiring Discrimination

17-21, 18-21

In the case of *Hazelwood School District vs. United States* (1977), the U.S. government sued the City of Hazelwood, a suburb of St. Louis, on the grounds that it discriminated against African Americans in its hiring of school teachers (Finkelstein and Levin, 1990). The statistical evidence introduced noted that of the 405 teachers hired in 1972 and 1973 (the years following the passage of the Civil Rights Act), only 15 had been African American. If you include the city of St. Louis itself, then 15.4% of the teachers in the county were African American; if you do not include the city of St. Louis, then 5.7% of the teachers in the county were African American.

a. Identify the parameter of interest here in words.
b. Conduct a significance test to assess whether the proportion of African American teachers hired by the school district is statistically significantly less than .154 (the proportion of county teachers who were African American). Use the .01 significance level. Along with your conclusion, report the null and alternative hypotheses, a sketch of the sampling distribution specified by the null hypothesis, a check of the technical conditions, the test statistic, and the *p*-value.
c. Conduct a significance test to assess whether the proportion of African American teachers hired by the school district is statistically significantly less than .057 (the proportion of county teachers who were African American if you exclude the city of St. Louis). Again use the .01 significance level and report the null and alternative hypotheses, test statistic, and *p*-value along with your conclusion.

d. Write a few sentences comparing and contrasting the conclusions of these tests with regard to the issue of whether the Hazelwood School District was practicing discrimination.

Activity 17-22: Marriage Ages

8-17, 9-6, 16-19, **17-22,** 23-1, 23-12, 26-4, 29-17, 29-18

Reconsider Activity 16-19, in which you analyzed sample data and found the sample proportion of marriages in which the bride was younger than the groom. Conduct a test of significance to address whether the sample data support the theory that the bride is younger than the groom in more than half of all the marriages in that particular county. Report the details of the test and write a short paragraph describing and explaining your findings.

Activity 17-23: Veterans' Marital Problems

17-23, 18-20

Researchers found that in a sample of 2101 Vietnam veterans, 777 had been divorced at least once (Gimbel and Booth, 1994). U.S. Census figures indicate that among all American men aged 30–44 when the study was conducted in 1985, 27% had been divorced at least once. Conduct a test of significance to assess whether the sample data from the study provide strong evidence that the divorce rate among all Vietnam veterans is greater than 27%. Report the null and alternative hypotheses (identifying the symbols you introduce), the test statistic, and the p-value. Also write a one-sentence conclusion and explain the reasoning process by which the conclusion follows from the test results.

Activity 17-24: Distinguishing Between Colas

13-13, **17-24,** 18-9

Reconsider a cola discrimination taste test in which a subject is presented with three cups, two of which contain the same brand of cola and one of which contains a different brand. The subject is to identify which one of the three cups contains a different brand of cola than the other two.

a. State the relevant null and alternative hypotheses for testing whether the subject would correctly identify the different brand more than one-third of the time in the long run.

Suppose a test consists of 60 trials.

b. Determine the test statistic and p-value if the subject identifies the different brand correctly on 21 trials.

c. Determine the test statistic and p-value if the subject identifies the different brand correctly on 30 trials.

d. Determine the smallest number of correct identifications that would lead to rejecting the null hypothesis at the $\alpha = .05$ significance level. Along with your answer, report the value of the test statistic and p-value corresponding to that answer. *Hint:* First draw a sketch of the sampling distribution of the sample proportion $\hat{p}$.

e. Repeat part d for the $\alpha = .01$ significance level.

f. Which answer (part d or e) is greater? Explain why this makes sense.

TOPIC

18 • • • More Inference Considerations

Do one-third of all American households own a pet cat? If not, is the actual proportion close to one-third? If a baseball player improves his batting success rate substantially, or if a new drug succeeds at alleviating pain more often than the standard drug, is a test of significance guaranteed to reveal the improvements? If not, what factors affect how likely the test is to show the improvements? Finally, if an alien landed on Earth and set out to estimate the proportion of human beings who are female, what might the alien do wrong in constructing its confidence interval? In this topic, you will examine such dissimilar, occasionally even silly, questions as you explore some of the finer points of confidence intervals and significance tests.

Overview

In the previous two topics, you explored and applied the two principal techniques of statistical inference: confidence intervals and tests of significance. This topic will give you more experience with applying these procedures, and you will also investigate their properties, including some fairly subtle ones, further. More specifically, you will consider the relationship between intervals and tests, learn to watch for ways in which these techniques are sometimes misapplied, and explore the important concept of power.

Preliminaries

1. Guess what proportion of American households includes a pet cat.

2. If 31.6% of a random sample of American households includes a pet cat, would you be fairly convinced that the proportion of all American households who have a cat is different than one-third?

3. If 31.6% of a random sample of American households includes a pet cat, would you be fairly convinced that the proportion of all American households who have a cat is much different than one-third?

4. Suppose a baseball player who has always been a .250 career hitter (meaning that his hitting success probability has been 1/4) suddenly improves over one winter to the point where he now has a 1/3 success probability of getting a hit during an at-bat. Do you think he would be likely to convince the team manager of his improvement in a trial consisting of 30 at-bats?

5. Do you think this baseball player would be more or less likely to convince the team manager of his improvement in a trial of 100 at-bats?

In-Class Activities

Activity 18-1: Generation M

3-8, 4-14, 13-6, 16-1, 16-3, 16-7, **18-1,** 21-11, 21-12

Recall from Activity 16-1 that a random sample of 2032 youths aged 8–18 revealed that 1381 (68%) have a television in their bedrooms.

a. Is 68% a parameter or a statistic? Explain.

b. Determine a 99% confidence interval for the population proportion of American 8–18-year-olds who have a television in their bedrooms (call this value π). (Feel free to use your calculator. Select **1-PropZInt** by pressing STAT, selecting the TESTS menu, and arrowing down to select it.)

c. Does the value .65 fall within this interval? What about .70, .6667, and .707?

.65: .70: .6667: .707:

d. Based solely on your answers to part c, would you expect a significance test to reject that the population proportion equals .65 at the $\alpha = .01$ level? Explain. What about testing that the population proportion equals .7? *Hint:* Remember that a confidence interval contains plausible values of the parameter.

.65: .70:

e. Use your calculator to conduct a two-sided test of whether the sample data provide strong evidence that the population proportion who have a television in their bedrooms differs from .65 at the .01 level. To do this, you will need to use **1-PropZTest** (press STAT and select the TESTS menu). Enter **.65** for P_0, **68%** of

2032 for n, and select the $\neq$ symbol. Your Home screen should look like the following:

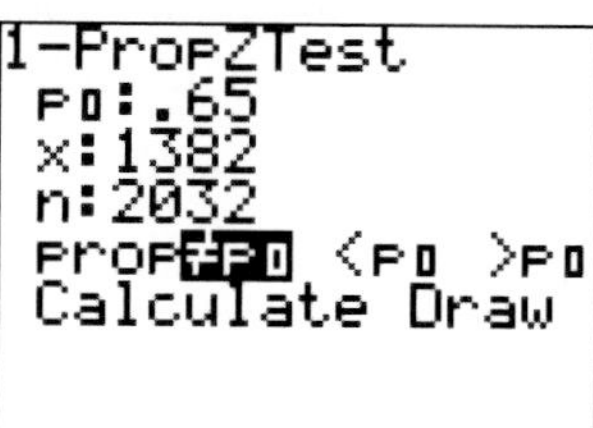

Select **Calculate** and then record the test statistic and p-value in the first row of the following table. Also indicate whether the result is significant at the .01 level. Was your prediction in part d correct?

f. Repeat part e for testing whether the population proportion differs from .7.

g. Repeat part e for two more hypothesized values of the population proportion: two-thirds (approximately .6667) and .707.

Hypothesized Value	Contained in 99% Confidence Interval?	Test Statistic	*p*-value	Significant at .01 Level?
.65				
.6667				
.7				
.707				

h. Do you notice any connection between whether a 99% confidence interval for π includes a particular value and whether the sample proportion differs significantly from that particular value at the $\alpha = .01$ level? Explain.

This activity reveals a **duality** between confidence intervals for estimating a population parameter and a two-sided test of significance regarding the value of that parameter. Roughly speaking, if a 99% confidence interval for a parameter does not include a particular value, then a two-sided test of whether the parameter equals that particular value will be statistically significant at the $\alpha = .01$ level. The same is true for a 90% confidence interval with the .10 significance level and for a 95% confidence interval with the .05 significance level and so on.

Confidence intervals and tests of significance are complementary procedures. Whereas tests of significance can establish strong evidence that a parameter differs from a hypothesized value, confidence intervals estimate the magnitude of that difference.

Watch Out

This duality does not always work exactly when the parameter is a population proportion, because the confidence interval (CI) uses the standard error of $\hat{p}$ and the test statistic uses the standard deviation of $\hat{p}$ based on the hypothesized value of π. But the principle still holds, and any departures from the rule will be very small except with small sample sizes.

Activity 18-2: Pet Ownership

13-9, 13-14, 13-15, **18-2,** 20-21

As you first saw in Activity 13-14, a sample survey of 80,000 households in 2001, conducted by the American Veterinary Medical Association and published in the *Statistical Abstract of the United States 2006,* found that 31.6% of households sampled owned a pet cat.

a. Is this number a parameter or a statistic? Explain, and indicate the symbol used to represent it.

b. Use your calculator to conduct a test of whether the sample data provide evidence that the population proportion of households who own a pet cat differs from one-third. State the hypotheses, and report the test statistic and p-value. State your test decision for the $\alpha = .001$ significance level. Summarize your conclusion, stating your comments in the context of this study.

c. Use your calculator to produce a 99.9% confidence interval for the population proportion of all American households that own a pet cat. Interpret this interval.

d. Is the confidence interval consistent with the test result? Explain.

e. Do the sample data provide *very* strong evidence that the population proportion of households who own a pet cat is not one-third? Explain whether the p-value or the confidence interval helps you to decide.

f. Do the sample data provide strong evidence that the population proportion of households who own a pet cat is *very* different from one-third? Explain whether the *p*-value or the confidence interval helps you to decide.

This activity illustrates that *statistical* significance is not the same thing as *practical* significance. A statistically significant result is simply one that is unlikely to have occurred by chance alone, which does not necessarily mean the result is substantial or important in a practical sense. Although there is strong reason to believe the actual population proportion of households owning a pet cat does indeed differ from one-third, that proportion is actually *quite close* to one-third—close enough to be not worth arguing over for most people. When you work with very large samples, an unimportant result can be considered "statistically significant." The *p*-value tells you whether the difference could have arisen by chance alone, not whether it is interesting or important. Confidence intervals are useful for estimating the size of the effect involved and should be used in conjunction with significance tests.

Activity 18-3: Racquet Spinning

11-9, 13-11, 17-3, 17-18, **18-3,** 18-12, 18-13

Consider again the racquet spinning exercise from Activities 11-9 and 17-3, and continue to suppose a goal is to determine whether the sample data provide evidence that the proportion of "up" results would differ from .5 in the long run.

a. Suppose you were to spin the racquet 200 times and obtain 113 "up" results. Use your calculator to determine the appropriate test statistic and *p*-value. Also report the value of the sample proportion of "up" landings and whether that sample proportion differs significantly from .5 at the $\alpha = .05$ level.

Sample proportion: Test statistic: *p*-value: Significant at .05?

b. Repeat part a supposing you obtained 115 "up" results in 200 spins.

Sample proportion: Test statistic: *p*-value: Significant at .05?

c. Repeat part a supposing that you obtained 130 "up" results in 200 spins.

Sample proportion: Test statistic: *p*-value: Significant at .05?

d. In which pair of cases (a and b, a and c, or b and c) are the sample results most similar?

e. In which pair of cases (a and b, a and c, or b and c) are the decisions about significance at the .05 level the same?

The moral here is that it is unwise to treat standard significance levels as "sacred." It is much more informative to consider the *p*-value of the test and to base your decision on the strength of evidence provided by the *p*-value. There is no sharp border between "significant" and "insignificant," only increasingly strong evidence as the *p*-value decreases. Reports of significance tests should include the sample information and p-value, not just a statement of statistical significance or a decision about rejecting a hypothesis.

Activity 18-4: Female Senators

6-12, **18-4,** 18-8

Suppose an alien lands on Earth, notices that the human species has two different sexes, and sets out to estimate the proportion of humans who are female. Fortunately, the alien had a good statistics course on its home planet, so it knows to take a sample of human beings and produce a confidence interval. Suppose the alien then happens upon the members of the 2007 U.S. Senate as its sample of human beings, so it finds 16 women and 84 men in its sample.

a. Use this sample information to form a 95% confidence interval for the actual proportion of all humans who are female.

b. Is this confidence interval a reasonable estimate of the actual proportion of all humans who are female?

c. Explain why the confidence interval procedure fails to produce an accurate estimate of the population parameter in this situation.

d. It clearly does not make sense to use the confidence interval in part a to estimate the proportion of women on Earth, but does the interval make sense for estimating the proportion of women in the 2007 U.S. Senate? Explain your answer.

Watch Out

This example illustrates some important limitations of inference procedures:

- First, inference procedures do not compensate for the problems of a biased sampling procedure. If the sample is collected from the population in a biased manner, the ensuing confidence interval will be a biased estimate of the population parameter of interest.
- Second, confidence intervals and significance tests use *sample* statistics to estimate *population* parameters. If the data at hand constitute the entire population of interest, then constructing a confidence interval from these data is meaningless. In this case, you know without doubt that the proportion of women in the population of the 2007 U.S. senators is exactly .16, so it is senseless to construct a confidence interval from these data.

Activity 18-5: Hypothetical Baseball Improvements

18-5, 18-16, 18-17

Suppose a baseball player who has always been a .250 career hitter works extremely hard during the winter off-season. He wants to convince the team manager that he has improved, and so the manager offers him a trial of 30 at-bats with which to demonstrate his improvement.

a. State the null and alternative hypotheses to be tested on the resulting data, in symbols and in words.

Two kinds of errors can be made with a test of significance: The null hypothesis can be rejected when it is actually true (called a **Type I error**), and the null hypothesis can fail to be rejected when it is actually false (called a **Type II error**). The significance level α of a test puts an upper bound on the probability of a Type I error.

b. Describe (in words) what a Type I error means in the context of the baseball player.

c. Describe (in words) what a Type II error means in this context.

18

A Type I error is sometimes referred to as a *false alarm* because the researcher mistakenly thinks that the parameter value differs from what was hypothesized. Similarly, a Type II error can be called a *missed opportunity* because the parameter really did differ from what was hypothesized, yet the researchers failed to realize it.

Now suppose that the player has genuinely improved to the point where he now has a .333 probability of getting a hit during an at-bat.

d. Open the `Power Simulation` applet. Set the hypothesized value of π to **.250** and the alternative value to **.333.** Set the sample size to **30,** and ask for **200** samples. Click Draw Samples. The top dotplot (Hypothesized) shows the distribution of the number of hits in the 30 trials for the 200 samples if the hitter's success probability is .250.

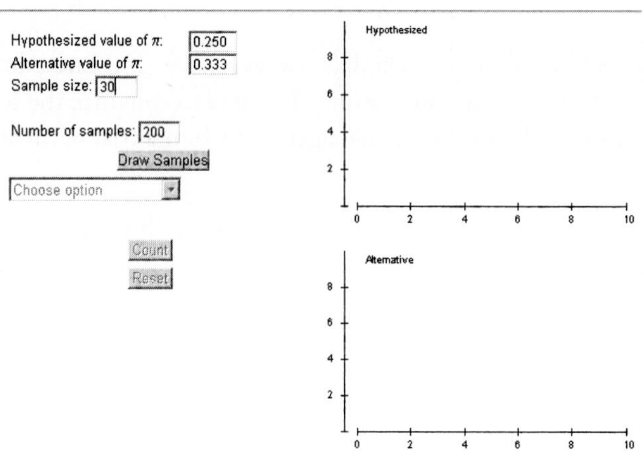

Describe this distribution.

e. Based on the dotplot of the simulated data, approximately how many hits would the player have to get in 30 at-bats so the probability of a .250 hitter doing that well by chance alone is less than .05? *Hint:* Use the pull-down menu to select Level of Significance and set α to **.05.** Click Count.

f. The bottom dotplot (Alternative) uses .333 as the success probability. Comment on the amount of overlap between the two distributions.

g. In what percentage of the 200 samples did the .333 hitter exceed the number of hits that you identified in part e? (This would be the percentage of the 200 samples in which the .333 hitter would do well enough in 30 at-bats to convince the manager that a .250 hitter would have been very unlikely to do that well simply by chance.)

h. Is it very likely that a .333 hitter will be able to establish that he is better than a .250 hitter in a sample of 30 at-bats? Explain. *Hint:* Refer to your answer to part g and to the overlap between these two distributions that you noted in part f.

The **power** of a statistical test is the probability that the null hypothesis will be rejected when it is actually false (and therefore should be rejected). Particularly with small sample sizes, a test might have low power, so it is important to recognize that failing to reject the null hypothesis does not mean accepting it as being true.

i. Use your simulation results to report the approximate power of this significance test involving the baseball player. *Hint:* Focus on your answer to part g.

j. Repeat parts d–i assuming that the player has a sample of **100** at-bats in which to establish his improvement. Write a paragraph summarizing your findings about the power of the test with this larger sample size.

Increasing the sample size is one way to obtain a more powerful test, i.e., one that is more likely to detect a difference from the hypothesized value when a difference is actually there. With very large sample sizes, even minor differences can be detected, which reinforces the distinction between statistical and practical significance.

k. If the player had improved to the point of being a .400 hitter, would you expect the test to be more or less powerful than when his improvement was at the .333 level? Explain.

l. If your analysis had used the $\alpha = .10$ rather than $\alpha = .05$ significance level, would you expect the test to be more or less powerful? Explain.

m. In addition to sample size, list two other factors that are directly related to the power of a test.

√

Activity 18-6: *West Wing* Debate

16-14, **18-6**

Recall from Activity 16-14 that the popular television drama *The West Wing* held a live debate between two (fictional) candidates for president on Sunday, November 6, 2005. Immediately afterward, an MSNBC/Zogby poll found that 54% (of real people) favored Democratic Congressman Matt Santos, played by Jimmy Smits, whereas 38% favored Republican Senator Arnold Vinick, played by Alan Alda. The poll was conducted online with a sample of 1208 respondents; the Zogby company screened the online respondents to try to ensure they were representative of the population of adult Americans.

a. Describe the relevant population and parameter.

b. Use your calculator to produce 90%, 95%, and 99% confidence intervals for the population proportion who favored Santos.

c. Comment on how these intervals compare (midpoints and widths).

d. Do these intervals suggest that more than half of the population favored Santos? Explain.

e. State the null and alternative hypotheses for testing whether the sample data provide strong evidence that more than half of the population favored Santos.

f. What do the intervals in part b say about the p-value for testing the hypotheses in part c? *Hint:* Be careful because you should have a one-sided alternative in part c, but the duality result holds for a two-sided alternative.

g. Describe what a Type I error would mean and what a Type II error would mean in the context of the hypotheses in part e.

h. Would a test of the hypotheses in part e have more power if 55% of the population actually favored Santos or if 52% of the population actually favored Santos? Explain.

i. Would the sample proportion of .54 favoring Santos be more impressive (i.e., more favorable to Santos) if the sample size were 10,000 or 100? Explain.

Solution

a. The population of interest is all adult Americans who are familiar with these fictional candidates. The parameter (call it π) is the proportion of this population who would have supported Santos if they had been asked.

b. The 90% CI for π is .54 ± .024, which is (.516, .564).
The 95% CI for π is .54 ± .028, which is (.512, .568).
The 99% CI for π is.54 ± .037, which is (.503, .577).

c. The midpoints are all the same, namely .54, the sample proportion of Santos supporters. The 99% CI is wider than the 95% CI, and the 90% CI is the narrowest.

d. Yes. All three intervals contain only values greater than .5, so they do suggest, even with 99% confidence, that more than half of the population favored Santos.

e. H_0: $\pi = .5$ (half of the population favored Santos)
H_a: $\pi > .5$ (more than half of the population favored Santos)

f. Because all three intervals fail to include the value0.5, you know that the *p*-value for a two-sided alternative would be less than .10, .05, and .01. Because you have a one-sided alternative in this case, you know that the *p*-value will be less than .01 divided by 2, or .005.

g. A Type I error occurs when the null hypothesis is really true but is rejected. In this case, a Type I error would mean that you conclude that Santos was favored by more than half of the population when in truth he was not favored by more than half. In other words, committing a Type I error means concluding that Santos was ahead (favored by more than half) when he wasn't really. A Type II error occurs when the null hypothesis is not really true but is not rejected (you continue to believe a false null hypothesis). In this case, a Type II error means that you conclude Santos was only favored by half of the population when in truth he was favored by more than half of the population. In other words, committing a Type II error means concluding that Santos was not ahead when he really was.

h. The test would be more powerful if Santos really was favored by 55% rather than 52%. The higher population proportion would make it more likely to reject the null hypothesis that only half of the population favored Santos, because the distribution of sample proportions would center around .55 rather than .52 (further from .5).

i. The larger sample (10,000) would produce stronger evidence that more than half of the population favored Santos. With less variability in the sampling distribution, the *p*-value would be much smaller.

Watch Out

This entire topic has been about issues to watch out for and the subtle ways in which confidence intervals and significance tests are often misinterpreted or misunderstood. Keep in mind that

- A statistically significant result might not be practically significant, especially with large sample sizes.
- Reporting a *p*-value is more informative than simply reporting a test decision at one particular significance level.
- A test might have insufficient power to reject a null hypothesis even when it is wrong, so you should never "accept" a null hypothesis, especially with small sample sizes.
- When you have data for the entire population, do not apply an inference technique (confidence interval or significance test).

Wrap-Up...........

This topic aimed to deepen your understanding of confidence intervals and tests of significance so you can better understand the relationship between them and avoid misinterpreting them. You discovered a **duality** between intervals and tests: when a confidence interval includes a particular value, then that value will not be rejected by the corresponding two-sided significance test. You also explored the distinction between practical and statistical significance. For example, a significance test based on a large sample of households convinced you that the proportion of American households with a pet cat is not one-third, but a confidence interval showed that the population proportion is actually quite close to one-third in practical terms. You also learned that common significance levels such as .05 and .01 are useful but not sacred.

You also investigated the concepts of power and types of error associated with statistical tests. In the baseball example, a **Type I error** means to decide that the player has improved when he really has not (a *false alarm*), and a **Type II error** means to decide that the player has not improved when he really has (a *missed opportunity*). **Power** is the probability of rejecting a null hypothesis that is actually false, and you saw that sample size plays a large role in determining how powerful a test is. Small samples typically lead to tests with low power, meaning you are unlikely to detect a difference or improvement even when there really is one.

Finally, the "alien" example reminded you that intervals and tests are bound to produce misleading findings when the sampling method is biased. In fact, statistical inference applies only when a sample has been drawn from a population in the first place. In cases where you have access to the entire population of interest (such as the 100 U.S. senators), you can describe the population but should not apply these statistical inference procedures to the data.

Some useful definitions to remember and habits to develop from this topic include

- Confidence intervals and significance tests have different goals. Intervals estimate the value of an unknown population parameter, also indicating the amount of uncertainty in the estimate. Significance tests assess how much evidence the sample data provide against a particular hypothesized value for the population parameter.
- Whenever a two-sided test rejects a particular hypothesized value at a certain significance level α, then the confidence interval at the corresponding confidence level (for example, 95% confidence for $\alpha = .05$) will not include that hypothesized value.

- Whenever a test gives a statistically significant result, it is useful to follow up with a confidence interval (CI). A CI can help determine whether the result is also of practical significance.
- Especially with large sample sizes, a statistically significant result might fail to be practically important.
- Reporting a *p*-value is much more informative than simply providing a yes/no statement of whether the sample result is significant at a certain α level.
- A Type I error occurs when you reject a null hypothesis that is actually true. A Type II error occurs when you fail to reject a null hypothesis that is actually false.
- Sample size affects the power of a test. All else being equal, larger samples produce more powerful tests than smaller samples.
- Power is also influenced by the significance level α and by the actual value of the parameter. Using a smaller (i.e., more stringent) significance level *reduces* the power of the test. Having an actual parameter value that differs more from the hypothesized value *increases* the power of the test.
- Never state a conclusion as "accept the null hypothesis." Failing to have enough evidence to reject the null hypothesis does not necessarily mean that you have proven the null hypothesis to be true.
- Be very cautious about generalizing to a larger population, with either a confidence interval or a significance test, if the sample is not selected randomly from that population.
- Do not apply these inference procedures when you have access to data from the entire population.

In the next two topics, you will learn a confidence interval and a test of significance for a population *mean* rather than a population *proportion.* In other words, you will analyze a quantitative variable rather than a categorical one. You will find that although the details of implementing the procedures necessarily change, the basic structure, reasoning, and interpretation do not.

Homework Activities

Activity 18-7: Charitable Contributions

16-18, **18-7**

Recall from Activity 16-18 that the 2004 General Social Survey found that 1052 people from a random sample of 1334 American adults claimed to have made a financial contribution to charity in the previous year.

a. Find a 90% confidence interval for the proportion of all American households that made a financial contribution to charity in the previous year.
b. Repeat part a with a 99% confidence interval.
c. Based on these confidence intervals, without carrying out a test of significance, indicate whether this sample proportion differs significantly from 75% at the $\alpha = .01$ level. Explain your reasoning.
d. Based on these confidence intervals, without doing a test of significance, indicate whether this sample proportion differs significantly from 80% at the $\alpha = .10$ level. Explain your reasoning.

Activity 18-8: Female Senators

6-12, 18-4, **18-8**

Recall from Activity 18-4 that the 2007 U.S. Senate consists of 16 women and 84 men.

a. Treat these numbers as sample data and calculate the test statistic for the significance test of whether the population proportion of women is less than .50.
b. Use the test statistic and the Standard Normal Probabilities Table (Table II) or your calculator to calculate the *p*-value of the test.
c. If the goal is to decide whether women constitute less than half of the entire U.S. Senate of 2007, does this test of significance have any meaning? Explain.

Activity 18-9: Distinguishing Between Colas

13-13, 17-24, **18-9**

Recall the cola taste test described in Activity 17-24. Each subject is presented with three cups, two of which contain the same brand of cola and one of which contains a different brand. The subject is to identify which one of the three cups contains a different brand of cola than the other two.

a. Report (in symbols and in words) the null hypothesis of the test, corresponding to the conjecture that a subject is simply guessing at identifying the one cup of soda that differs from the other two.
b. Suppose one particular subject (Randy) is actually able to identify the different brand 50% of the time in the long run. Would this subject's sample data necessarily lead to rejecting the null hypothesis? Explain.
c. Suppose that Randy participates in $n = 50$ trials. Verify that the null hypothesis would be rejected at the .05 significance level if the subject obtained a sample proportion of correct identifications of .46 or greater.
d. Is there greater than a 50/50 chance that Randy, with his actual 50% success rate, will get a sample proportion of .46 or greater? Explain, based on a CLT calculation.
e. Would the probability in part d increase, decrease, or remain the same if the sample size were 100 instead of 50? Explain.
f. Would the probability in part d increase, decrease, or remain the same if Randy's probability of a correct identification were 2/3 instead of 1/2? Explain.

Activity 18-10: Voter Turnout

4-19, **18-10**

A random sample of 2613 adult Americans in 1998 revealed that 1783 claimed to have voted in the 1996 presidential election.

a. Use these sample data to construct a 99.9% confidence interval for π, the proportion of all adult Americans who voted in that election.
b. Even though this truly was a random sample, do you really have 99.9% confidence that this interval captures the actual proportion who voted in 1996? Explain.
c. The Federal Election Commission reported that 49.0% of those eligible to vote in the 1996 election had actually voted. Is this value included within your interval?

d. Do you think this interval succeeds in capturing the proportion of all adult Americans who would claim to have voted in 1996? Explain how this parameter differs from that in part c.

e. Explain why the confidence interval renders it unnecessary to conduct a significance test of whether π differs from .49 at the .001 level.

Activity 18-11: Phone Book Gender

4-17, 16-16, **18-11**

Recall from Activity 4-17 the sample data collected from a random page of the San Luis Obispo County telephone book: 36 listings had both male and female names, 77 had male names, 14 had female names, 34 had initials only, and 5 had pairs of initials. Before collecting the data, the authors conjectured that less than half of the names in the phone book would be female.

a. A total of how many first names were studied (ignore listings with only initials)? How many of them were female names? What proportion of the names were female?

b. Identify the observational units in this study.

c. Identify the population and the parameter of interest.

d. Do the sample data support the authors' conjecture that less than half of the names in the phone book would be female at the $\alpha = .05$ level? Report the details of the test and write a short paragraph describing your findings and explaining how your conclusions follow from the test results. Also discuss the technical conditions.

e. Accompany your test with a 95% confidence interval. Interpret the interval, and comment on how it relates to the test result.

f. Do the sample data provide evidence that less than half of the residents of San Luis Obispo County are female? Explain your answer, and be sure to discuss how this question differs from the one you addressed in part d.

Activity 18-12: Racquet Spinning

11-9, 13-11, 17-3, 17-18, 18-3, **18-12**, 18-13

Recall from Activities 11-9 and 17-3 that 100 spins of a tennis racquet produced 46 "up" results and 54 "down" results. Suppose you want to test whether a spun tennis racquet is equally likely to land "up" or "down," i.e., whether it would land "up" 50% of the time in the long run.

a. Identify with a symbol and in words the *parameter* of interest in this experiment.

b. Considering the stated goal of the study, is the alternative hypothesis one-sided or two-sided? Explain.

c. Specify the null and alternative hypotheses for this study, both in symbols and in words.

d. Use your calculator to calculate the relevant test statistic and p-value.

e. Use your calculator to calculate how the test statistic and p-value would have been different if the 100 spins had produced 54 "up" and 46 "down" results.

f. In either of these cases (46 of one outcome and 54 of the other), do the sample data provide strong evidence that a spun tennis racquet would *not* land "up" 50% of the time in the long run? Would you reject the null hypothesis at the .05 significance level?

Activity 18-13: Racquet Spinning

11-9, 13-11, 17-3, 17-18, 18-3, 18-12, **18-13**

Reconsider the previous activity. Now suppose that the goal of the study is to investigate whether a spun tennis racquet tends to land "up" less than 50% of the time.

a. Restate the null and alternative hypotheses (in symbols).
b. Use your calculator to determine the test statistic and p-value of the one-sided test, assuming that 46 of the 100 sample spins landed "up." Report the test statistic and p-value of the test, and comment on how they compare to the ones found with the two-sided test.
c. Still supposing that the goal of the experiment is to investigate whether a spun tennis racquet tends to land "up" less than 50% of the time, use your calculator to determine the test statistic and p-value of the one-sided test assuming that 54 of the 100 sample spins landed "up." Again report the test statistic and p-value of the test, and comment on how they compare to the values found with the two-sided test.
d. For the sample results in part c, explain why the formal test of significance is unnecessary.

Activity 18-14: Penny Activities

16-9, 16-10, 16-11, 17-9, **18-14,** 18-15

Recall Activity 16-9 on page 334 and the data you gathered on penny flipping, spinning, and tilting.

a. For each of those three activities, use your sample data to test whether the population proportion of heads differs significantly from one-half at the .05 significance level.
b. Comment on whether your test results are consistent with your 95% confidence intervals found in Activity 16-9.

Activity 18-15: Penny Activities

16-9, 16-10, 16-11, 17-9, 18-14, **18-15**

Statistics professor Robin Lock has asked his students to flip, spin, and tilt pennies for many years, and he has compiled a running total of the results:

- 14,709 heads in 29,015 flips
- 9,197 heads in 20,422 spins
- 10,087 heads in 14,611 tilts

a. For each of the three activities, determine a 95% confidence interval for the actual probability (long-run proportion) of heads.
b. Explain why it makes sense that these intervals are so narrow.
c. Based on these intervals, for which activities would you reject at the .05 significance level that the probability of heads is .5? Explain.
d. For each of the three penny activities, conduct the test of whether the probability of heads differs significantly from one-half. Report the test statistics and p-values. Do the results agree with your answer to part c?
e. With regard to penny *flipping,* would you regard the difference from one-half as *practically* significant even if it is statistically significant? Explain.

Activity 18-16: Hypothetical Baseball Improvements

18-5, **18-16,** 18-17

Reconsider the baseball player from Activity 18-5. Suppose that he actually became a .400 hitter (i.e., had a .4 probability of getting a hit during an at-bat). Investigate whether the power for a significance test with a sample size of 30 at-bats is greater or less than it was when he became a .333 hitter. To do this, repeat your simulation analysis of Activity 18-5 for a **.400** hitter. Write a paragraph reporting your findings concerning the change in the test's power.

Activity 18-17: Hypothetical Baseball Improvements

18-5, 18-16, **18-17**

Consider once again the baseball player from Activity 18-5 who became a .333 hitter. Investigate whether the test is more or less powerful at higher significance levels by repeating your simulation analysis of Activity 18-5 using α = **.10** rather than α = .05 as the significance level. Write a paragraph reporting your findings concerning the change in the test's power resulting from this change in significance level.

Activity 18-18: Flat Tires

17-1, 17-2, 17-4, 17-10, 17-16, **18-18,** 24-18

Reconsider the "flat tire" scenario from Activities 17-1, 17-2, and 17-4.

a. Report the null and alternative hypotheses being tested.
b. Describe what a Type I error and a Type II error mean in this context.
c. Perform a simulation using the `Power Simulation` applet to approximate the power of this test with a sample size of **100** and a significance level of **.05,** assuming that the right-front tire is actually chosen 50% of the time.
d. Would the power in part c increase, decrease, or remain the same if the sample size were 200 and all else remained unchanged? Explain.
e. Would the power in part c increase, decrease, or remain the same if the significance level were .10 and all else remained unchanged? Explain.
f. Would the power in part c increase, decrease, or remain the same if the right-front tire were actually chosen 40% of the time and all else remained unchanged? Explain.

Activity 18-19: Emotional Support

4-15, **18-19**

Recall from Activity 4-15 that sociologist Shere Hite received mail-in questionnaires from 4500 women, 96% of whom claimed that they give more emotional support than they receive from their husbands or boyfriends (Moore, 1995). Also recall that an ABC News/*Washington Post* poll of a random sample of 767 women found that 44% claimed to give more emotional support than they receive.

a. Determine the margin of error for each of these surveys. Also report each survey's 95% confidence interval for the proportion of all American women who feel they give more emotional support than they receive.
b. Are these two confidence intervals similar? Do they overlap at all?
c. Which survey has the smaller margin of error, i.e., the narrower confidence interval?
d. Which of these two confidence intervals do you have more confidence in? Explain.

Activity 18-20: Veterans' Marital Problems

17-23, **18-20**

Recall from Activity 17-23 the study that investigated whether Vietnam veterans tend to get divorced at a higher rate than men aged 30–44 at that time (Gimbel and Booth, 1994).

a. State the null and alternative hypotheses for this study.
b. Describe what a Type I error would mean in this context.
c. Describe what a Type II error would mean in this context.

Activity 18-21: Hiring Discrimination

17-21, **18-21**

Recall from Activity 17-21 the legal case in which the U.S. government sued the Hazelwood School District on the grounds it discriminated against black teachers in its hiring practices. Describe what a Type I error and Type II error would mean in this context. Would you consider one of these to be a more serious error than the other? Explain.

TOPIC

19 ••• Confidence Intervals: Means

How much money do Americans spend on average on Christmas presents? What is the body temperature of an "average" healthy adult? For how long did an "average" student at your school sleep last night? How much weight does a typical college student carry in his or her backpack? Although these questions concern different contexts, they nevertheless sound quite similar. In this topic, you will learn how to construct a confidence interval to address these and other questions.

Overview

In Topic 16, you explored confidence intervals for a population *proportion.* In this topic, you will turn from binary categorical variables to quantitative variables, so now the population *mean* is the parameter of interest. You will investigate and apply confidence interval procedures for estimating a population mean. Although some of the procedure's details are different, and you will work with a new probability model called a *t*-distribution, you will find that the reasoning, structure, and interpretation of a confidence interval remain unchanged.

Preliminaries

1. Guess the average amount of sleep that students in your class got last night. Then make guesses for the least and most hours of sleep that a student in your class got last night.

Average: Least: Most:

2. Mark on the following scale an interval that you believe with 90% confidence to include the mean amount of sleep (in hours) that students at your school got last night.

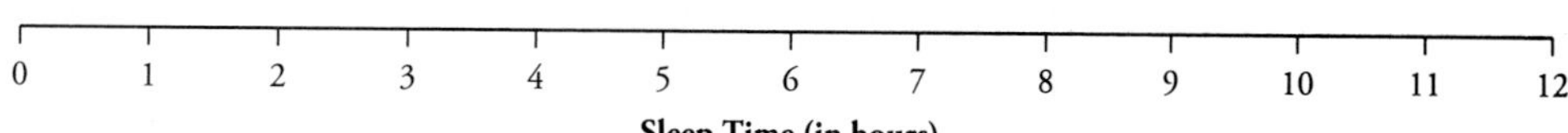

3. Mark on the following scale an interval that you believe with 99% confidence to include the mean amount of sleep (in hours) that students at your school got last night.

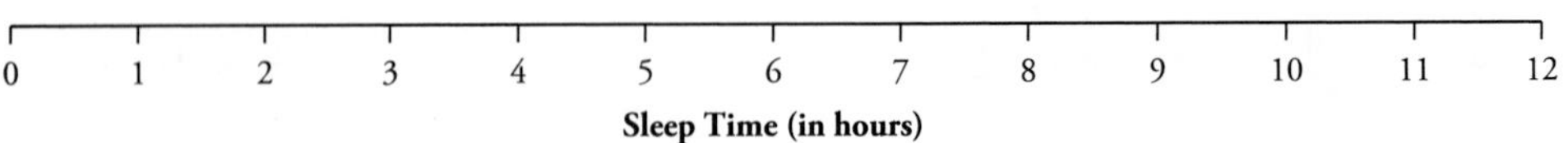

4. Which of these two intervals is wider?

5. Record the times that the students in your class went to bed last night and woke up this morning. Also calculate and record the amount of sleep time, in hours (e.g., 2 hours and 15 minutes = 2.25 hours).

Student	Bedtime	Wake Time	Sleep Time
1			
2			
3			
...			

6. Guess the average number of hours per week that third- or fourth-graders spend watching television.

In-Class Activities

Activity 19-1: Christmas Shopping

14-3, 14-7, 15-11, **19-1**

Recall from Activity 14-3 on page 290 that the Gallup organization conducted a survey about how much money people expected to spend on Christmas presents in 1999. The sample mean of the 922 responses was $857.

a. Is the variable *amount expected to be spent on Christmas presents in 1999* categorical or quantitative?

b. Is $857 a parameter or a statistic? Explain. What symbol represents it?

c. Identify (in words) the parameter of interest in this study. What symbol represents this parameter?

d. Do you know the value of the parameter in this study? Is it more likely to be close to $857 than far from it?

As you have seen, you can form an interval estimate of a parameter by starting with the sample statistic and going two standard deviations on either side of it. The Central Limit Theorem for a sample mean tells you the standard deviation of a sample mean $\bar{x}$ is $\sigma/\sqrt{n}$, where σ represents the population standard deviation and n the sample size.

e. Suppose for these expected shopping expenditures that the population standard deviation is $\sigma = 250$. Calculate the standard deviation of the sample mean $\bar{x}$.

f. Add and subtract two standard deviations to/from the observed sample mean to form a reasonable interval estimate of the population mean μ.

When the population standard deviation σ is known, a confidence interval for a population mean is given by $\bar{x} \pm z^* \frac{\sigma}{\sqrt{n}}$. However, a major drawback to using this procedure is that it requires you to know the value of the population standard deviation σ, which you almost never know. (After all, if you knew the population standard deviation, isn't it likely you would also know the population mean and do not need to estimate it?)

g. What is a reasonable substitute for σ that you can calculate from sample data?

> To estimate the standard deviation of a sample mean $\bar{x}$, replace the population standard deviation σ with the sample standard deviation s, producing $s/\sqrt{n}$. This is known as the **standard error** of the sample mean.

It seems reasonable to form a confidence interval for a population mean μ with $\bar{x} \pm z^* \frac{s}{\sqrt{n}}$. But a problem emerges with this procedure, especially with small sample sizes, as you will discover through a simulation analysis.

h. Open the `Simulating Confidence Intervals` applet. Use the pull-down menus to switch to means and then select the method called "z with sigma." Set the population mean to μ = **1000** and the population standard deviation to σ = **250.** Set the sample size to n = **10** and the confidence level to **95%.** Ask for **100** intervals, and click Sample.

Simulating Confidence Intervals

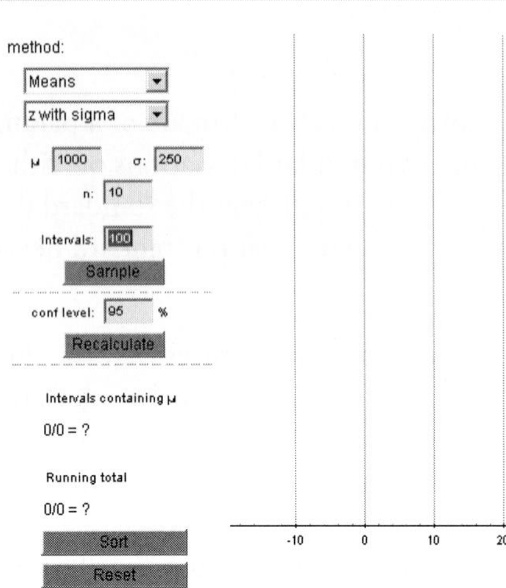

Notice that the applet generates 100 samples, calculates a 95% confidence interval for μ from each sample, and colors the intervals green when they succeed in capturing the population mean and red when they fail. What percentage of these 100 intervals succeed in capturing the value of μ (indicated by the vertical line) between the two endpoints?

i. Continue to click Sample until you have produced 1000 intervals. What is the running total percentage of intervals that succeed in capturing the population mean? Is this value close to what you expected? Explain.

j. Now change the method to "*z* with *s*." This method estimates σ with each sample's standard deviation *s* in calculating each interval. Click Sample and continue to click until you have generated 1000 intervals. What percentage of these 1000 intervals succeed in capturing the population mean? Is this value very close to 95%?

The problem with simply replacing σ with *s* is that noticeably fewer than 95% of the intervals succeed in capturing the population mean, even though they are being called 95% confidence intervals. (This happens because there will be some samples where two unfortunate things happen: the sample mean is a little farther than expected from the population mean, and the sample standard deviation is a little less than the population standard deviation.) To compensate for this, you need to make the intervals a bit longer, so more of them will succeed and the overall percentage will be closer to the stated 95%. To do this, you use a different multiplier than the z^* critical value. You use what is called a ***t*-distribution.** The resulting t^* critical values will reflect the additional uncertainty introduced by estimating σ with *s*.

Confidence interval for a population mean (*t*-interval):

$$\bar{x} \pm t^* \frac{s}{\sqrt{n}}$$

where t^* is the appropriate critical value from the t-distribution with $n - 1$ degrees of freedom for the desired confidence level.

Two **technical conditions** must be satisfied for this *t*-interval procedure to be valid:

- The sample is a simple random sample from the population of interest.
- *Either* the sample size is large ($n \geq 30$ as a guideline) *or* the population is normally distributed.

The second technical condition stems from what you learned in Topic 14: The sampling distribution of the sample mean will be normal if the population itself is normal, or this sampling distribution will be approximately normal for any population shape as long as the sample size is large. A sample size of at least 30 is generally regarded as large enough for the procedure to be valid. If the sample size is less than 30, examine visual displays of the sample data to see whether they appear to follow a normal distribution. If the sample appears to be roughly normal, then you can assume that the population follows a normal distribution.

The *t*-procedure is fairly *robust* in that it tends to give reasonable results even for small sample sizes as long as the population is not severely skewed and does not have extreme outliers.

The reasoning and interpretation of these confidence intervals are the same as always: If you were to repeatedly take random samples from the population and apply this procedure over and over, then in the long run, 95% of the intervals generated would succeed in containing the population mean. This allows you to say that you are "95% confident" that the one interval you actually constructed contains the actual value of the population mean.

Watch Out

- As mentioned in the last topic, it really is crucial to keep notation and terminology straight in your mind. Remember that μ denotes a population mean and $\bar{x}$ a sample mean. The whole point of the confidence interval is to estimate the *unknown* value of μ, based on the *observed* value of $\bar{x}$.
- Similarly, remember that σ stands for a population standard deviation and s for a sample standard deviation. Be especially careful with the phrase "standard deviation." Not only is the population standard deviation σ different from the sample standard deviation s, but also the standard deviation of the sample mean ($\sigma/\sqrt{n}$) and the standard error of the sample mean ($s/\sqrt{n}$) are all different quantities as well.

19

Activity 19-2: Exploring the *t*-Distribution

19-2, 20-5, 20-6

The *t*-distribution is actually an entire family of distribution curves, similar to the normal distributions. Whereas a normal distribution is identified by its mean and standard deviation, a *t*-distribution is characterized by an integer number called its **degrees of freedom** (abbreviated df). These *t*-distributions are mound-shaped and centered at zero, but they are more spread out (i.e., they have wider, fatter tails) than standard normal distributions. As the number of degrees of freedom increases, the tails get lighter and the *t*-distribution gets closer and closer to a normal distribution.

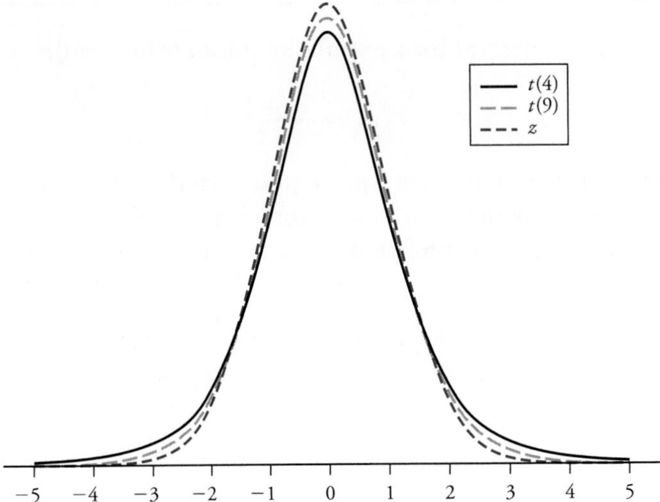

Previously, you have used the Standard Normal Probabilities Table to find critical values; now you will learn how to use a *t*-table. A *t*-table (Table III) can be found on page 625. Notice that each line of the table corresponds to a different value for the degrees of freedom. Always start by going to the relevant line for the degrees of freedom with which you are working. In conducting inferences about a population mean, the degrees of freedom can be found by subtracting 1 from the sample size (df $= n - 1$). Next, note that across the top of the table are various values for "area to the right." Finally, observe that the body of the table gives values such that the probability of lying to the right of that value (equivalent to the area to the right of that value under the *t*-distribution) is given at the top of each column.

Suppose you need to find the critical value t^* for a 95% confidence interval based on a sample size of $n = 10$.

a. Draw a rough sketch of the *t*-distribution with 9 degrees of freedom. *Hint:* It should look very much like a standard normal curve.

b. The critical value t^* for a 95% confidence interval is the value such that 95% of the area under the curve is between $-t^*$ and t^*. Shade this area on your sketch.

c. What is the area to the right of t^* under the curve? *Hint:* This area is not .05.

d. Look at the *t*-table (Table III) to find the value of t^* with that area to its right under a *t*-distribution with 9 degrees of freedom. Report this value.

e. Is this critical value less than or greater than the critical value z^* from the standard normal distribution for a 95% confidence interval? Explain why this is helpful, based on your motivation for needing the t-distribution instead of the z-distribution.

Critical values t^* from the t-distribution are always greater than their counterparts from the z- (normal) distribution, reflecting the greater uncertainty introduced by estimating σ with s. This larger multiplier makes the intervals just long enough that 95% of them (or whatever the confidence level) succeed in capturing the value of the population mean.

f. Find the critical value t^* for a 95% confidence interval based on a sample size of $n = 30$. How does this value compare to the previous t^* value? Explain why this is appropriate for the interval procedure as well. *Hint:* Think about whether a larger sample size would increase or decrease the uncertainty in estimating σ by s.

g. Find the critical value t^* for a 90% and 99% confidence interval, based on a sample size of $n = 30$. Which is greater? Explain why this is appropriate.

h. Find the critical value t^* for a 95% confidence interval based on a sample size of $n = 130$. *Hint:* When the desired value for degrees of freedom is not shown in the table, always round down to be *conservative,* because it's better to have a longer interval than necessary to ensure the achieved confidence level is at least as large as the stated confidence level.

Activity 19-3: Body Temperatures

12-1, 12-19, 15-3, 15-18, 15-19, **19-3,** 19-7, 20-11, 22-10, 23-3

Recall from Activities 12-1 and 15-3 that the body temperatures (in degrees Fahrenheit) have been recorded for a sample of 130 healthy adults (Shoemaker, 1996). The sample mean body temperature is 98.249 degrees, and the sample standard deviation is 0.733 degrees.

a. Calculate a 95% confidence interval for the population mean body temperature, based on the sample results for these 130 healthy adults.

b. Write a sentence interpreting this interval. *Hint:* Ask yourself what you believe to be in this interval with 95% confidence. Then write a separate sentence interpreting what the phrase "95% confidence" means. *Hint:* Refer back to the green and red intervals in the applet used in Activity 19-1.

c. Create and examine a dotplot or histogram of the sample data (use the list named **BDTMP** in the grouped file BODYTEMPS). Also create and examine a normal probability plot. Does the *sample* distribution of body temperatures appear to be roughly normal?

d. Is normality of the *population* of body temperatures required for this *t*-procedure to be valid with these data? Explain.

e. Comment on whether the other technical condition required for the validity of this *t*-interval is satisfied.

f. Use your calculator's **T-Interval** command to confirm your calculation of the 95% confidence interval (CI) in part a. The **T-Interval** can be found by pressing [STAT] and selecting the **TESTS** menu. The **T-Interval** test gives you a choice of inputs, either **Data** or **Stats.** For this activity, select **Data** and then input the list named **BDTMP.** The frequency should be set to **1** and the confidence level (**C-Level**) set to the level of interest. (*TI hint:* If you select the **Stats** option for the **T-Interval,** then you will need to enter the sample mean, standard deviation, and the sample size.) Then use your calculator to produce a 90% CI and a 99% CI for the population mean body temperature. Record these here.

90% CI for μ: 99% CI for μ:

g. Comment on how the midpoints and widths of these intervals compare.

h. Based on these three confidence intervals (90%, 95%, and 99%), and assuming you have a representative sample, does it appear that 98.6 degrees Fahrenheit is a plausible value for the mean body temperature for the population of all healthy adults? Explain.

i. Suppose the sample size had been 13 rather than 130, but the sample statistics turned out exactly the same. How would you expect a 95% CI to differ in this case from the 95% interval in part a? Explain.

j. Produce the CI mentioned in part i, and describe how the interval has changed. *Hint:* Comment on both its midpoint and width.

Watch Out

When checking the second technical condition for the *t*-interval, normality of the population is not necessary if the sample size is large. So, even if the sample data have a skewed distribution, the *t*-interval can still be used if the sample size is large.

Activity 19-4: Sleeping Times

8-1, **19-4,** 19-5, 19-12, 19-19, 20-2, 20-7

Suppose you want to estimate the mean sleep times of all students at your school last night. Consider the four different (hypothetical) samples of sleep times presented in the dotplots and boxplots shown here.

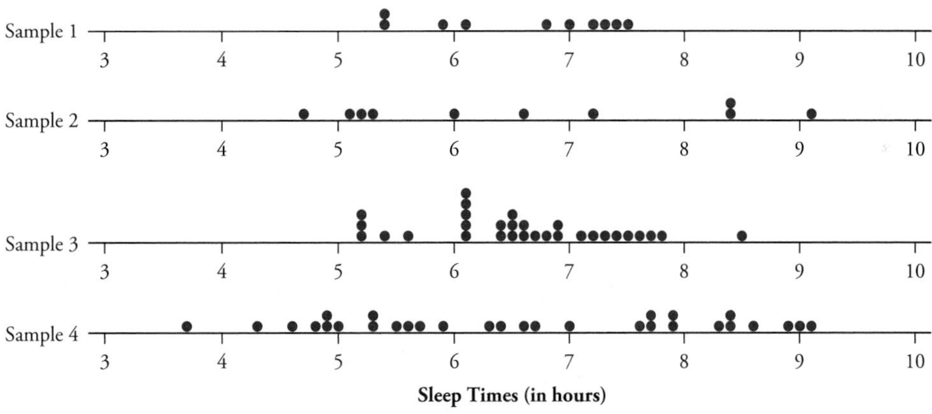

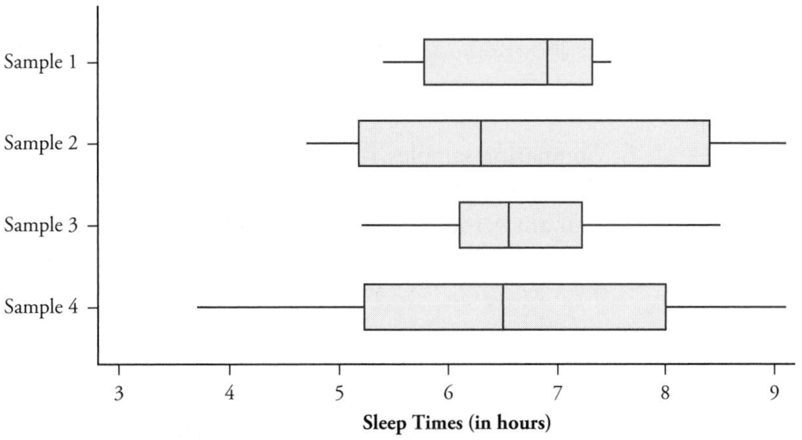

a. The following descriptive statistics were calculated from these sample data. Fill in the Sample Number column by figuring out which statistics go with which plots.

Sample Number	Sample Size	Sample Mean	Sample SD
	30	6.6	0.825
	10	6.6	0.825
	10	6.6	1.597
	30	6.6	1.597

b. What do all of these samples have in common?

c. What strikes you as the most important difference between the distribution of sleep times in sample 1 and in sample 2?

d. What strikes you as the most important difference between the distribution of sleep times in sample 1 and in sample 3?

The following table gives a 95% confidence interval for the population mean sleep time, based on each sample's data (presented in the same order as in the previous table):

Sample Number	Sample Size	Sample Mean	Sample SD	95% CI
3	30	6.6	0.825	(6.29, 6.91)
1	10	6.6	0.825	(6.01, 7.19)
2	10	6.6	1.597	(5.46, 7.74)
4	30	6.6	1.597	(6.00, 7.20)

e. Comparing samples 1 and 2, which produces a more precise estimate of μ (i.e., a narrower confidence interval for μ)? Explain why this makes sense. *Hint:* Refer to your answer in part c.

f. Comparing samples 1 and 3, which produces a more precise estimate of μ (i.e., a narrower confidence interval for μ)? Explain why this makes sense. *Hint:* Refer to your answer in part d.

In addition to sample size and confidence level, the sample standard deviation plays a role in determining the width of a confidence interval for a population mean. Samples with more variability (less precise measurements) produce wider confidence intervals.

Activity 19-5: Sleeping Times

8-1, 19-4, **19-5,** 19-12, 19-19, 20-2, 20-7

Now consider the data on sleep times collected from the students in your class in the Preliminaries section.

a. Use your calculator to produce graphical displays of the distribution. Write a few sentences commenting on key features of this distribution.

b. Use your calculator to calculate the sample size, sample mean, and sample standard deviation. Record these values here, and indicate the symbol used to represent each value.

c. State the technical conditions required for the *t*-interval to be a valid procedure. For each condition, comment on whether it appears to be satisfied.

d. Use your calculator to construct a 90% confidence interval for estimating the mean sleep time for *all* students at your school on that particular night. Also write a sentence interpreting this interval in context.

e. Count how many of the sample sleep times fall within this confidence interval. What percentage of the sample does this represent?

f. Is the percentage in part e close to 90%? Should it be? Explain. *Hint:* The issue here is not whether the sample was selected randomly from the population.

Watch Out

Confidence intervals of this type estimate the value of a population *mean.* They do not estimate the values of *individual* observations in the population or in the sample. You do not expect 90% of the sample data to be in the interval, nor are you claiming that 90% of the population's sleep times are contained in this interval. Especially with large samples, it is possible that very few sample (or population) values might fall inside the interval. In this case, you are simply 90% confident that the value of the *population mean* sleeping time among all students at your school is somewhere inside this interval. (Of course, this claim is only valid if you are willing to regard your sample as representative of the entire population on this issue, which would be problematic if, say, this was an early morning class).

Activity 19-6: Backpack Weights

2-13, 10-12, **19-6,** 20-17

Refer to Activities 2-13 and 10-12, which described a study (Mintz, Mintz, Moore, and Schuh, 2002) in which student researchers recorded the body weights and backpack weights for a sample of 100 students on the Cal Poly campus (BACKPACK). The students wanted to see how well Cal Poly students adhered to recommendations to carry less than 10% of their body weight in their backpacks.

a. Identify the observational units, variable (and its type), sample, and population in this study.

b. Use your calculator to create the *ratio of backpack weight to body weight* (use the BPKWT and BDYWT lists, respectively) variable for each student in the sample. Produce and comment on graphical displays and numerical summaries of the distribution of this ratio for these 100 students.

c. Determine a 99% confidence interval for the population mean weight ratio among all Cal Poly students at the time this study was conducted.

d. Interpret this interval, and also explain what the phrase "99% confidence" means.

e. Comment on whether the technical conditions required for the validity of this interval are satisfied (as well as you can with the information provided).

f. Would you expect 99% of the students in the sample to have a weight ratio in this interval? Would you expect this of 99% of the students in the population? Explain.

Solution

a. The observational units are the students. The variable is the *ratio of backpack weight to body weight,* which is quantitative. The sample is the 100 Cal Poly students whose weights were recorded by the student researchers. The population is all Cal Poly students at the time the study was conducted.

b. The following dotplot reveals that the distribution of these weight ratios is a bit skewed to the right. The center is around .07 or .08 (mean $\bar{x}$ = .077, median = .071). The five-number summary is (.016, .050, .071, .096, .181), so students in the sample carried as little as 1.6% of their weight in their backpacks and as much as 18.1% of their weight in their backpacks. The standard deviation of these ratios is s = .037.

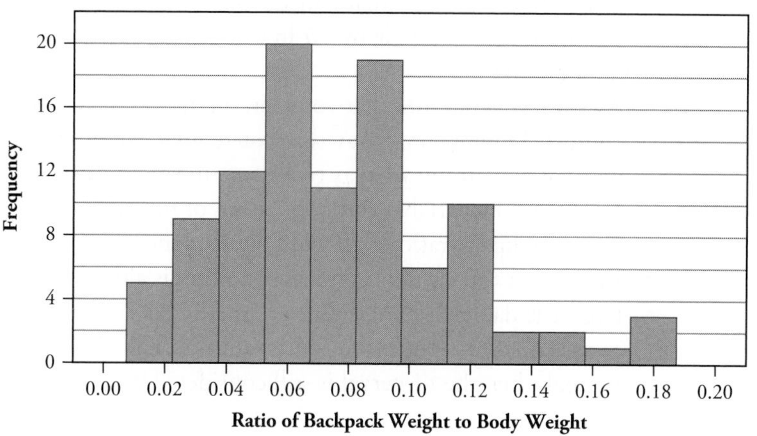

c. Calculating a 99% CI for the population mean by hand using

$$\bar{x} \pm t^* \frac{s}{\sqrt{n}}$$

gives .0771 ± 2.639 × .0366/$\sqrt{100}$, which is .0771 ± .0097, which corresponds to the interval from 0.0674 through 0.0868. (This t^* value should be based on 99 df, but we used 80 df here, the closest value less than 99 that appears in Table III.) Using your calculator gives a slightly more accurate 99% CI for μ of .0675 through .0867.

d. You are 99% confident that the mean weight ratio of backpack-to-body weights among all Cal Poly students at the time of this study is between .0674 and .0868. In other words, you are 99% confident that the average Cal Poly student wears between 6.74% and 8.68% of his/her body weight in their backpack. By "99% confidence," you mean that 99% of all intervals constructed with this method would succeed in capturing the actual value of the population mean weight ratio.

e. The first condition is that the sample be randomly selected from the population. This is not literally true in this case, because the student researchers did not obtain a list of all students at the university and select randomly from that list, but they did try to obtain a representative sample. The second condition is either that the population of weight ratios is normal or that the sample size is large. In this case, the sample size is large (n = 100, which is much greater than 30), so this condition is satisfied; even though, the distribution of ratios in the sample is somewhat skewed (and so presumably is the population).

f. You do not expect 99% of the sample, nor 99% of the population, to have a weight ratio between .0673 and .0869. You are 99% confident that the population *mean* weight ratio is between these two endpoints. In fact, only 18 of the 100 students in the sample have a weight ratio in this interval.

Watch Out

- Now that we have discussed two confidence interval procedures, the first question to ask when constructing a confidence interval is whether the underlying variable of interest is categorical or quantitative because this determines whether you are estimating a population proportion or a population mean. In the previous example, the *ratio* variable is quantitative. This might seem a little strange because this *ratio* variable gives numbers between 0 and 1, but it is not categorical (a numerical value is observed for each person). Another unsubtle clue is that part c clearly asks you to estimate a population *mean.*
- Be sure to carry many decimal places of accuracy in your intermediate calculations. If you round off too much in intermediate steps, your final calculations can turn out to be quite wrong. Go ahead and round your final answer to a few decimal places, but remember that a mean need not be an integer.
- Do not forget to divide by $\sqrt{n}$ in calculating the standard error portion of the CI for a population mean.
- As always, make sure that your interpretation of the calculated interval refers to context. Do not simply say that you're 99% confident that μ is within the interval without explaining what μ represents in the context of the study. This entails clarifying the variable and population of interest, as well as the parameter (e.g., mean backpack ratio of all Cal Poly students).
- Remember the second technical condition is either/or, not both. If the sample size is large, the data do not have to be normally distributed. On the other hand, a small sample does not render the *t*-interval invalid; if the sample data appear to be roughly normal, then the *t*-interval is still considered valid even with a small sample size.

Wrap-Up...........

This topic introduced you to confidence intervals for a population *mean.* You found that the reasoning, structure, and interpretation of these intervals are the same as for a population proportion. For instance, these intervals have the basic form: *sample statistic* ± (*critical value*) × (*standard error of statistic*). And you still interpret "95% confidence" to mean that 95% of all intervals generated by this procedure succeed in capturing the population parameter (in this case, population *mean*) of interest. One difference is that the sample mean is the statistic of interest. Another is that the standard error of the statistic ($s/\sqrt{n}$) involves the sample standard deviation. Perhaps the most important difference is that the critical value is based on the ***t*-distribution** rather than the standard normal (*z*-) distribution.

You also studied the properties of these intervals. For example, you again found that a larger sample size produces a narrower interval, and a higher confidence level generates a wider interval (if all else remains the same). You also learned that more variability in the sample data leads to a wider confidence interval.

One common misunderstanding of these intervals is to think (wrongly) that 95% of the *data* should fall within the interval. The interval estimates the population *mean,* not individual data values. Especially with a large sample that produces a narrow interval, it is not uncommon for the interval to capture only a small percentage of the individual data values. You should not expect 95% of the sample values, or 95% of the population values, to fall within a 95% confidence interval for μ.

Some useful definitions to remember and habits to develop from this topic include

- Remember to check the technical conditions before applying the procedure. This procedure requires either a large sample size *or* a normally distributed population. If the sample size is less than 30, examine visual displays of the sample data to see whether they appear to follow a normal distribution.

- Be careful to consider whether the sample was randomly selected from the population of interest before generalizing the confidence interval to that population.
- A larger sample produces a narrower interval.
- A higher confidence level produces a wider interval.
- More sample variability produces a wider interval.
- Do not forget that this confidence interval procedure applies to a *quantitative* variable, not a categorical variable.
- Remember this confidence interval estimates a population *mean,* not individual data values.

In the next topic, you will continue to work with quantitative variables, turning your attention to a test of significance about a population mean.

Homework Activities

Activity 19-7: Body Temperatures

12-1, 12-19, 15-3, 15-18, 15-19, 19-3, **19-7,** 20-11, 22-10, 23-3

Reconsider the data on body temperatures from Activity 19-3 (BODYTEMPS). Now consider the body temperatures of men and women separately (use the **BTMPM** and **BTMPF** lists, respectively).

a. Produce a 95% confidence interval for the population mean body temperature of a healthy male. Then do the same for the population of healthy females.
b. Interpret these intervals.
c. Comment on how the intervals compare to each other, addressing whether they suggest that men and women have different mean body temperatures.
d. Report the half-width of these two intervals. Which is wider? What aspect of the data causes the interval to be wider?
e. Compare the half-width of these intervals to the half-width of the interval based on the entire sample from Activity 19-3. Explain why these two intervals are wider than that one.
f. Investigate and comment on whether the technical conditions required for the validity of this procedure appear to be satisfied for each gender.

Activity 19-8: Credit Card Usage

1-9, 16-12, 17-11, **19-8,** 20-10

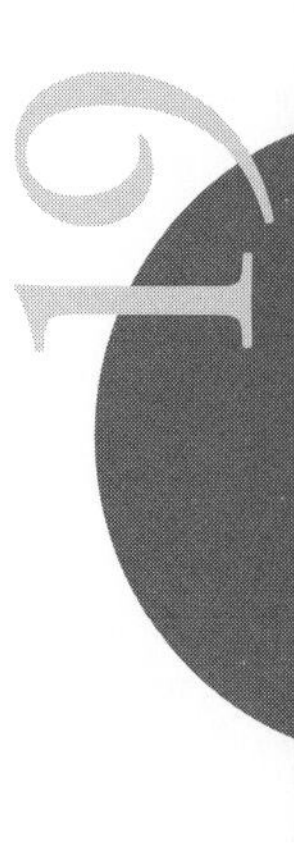

Refer to the study described in Activity 16-12. The Nellie Mae organization found that in a random sample of undergraduate students taken in 2004, the average credit card balance was $2169. Take the sample size to be 1074, which corresponds to 76% of the sample of 1413 students who hold a credit card.

a. What additional information is needed to produce a 95% confidence interval for the mean credit card balance in the population of all undergraduate students who held a credit card in 2004?
b. The Nellie Mae report did not mention the sample standard deviation of these credit card balances. Suppose this standard deviation was $1000. Produce and interpret a 95% confidence interval for the population mean in this case.
c. Would you expect 95% of the population to have a credit card balance in this interval? Explain.
d. Repeat part b, supposing the standard deviation to be $2000.
e. Which interval is longer? Explain why this makes sense.
f. Which standard deviation value ($1000 vs. $2000) would you suggest is more plausible for these data? Explain your reasoning.

Activity 19-9: Social Acquaintances

9-8, 9-9, 10-13, 10-14, **19-9**, 19-10, 20-12

Reconsider the data that you collected in Topic 9, based on the "social acquaintance" exercise described by Malcolm Gladwell in *The Tipping Point.*

a. Use these sample data to produce a 90% confidence interval for the population mean number of acquaintances from this list among all students at your school.
b. Interpret what this interval says. Also include an explanation of what "90% confidence" means in this context.
c. Determine how many and what proportion of the sampled students have an acquaintance number that falls within this interval.
d. Is this proportion close to 90%? Should it be? Explain.

Activity 19-10: Social Acquaintances

9-8, 9-9, 10-13, 10-14, 19-9, **19-10**, 20-12

Reconsider the previous activity. The file ACQUAINTANCESCP (ACQNT) contains data from 99 undergraduate students at Cal Poly who engaged in this exercise during the winter of 2006.

a. Determine a 90% confidence interval for the population mean number of acquaintances from this list among all students at Cal Poly in the winter of 2006.
b. Compare your class results with those from Cal Poly. Write a paragraph or two summarizing your comparison, including graphical displays and numerical summaries, as well as commenting on the confidence intervals.

Activity 19-11: Nicotine Lozenge

1-16, 2-18, 5-6, 9-21, **19-11**, 20-15, 20-19, 21-6, 22-8

Recall from Activity 1-16 the experiment that investigated the effectiveness of a nicotine lozenge for subjects who wanted to quit smoking (Shiffman et al., 2002). Before the treatments began, subjects answered background questions including how many cigarettes they smoked per day. Among the 1818 subjects in the study, the average was 22.0 cigarettes per day, and the standard deviation was 10.8 cigarettes per day.

a. Identify the population and parameter of interest.
b. Produce a 99% confidence interval for the population mean number of cigarettes smoked per day.
c. Based on this interval, does it seem plausible to assert that the population mean is 20 cigarettes (one pack) per day? Explain.

Activity 19-12: Sleeping Times

8-1, 19-4, 19-5, **19-12**, 19-19, 20-2, 20-7

Consider again the sleeping times collected in class and the confidence interval that you produced from that data. Describe how the confidence interval would have been different if the only change had been

a. A larger standard deviation among the sample sleeping times
b. A smaller sample size
c. A larger sample mean by 0.5 hours
d. Each person's sleep time had been 15 minutes longer than reported

Activity 19-13: Critical Values

12-18, 16-2, 16-20, **19-13**

a. Use the t-table to find the critical values t^* corresponding to the following confidence levels and degrees of freedom, filling in a table like the one shown here with those critical values:

	Confidence Levels			
Degrees of Freedom	80%	90%	95%	99%
4				
11				
23				
80				
Infinity				

b. Does the critical value t^* get larger or smaller as the confidence level increases (if the number of degrees of freedom remains the same)?

c. Does the critical value t^* get larger or smaller as the number of degrees of freedom increases (if the confidence level remains the same)?

d. Do the critical values from the t-distribution corresponding to infinitely many degrees of freedom look familiar? Explain. (Refer back to Topic 16 if they do not.)

Activity 19-14: Sentence Lengths

The following data are the lengths (measured as numbers of words) in a sample of 28 sentences from Chapter 3 of John Grisham's novel *The Testament:*

17	21	8	32	13	16	17	37	27	20	30	15	64	34
18	26	23	17	5	10	29	9	22	18	7	16	13	10

a. Create a graphical display of these data. Write a few sentences commenting on key features of the distribution.

b. Use these sample data to produce a 95% confidence interval for the mean length among all sentences in this book.

c. Comment on whether the technical conditions necessary for the validity of this procedure seem to be satisfied.

d. Remove the outlier, and recalculate the 95% confidence interval. Comment on how this interval has changed. Did removing the outlier have much impact on the interval?

Activity 19-15: Coin Ages

12-16, 14-1, 14-2, **19-15**

Recall from Activity 14-1 on page 281 the population of 1000 coins from which you selected a random sample.

a. As you did in Activity 14-1, use a table of random digits to select a random sample of ten pennies from this population. Record their ages.
b. Use this sample to construct a 90% confidence interval for the mean age of the pennies in this population.
c. Do you think that the technical conditions for this procedure have been met? Explain.
d. The population mean age among these 1000 pennies is 12.26 years. Does the interval in part b succeed in capturing this population mean?
e. If you had constructed a 95% confidence interval, would this procedure have been more likely to capture the population mean than your interval in part b?
f. If you had taken a random sample of 40 pennies and calculated a 90% interval, would this procedure have been more likely to lead to an interval that captures the population mean than your interval in part b? Explain.

Activity 19-16: Children's Television Viewing

1-15, **19-16,** 20-4, 22-14, 22-15

Recall from Activity 1-15 the study that investigated a relationship between watching television and obesity in third- and fourth-grade children (Robinson, 1999). Prior to assigning children to treatment groups, researchers gathered baseline data on their television viewing habits. Children were asked to report how many hours of television they watched in a typical week. The 198 responses had a mean of 15.41 hours and a standard deviation of 14.16 hours.

a. Are these values (15.41, 14.16) parameters or statistics? Explain.
b. Do you think that the technical conditions for the confidence interval for μ have been met? Explain.
c. Use this sample information to determine 90%, 95%, and 99% confidence intervals for the mean hours of television watched per week among all third- and fourth-grade children.
d. Do any of these intervals include your guess from the Preliminaries section?
e. In this situation would it make much difference if you used the z^* critical values rather than t^*? Explain.

Activity 19-17: Close Friends

19-17, 19-18, 22-1, 22-5, 22-22

The 2004 General Social Survey (GSS) interviewed a random sample of adult Americans. For one question the interviewer asked: "From time to time, most people discuss important matters with other people. Looking back over the last six months–who are the people with whom you discussed matters important to you? Just tell me their first names or initials." The interviewer then recorded how many names or initials the respondent mentioned. Results are tallied in the following table:

Number of Close Friends	0	1	2	3	4	5	6	Total
Count (number of respondents)	397	281	263	232	128	96	70	1467

a. Identify the observational units and variable in the study. Is the variable categorical or quantitative?
b. This distribution is sharply skewed to the right, but a t-interval is still valid. Explain why.

c. Use your calculator to produce a 90% confidence interval for the mean number of close friends in the population of American adults. For this activity, you will need to create two lists: one containing the numbers 0 through 6 and the second with the counts. You will use the first list as the data list and the second list as the frequency list (replace 1 with the list name).

d. Which two of the following are reasonable interpretations of this confidence interval and its confidence level:

- You can be 90% confident that the mean number of close friends in the population is between the endpoints of this interval.
- Ninety percent of all people in this sample reported a number of close friends within this interval.
- If you took another sample of 1467 people, there is a 90% chance that its sample mean would fall within this interval.
- If you repeatedly took random samples of 1467 people, this interval would contain 90% of your sample means in the long run.
- If you repeatedly took random samples of 1467 people and constructed *t*-intervals in this same manner, 90% of the intervals in the long run would include the population mean number of close friends.
- This interval captures the number of close friends for 90% of the people in the population.

e. For one of the incorrect interpretations in part d, explain why it is incorrect.

f. Describe how the interval would change if all else remained the same except

- The sample size were larger
- The sample mean were larger
- The sample values were less spread out
- Every person in the sample reported one more close friend

Activity 19-18: Close Friends

19-17, **19-18,** 22-1, 22-5, 22-22

Refer to the previous activity.

a. Produce and interpret a 90% confidence interval for the population proportion of people who would report having 0 close friends. *Hint:* Note that the parameter being estimated is not a mean.

b. Repeat part a for the proportion of people who would report having five or more close friends.

c. What information is gained in reporting a *t*-interval for the population mean rather than for these intervals?

Activity 19-19: Sleeping Times

8-1, 19-4, 19-5, 19-12, **19-19,** 20-2, 20-7

Consider your analysis of students' sleep times from Activity 19-5. Let μ represent the mean sleep time in the population of all students at your school. Identify each of the following statements as legitimate or illegitimate interpretations of the 95% confidence interval that you produced. For each illegitimate interpretation, explain why it is incorrect.

a. You can be 95% confident that the interval contains the true value of μ.

b. If you repeatedly took random samples of college students and generated 95% confidence intervals in this manner, then in the long run 95% of the intervals so generated would contain the actual value of μ.

c. The probability is .95 that μ lies within the interval.
d. You can be 95% confident that the sleep time for any particular student falls within the interval.
e. Ninety-five percent of the students in the population have sleep times that fall within the interval.

Activity 19-20: Planetary Measurements

8-12, 10-20, **19-20**, 27-13, 28-20, 28-21

Reconsider the data presented in Activity 8-12 that listed (among other things) the distance from the sun for each of the nine planets in our solar system. The mean of the distances turns out to be 1102 million miles, and the standard deviation is 1341 million miles.

a. Use these numbers to construct a 95% confidence interval.
b. Does this interval make any sense at all? If so, what population parameter does it estimate? Do you know the exact value of that parameter? Explain.

Activity 19-21: Hypothetical ATM Withdrawals

9-24, **19-21**, 22-25

The following dotplots display samples of withdrawal amounts from three different automatic teller machines (HYPOATM):

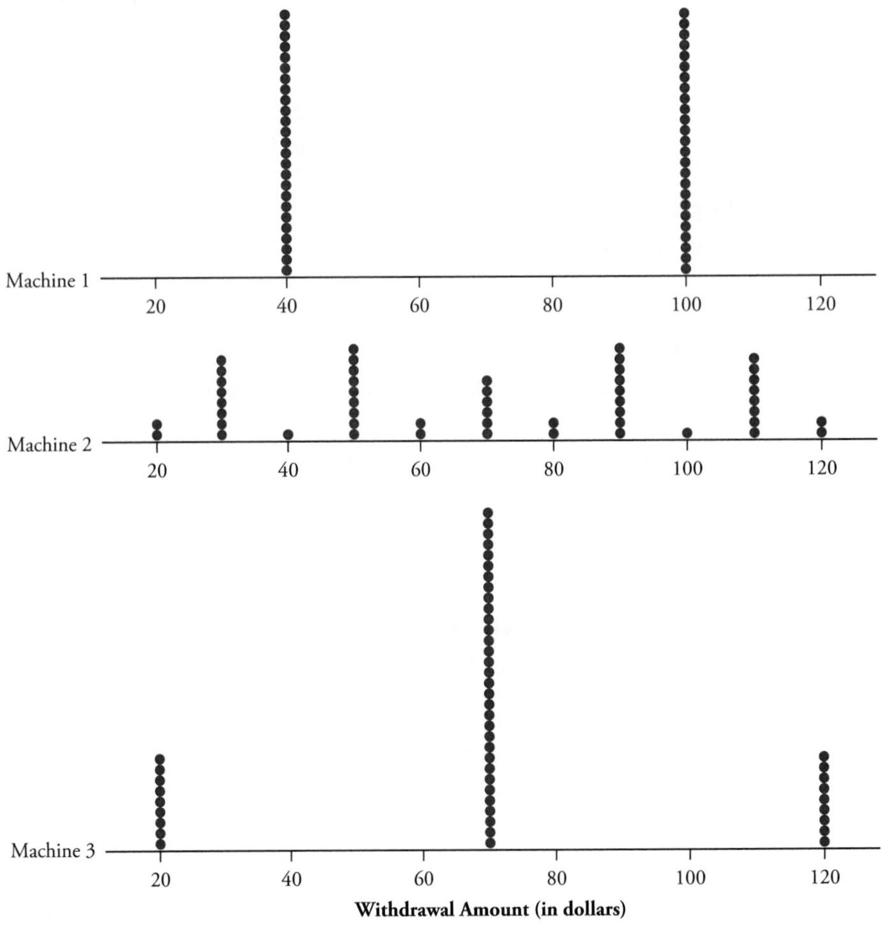

a. Write a paragraph comparing and contrasting the three distributions of ATM withdrawal amounts.

b. Use your calculator to compute the sample size, sample mean, and sample standard deviation of the withdrawal amounts for each machine (the MACH1, MACH2, and MACH3 lists, respectively). Also use your calculator to determine a 95% confidence interval for the population mean withdrawal amount among all withdrawals for each machine. Record the results in the following table.

	Sample Size	Sample Mean	Sample SD	95% CI for μ
Machine 1				
Machine 2				
Machine 3				

c. Summarize what this exercise reveals about whether a confidence interval for a mean describes all aspects of a dataset.

Activity 19-22: Your Choice

6-30, 12-22, **19-22,** 21-28, 22-27, 26-23, 27-22

a. Think of a situation in which you would be interested in producing a confidence interval to estimate a population *proportion.* Describe precisely the observational units, population, and parameter involved. Also describe how you might select a sample from the population. *Hint:* Be sure to think of a *categorical* variable so dealing with a proportion is sensible.

b. Repeat part a for a population mean rather than a proportion. *Hint:* Be sure to think of a *quantitative* variable.

TOPIC 20 • • • Tests of Significance: Means

The golden ratio is a famous number not only in mathematics but also in art, music, and architecture. It has been suggested since ancient times that rectangles for which the ratio of width to length equals this golden value are aesthetically pleasing. In this topic, you will investigate whether beaded artwork of Shoshoni Indians appears to adhere to this famous number. Other questions that you will investigate include: Did recent rule changes made by a sports league serve the intended purpose of increasing scoring? Do college students tend to get the recommended eight hours of sleep per night? Do schoolchildren watch more than two hours of television per day on average?

Overview

In the last topic, you studied confidence intervals for a population mean. Now you will turn to the other major type of statistical inference procedure, tests of significance, and apply these procedures to *quantitative* data and, therefore, a population *mean*. This procedure will lead you to one of the most famous of statistical techniques: the *t*-test.

Preliminaries

1. Guess how many points are scored (by the two teams combined) in a typical professional basketball game.

2. If you were looking to find evidence that rule changes have increased basketball scoring from the previous season, do you think that a sample of one day's games or a sample of one week's games would be more informative?

3. Guess your instructor's age.

4. Gather and record data on your classmates' guesses of your instructor's age.

In-Class Activities

Activity 20-1: Basketball Scoring

20-1, 20-9, 20-20

Prior to the 1999–2000 season, the National Basketball Association made several rule changes designed to increase scoring. The average number of points scored per game in the previous season had been 183.2. Let μ denote the mean number of points scored per game in the 1999–2000 NBA season.

a. If the rule changes had no effect on scoring, what value would μ have? Is this a null or an alternative hypothesis?

b. If the rule changes had the desired effect on scoring, what would be true about the value of μ? Is this a null or an alternative hypothesis?

The following sample data are the number of points scored in the 25 NBA games played from December 10–12, 1999:

196	198	205	163	184	224	206	190	140	204	200	190	195
180	200	180	198	243	235	200	188	197	191	194	196	

c. Enter the data in the table into your calculator and then create and examine visual displays of this distribution. Write a few sentences commenting on the distribution of points scored per game, particularly addressing the issue of whether scoring seems to have increased over the previous season's mean of 183.2 points per game.

d. Use your calculator to calculate the mean and standard deviation of this sample of points per game.

Sample mean $\bar{x}$: Sample standard deviation s:

e. Is the sample mean in the direction specified in the alternative hypothesis? In other words, is the mean of the points scored per game in this sample greater than the 1998–1999 season's mean?

f. Is it *possible* to have gotten such a large sample mean even if the new rules had no effect on scoring? Explain.

The structure, reasoning, and interpretation of significance tests regarding a population mean are the same as for a population proportion. The null hypothesis asserts that the population mean equals some value of interest, whereas the alternative hypothesis reflects the researcher's conjecture. The test statistic is calculated by taking the difference between the observed sample mean and the hypothesized population mean and dividing by the standard error of the sample mean. When the null hypothesis is true and the population has a normal distribution, the test statistic follows a t-distribution with $n - 1$ degrees of freedom, so the p-value is calculated from the t-distribution. This p-value again reports the probability of getting such an extreme sample, or even more extreme, if the null hypothesis were true. Thus, the smaller the p-value, the stronger the evidence against the null hypothesis supplied by the sample data.

Test of significance for a population mean μ (t-test):

$H_0: \mu = \mu_0$

i. $H_a: \mu < \mu_0$ *or*
ii. $H_a: \mu > \mu_0$ *or*
iii. $H_a: \mu \neq \mu_0$

Test statistic: $t = \dfrac{\bar{x} - \mu_0}{s/\sqrt{n}}$

p-value:

i. $\Pr(T_{n-1} \leq t)$ *or*
ii. $\Pr(T_{n-1} \geq t)$ *or*
iii. $2 \times \Pr(T_{n-1} \geq |t|)$

where μ_0 represents the hypothesized value of the population mean and T_{n-1} represents a t-distribution with $n - 1$ degrees of freedom.

The **technical conditions** for this procedure to be valid are

- The data are gathered from a simple random sample from the population of interest.
- Either the population values follow a normal distribution or the sample size is large ($n \geq 30$ as a guideline).

Watch Out

- Remember hypotheses are always about *parameters,* not statistics. In this case, the relevant parameter is a population mean, denoted by μ, because the variable (*points scored*) is quantitative.
- The alternative hypothesis should always be formulated before collecting the sample data, based on the research question.
- Note that the denominator of the test statistic is the standard error of the sample mean, which you also encountered in Topic 19 when forming a confidence interval for μ. Do not forget the $\sqrt{n}$ term in this expression.
- When calculating the p-value for a two-sided alternative, include the total area in *both* tails of the t-distribution beyond the value of the test statistic. But take advantage of the symmetry of the t-distribution and calculate this total area by doubling the area in the right tail.

g. Translate the statements in part a and part b into null and alternative hypotheses about μ by completing the following:

$H_0: \mu =$

$H_a: \mu$

h. Comment on whether the technical conditions for the validity of this *t*-test have been met.

i. Regardless of your answer to part h, use the sample statistics you found in part d to compute the value of the *t*-test statistic from this sample.

j. Draw a rough sketch of the *t*-distribution, with $n - 1$ degrees of freedom representing the sampling distribution of the *t*-test statistic when the null hypothesis is true (remember to label the horizontal axis). Shade the area under this curve lying to the right of the test statistic value you calculated in part b.

As you used the *t*-table (Table III) in the previous topic to find critical values for confidence intervals, you will also use it to find *p*-values for significance tests.

k. Use the *t*-table to approximate the area to the right of your test statistic value with $n - 1$ degrees of freedom. *Note:* The *p*-value in this example is equal to the area to the right of the test statistic value that you calculated. Use the appropriate row of the *t*-table to find two values on either side of your test statistic. Read off the probabilities from the top of the table that correspond to these two values. You can report that the *p*-value is between these two probabilities. If the test statistic value is off the chart, determine the bound on the *p*-value from the last probability listed. Your sketches should be very helpful.

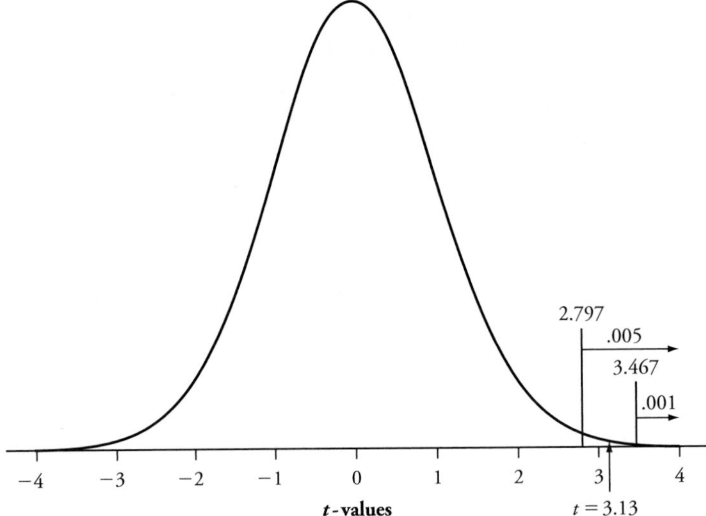

l. Use the `Test of Significance Calculator` applet, as in part b of Activity 17-2 on page 347, to verify your calculations for this test and to calculate the p-value more exactly. *Hint:* Use the pull-down menu to change the procedure from One Proportion to One Mean.

Applet p-value:

m. Interpret the p-value in the context of these data and hypotheses.

n. Would you reject the null hypothesis at the .10 level? At the .05 level? At the .01 level? At the .005 level? Explain.

o. Write a sentence or two summarizing your conclusion about whether the sample data provide convincing evidence that the mean points per game in the 1999–2000 season is greater than in the previous season. Include an explanation of the reasoning process by which your conclusion follows from the test result. Also mention any reservations you have about the technical conditions for this study.

Watch Out

Think through the reasoning process of tests and p-values as you learned in the previous topic. A small p-value indicates that you are unlikely to obtain such extreme sample data if the null hypothesis is true, which provides evidence against the null hypothesis and in favor of the alternative hypothesis. The smaller the p-value, the stronger the evidence against the null hypothesis.

Activity 20-2: Sleeping Times

8-1, 19-4, 19-5, 19-12, 19-19, **20-2,** 20-7

Reconsider the data that you collected and analyzed on the sleep times for yourself and your classmates in Activity 19-5 on page 389.

a. Conduct a significance test of whether your class data provide strong evidence that the mean sleep time among all students at your school that particular night differs from seven hours. Clearly define the parameter of interest and report the hypotheses, test statistic, and *p*-value. Would you reject the null hypothesis at the .05 level? *Hint:* You might need to use the symmetry of the *t*-distribution in finding the *p*-value.

b. Do you think the technical conditions for the validity of this procedure have been met? Explain.

Now consider the four samples of hypothetical sleep times presented in Activity 19-4. Dotplots and summary statistics are reproduced here. (The data are stored in the grouped file HYPOSLEEP, containing the lists SLP1, SLP2, SLP3, and SLP4, respectively.)

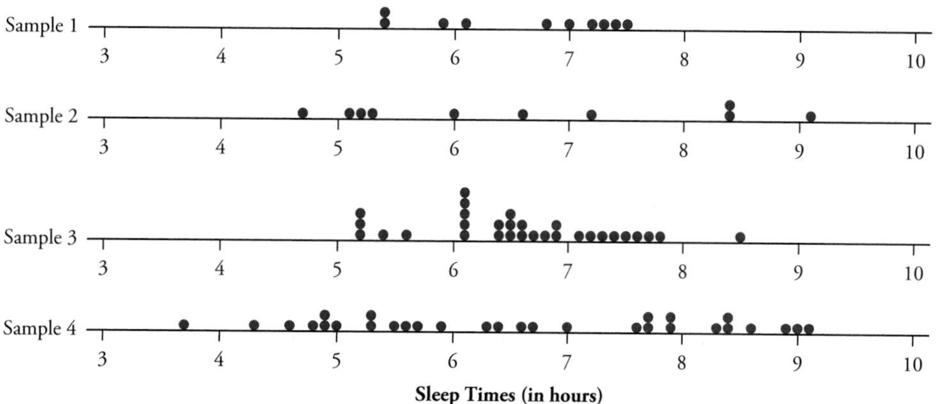

Sample Number	Sample Size	Sample Mean	Sample SD	Test Statistic	p-value
1	10	6.6	0.825		
2	10	6.6	1.597		
3	30	6.6	0.825		
4	30	6.6	1.597		

20

c. Comparing samples 1 and 2, which sample do you think supplies stronger evidence that $\mu \neq 7$ (i.e., that the population mean sleep time differs from 7 hours)? In other words, which sample (1 or 2) would produce a smaller *p*-value for the appropriate test of significance? Explain.

d. Comparing samples 1 and 3, which sample do you think supplies stronger evidence that $\mu \neq 7$ (i.e., that the population mean sleep time differs from 7 hours)? In other words, which sample (1 or 2) would produce a smaller *p*-value for the appropriate test of significance? Explain.

e. For each of these four samples, use your calculator's **T-Test** command to compute the test statistic and *p*-value for testing that the population mean differs from 7 hours. Record these in the table following part b. The **T-Test** can be found by pressing [STAT] and selecting the TESTS menu. For the **T-Interval,** select **Data** (lists **SLP1, SLP2, SLP3,** and then **SLP4**) as input. Set the frequency to **1.**

f. Which of the samples give(s) you enough evidence to reject the null hypothesis at the .05 level and conclude that the mean sleep time is, in fact, different than seven hours?

g. Comment on whether your conjectures in part c and part d are confirmed by the test results.

This activity should reinforce what you discovered in Activities 17-4 and 19-4 about effects of sample size and sample variation. An observed difference between a sample mean and a hypothesized mean is more statistically significant (i.e., less likely to occur by chance, indicated by a smaller *p*-value) with a larger sample than with a smaller one. Also, that difference is more significant if the sample data values themselves are less spread out (less variable, more consistent) than if they are more spread out. Naturally, the farther the sample mean is from the hypothesized mean, the more statistically significant the result.

Activity 20-3: Golden Ratio

The ancient Greeks made extensive use of the "golden ratio" in art and literature, for they believed that a width-to-length ratio of 0.618 was aesthetically pleasing. Some have conjectured that American Indians also used the same ratio (Hand et al., 1993). The following data are width-to-length ratios for a sample of 20 beaded rectangles used by the Shoshoni Indians to decorate their leather goods.

0.693	0.662	0.690	0.606	0.570	0.749	0.672	0.628	0.609	0.844
0.654	0.615	0.668	0.601	0.576	0.670	0.606	0.611	0.553	0.933

a. Produce a histogram and comment on the distribution of these ratios.

b. Conduct a *t*-test of whether these sample data lead to rejecting the hypothesis that the population mean width-to-length ratio equals 0.618. Use the .10 significance level and a two-sided alternative. Report all aspects of the test, including a check of technical conditions, and summarize your conclusion in context.

Activity 20-4: Children's Television Viewing

1-15, 19-16, **20-4,** 22-14, 22-15

Activity 1-15 described a study on children's television viewing conducted by Stanford researchers (Robinson, 1999). At the beginning of the study, parents of third- and fourth-grade students at two public elementary schools in San Jose were asked to report how many hours of television the child watched in a typical week. The 198 responses had a mean of 15.41 hours and a standard deviation of 14.16 hours.

Conduct a test of whether these sample data provide evidence at the .05 level for concluding that third- and fourth-grade children watch an average of more than two hours of television per day. Include all the components of a significance test, and explain what each component reveals. Start by identifying the observational units, variable, and population.

Solution

The observational units are third- and fourth-grade students. The sample consists of the 198 students at two schools in San Jose. The population could be considered all American third- and fourth-graders, but it might be more reasonable to restrict the population to be all third- and fourth-graders in the San Jose area at the time the study was conducted.

The variable measured here is the *amount of television the student watches in a typical week,* which is quantitative. The parameter is the mean number of hours of television watched per week among the population of all third- and fourth-graders. This population mean is denoted by μ. The question asked about watching an average of two hours of television per day, so convert that to be 14 hours per week.

The null hypothesis is that third-and fourth-graders in the population watch an average of 14 hours of television per week (H_0: $\mu = 14$). The alternative hypothesis is that these children watch more than 14 hours of television per week on average (H_a: $\mu > 14$).

20

You should check the technical conditions for the *t*-test before you proceed:

- The sample of children was not chosen randomly; they all came from two schools in San Jose. You might still consider these children to be representative of third- and fourth-graders in San Jose, but you might not be willing to generalize to a broader population.
- The sample size is large enough (198 is far greater than 30) that the second condition holds regardless of whether the data on television watching follow a normal distribution. You do not have access to the child-by-child data in this case, so you cannot examine graphical displays; however, the large sample size assures you that the *t*-test is nevertheless valid to employ.

The sample size is $n = 198$; the sample mean is $\bar{x} = 15.41$ hours; and the sample standard deviation is $s = 14.16$ hours. The test statistic is

$$t = \frac{15.41 - 14}{14.16/\sqrt{198}} \approx 1.401$$

indicating the observed sample mean lies 1.401 standard errors above the conjectured value for the population mean. Looking in Table III, using the 100 df line (rounded down from the actual df of $198 - 1 = 197$), reveals the *p*-value (probability to the right of $t = 1.401$) to be between .05 and .10. The calculator gives the *p*-value more exactly to be .081.

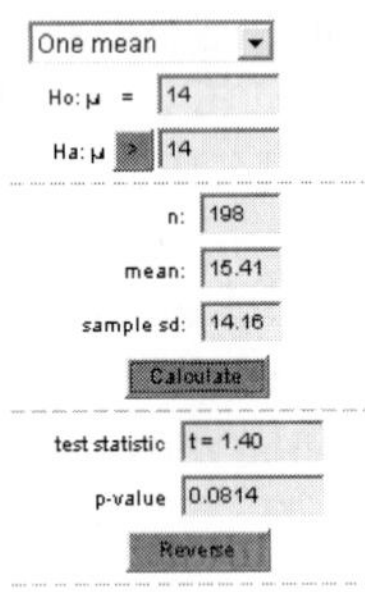

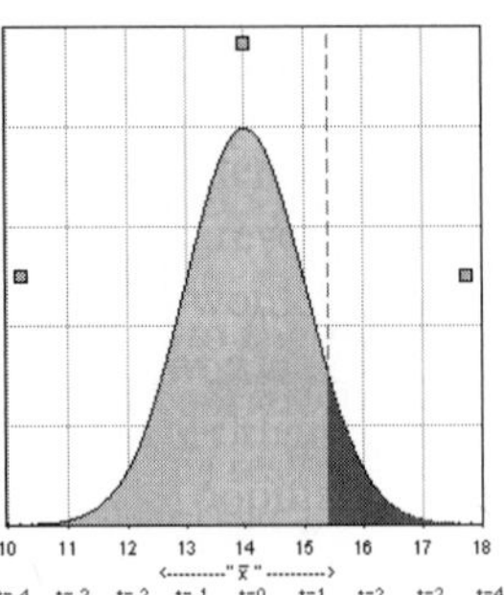

This *p*-value is not less than the .05 significance level. The sample data, therefore, do not provide sufficient evidence to conclude that the population mean is greater than 14 hours of television watching per week. This conclusion stems from realizing that obtaining a sample mean of 15.41 hours or greater would not be terribly uncommon when the population mean is really 14 hours per week. If you had used a greater significance level (such as .10), which requires less compelling evidence in order to reject a hypothesis, then you would have concluded that the population mean exceeds 14 hours per week.

Watch Out

- The first step is to decide whether your test is about a proportion or a mean. One way to judge is to ask whether the variable (as measured on the individual observational units) is categorical or quantitative. Also be on the lookout for more obvious clues, such as the word *average* appearing in the statement of this problem.
- Remember, the hypotheses are about parameters, not statistics. Also remember the hypothesized value comes from the research question, not from the sample data. Also, the direction of the alternative hypothesis should be based on the research question and not on the data.

- When a sample is not chosen randomly, be cautious about generalizing the study results to a larger population. Try instead to think about what population the sample might be representative of, and even then be wary.
- Do not forget to include the $\sqrt{n}$ factor in calculating the value of the test statistic.

Wrap-Up...........

This topic introduced you to tests of significance concerning a population *mean*. You found that the structure, reasoning, and interpretation of these tests are the same for a mean as they were for a proportion. As always, the *p*-value is the key, because it reports how likely you would have been to observe such extreme sample data if the null hypothesis were true. So, a small *p*-value provides evidence against the null hypothesis that the population mean equals a particular hypothesized value.

Two ways in which the ***t*-test** for a mean is different from the *z*-test for a proportion are that the *p*-value is calculated from the *t*-distribution instead of the normal distribution, and the variability in the sample data also plays a role in determining the *p*-value. For example, learning that the average sleep time in a sample of students is 6.6 hours is not enough information to assess whether the population average sleep time is less than 7 hours, because you also need to know the sample size and how variable (as measured by the sample standard deviation) those sleep times are.

Some useful definitions to remember and habits to develop from this topic include

- Specify the null and alternative hypotheses based on the research question, prior to seeing the sample data. With a quantitative variable, the hypotheses concern a population mean.
- Check the technical conditions before applying the procedure. As with a *t*-interval, the *t*-test requires either a large sample size or a normally distributed population. You can generally regard a sample of at least 30 as large enough for the procedure to be valid. If the sample size is less than 30, examine visual displays of the sample data to see whether they appear to follow a normal distribution.
- Consider whether the sample was randomly selected from the population of interest before generalizing the test conclusion to that population.
- Do not forget to begin your analysis with graphical and numerical summaries of the sample data. Do not jump into a *t*-test before examining the sample data first.

Thus far you have studied confidence intervals and tests of significance for both categorical and quantitative variables, but you have only considered cases with a *single* sample. In the next unit, you discover how to apply these inference procedures to research questions that call for comparing two groups. These procedures will be especially useful for analyzing data from randomized experiments.

● ● ● Homework Activities

Activity 20-5: Exploring the *t*-Distribution

19-2, **20-5**, 20-6

a. Use the *t*-table to find the *p*-values (as accurately as possible) corresponding to the following test statistic values and degrees of freedom, filling in a table like the one shown here with those *p*-values:

df	Pr($T \geq$ 1.415)	Pr($T \geq$ 1.960)	Pr($T \geq$ 2.517)	Pr($T \geq$ 3.168)
4				
11				
23				
80				
Infinity				

b. Does the *p*-value get larger or smaller as the value of the test statistic increases (if the number of degrees of freedom remains the same)?

c. Does the *p*-value get larger or smaller as the number of degrees of freedom increases (if the value of the test statistic remains the same)?

Activity 20-6: Exploring the *t*-Distribution

19-2, 20-5, **20-6**

a. Use the *t*-table and/or your answers to Activity 20-5 to find the following *p*-values:

df	Pr($T \leq -1.415$)	Pr($T \leq -1.960$)	$2 \times$ Pr($T \geq \|-2.517\|$)	$2 \times$ Pr($T \geq \|-3.168\|$)
4				
11				
23				
80				
Infinity				

b. Describe how the *p*-values in the first two columns of the table compare with those in the first two columns of the table from Activity 20-5. Explain why this makes sense.

c. Describe how the *p*-values in the last two columns of the table compare with those of the last two columns of the table from Activity 20-5. Explain why this makes sense.

Activity 20-7: Sleeping Times

8-1, 19-4, 19-5, 19-12, 19-19, 20-2, **20-7**

Reconsider the hypothetical samples of sleep times (HYPOSLEEP) presented in Activity 19-4 and analyzed in Activity 20-2. Suppose now that you are interested in testing whether the sample data provide evidence that the population mean sleep time is *less than* seven hours. Use your calculator to conduct the test of significance for each of the four samples with a *one-sided* alternative hypothesis. Record the *p*-value for each sample and comment on how these *p*-values compare with the one-sided values found in Activity 20-2.

Activity 20-8: UFO Sighters' Personalities

20-8, 22-17

In a 1993 study, researchers took a sample of people who claimed to have had an intense experience with an unidentified flying object (UFO) and a sample of people who did not claim to have had such an experience (Spanos et al., 1993). They then compared the two groups on a wide variety of variables, including IQ. Suppose you want to test whether the average IQ of those who have had such a UFO experience is higher than 100, so you want to test H_0: $\mu = 100$ vs. H_a: $\mu > 100$.

a. Identify clearly what the symbol μ represents in this context.
b. Is this a one-sided or a two-sided test? Explain how you can tell.

The sample mean IQ of the 25 people in the study who claimed to have had an intense experience with a UFO was 101.6; the standard deviation of these IQs was 8.9.

c. Does this information enable you to check the technical conditions completely? What needs to be true for this procedure to be valid?
d. Calculate the test statistic and draw a sketch with shaded area corresponding to obtaining a test statistic as extreme or more extreme than this one observed for the sample of 25 UFO observers.
e. Use the *t*-table or your calculator to determine (as accurately as possible) the *p*-value.
f. Write a sentence interpreting the *p*-value in the context of this sample and these hypotheses. Summarize the conclusion of your test in context.

Activity 20-9: Basketball Scoring

20-1, **20-9,** 20-20

Recall from Activity 20-1 that the NBA made rule changes prior to the 1999–2000 season that were intended to boost scoring from the average of 183.2 points per game in the previous season. The November 29, 1999, issue of *Sports Illustrated* reported that for the first 149 games of the 1999–2000 season, the mean number of points per game was 196.2.

a. State in words and in symbols the hypotheses for testing whether the sample data provide strong evidence that the mean for the entire 1999–2000 season exceeds 183.2.
b. Do you have enough information to calculate the test statistic? Explain.
c. How large would the test statistic have to be in order to reject the null hypothesis at the .01 level?
d. If the sample standard deviation for these 149 games were close to the standard deviation for the 25 games that you analyzed in Activity 20-1, would the test statistic exceed the rejection value in part c? By a lot? Explain.
e. Even though the magazine did not provide all the information necessary to conduct a significance test, can you reasonably predict whether the test result would be significant at the .01 level? Explain, based on your answers to parts c and d.
f. Does the validity of this test procedure depend on the scores being normally distributed? Explain.

Activity 20-10: Credit Card Usage

1-9, 16-12, 17-11, 19-8, **20-10**

Refer to the study described in Activities 16-12 and 19-8. The Nellie Mae organization found that in a random sample of undergraduate students taken in 2004, the average credit card balance was \$2169. Again take the sample size to be 1074, which corresponds to the 76% of the sample of 1413 students who hold a credit card. Suppose (for now) the sample standard deviation of these credit card balances is \$1000.

a. Conduct a significance test of whether the sample data provide strong evidence (at the $\alpha = .05$ level) that the population mean credit card balance exceeds \$2000. Report all aspects of the test, including a check of the technical conditions.
b. Repeat part a, assuming that the sample standard deviation of these credit card balances is \$2000.
c. Which scenario produces a greater p-value? Explain why this makes sense.

Activity 20-11: Body Temperatures
12-1, 12-19, 15-3, 15-18, 15-19, 19-3, 19-7, **20-11,** 22-10, 23-3

Reconsider the data from Activity 19-3 on body temperatures for a sample of 65 healthy adult males and 65 healthy adult females (BODYTEMPS). For each gender, use your calculator to produce graphical and numerical summaries of the body temperatures. Then conduct a t-test of whether these sample data provide strong evidence that the population mean body temperature differs from 98.6 degrees. Provide all components of a hypothesis test, and summarize your conclusions. Also comment on how your conclusions are similar or different between the two genders.

Activity 20-12: Social Acquaintances
9-8, 9-9, 10-13, 10-14, 19-9, 19-10, **20-12**

Reconsider the data that you collected in Topic 9, based on the "social acquaintance" exercise described by Malcolm Gladwell in *The Tipping Point.*

a. Use the sample data from your class to test whether the population mean number of acquaintances differs from 30. Report the hypotheses, test statistic, and p-value. Also state the test decision using the $\alpha = .025$ significance level, and summarize your conclusion.
b. To what population would you be willing to generalize the result of your significance test in part a? Explain.

Activity 20-13: Age Guesses
8-20, **20-13,** 20-14

Consider the data collected in the Preliminaries section concerning guesses of your instructor's age. Conduct a full analysis of the data with regard to the question of whether the guesses tend to average out to the actual age. (If your instructor prefers not to reveal his or her actual age, address whether the guesses tend to average out to the age that you personally guessed.) Include graphical and numerical summaries as well as an appropriate test of significance, and be sure to identify the population and parameter of interest very clearly. Write a paragraph or two describing and explaining your analysis and findings.

Activity 20-14: Age Guesses
8-20, 20-13, **20-14**

After five weeks of a ten-week term, a 44-year-old statistics instructor asked his students to guess his age. The responses are displayed in the following dotplot:

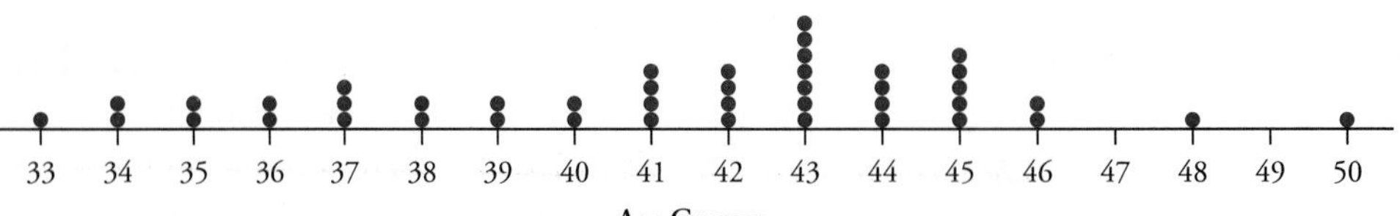

a. Comment on the distribution of age guesses. Be sure to address whether students tended to misjudge his age on the high side or low side or whether their guesses averaged out to approximately the correct age. *Hint:* Remember to mention center, spread, shape, and outliers.
b. The sample size is 44; the sample mean is 41.182 years; and the sample standard deviation is 3.996 years. Are these parameters or statistics? Identify the appropriate symbol for each of these numbers.
c. Identify the relevant population and parameter of interest.
d. Conduct a significance test of whether these sample data provide evidence that the population mean age guess differs from the instructor's actual age of 44 years. Include all components of a test, including a check of technical conditions. State the test decision that you would make at the $\alpha = .01$ significance level, and summarize your conclusion in context.

Activity 20-15: Nicotine Lozenge

1-16, 2-18, 5-6, 9-21, 19-11, **20-15**, 20-19, 21-6, 22-8

Recall from Activities 1-16 and 19-11 the experiment that investigated the effectiveness of a nicotine lozenge for subjects who wanted to quit smoking (Shiffman et al., 2002). Before the treatments began, subjects answered background questions, including how many cigarettes they smoked per day. Among the 1818 subjects in the study, the average was 22.0 cigarettes per day, and the standard deviation was 10.8 cigarettes per day.

a. Use these sample statistics to test whether the population mean differs from 20 cigarettes (one pack) per day.
b. Suppose that a smaller study produced the same sample mean and standard deviation, but with a sample size of only 100 subjects. Conduct the test from part a in this case.
c. Describe how your test results differ between part a and part b, and explain why this makes sense.

Activity 20-16: Random Babies

11-1, 11-2, 14-8, 16-17, **20-16**, 20-22

Recall the simulated data that you collected by shuffling and dealing cards to represent assigning babies to mothers at random in Activity 11-1 on page 216. Consider the variable *number of matches*.

a. Report (by either recalling or recalculating) the sample size (number of repetitions conducted by yourself and your classmates), the sample mean number of matches per repetition, and the sample standard deviation of those numbers of matches.
b. Recall that your theoretical analysis in part g of Activity 11-2 on page 220 revealed the long-run average number of matches to be exactly 1. Use the sample statistics from part a to test whether the simulated data provide strong reason to doubt that the population mean equals 1. Report the hypotheses (in words and in symbols), the sampling distribution specified by the null hypothesis, the test statistic, and the *p*-value. Also comment on the technical conditions and write a conclusion.

Activity 20-17: Backpack Weights

2-13, 10-12, 19-6, **20-17**

Recall the data on backpack weights and body weights of college students from Activity 19-6 (BACKPACK). Consider the ratio of backpack weight to body weight.

a. Is this ratio a categorical or quantitative variable?
b. Conduct a test of whether the sample data provide evidence that the population mean ratio differs from .10. Report all aspects of the test, including a check of technical conditions. Also summarize your conclusion.
c. Comment on whether your test result is consistent with the findings of the confidence interval found in Activity 19-6 on pages 390 and 391.

Activity 20-18: Looking Up to CEOs

14-4, **20-18**

Recall from Activity 14-4 the study that took a random sample of male chief executive officers (CEOs) of American companies to test whether their average height exceeds the 69-inch average height for adult American males (Gladwell, 2005).

a. State the null and alternative hypotheses for this study.
b. Describe what a Type I error would mean in this context.
c. Describe what a Type II error would mean in this context.

Activity 20-19: Nicotine Lozenge

1-16, 2-18, 5-6, 9-21, 19-11, 20-15, **20-19**, 21-6, 22-8

Recall the data from Activity 19-11 that you analyzed on number of cigarettes smoked per day by subjects in an experiment that investigated the effectiveness of a nicotine lozenge. Among the 1818 subjects in the study, the average was 22.0 cigarettes per day, and the standard deviation was 10.8 cigarettes per day.

a. Based on the 99% confidence interval for the population mean μ, what potential values of μ would be rejected at the .01 significance level? Explain.
b. Conduct a test of whether the population mean differs from 22.0 at the .05 significance level.
c. Based on the *p*-value in part b, what can you say regarding a 95% confidence interval for μ? Explain.

Activity 20-20: Basketball Scoring

20-1, 20-9, **20-20**

Reconsider the data that you analyzed in Activity 20-1 concerning whether the mean points scored per NBA game was higher in the 1999–2000 season than it had been in the previous season. Suppose your sample had consisted of only the ten games played on December 10, 1999, whose total points scored were as follows:

196	198	205	163	184	224	206	190	140	204

a. Examine visual displays and summary statistics for these data. Write a few sentences describing the distribution with respect to the question of whether the games average more than 183.2 points, which had been the previous season's average.
b. Use these sample data to test whether there is reason to believe that the mean points per game during the 1999–2000 NBA season exceeds 183.2. Report your hypotheses, test statistic, and *p*-value.
c. Would you reject the null hypothesis at the $\alpha = .05$ level? Write a sentence indicating what this signifies in this context.

d. Would you *accept* the null hypothesis and conclude that the mean points scored per game in the 1999–2000 season is, in fact, 183.2? Explain.
e. Remove the outlier from the sample data and repeat this analysis. Report your results, and also comment on what effect removing the outlier has on your findings.

Activity 20-21: Pet Ownership

13-9, 13-14, 13-15, 18-2, **20-21**

Recall from Activity 18-2 that in a survey of 80,000 households conducted by the American Veterinary Medical Association in 2001, 31.6% of households reported that they owned a pet cat. Of those 25,280 households that did own a pet cat, the mean number of cats per household was 2.1.

a. State (in words and in symbols) the hypotheses for testing whether the sample data provide strong evidence that the mean number of cats per cat-owning household exceeds two.
b. What additional sample information do you need to conduct this test?
c. Supply a reasonable estimate for the missing sample statistic, and calculate the test statistic and *p*-value. Does the sample provide strong evidence that the mean number of cats exceeds two? Explain.
d. As a check on the sensitivity of this test, double your estimate for the missing sample statistic and repeat part c. Does your conclusion change substantially? Explain.
e. Using your estimate from part c, find a 99% confidence interval for the population mean number of cats per cat-owning household. Would you say that this mean value greatly exceeds two cats in a practical sense? Explain.

Activity 20-22: Random Babies

11-1, 11-2, 14-8, 16-17, 20-16, **20-22**

Recall again from Activity 11-1 on page 216 and Activity 11-2 on page 220 the class simulation of the "random babies" activity. Let π be the long-term proportion of repetitions that result in no matches, and let μ be the long-term mean number of matches per repetition.

a. Using the simulated sample data, conduct a significance test of whether π differs from .4. Report the hypotheses, test statistic, and *p*-value. Is the result significant at the $\alpha = .05$ level?
b. If the test result in part a is not significant, does it follow that you accept π equals .4 exactly?
c. Recall from your theoretical analysis in Activity 11-2 the exact value of π. Explain how this relates to part b.
d. Using the simulated sample data, conduct a significance test of whether μ differs from 1.1. Report the hypotheses, test statistic, and *p*-value. Is the result significant at the $\alpha = .05$ level?
e. If the test result in part a is not significant, does it follow that you accept μ equals 1.1 exactly?
f. Recall from your theoretical analysis in Activity 11-2 the exact value of μ. Explain how this relates to part e.

UNIT 5 • • •

Inference from Data: Comparisons

TOPIC

21 Comparing Two Proportions

Would you feel more comfortable performing some skilled task, such as playing a video game, if an observer with a vested interest were watching you, or would you rather not have someone else's potential payoff relying on your performance? What do psychology experiments reveal about this, and how can statistics help to settle the issue? Or suppose you want to know whether men and women differ with regard to how satisfied they are with their own appearance. In this topic, you will learn how to apply the inference principles that you learned in the previous unit to address questions of comparisons like these.

Overview

In the previous unit, you discovered and explored techniques of statistical inference for drawing conclusions about population parameters on the basis of sample statistics. All of those procedures applied to a *single* parameter (proportion or mean) from a *single* population. In the next two topics, you will investigate and apply inference procedures for comparing parameters between *two* populations or treatment groups. These techniques are especially important due to the crucial role of comparison in experimental design. The randomization involved in experiments enables you to make inferences about the effects of the explanatory variable on the response variable. These procedures are similarly applicable for comparing two groups that have been randomly selected from two populations. In this topic, you will investigate inference procedures for comparing proportions between two experimental groups or two samples. You will find that what you have already learned about significance tests and confidence intervals will serve you well, because the basic reasoning, structure, and interpretation of these tests and intervals are the same whether you're making inferences about a single population parameter or comparing two population parameters.

Preliminaries

1. Would you expect to perform better on a cognitive task if you were being observed by someone with a vested interest in your performance or by a neutral observer?

2. Suppose that 10 patients are randomly assigned to receive a new medical treatment, while another 10 patients are randomly assigned to receive the old standard treatment. If 7 patients in the new group and 5 patients in the old group improve, would you be strongly convinced, somewhat convinced, or not convinced that the new treatment is superior to the old treatment?

3. Suppose that 10 patients are randomly assigned to receive a new medical treatment, while another 10 patients are randomly assigned to receive the old standard treatment. If 9 patients in the new group and 3 patients in the old group improve, would you be strongly convinced, somewhat convinced, or not convinced that the new treatment is superior to the old treatment?

4. Guess what proportion of American adults say that they are satisfied with their physical appearance.

5. Would you expect the proportion of adults who are satisfied with their physical appearance to be higher among men or among women?

In-Class Activities

Activity 21-1: Friendly Observers

5-27, **21-1,** 23-11

Recall from Activity 5-27 that in a study published in the *Journal of Personality and Social Psychology* (Butler and Baumeister, 1998), researchers investigated a conjecture that having an observer with a vested interest would decrease subjects' performance on a skill-based task. Subjects were given time to practice playing a video game that required them to navigate an obstacle course as quickly as possible. They were then told to play the game one final time with an observer present. Subjects were randomly assigned to one of two groups. One group (call it A) was told that the participant and observer would each win $3 if the participant beat a certain threshold time, and the other group (B) was told only that the participant would win the prize if the threshold were beaten. The following table summarizes results similar to those found in this study:

	A: Observer Shares Prize	B: No Sharing of Prize	Total
Beat Threshold	3	8	11
Did Not Beat Threshold	9	4	13
Total	12	12	24

a. Identify the explanatory and response variables in this study. Also classify each variable as categorical (also binary) or quantitative.

Explanatory: Type:

Response: Type:

b. Is this an observational study or an experiment? Explain how you know.

c. Calculate the sample proportions of success for each group. (Denote these by $\hat{p}_A$ and $\hat{p}_B$.) Also display the results in a segmented bar graph.

d. Do these sample proportions differ in the direction conjectured by the researchers?

e. Even if there were absolutely *no effect* due to the observers' incentive, is it possible to have obtained such a big difference between the two groups simply because of chance variation (i.e., due to the random assignment)?

In keeping with the reasoning of significance tests, you will now consider how likely it would be to get such sample results simply by random assignment, if in fact the observers' incentive had no effect on the subjects' performance. One way to analyze this question is to assume that there actually is no treatment effect and so the 11 successes would have been successes regardless of which group these subjects had been assigned to, and similarly for the 13 failures. You can then *simulate* the process of assigning the 24 subjects at random to the two groups, observing how often you obtain a result at least as extreme (3 or fewer of the successes assigned to group A) as in the actual study. Repeating this process a large number of times will give you a sense for how unusual it would be for this sample result to occur by chance alone, still assuming the observers' incentive had no effect on subject performance.

f. Select 24 cards (index cards or playing cards) to represent the 24 subjects in this study. Mark 11 cards as "success" and 13 as "failure," shuffle them well, and randomly deal out 12 cards to represent the cases assigned to group A. How many of these 12 cards are successes? Is this result at least as extreme as the result in the actual study (3 or fewer successes in group A)?

g. Repeat this simulation a total of five times, recording your results in the table:

Repetition #	1	2	3	4	5
Number of "Successes" Assigned to Group A					
Is Result as Extreme as in Actual Study?					

h. Combine your results with the rest of the class's results, creating a dotplot of the number of successes randomly assigned to group A.

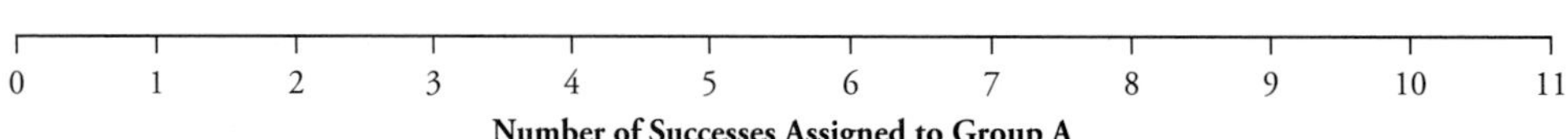

i. What value or values for the number of successes in group A are the most common? Explain why this makes sense. *Hint:* Remember the assumption behind your simulation is that the observers' incentive has no effect on subjects' performance; the only differences between the two groups arise from random assignment.

j. How many repetitions were performed by the class as a whole? How many repetitions gave a result at least as extreme as the actual sample (3 or fewer successes in group A)? What proportion of the repetitions does this value represent?

k. Remember that your random shuffling and dealing assumed the observers' incentive had no effect on subjects' performance. Based on these simulated results, does it appear *very* unlikely, *somewhat* unlikely, or *not* unlikely for random assignment to produce a result as extreme as the actual sample when the observer has no effect on subjects' performance?

l. In light of your answer to the previous question, considering that the actual sample is what the researchers found (3 or fewer successes in group A), would you say the data provide reasonably strong evidence in support of the researchers' conjecture? Explain.

This activity should reinforce the important idea that *statistical significance* assesses the likelihood of an observed result by asking how often such an extreme result would occur *by chance alone*. When the sample result is unlikely to occur by chance alone, it is said to be statistically significant. As always, the probability of obtaining a result at least as extreme as the sample by chance alone is known as the *p-value* of the test. Your class simulation has approximated this *p*-value.

You can approximate the *p*-value more accurately by performing more repetitions of the random assignment. The calculator facilitates this.

m. Open the applet `Friendly Observer Simulation`. Make sure the Number of repetitions is set to **1,** and then click Randomize. Click Randomize four more times. Does the number of successes randomly assigned to group A vary? Record these five values.

Friendly Observer Simulation

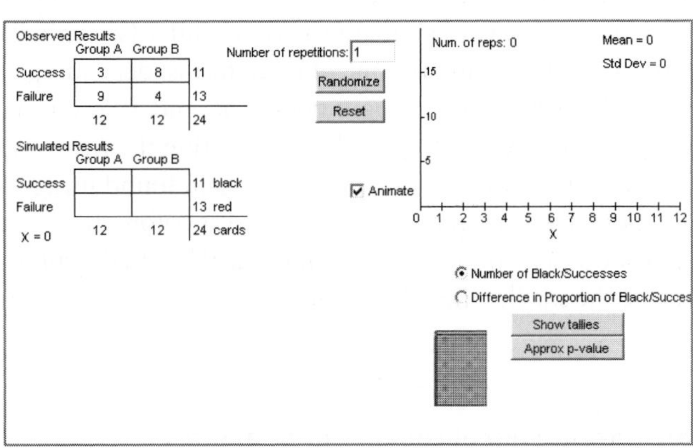

n. Unclick the Animate box, ask for **495** more repetitions, and click Randomize. Describe the resulting distribution of the number of successes randomly assigned to group A.

o. Click Show tallies (which counts the number of repetitions that resulted in each value for *x*, the number of successes in group A), and then click Approx *p*-value (which determines the proportion of these 500 repetitions that produced a result as least as extreme as in the actual study). Report this approximate *p*-value as determined from your simulation.

p. Do you need to reconsider or refine your answers to part k and part l based on this larger simulation?

Rather than rely on simulations, you can use a formal test of significance to compare success proportions between two groups. The structure and reasoning of this test are the same as for a single population proportion. Two-sample procedures arise from two different scenarios:

- Scenario 1 applies when subjects have been *randomly assigned* between two treatment groups (i.e., in a randomized experiment). This scenario allows for drawing a *cause-and-effect conclusion* between the explanatory and response variables if the difference between the groups is judged to be statistically significant.

- Scenario 2 applies when observational units have been *randomly sampled* from two populations. This scenario allows for *generalizing* conclusions about the samples to the larger populations.

(Of course, some studies employ both random sampling from a population and then random assignment to treatment groups.)

We will denote the underlying population proportion in the two groups by π_1 and π_2. As outlined below, the null hypothesis asserts that these population proportions are equal—i.e., that the explanatory variable has no effect on the response variable (scenario 1)—or that two populations do not differ (scenario 2). Once again, the alternative hypothesis can take one of three forms, depending on the researchers' conjecture about the relative magnitudes of π_1 and π_2 prior to conducting the study. The test statistic is calculated by comparing the two sample proportions and standardizing appropriately. The *p*-value is again found using the Standard Normal Probabilities table. You interpret this *p*-value as before: It represents the probability of obtaining a test statistic at least as extreme as the actual result if the null hypothesis is true. Thus, the smaller the *p*-value, the stronger the evidence against the null hypothesis.

Significance test of equality of π_1 and π_2:
Null hypothesis: H_0: $\pi_1 = \pi_2$
Alternative hypothesis:

i. H_a: $\pi_1 < \pi_2$ *or*
ii. H_a: $\pi_1 > \pi_2$ *or*
iii. H_a: $\pi_1 \neq \pi_2$

Test statistic:

$$z = \frac{\hat{p}_1 - \hat{p}_2}{\sqrt{\hat{p}_c(1 - \hat{p}_c)\left(\frac{1}{n_1} + \frac{1}{n_2}\right)}}$$

where $\hat{p}_c$ represents the *combined* sample proportion of successes if the two groups are combined into one sample.

p-value:

i. $\Pr(Z \leq z)$ *or*
ii. $\Pr(Z \geq z)$ *or*
iii. $2 \times \Pr(Z \geq |z|)$

The **technical conditions** required for this procedure to be valid are that

- The data arise from randomly assigning subjects to two treatment groups *or* from independent random samples from two populations.
- The sample sizes are large enough, relative to the success proportions involved, so that $n_1\hat{p}_c \geq 5$ and $n_1(1 - \hat{p}_c) \geq 5$ and $n_2\hat{p}_c \geq 5$ and $n_2(1 - \hat{p}_c) \geq 5$.

Watch Out

- The calculation details are identical for the two scenarios (random assignment or random sampling), so it is easy to avoid thinking about the differences. But these two scenarios, based on different uses of randomness, permit very different scopes of conclusions to be drawn, as you first studied in Topic 3. Always pay attention to which of these two scenarios pertains when conducting such a test, and draw your conclusions accordingly.

- A minor variation on scenario 2 is that *one* random sample is taken from a population and then separated by the explanatory variable (such as gender), rather than taking separate independent samples of men and women. As long as two groups can be considered independent, we will consider this as behaving like scenario 2.
- If neither of these scenarios applies, for example, in an observational study where subjects are not randomly selected, then be cautious about drawing any conclusions from the test. Confounding variables can prevent drawing a cause-and-effect conclusion, and biased sampling can prevent generalizing to a larger population. But the significance test still might be useful for possibly eliminating random chance as a plausible explanation for the observed difference between the groups.

Activity 21-2: Back to Sleep

4-5, 6-5, **21-2**

Recall from Activity 6-5 on page 106 that the National Infant Sleep Positioning Study began conducting national surveys in 1992 to examine how well parents were heeding recommendations to place sleeping babies on their backs or sides. For each year through 1996, researchers interviewed random samples of about 1000 parents of infants younger than eight months old, asking about the position in which the infant regularly slept (Willinger et al., 1998).

a. Does this study involve random assignment to groups or random sampling from populations?

b. Suppose you want to analyze the sample data to test whether the proportion of infants who sleep on their stomachs (the *discouraged* position) decreased between 1992 and 1996. State the relevant null and alternative hypotheses to be tested, in both symbols and words. (Use π_{1996} and π_{1992} as the symbols for the population proportions.)

The sample data revealed that 70% of infants slept on their stomachs in 1992 and 24% did so in 1996. Assume that the sample size was 1000 for each year.

c. Use this information to determine $\hat{p}_c$, the combined sample proportion of infants who slept on their stomachs.

d. Calculate the test statistic for testing the hypotheses of part b. Does this value impress you as a large z-score?

e. Based on this test statistic, and also considering the inequality direction in your alternative hypothesis, use Table II (Standard Normal Probabilities) to calculate the p-value of the test.

f. Write a sentence or two interpreting what this p-value means in the context of this study.

g. What test decision (reject or fail to reject the null hypothesis) would you make at the $\alpha = .01$ significance level?

h. Summarize the conclusion you would draw about whether the sample data provide strong evidence that the proportion of infants who slept on their stomachs decreased between 1992 and 1996.

Although this test of significance addresses the question about the *extent of evidence* to which the population proportions differ, it does not provide information about the *magnitude* of this difference. You can address this issue with a confidence interval for the difference in population proportions.

Confidence interval for $\pi_1 - \pi_2$:

$$(\hat{p}_1 - \hat{p}_2) \pm z^* \sqrt{\frac{\hat{p}_1(1-\hat{p}_1)}{n_1} + \frac{\hat{p}_2(1-\hat{p}_2)}{n_2}}$$

where z^* is the appropriate (for the desired confidence level) critical value from the standard normal distribution.

The **technical conditions** necessary for the validity of this confidence interval procedure are that

- The data arise from randomly assigning subjects to two treatment groups or from independent random samples from two populations.
- There are at least five "successes" and five "failures" in each sample group.

i. Calculate a 95% confidence interval for estimating the difference in population proportions of infants who slept on their stomachs ($\pi_{1992} - \pi_{1996}$).

j. Describe what this interval reveals about the difference in population proportions. In particular, comment on the importance of whether the interval includes the value zero, contains only negative values, or contains only positive values. *Hint:* Remember the interval estimates the *difference* in population proportions. As always, relate your comments to this context of infant sleeping positions.

k. Use your calculator's **2-PropZTest** command (found by pressing [STAT] and selecting the TESTS menu) to confirm your test and interval calculations. In this case, **x1 = 700, n1 = 1000, x2 = 240,** and **n2 = 1000.**

l. Suppose you had instead calculated a 95% confidence interval for $\pi_{1996} - \pi_{1992}$ (notice the order of subtraction has been switched). Before performing the calculation, how do you expect this interval to compare to the one in part i? Explain.

m. Select **2-PropZInt** by pressing [STAT] and arrowing over to the TESTS menu to compute a 95% confidence interval for $\pi_{1996} - \pi_{1992}$. (*TI hint:* For this exercise, you will need to select **Stats** rather than **Data** for **Inpt.** The **Data** selection is for when you want to input information from a list.) Was your prediction in part l correct? What does this interval reveal?

This last question should convince you that the group labels (i.e., which group you label as 1 and which group you label as 2) and the order of subtraction do not change your conclusions at all. The intervals should turn out to be negatives of each other, so the interpretation and conclusion are identical in both cases. Similarly, in a test of significance, switching the group labels leads to the same *p*-value and conclusion. As long as you are consistent throughout a question (meaning as long as you perform the calculations consistently with how you set up the hypotheses), it does not matter which group you call group 1 and which you call group 2.

Watch Out

- In calculating the combined sample proportion ($\hat{p}_c$), notice that it's not generally correct to take the average of the two group proportions. This shortcut works if the group sample sizes are identical, but not otherwise.
- Notice the standard error (i.e., the square root expression) is calculated slightly differently for the significance test and the confidence interval. The significance test makes use of the combined sample proportion $\hat{p}_c$, whereas the confidence interval does not. This is because the null hypothesis asserts that the population proportions are equal, so information from both samples is combined to estimate that common population proportion value, whereas no such assertion is made with the confidence interval procedure.

Activity 21-3: Preventing Breast Cancer

6-14, 6-15, **21-3,** 21-18, 25-22

Recall from Activity 6-14 that the Study of Tamoxifen and Raloxifene (STAR) enrolled more than 19,000 postmenopausal women who were at increased risk for breast cancer. Women were randomly assigned to receive one of the drugs (tamoxifen or raloxifene) daily for five years. Researchers kept track of which women developed invasive breast cancer, and which women did not, during the course of the study.

a. Is this an observational study or an experiment? Explain how you know.

b. Identify the explanatory and response variable in this study.

Explanatory:

Response:

c. Let π_T represent the proportion of *all potential* tamoxifen-takers who develop invasive breast cancer within five years, and let π_R denote the proportion of *all potential* raloxifene-takers who develop invasive breast cancer within five years. Use these symbols to write the appropriate null and alternative hypotheses for testing whether these drugs differ with regard to their effectiveness for preventing breast cancer.

Initial results released in April 2006 revealed that 163 of the 9726 women in the tamoxifen group had developed invasive breast cancer, compared to 167 of the 9745 women in the raloxifene group.

d. Organize this information in a 2×2 table. *Hint:* Put the explanatory variable groups in columns.

e. Calculate the sample proportions of women who developed invasive breast cancer in each of the two experimental groups. Record these values along with the symbols used to represent them.

f. Calculate the test statistic for testing the hypotheses in part c. (Feel free to use your calculator .)

g. Would you say this is a large *z*-score? Explain why this value is not surprising, in light of your answers to part e.

h. Calculate the *p*-value. (Feel free to use your calculator.) *Hint:* Keep in mind that you should have a two-sided alternative in this case.

i. Summarize the conclusion you would draw from these data in the context of this study. Be sure to address issues of causation and generalizability.

Activity 21-4: Perceptions of Self-Attractiveness

21-4, 21-9, 21-10

A survey conducted by the Gallup organization in 1999 asked American adults whether they were satisfied with their physical attractiveness or wished they could be more attractive. The survey revealed that 71% of the women and 81% of the men said that they were satisfied with their appearance.

a. What additional information is necessary to calculate a *p*-value and determine whether a significantly higher proportion of men than women are satisfied with their appearance?

b. Describe a circumstance (related to the missing information) in which these sample proportions would not convince you at all that the two genders differ with regard to satisfaction with their appearance.

c. Describe a circumstance in which these sample proportions would *strongly* convince you that the two genders differ with regard to satisfaction with their appearance.

d. Use your calculator to determine the test statistic and *p*-value of the appropriate significance test for the different sample sizes listed in this table. Also record whether the difference in sample proportions is statistically significant at the .10, .05, and .01 levels.

					Significant at:		
Sample Size (each group)	Satisfied Women	Satisfied Men	Test Statistic	p-value	$\alpha = .10$?	$\alpha = .05$?	$\alpha = .01$?
100	71	81					
200	142	162					
500	355	405					

e. Write a sentence or two commenting on the role of sample size in determining whether a difference between two sample proportions is statistically significant.

This activity should reinforce what you learned earlier about the effect of sample size on inference procedures. An observed difference in sample proportions is more statistically significant (unlikely to have occurred by chance alone) with larger sample sizes. Even a small difference can be statistically significant with a large sample size.

Activity 21-5: Graduate Admissions Discrimination

6-26, **21-5**

Recall from Activity 6-26 on page 117 that the University of California at Berkeley was alleged to have discriminated against women in its graduate admissions practices in 1973 (Bickel et al., 1975). When you analyzed the data from the six largest graduate programs, you found that 1195 of the 2681 male applicants had been accepted and that 559 of 1835 female applicants had been accepted.

a. Use your calculator to conduct the appropriate test of significance for assessing whether this difference in sample proportions is statistically significant. Report the null and alternative hypotheses (in words and in symbols), the test statistic, and the *p*-value of the test. Are these sample results statistically significant at commonly used levels of significance?

b. Referring back to the analysis you performed in Activity 6-26 (or answering those questions now if you have not done so), do you regard this statistically significant difference as evidence of discrimination? Explain.

These inference procedures for comparing two proportions can be used with data from observational studies as well as from controlled experiments. No matter how statistically significant a difference might be, however, you *cannot* draw conclusions about *causation* from an observational study. In this case, you do not have random samples from populations either, so the significance test only tells you that the observed difference in acceptance rates between men and women cannot reasonably be explained by random chance.

√

Activity 21-6: Nicotine Lozenge

1-16, 2-18, 5-6, 9-21, 19-11, 20-15, 20-19, **21-6**, 22-8

Recall from Activity 5-6 the study in which smokers who wanted to quit were randomly assigned to take either a nicotine lozenge or a placebo lozenge (Shiffman et al., 2002). The subjects were then monitored over the course of a year for whether they successfully abstained from smoking for that year.

a. Identify the explanatory and response variables.

b. Is this an experiment or an observational study? Explain.

c. State the appropriate null and alternative hypotheses, in symbols and in words, for testing whether the nicotine lozenge is helpful in terms of quitting smoking, as compared with a placebo.

Of the 459 subjects assigned to take the nicotine lozenge, 82 successfully abstained from smoking for the year, compared to 44 successful abstainers among the 458 subjects in the placebo group.

d. Organize these data into a 2 × 2 table, and construct a segmented bar graph to display these data. Comment on what the graph reveals.

e. Conduct the appropriate test of significance. Report the test statistic and *p*-value. Also include a check of technical conditions and summarize your conclusion. Be sure to comment on the appropriate scope of conclusions you can draw from this study.

f. Determine and interpret a 95% confidence interval for estimating the difference in population proportions showing successful abstinence between the two treatment groups.

g. Calculate a 95% confidence interval for the proportion of smokers who successfully abstain for a year with the nicotine lozenge. Comment on what additional information this interval provides about the effectiveness of the nicotine lozenge.

Solution

a. The explanatory variable is the type of lozenge (nicotine or placebo). The response variable is whether the smoker successfully abstains from smoking for the year. Both variables are categorical and binary.

b. This is an experiment because the researchers *assigned* the subjects to take a particular kind of lozenge (nicotine or placebo).

c. The null hypothesis is that the nicotine lozenge is no more (or less) effective than the placebo, where effectiveness is measured by the proportion of smokers who successfully abstain from smoking for a year. The alternative hypothesis is that the nicotine lozenge is *more* effective than the placebo, meaning a higher proportion of smokers (who are interested in and might potentially use such a product) would successfully quit with a nicotine lozenge than with a placebo lozenge. In symbols, the hypotheses are H_0: $\pi_{nicotine} = \pi_{placebo}$ vs. H_a: $\pi_{nicotine} > \pi_{placebo}$, where π represents the population proportion of smokers who successfully abstain for a year if given the nicotine lozenge or the placebo.

d. The 2 × 2 table is shown here:

	Nicotine Lozenge	Placebo Lozenge	Total
Successfully Abstained	82	44	126
Resumed Smoking	377	414	791
Total	459	458	917

The following segmented bar graph shows those smokers taking the nicotine lozenge had a higher success rate (proportion) in this study than those smokers taking the placebo lozenge, almost twice as high (.179 vs. .096). But the graph also reveals that in both groups, many more smokers resumed smoking than were able to abstain successfully.

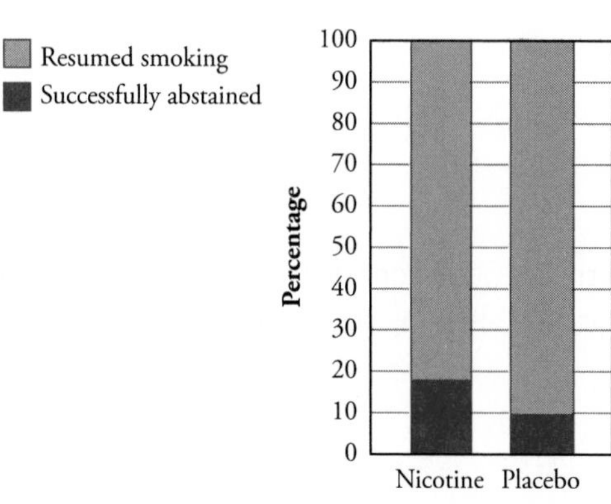

e. The sample proportions of smokers who successfully abstained from smoking in each group are

$$\hat{p}_{\text{nicotine}} = \frac{82}{459} \approx .179, \; \hat{p}_{\text{placebo}} = \frac{44}{458} \approx .096$$

The combined sample proportion of smokers who abstained is

$$\hat{p}_c = \frac{82 + 44}{459 + 458} = \frac{126}{917} \approx .137$$

The test statistic is

$$z = \frac{.179 - .096}{\sqrt{(.137)(1 - .137)\left(\frac{1}{459} + \frac{1}{458}\right)}} \approx 3.65$$

The p-value is the area to the right of 3.65 under the standard normal curve, which Table II reveals to be .0001. This test is valid because random assignment was used to put subjects in groups and because the sample size condition is also met: $459(.137) = 62.88$ is greater than 5, as are $458(.137) = 62.75$, $459(1 - .137) = 396.12$, and $458(1 - .137) = 395.25$.

This very small p-value indicates the experimental data provide very strong evidence against the null hypothesis, which means very strong evidence that the population proportion of smokers who would successfully abstain from smoking is higher with the nicotine lozenge than with the placebo lozenge. Because this was a randomized experiment, you can conclude the nicotine lozenge *causes* an increase in the proportion of smokers who would successfully abstain, as compared to the placebo group. You are not told how the subjects were selected for the study, but you do know that they were hoping to quit smoking, so this conclusion should be limited to smokers hoping to quit. They were probably chosen from a particular geographic area, so you might not want to generalize beyond that area, but it is hard to imagine why smokers in one area would respond differently to these lozenges than smokers in another area.

f. A 95% confidence interval for the difference $\pi_{\text{nicotine}} - \pi_{\text{placebo}}$ is

$$(.179 - .096) \pm 1.96\sqrt{\frac{(.179)(1 - .179)}{459} + \frac{(.096)(1 - .096)}{458}}$$

which is $.083 \pm 1.96(.023)$, which is $.083 \pm .044$, which is the interval from .039 through .127. The researchers are 95% confident that the proportion of smokers using a nicotine lozenge who successfully quit would be higher than the successful proportion of smokers using a placebo lozenge by somewhere between .039 and .127. This interval procedure is valid because of random assignment and because the number of successes and failures in both groups exceeds 5 (the smallest of these numbers is 44 successes in the placebo group) with the sampling issues cautioned about in part e.

g. Notice this question calls for a confidence interval for a single proportion: π_{nicotine}, so you need to use the procedure from Topic 16:

$$\hat{p}_{\text{nicotine}} \pm z^* \sqrt{\frac{\hat{p}_{\text{nicotine}}\left(1 - \hat{p}_{\text{nicotine}}\right)}{n_{\text{nicotine}}}}$$

which is $.179 \pm 1.96(.018)$, which is $.179 \pm .035$, which is the interval from .144 through .214. The researchers can be 95% confident that if the population of all smokers who wanted to quit were to use the nicotine lozenge, the proportion who would successfully abstain from smoking for one year would be between .144

and .214. So, even though the experiment provides strong evidence the nicotine lozenge works better than a placebo, most smokers (more than 3/4) would be unable to quit even with the nicotine lozenge.

Watch Out

- Be sure to plug in proportions (which must be between 0 and 1), not percentages or counts, in these calculations.
- Be especially careful about rounding errors with these calculations. During intermediate steps of the calculation, carry as many decimal places of accuracy as you can. Then, at the end, round your final answer. If you round too much in the early stages, the final answer can be considerably different.
- When you interpret a confidence interval, always start by asking yourself what parameter the interval is estimating. For example, notice you have two different confidence intervals in the previous activity. Part f estimates the *difference* in proportions $\pi_{\text{nicotine}} - \pi_{\text{placebo}}$. The key question with such an interval is whether it includes zero or is entirely positive or entirely negative. On the other hand, part g estimates only π_{nicotine}. Such an interval estimates the actual value of that population proportion.
- We emphasize that random assignment and random sampling serve two very different purposes and permit very different kinds of conclusions. This study involved random assignment, but not random sampling, so you can draw a cause-and-effect conclusion between the nicotine lozenge and the increased rate of abstinence, but you should be cautious about generalizing this result to a larger population.

Wrap-Up...........

In this topic, you investigated tests of significance for comparing proportions between two groups. You used physical (card shuffling) and computer simulations to study the reasoning behind the test procedure. The *p*-value of the test represents the probability of getting such an extreme sample or experimental result by random chance alone.

You applied this test in two different scenarios: random assignment of subjects to treatment groups and random sampling from two populations. Examples of random assignment include the "friendly observers" psychology study and the medical experiment comparing tamoxifen with raloxifene. Examples of random sampling include the "back to sleep" study and the survey of people's impressions of their own physical attractiveness. The mechanics of the test procedure are identical in both settings, but the scope of conclusions that can be drawn differ in important ways. As you first learned in Unit 1, random assignment in an experiment allows for a cause-and-effect conclusion if the difference between the groups is statistically significant. On the other hand, random sampling enables you to generalize from the samples at hand to the larger populations from which the samples were selected. Some studies employ both types of randomness, so causal conclusions can be drawn about the larger populations.

In addition to the test of significance, you learned a confidence interval procedure for estimating the difference in population proportions between two groups. You also explored the effects of sample sizes on the significance test and confidence interval. All else being equal, larger sample sizes produce more significant results (i.e., smaller *p*-values) and narrower confidence intervals. For example, the difference between 71% and 81% (the sample results with the self-attractiveness survey) is statistically significant with large sample sizes, but not with more moderate sample sizes.

Some useful definitions to remember and habits to develop from this topic include

- Even now that you are studying more sophisticated inference procedures, always begin any analysis of data with graphical displays (e.g., segmented bar graphs) and numerical summaries (e.g., conditional proportions).
- Remember to state the hypotheses based on the research question, not based on how the sample results turn out.
- Remember to check the technical conditions before applying any inference procedure, or at least before interpreting the results.
- When a test reveals that a sample result is statistically significant, follow up with a confidence interval to estimate the magnitude of the difference.
- Remember a cause-and-effect conclusion can only be drawn from a significant test result if that data were collected according to a randomized experiment.

In the next topic, you will turn from categorical variables to quantitative ones, and you will study inferential techniques for comparing two means.

Homework Activities

Activity 21-7: Botox for Back Pain

5-4, 6-19, 6-20, **21-7**

Recall from Activity 6-19 that a randomized, double-blind experiment investigated whether botulinum toxin A (botox) is effective for treating patients with chronic low-back pain (Foster et al., 2001). At the end of eight weeks, 9 of 15 subjects in the botox group had experienced substantial relief, compared to 2 of 16 subjects in the group who had received an ordinary saline injection.

a. Identify the explanatory and response variables in this study.
b. Organize the data in a 2 × 2 table.

The following histogram displays the results of 1000 random assignments, assuming that botox has no effect and the study would have always produced 11 subjects with substantial pain reduction and 20 subjects without such pain reduction anyway. Notice the variable displayed is the *number of successes randomly assigned to botox group.*

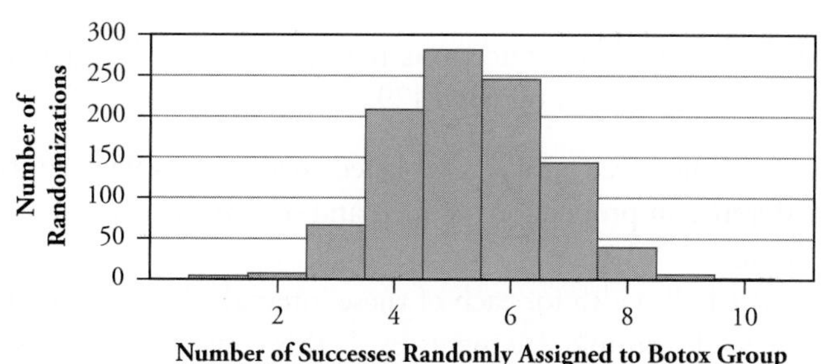

c. Use the results of this simulation to determine an approximate p-value. Describe how you calculated this value.
d. Interpret what this approximate p-value means. Also describe the conclusion that you could draw about whether botox is a more effective treatment for back pain than an ordinary saline injection. Explain the reasoning process by which the approximate p-value leads to this conclusion.
e. Check whether the technical conditions of the two-proportion z-test are satisfied in this case.

Activity 21-8: "Hella" Project

6-7, **21-8**

Reconsider the "hella" project described in Activity 6-7 (Mead et al., 2005). The student researchers took a random sample of 40 first-year students at Cal Poly and asked whether they were from northern or southern California and whether they used the slang word "hella" regularly in conversation. Before collecting the data, the student researchers conjectured that those students from northern California would be more likely to use the word. The following table summarizes the results:

	Northern Californians	Southern Californians	Total
Uses "Hella" Regularly	10	3	13
Does Not Use "Hella" Regularly	5	22	27
Total	15	25	40

a. Calculate the sample proportion of students in each group who use the word "hella" regularly.
b. Construct and comment on a segmented bar graph for comparing the sample proportions of students who use the word "hella" regularly.
c. Use the `Two-Way Table Simulation` applet to conduct a simulation analysis and determine an approximate p-value. Reproduce a histogram of the simulation results, and report the approximate p-value.
d. Interpret what this approximate p-value means. Also describe the conclusion you draw about whether students from northern California are really more likely to use the word "hella" than those students from southern California. Explain the reasoning process by which the approximate p-value leads to this conclusion.
e. Check whether the technical conditions of the two-proportion z-test are satisfied in this case.

Activity 21-9: Perceptions of Self-Attractiveness

21-4, **21-9**, 21-10

Reconsider the survey result concerning perceptions of self-attractiveness that you investigated in Activity 21-4 on page 429.

a. For each of the three sample sizes listed there, find a 95% confidence interval for the difference in proportions of men and of women who are satisfied with their appearance.
b. Report the half-width for each of these intervals. Does the half-width increase or decrease as the sample size increases? Is this consistent with what happens with confidence intervals for a single proportion?
c. With which sample sizes does the 95% confidence interval include the value zero and with which does it not? Explain how this result is consistent with your test results in Activity 21-4.

Activity 21-10: Perceptions of Self-Attractiveness

21-4, 21-9, **21-10**

Reconsider again the survey result that you investigated in Activity 21-4 and in the previous activity. Suppose the sample proportions of men and women who indicated satisfaction with their appearance had been 76% for men and 75% for women. Further suppose the sample size had been the same for each group.

a. Find a sample size for which this difference would be statistically significant at the $\alpha = .01$ level. Report the test statistic and *p*-value of the test.
b. Construct a 99% confidence interval for the difference in the population proportions using the sample size you found in part a.
c. Repeat parts a and b if the sample proportions had been 75.6% and 75.5%, respectively.
d. Would you regard the difference as *practically* significant in the situation described in part c? Explain.

Activity 21-11: Generation M

3-8, 4-14, 13-6, 16-1, 16-3, 16-7, 18-1, **21-11,** 21-12

Recall from Activities 16-1 and 16-3 that the Kaiser Family Foundation commissioned an extensive survey in 2004 that investigated the degree to which American youths aged 8–18 have access to various forms of media. Of the 1036 girls in the sample, 64% had a television in their bedrooms, compared to 72% of the 996 boys in the sample.

a. Suppose you want to use these sample results to produce a 95% confidence interval for $\pi_g - \pi_b$. Describe in words what $\pi_g - \pi_b$ represents.
b. Calculate this interval, and interpret what it reveals. Be sure to mention whether the interval contains all negative values, all positive values, or some of each.
c. Calculate a 99% confidence interval for $\pi_g - \pi_b$. Comment on how its midpoint and width compare to the 95% interval.

Activity 21-12: Generation M

3-8, 4-14, 13-6, 16-1, 16-3, 16-7, 18-1, 21-11, **21-12**

Reconsider the previous activity. The Kaiser survey also asked about whether the teenager's bedroom contained many other types of media in addition to television. The following table reports the sample results for boys and for girls.

	Radio	Video Game Player	Internet	Telephone
Boys	82%	63%	24%	39%
Girls	86%	33%	17%	42%

Conduct the appropriate tests to determine for which media device boys and girls differ significantly. Then, for the devices for which they do differ significantly, estimate the magnitude of the difference with a 95% confidence interval. Write a paragraph or two summarizing your findings. (Feel free to use your calculator, and recall that the sample sizes are 1036 girls and 996 boys.)

Activity 21-13: AZT and HIV

5-12, 6-2, **21-13**

Recall from Activity 6-2 the study of whether AZT helps to reduce transmission of AIDS from mother to baby (Connor et al., 1994). Of the 180 babies whose mothers had been randomly assigned to receive AZT, 13 babies were HIV-infected, compared to 40 of the 183 babies in the placebo group.

a. Create a segmented bar graph to display these results. Comment on what the graph reveals.
b. Check the technical conditions for whether a *z*-test can be applied to these data. Be sure to mention whether the study involves random sampling from populations or random assignment to treatment groups.
c. Conduct the appropriate test of whether the data provide evidence that AZT is more effective than a placebo for reducing mother-to-infant transmission of AIDS. Report the hypotheses, test statistic, and *p*-value. Also indicate the test decision using the $\alpha = .01$ significance level.
d. Estimate how much better (lower transmission rates) AZT is than a placebo with a 99% confidence interval. Also be sure to interpret this interval.
e. Summarize the conclusion that you could draw from this study. Also explain the reasoning process by which this conclusion follows from your test result.

Activity 21-14: Flu Vaccine

6-16, **21-14**

Recall from Activity 6-16 that in January 2004, the Centers for Disease Control and Prevention published the results of a study that looked at workers at Children's Hospital in Denver, Colorado. Of the 1000 people who had chosen to receive the flu vaccine (before November 1, 2003), 149 still developed flu-like symptoms. Of the 402 people who did not get the vaccine, **68** developed flu-like symptoms.

a. Calculate conditional proportions and create a segmented bar graph to display these results.
b. Conduct a test of whether these data provide strong evidence (at the $\alpha = .05$ level) that the flu vaccine is effective. Report all aspects of the test, including a check of technical conditions. Also summarize your conclusion.
c. Comment on the issues of generalizability and causation as they pertain to this study.

Activity 21-15: Suitability for Politics

6-8, 6-9, **21-15,** 25-19, 25-20

Recall from Activity 6-8 that the 2004 General Social Survey asked a random sample of American adults whether they agree with the proposition that men are better suited emotionally for politics than women are. The sample results by gender are summarized in the following table:

	Men	Women
Agree	109	103
Disagree	276	346

a. State the hypotheses for testing whether men tend to agree with this proposition more often than women do.
b. Comment on whether the technical conditions for conducting the test are satisfied.
c. Calculate the test statistic and *p*-value.
d. Interpret the *p*-value in the context of this issue.
e. Construct a 95% confidence interval for the difference in population proportions of agreement between men and women.
f. Write a sentence or two summarizing your findings.

Activity 21-16: Hand Washing

2-2, **21-16,** 25-16

Recall from Activity 2-2 that researchers monitored the behavior of more than 6300 users of public restrooms in venues such as Turner Field in Atlanta and Grand Central Station in New York City. They found that 2393 of 3206 men washed their hands, compared to 2802 of 3130 women. Analyze these data, with graphical displays as well as a significance test and confidence interval. Summarize and explain your conclusions about whether men and women differ with regard to how often they wash their hands in public restrooms.

Activity 21-17: Volunteerism

13-16, 15-16, **21-17**

Recall from Activity 13-16 that the Bureau of Labor Statistics conducts an extensive survey about volunteerism, using a random sample of about 60,000 American households. For the sample taken in 2005, 25.0% of the men sampled had done volunteer work, compared to 32.4% of the women sampled.

a. Are .250 and .324 parameters or statistics? Explain.
b. What additional information do you need to make inferences about the population from this study?
c. Suppose the survey questioned 30,000 men and 30,000 women. Estimate the difference in population proportions of men and women who did volunteer work with a 99% confidence interval. Also interpret this interval.
d. Explain why this interval is so narrow.
e. Would you expect a 99.9% confidence interval to be wider or narrower? Explain.

Activity 21-18: Preventing Breast Cancer

6-14, 6-15, 21-3, **21-18,** 25-22

Reconsider the Study of Tamoxifen and Raloxifene (STAR), particularly the issues raised in Activity 6-15 on page 113. The study hoped to show that raloxifene would reduce the risk of dangerous side effects involving blood clots, as compared to tamoxifen (with a similar cancer incidence rate). The initial report also revealed that 53 of the 9726 women in the tamoxifen group had a blood clot in a major vein, compared to 65 of the 9745 women taking raloxifene. Further, 54 of the 9726 women taking tamoxifen developed a blood clot in the lung, compared to 35 of the 9745 women taking raloxifene. Compare the two drugs' performance for each of these side effects with appropriate significance tests and confidence intervals. Report all aspects of these procedures, and write a paragraph or two summarizing your findings.

Activity 21-19: Underhanded Free Throws

Sports Illustrated columnist Rick Reilly wrote about free-throw shooting in the December 11, 2006, issue of the magazine. He said that he went to his neighborhood gym and successfully made 63% of his free-throw attempts, using a conventional technique. Then he received instruction from Rick Barry, a former player famous for the underhand method of shooting free throws. Reilly went back to the gym and shot underhanded, successfully making 78% of his attempts.

a. State the hypotheses for testing whether Reilly's success proportion was significantly higher with the underhand method than with the conventional technique.
b. What additional information do you need to conduct the significance test?

c. Would you expect Reilly's results to be more significant if he made 100 attempts with each method or if he made 500 attempts with each method?
d. Determine the test statistic and *p*-value, assuming that he made 100 attempts with each method.
e. Repeat part d, assuming that he made 500 attempts with each method.
f. Summarize the conclusion that you would draw in each of these cases.
g. Would you feel comfortable generalizing these results to all players who could be taught by Rick Barry? Explain.

Activity 21-20: Solitaire

11-22, 11-23, 15-4, 15-14, **21-20,** 27-18

Recall from Activities 11-22 and 11-23 that author A won 25 times in 217 games of solitaire and that author B won 74 times in 444 games.

a. Can you analyze these data even though the two people played a different number of games?
b. Calculate the proportion of solitaire games won by each author, and construct a segmented bar graph to display the distributions.
c. Do these data suggest that the winning probabilities for these two people differ at the .10 significance level? At the .05 level? Show the details of your test procedure.
d. Find a 90% confidence interval to estimate the difference in the probabilities ($\pi_A - \pi_B$).
e. Are the technical conditions satisfied for the inference procedures in parts b and c? Explain.
f. How would the interval differ if you found a confidence interval for $\pi_B - \pi_A$?
g. Explain how the test result reveals whether a 95% confidence interval for the difference in probabilities would include the value zero.
h. Explain how the *p*-value would have differed if you had tested that author B's success probability is higher than author A's.

Activity 21-21: Magazine Advertisements

16-15, 17-20, **21-21**

Recall from Activity 16-15 that the September 13, 1999, issue of *Sports Illustrated* had 54 pages with advertisements among its 116 pages, whereas the September 14, 1999, issue of *Soap Opera Digest* had 28 pages with advertisements among its 130 pages.

a. Calculate the sample proportions of pages with ads in these magazines.
b. Construct a segmented bar graph to compare the proportions of pages with ads between the two magazines.
c. Conduct the appropriate test of significance to assess whether the magazines' proportions of pages with ads differ significantly at the $\alpha = .01$ level. Report the hypotheses (in symbols and in words), the test statistic, and the *p*-value.
d. Write a sentence or two summarizing your conclusion and how it follows from this test result.
e. What assumption must you make in order for this test procedure to be valid?

Activity 21-22: Wording of Surveys

21-22, 21-23

Much research goes into identifying factors that can unduly influence people's responses to survey questions. In a 1974 study, researchers conjectured that people are prone to acquiesce—in other words, to agree with attitude statements presented to them

(Schuman and Presser, 1981). They investigated their claim by asking subjects whether they agree or disagree with the following statement. Some subjects were presented with form A and others with form B:

- Form A: "Individuals are more to blame than social conditions for crime and lawlessness in this country."
- Form B: "Social conditions are more to blame than individuals for crime and lawlessness in this country."

The responses are summarized in the table:

	Blame Individuals	Blame Social Conditions
Form A	282	191
Form B	204	268

a. Let π_A represent the population proportion of all potential form A subjects who contend that individuals are more to blame, and let π_B denote the population proportion of all potential form B subjects who contend that individuals are more to blame. If the researchers' claim about acquiescence is valid, should π_A be greater than π_B or vice versa?

b. Calculate the sample proportion of form A subjects who contend that individuals are more to blame. Then calculate the sample proportion of form B subjects who contend that individuals are more to blame.

c. Conduct the appropriate test of significance to assess the researchers' conjecture. Report the null and alternative hypotheses, the test statistic, and *p*-value. Also check technical conditions and write a one-sentence conclusion in context.

Activity 21-23: Wording of Surveys

21-22, **21-23**

Researchers have conjectured that the use of the words "forbid" and "allow" can affect people's responses to survey questions (Schuman and Presser, 1981). In a 1976 study, one group of subjects was asked, "Do you think the United States should forbid public speeches in favor of communism?" while another group was asked, "Do you think the United States should allow public speeches in favor of communism?" Of the 409 subjects asked the "forbid" version of the question, 161 favored the forbidding of communist speeches. Of the 432 subjects asked the "allow" version of the question, 189 favored allowing the speeches.

a. Calculate the sample proportion of "forbid" subjects who oppose communist speeches (i.e., favor forbidding them) and the sample proportion of "allow" subjects who oppose communist speeches (i.e., do *not* favor allowing them).

b. Conduct the appropriate two-sided test of significance to test the researchers' conjecture. Report the null and alternative hypotheses (explaining any symbols that you use) as well as the test statistic and *p*-value. Check technical conditions, and indicate whether the difference in sample proportions is statistically significant at the .10, .05, and .01 significance levels.

c. A 1977 study asked 547 people, "Do you think the government should forbid the showing of X-rated movies?" A total of 224 answered in the affirmative (i.e., to forbid). At the same time, a group of 576 people were asked, "Do you think the government should allow the showing of X-rated movies?" A total of

309 answered in the affirmative (i.e., to allow). Repeat parts a and b for the results of this experiment. *Hint:* Be very careful to calculate and compare relevant sample proportions. Do not simply calculate proportions of "affirmative" responses.

d. A 1979 study asked 607 people, "Do you think the government should forbid cigarette advertisements on television?" A total of 307 people answered in the affirmative (i.e., to forbid). At the same time, a group of 576 people were asked, "Do you think the government should allow cigarette advertisements on television?" A total of 134 answered in the affirmative (i.e., to allow). Repeat parts a and b for the results of this experiment, heeding the same caution as in part c.

e. Write a paragraph summarizing your findings about the impact of forbid/allow distinctions on survey questions.

Activity 21-24: Questioning Smoking Policies

An undergraduate researcher at Dickinson College examined the role of social fibbing (the tendency of subjects to give responses they think the interviewer wants to hear) with the following experiment. Students were asked, "Would you favor a policy to eliminate smoking from all buildings on campus?" For a randomly assigned half of the subjects, the interviewer smoked a cigarette when asking the question; the other half were interviewed by a nonsmoker. Prior to conducting the experiment, the researcher suspected that students interviewed by a smoker would be less inclined to indicate they favored the ban. It turned out that 43 of the 100 students interviewed by a smoker favored the ban, compared to 79 of the 100 interviewed by a nonsmoker. Carry out the appropriate test of significance to assess the researcher's hypothesis. Write a brief conclusion as if to the researcher, addressing in particular the question of how likely her experimental results would have occurred by chance alone.

Activity 21-25: Teen Smoking

21-25, 21-26

A newspaper account of a medical study claimed the daughters of women who smoked during pregnancy were more likely to smoke themselves. The researchers surveyed teenagers, asking them whether they had smoked in the last year and then asking the mother whether she had smoked during pregnancy. Only 4% of the daughters of mothers who did not smoke during pregnancy had smoked in the past year, compared to 26% of girls whose mothers had smoked during pregnancy.

a. What further information do you need to determine whether this difference in sample proportions is statistically significant?

b. Suppose there had been 50 girls in each group. Use your calculator to conduct a two-sided significance test. Report the p-value and whether the difference in sample proportions is statistically significant at the .05 level.

c. Repeat part b supposing there had been 50 girls whose mothers had smoked and 200 whose mothers had not.

d. Repeat part b supposing there had been 200 girls in each group.

e. Construct a segmented bar graph to compare the smoking habits of girls whose mothers smoked and girls whose mothers did not smoke during pregnancy. Does the appearance of this graph change as the sample size increases? Explain.

f. Is this an experiment or an observational study? Explain.

g. Even if the difference in sample proportions is statistically significant, can this study establish that the pregnant mother's smoking caused the daughter's tendency to smoke? Explain. If appropriate, suggest a potentially confounding variable.

Activity 21-26: Teen Smoking
21-25, **21-26**

Reconsider the previous activity. The researchers also studied sons and found that 15% of the sons of mothers who had not smoked during pregnancy had smoked in the past year, compared to 20% of the sons of mothers who had smoked during pregnancy.

a. Suppose there had been 60 boys in each group. Use your calculator to conduct a two-sided significance test. Report the p-value and whether the difference in sample proportions is statistically significant at the .05 level.
b. Repeat part a supposing there had been 200 boys in each group.
c. Repeat part a supposing there had been 500 boys in each group.
d. Suppose the two groups had the same number of boys. Try to find the smallest number for this sample size that would make the difference in sample proportions statistically significant at the .05 level. You may either use trial and error with your calculator or solve the problem analytically by hand.

Activity 21-27: Candy and Longevity
3-3, **21-27**

Recall from Activity 3-3 that a study of male Harvard graduates revealed that 267 of 4529 candy consumers died during the course of the five-year study, compared to 247 of the 3312 nonconsumers.

a. Conduct a significance test of whether the proportion of candy consumers who died is statistically significantly lower, at the $\alpha = .05$ level, than the proportion of nonconsumers who died.
b. Does the way this study was conducted enable you to conclude a cause-and-effect relationship between candy consumption and increased survival? Explain.
c. Does the way this study was conducted enable you to generalize your conclusions to all adults, or even to all males? Explain.

Activity 21-28: Your Choice
6-30, 12-22, 19-22, **21-28**, 22-27, 26-23, 27-22

Write a paragraph detailing a real situation about which you would be interested in performing a test of significance to compare two *proportions*. Describe precisely the context involved and explain whether the study would be a controlled experiment or an observational study. Also identify carefully the observational units and variables involved. Finally, suggest how you might go about collecting the sample data.

TOPIC

22 • • • Comparing Two Means

Can a waitress earn higher tips simply by giving her name when she greets her customers? How can she design an experiment to investigate this? After she collects the data, how can she decide whether the results are statistically significant enough to convince her that telling customers her name really helps? And if she decides that giving her name helps, how can she estimate how much higher her tips will be, on average, when telling customers her name? In this topic, you will learn a statistical inference procedure for answering such questions.

Overview

In the previous topic, you studied the application of inference techniques to the comparison of two proportions. In this topic, you will examine the case of comparing two sample means. These inference procedures will again be based on the *t*-distribution. You will explore the effects of factors such as sample size and sample variability on these procedures. As you progress through this topic, you will find that the reasoning behind and interpretation of the procedures remain the same, and that an initial examination of the data, graphical and numerical, prior to applying formal inference procedures is, as always, important. You will again see that the method used to collect the data determines the scope of conclusions that can be drawn.

Preliminaries

1. One of the questions asked on the 2004 General Social Survey (GSS) was, "From time to time, most people discuss important matters with other people. Looking back over the last six months—who are the people with whom you discussed matters important to you? Just tell me their first names or initials." How many people would you mention in responding to this question?

2. How would you expect responses to this question to compare between men and women: Would you expect men to respond with more names on average, or women to respond with more names, or the responses to be very similar between men and women?

3. If one commuting route has a sample mean travel time of 32 minutes and another route has a sample mean commuting time of 28 minutes, what else would you want to know in order to assess whether the first route is really slower than the second?

4. Do you think a waitress can increase her tips by giving her name when she greets customers?

5. On a dinner check of approximately $20, about how much higher would you expect a waitress' tip to be (if at all) if she gave customers her name when she greeted them?

In-Class Activities

Activity 22-1: Close Friends

19-17, 19-18, **22-1,** 22-5, 22-22

Recall from Activity 19-17 that one of the questions asked of a random sample of adult Americans in the 2004 General Social Survey was, "From time to time, most people discuss important matters with other people. Looking back over the last six months—who are the people with whom you discussed matters important to you? Just tell me their first names or initials." The interviewer then recorded how many names or initials the respondent mentioned. Suppose you want to examine whether men and women differ with regard to how many names they tend to mention. (For convenience, we will refer to those named as "close friends.")

a. Is this an observational study or an experiment? Explain.

b. Identify the explanatory and response variable for this study. Also classify each variable as categorical (also binary) or quantitative.

Explanatory: Type:

Response: Type:

c. State the null and alternative hypotheses, in symbols and in words, for testing whether the sample data provide evidence that men and women differ with regard to the average number of close friends they tend to mention in response to this question.

d. Explain why the two-sample z-test procedure from Topic 21 does not apply in this situation. Also explain why the one-sample t-test procedure from Topic 20 does not apply.

Before you learn a new test procedure for handling this situation, let's begin with a preliminary analysis of the data. Sample responses by gender are tallied in the following table:

Number of Close Friends	0	1	2	3	4	5	6	Total
Number of Respondents (male)	196	135	108	100	42	40	33	654
Number of Respondents (female)	201	146	155	132	86	56	37	813

Some descriptive summaries follow:

	Sample Size	Mean	SD	Minimum	Lower Quartile	Median	Upper Quartile	Maximum
Male	654	1.861	1.777	0	0	1	3	6
Female	813	2.089	1.760	0	1	2	3	6

e. Are these values parameters or statistics? Explain.

f. Produce boxplots (on the same scale) to compare the distributions of the number of close friends between males and females.

g. Comment on any differences that you observe in the distributions of the number of close friends between males and females.

h. Would it be possible to obtain sample means this far apart even if the population means were equal? Explain.

Once again, because of sampling variability, you cannot conclude that simply because these sample means differ, the means of the respective populations must differ as well. As always, you can use a test of significance to establish whether a sample result (in this case, the observed difference in sample mean number of close friends) is "significant" in the sense of being unlikely to have occurred by chance (from random sampling) alone. Also, you can use a confidence interval to estimate the magnitude of the difference in the population means. You cannot use the procedure from Topic 21, however, because now you have a *quantitative* response variable and so are interested in comparing *means* rather than *proportions*.

Inference procedures for comparing the population means of two different groups are similar to those for comparing population proportions in that they take into account sample information from both groups. These procedures are similar to those for a *single* population mean in that they use the *t*-distribution, and the sample sizes, sample means, and sample standard deviations are the relevant summary statistics.

The details for conducting confidence intervals and significance tests concerning the difference between two population means, which will be denoted by μ_1 and μ_2, are presented here.

Significance test of equality of μ_1 and μ_2:
Null hypothesis: H_0: $\mu_1 = \mu_2$
Alternative hypothesis:

i. H_a: $\mu_1 < \mu_2$ *or*
ii. H_a: $\mu_1 > \mu_2$ *or*
iii. H_a: $\mu_1 \neq \mu_2$

Test statistic:

$$t = \frac{\bar{x}_1 - \bar{x}_2}{\sqrt{\frac{s_1^2}{n_1} + \frac{s_2^2}{n_2}}}$$

p-value:

i. $\Pr(T_k \leq t)$ *or*
ii. $\Pr(T_k \geq t)$ *or*
iii. $2 \times \Pr(T_k \geq |t|)$

where T_k represents a *t*-distribution with degrees of freedom *k* equal to the smaller of $n_1 - 1$ and $n_2 - 1$.

Confidence interval for $\mu_1 - \mu_2$:

$$(\bar{x}_1 - \bar{x}_2) \pm t_k^* \sqrt{\frac{s_1^2}{n_1} + \frac{s_2^2}{n_2}}$$

where t_k^* is the appropriate critical value (for the desired confidence level) from the *t*-distribution with degrees of freedom *k* equal to the smaller of $n_1 - 1$ and $n_2 - 1$.

Technical conditions:

- The data arise from random assignment of subjects to two treatment groups *or* from independent random samples from two populations.
- *Either* both sample sizes are large ($n_1 \geq 30$ and $n_2 \geq 30$ as a convention) *or* both populations are normally distributed (as judged by examining the sample data).

Notes

- As always, the symbols $\bar{x}$ and s represent a sample mean and a sample standard deviation, respectively. The subscripts indicate the population from which the observational units are randomly selected or the treatment group to which they are randomly assigned.

- The structure, reasoning, and interpretation of this test and interval procedure are the same as for other tests and intervals that you have studied.
- As with the two-sample z-procedures from Topic 21, these two-sample t-procedures apply to scenarios involving random sampling from two populations and/or random assignment to two treatment groups. As before, the calculations are identical, but the scope of conclusions is very different for these two scenarios.
- The degrees of freedom convention being used is a *conservative* approximation, meaning the degrees of freedom is on the low side, so the critical value will be slightly greater than it needs to be; thus the interval will be slightly wider and, therefore, will succeed in capturing $\mu_1 - \mu_2$ slightly more often than the confidence level indicates. When using your calculator, a more exact critical value will be computed for you. The p-value calculation can also differ a bit with your calculator, again based on the degrees of freedom.

i. Use the summary statistics provided after part d to calculate the test statistic for testing the hypotheses you stated in part c.

j. Use Table III (t-Distribution Critical Values) to find (as accurately as possible) the p-value of the test.

k. Which of the following is a correct interpretation of the p-value?

- The p-value is the probability that males and females have the same mean number of close friends in these samples.
- The p-value is the probability that males and females have the same mean number of close friends in the populations.
- The p-value is the probability that males have a higher mean number of close friends than females do.
- The p-value is the probability of getting sample data so extreme if, in fact, males and females have the same mean number of close friends in the populations.

l. Is this p-value small enough to reject the null hypothesis that these population means are equal at the $\alpha = .05$ significance level?

m. Is the observed difference in sample means statistically significant at the $\alpha = .01$ level?

n. State the technical conditions necessary for this procedure to be valid. Does the strong skewness in the sample data provide any reason to doubt the validity of this test?

o. Determine and interpret a 95% confidence interval for the difference in population means $\mu_f - \mu_m$. *Hint:* Be sure to comment on whether the interval is entirely negative, entirely positive, or contains zero. Also explain the importance of whether the interval includes zero.

p. Use your calculator's **2-SampTTest** command (found by pressing STAT and arrowing over to the **TESTS** menu) to verify your test and interval calculations. Enter the summary statistics you found in part d, and leave the **Pooled** option set to **No.** You can then select either **Calculate** or **Draw.**

Activity 22-2: Hypothetical Commuting Times

7-13, **22-2,** 22-6, 22-7

Suppose Alex wants to determine which of two possible driving routes gets him to school more quickly. Also suppose that over a period of 20 days, he randomly decides which route to drive each day. He then records the commuting times (in minutes) and displays them as follows:

Route 1	19.3	20.5	23.0	25.8	28.0	28.8	30.6	32.1	33.5	38.4
Route 2	23.7	24.5	27.7	30.0	31.9	32.5	32.6	35.5	38.7	42.9

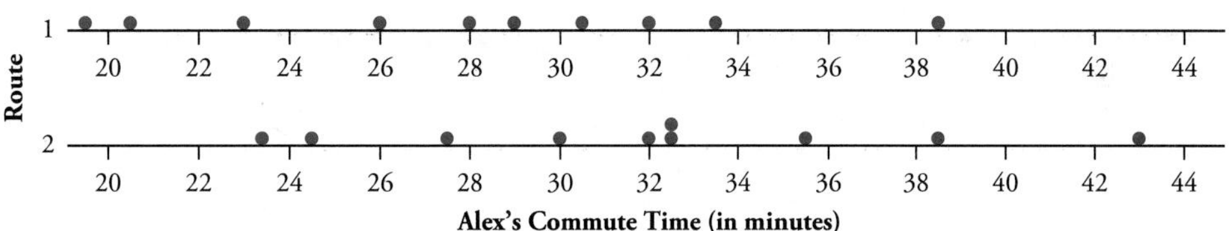

a. Does one route *always* get Alex to school more quickly than the other?

b. Do the data suggest that one route *tends* to get Alex to school more quickly than the other? If so, which route appears to be quicker?

c. The data for Alex's commute times (and several other commuters) are stored in the grouped file HYPOCOMMUTE, in the lists **ALEX1** and **ALEX2**. Use your calculator to determine the sample means and sample standard deviations of Alex's commuting times for each route; record them in the first three columns of the following table:

	Sample Size	Sample Mean	Sample SD	*p*-value
Alex: Route 1				
Alex: Route 2				

d. Use your calculator to conduct a significance test of whether Alex's sample commuting times provide evidence that the mean commuting times with these two routes differ. Record the *p*-value of the test in the table.

e. Are Alex's data statistically significant at any of the commonly used significance levels? Can Alex reasonably conclude that one route is faster than the other route for getting to school? Explain.

f. Use your calculator's **2-SampTInt** command (located by pressing STAT and arrowing over to the TESTS menu) to compute a 90% confidence interval for the difference in Alex's mean commuting times between route 1 and route 2. You can either enter the summary statistics you found in part c or enter the lists ALEX1 and ALEX2. Leave the **Pooled** option set to **No.** Record the confidence interval.

g. Does this interval include the value zero? Explain the importance of this.

Now consider three other commuters who conduct similar experiments to compare travel times for two different driving routes (the data are stored in the grouped file HYPOCOMMUTE, containing the lists ALEX1, ALEX2, BARB1, BARB2, CARL1, CARL2, DONN1, and DONN2). For the sake of comparison, Alex's results are also reproduced here.

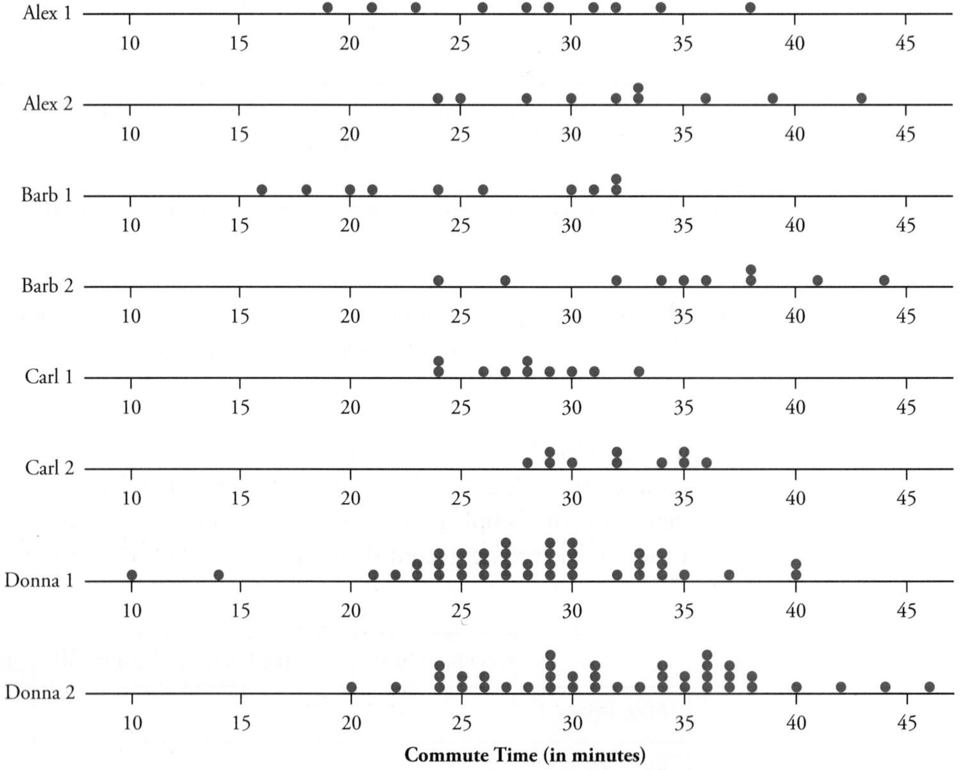

h. Based on your visual analysis of these pairs of dotplots, what strikes you as the most important difference between Alex's and Barb's results?

i. Based on your visual analysis of these pairs of dotplots, what strikes you as the most important difference between Alex's and Carl's results?

j. Based on your visual analysis of these pairs of dotplots, what strikes you as the most important difference between Alex's and Donna's results?

k. For each commuter (Barb, Carl, and Donna), use your calculator's **2-SampTTest** to conduct a significance test of whether the difference in his or her sample mean commuting times is statistically significant. Record the *p*-values of these tests in the following table, along with the appropriate sample statistics:

	Sample Size	Sample Mean	Sample SD	*p*-value
Barb: Route 1				
Barb: Route 2				
Carl: Route 1				
Carl: Route 2				
Donna: Route 1				
Donna: Route 2				

l. For each of these other commuters, explain why his or her commute times differ significantly between the two routes, whereas Alex's do not.

Barb:

Carl:

Donna:

These comparisons should help you to see the roles of sample sizes, means, and standard deviations in the two-sample *t*-test. All else being the same, the test result becomes more statistically significant (i.e., the *p*-value becomes smaller) as

- The difference in sample means increases
- The sample sizes increase
- The sample standard deviations decrease

Note that the researcher has no control over the sample means but can determine the sample sizes. Larger samples are better, but they require more time and expense. The researcher seems to have no control over the standard deviations, but this third bullet reveals why statisticians like to reduce variability as much as possible (for example, using better measurement tools). Topic 23 will describe a useful study design technique for reducing variability.

Watch Out

Remember that failing to reject a null hypothesis is not the same as accepting it. You do not have enough evidence to conclude that one route is faster than the other on average for Alex, but you should not conclude that the average commuting times are identical for the two routes. Larger samples might produce a statistically significant difference.

Activity 22-3: Memorizing Letters

5-5, 7-15, 8-13, 9-19, 10-9, **22-3,** 23-3

Recall from Activity 5-5 that you collected data on how well you and your classmates could memorize a sequence of letters. All students received the same 30 letters to memorize. For some students, these letters appeared in convenient three-letter groupings (JFK-CIA-FBI-. . .), but for the other students, the letters appeared in inconvenient groupings (JFKC-IAF-. . .). We suspected that those students who received the convenient three-letter chunks would tend to score higher (memorize more letters correctly) than those students who received the less convenient groupings.

a. Identify the observational units in this study.

b. Identify the explanatory and response variables. Also classify them.

Explanatory: Type:

Response: Type:

c. Does this study involve random sampling from populations, random assignment to experimental groups, both, or neither? Explain.

d. State the relevant null and alternative hypotheses, in words and in symbols.

e. Examine comparative visual displays (dotplots and boxplots) of your class data. Comment on what this preliminary analysis reveals about the question of interest.

f. Do the technical conditions of the two-sample *t*-procedures appear to be satisfied? Explain.

g. Use your calculator to compute the test statistic and *p*-value. Record these results. Also summarize your conclusion and explain the reasoning process by which the conclusion follows from your test.

h. Use your calculator to compute a 90% confidence interval for estimating the difference in mean population scores between the two groups.

i. Is it appropriate to conclude from this study that the convenient three-letter groupings cause higher memory scores on average? Explain why or why not.

This activity should remind you of the important role of *randomization* in designing experiments. Randomization serves to make the groups similar prior to the imposition of the treatment, so any significant differences observed after the experiment can reasonably be attributed to that treatment. With this memory experiment, if your class results produce a small *p*-value, then you can reasonably conclude that the convenient three-letter groupings were responsible for the higher memory scores. But, because the study design did not involve true random samples from a larger population, you might hesitate before generalizing the results to individuals, even other students, beyond your class.

√ Activity 22-4 : Got a Tip?

1-10, 5-28, 14-12, 15-17, **22-4,** 22-9, 25-3

Can waitresses increase their tips simply by introducing themselves by name when they greet customers? Garrity and Degelman (1990) report on a study in which a waitress collected data on two-person parties whom she waited on during Sunday brunch (with a fixed price of $23.21) at a Charley Brown's restaurant in southern California. For each party, the waitress used a random mechanism to determine whether to give her name as part of her greeting. Then she kept track of how much the party tipped at the end of their meal.

a. Is this an observational study or a randomized experiment? Explain.

b. Identify and classify the explanatory and response variables.

c. State the null and alternative hypotheses, in symbols and in words, for testing the waitress' conjecture.

d. Describe what a Type I error and a Type II error would mean in this study. Would you consider one of these errors to be more serious than the other one? Explain. *Hint:* Refer to Activity 18-5 on page 367 for a reminder of what these types of errors mean.

The sample mean tip amount for the 20 parties to whom the waitress gave her name was \$5.44, with a standard deviation of \$1.75. These statistics were \$3.49 and \$1.13, respectively, for the 20 parties to whom the waitress did not give her name.

e. Use this information to compute the test statistic and *p*-value (by hand or with your calculator).

f. What test decision would you reach at the $\alpha = .05$ level? What does this decision mean in the context of this study?

g. Do you have enough information to check whether the technical conditions of the two-sample *t*-test are satisfied here? If so, check them. If not, explain what additional information you would request from the waitress.

h. Calculate a 95% confidence interval for the difference in population mean tip amounts between the two experimental treatments (giving name or not). Also interpret what the interval means in this context.

i. Regardless of whether the technical conditions are met, summarize your conclusions from this test. Be sure to comment on issues of causation and generalizability.

Solution

a. This is a randomized experiment, because the waitress randomly assigned the two-person parties to be told her name or not as part of her greeting. We do have to consider that the waitress was not blind to the treatment condition and may have subconsciously provided better service for the customers she expected to give her a larger tip.

b. The explanatory variable is whether the waitress included her name as part of her greeting; this variable is categorical and binary. The response variable is the amount of the tip, which is quantitative.

c. The null hypothesis is that there is no effect from the waitress using her name in her greeting to customers. In other words, the null hypothesis says that the population mean tip amount will be the same when she uses her name as when she does not. The alternative hypothesis is that using her name has a positive effect, that the population mean tip amount is greater when she uses her name than when she does not. In symbols: H_0: $\mu_{\text{name}} = \mu_{\text{no name}}$ vs. H_a: $\mu_{\text{name}} > \mu_{\text{no name}}$.

d. A Type I error means the waitress decides using her name helps when it really doesn't, so she would waste the minimal effort of giving her name and reap no benefit from it. A Type II error means the waitress decides using her name is not helpful even though it actually is helpful, so she would not bother to give customers her name and would lose out on that benefit. Because the cost of giving her name is minimal, losing out on potential tips is probably more of a concern.

e. The test statistic is

$$t = \frac{5.44 - 3.49}{\sqrt{\frac{(1.75)^2}{20} + \frac{(1.13)^2}{20}}} \approx \frac{1.95}{0.466} \approx 4.19$$

Looking in Table III (*t*-Distribution Critical Values), with 19 degrees of freedom, reveals that this test statistic is off the chart, so the *p*-value is less than .0005.

f. Because the *p*-value is less than .05, you reject the null hypothesis at the $\alpha = .05$ level. Indeed, you would also reject the null hypothesis at the $\alpha = .01$ and even at the $\alpha = .001$ levels. The data and experimental design provide very strong evidence that giving her name to customers as part of her greeting does lead to higher tips on average.

g. You do not have enough information to check the technical conditions thoroughly. You do know the parties were randomly assigned to one group or the other. But the sample sizes are not large, so you should check whether the tip data could reasonably have come from normal distributions; however, you have only the summary statistics, not the actual tip amounts from each party, so you cannot check this condition. You should ask the waitress to provide the actual party-by-party tip amounts to help you access the shape of the distribution of tip amounts.

h. A 95% confidence interval for $\mu_{\text{name}} - \mu_{\text{no name}}$ is

$$(5.44 - 3.49) \pm 2.093\sqrt{\frac{(1.75)^2}{20} + \frac{(1.13)^2}{20}}$$

which is $1.95 \pm 2.093(0.466)$, which is 1.95 ± 0.97, which is the interval (0.98, 2.92). You can be 95% confident that the waitress would earn between \$0.98 and \$2.92 more per party with a \$23.21 check, on average, by giving her name as part of her greeting (assuming that the tip amounts are roughly normally distributed).

i. You can conclude a causal link between the waitress giving her name and receiving higher tips on average. Random assignment should have assured that the only difference between the groups was whether the party was given the waitress's name. Because the group who was told her name gave significantly higher tips on average (p-value $< .0001$), you can attribute that to being told her name in her greeting unless the waitress gave better service to customers to whom she gave her name. The confidence interval enables you to say more: that giving her name to customers increases the waitress' tips by an average of about 1–3 dollars per dining party at Sunday brunch in this restaurant. But you must be cautious about generalizing this result to other waitresses because only one waitress participated in this study. Even for this particular waitress, you should be cautious about generalizing the results to customers beyond those who partake of Sunday brunch at that particular Charley Brown's restaurant in southern California. You should also remember that these p-value and confidence interval calculations are only valid if the tip amounts roughly follow a normal distribution.

Watch Out

The cautions in this topic are not substantially new; however, we want to remind you of some good habits that are easy to forget:

- Be sure to choose the correct procedure in the first place. A good first step, as we've emphasized since Topic 1, is to identify the observational units and variables. In this study, the observational units are the dining parties; the explanatory variable is whether the waitress gives her name or not (categorical, binary); and the response variable is the amount of tip (quantitative). A two-sample t-procedure is appropriate in situations like this, when the explanatory variable is binary and the response variable is quantitative. Another way to think of this situation is that you want to compare two groups on a quantitative response variable.
- Remember the hypotheses are always about parameters, denoted by Greek letters. It's easy to forget and put symbols for sample statistics in the hypotheses. But you don't need to test whether the sample means are equal in the two groups, because you *know* the values of the sample mean tip amounts. The question is what you can *infer* about the unknown *population* means.
- Again be careful not to let rounding errors creep into your calculations. Also be aware your answers might differ very slightly if you use your calculator to conduct these tests and construct these intervals.
- Always remember to check technical conditions before taking a test or interval result seriously. Also notice that the second technical condition here is an either/or statement: the distributions do not have to be normal if the sample sizes are large.
- Remember the scope of conclusions, with regard to causation and generalizability, depends on how the study is conducted.
- Do not forget to relate your conclusions to the context of the study. Do not simply say "reject H_0" and leave it at that. Such a conclusion would not be very helpful to this waitress. Rather, say the data provide very strong evidence that giving her name to customers increases her tips on average by about 1–3 dollars per dining party at Sunday brunch in this restaurant.

Wrap-Up...........

This topic has extended your knowledge of inference techniques to include the goal of comparing two *means*. You also saw that the structure and reasoning of inference procedures are the same as in previous situations you have examined.

You discovered the roles played by the sample sizes, means, and standard deviations in these inference procedures. The most obvious, but important, relationship here is that the difference between the two groups is more significant, as reflected by a smaller

p-value, when the difference between the sample means is greater. As you saw with the commuting times activity, the difference between two groups becomes more significant as the sample sizes increase and also as the standard deviations decrease, indicating less variability in the response variable measurements (if all else remains the same).

You were reminded again that the scope of conclusions you can draw depends on how the data are collected. When the data are obtained from independent random samples of the two populations, as with the GSS question about close friends, then you can generalize the results to the larger populations. When the data are gathered after randomly assigning subjects to treatment groups, as with the memory experiment and the waitress tip study, then you can draw a cause-and-effect conclusion between the variables.

Some useful definitions to remember and habits to develop from this topic include

- Consider applying these two-sample t-procedures when you want to compare a quantitative response variable between two groups.
- The sample means, sample standard deviations, and sample sizes are needed to apply these procedures.
- When the sample sizes are small, you also need to examine the raw data and judge whether both distributions are roughly normal in order to determine whether the technical conditions are satisfied.
- If a two-sample t-test reveals that the difference between the groups is statistically significant, follow up with a confidence interval to estimate the magnitude of that difference.

The procedures described in this topic apply when the two samples are drawn *independently* from the two populations or when subjects are assigned randomly to treatment groups. In the next topic, you will learn about a different way to design an experiment, or collect data, called a matched pairs design. You will study the advantages of such a design and learn how to analyze the resulting data.

Homework Activities

Activity 22-5: Close Friends

19-17, 19-18, 22-1, **22-5**, 22-22

Consider the data you analyzed in Activity 22-1 on how many close friends a person has. Describe the effect (decrease, increase, or remain the same) on the p-value of the significance test, and explain your reasoning in each case, if everything remains the same except

a. All of the women sampled have one more close friend than they originally reported.
b. All of the men sampled have one more close friend than they originally reported.
c. Every man and every woman sampled have one more close friend than they originally reported.
d. Both sample standard deviations are larger.
e. Both sample sizes are larger.

Activity 22-6: Hypothetical Commuting Times

7-13, 22-2, **22-6**, 22-7

Reconsider the commuting time experiments presented in Activity 22-2 on page 449. For each of the following people (Earl, Fred, Grace, Harry, and Ida), determine whether his or her sample results are statistically significant. Do not perform any calculations; simply compare the sample results to those of Alex, Barb, Carl, or Donna. In each case, indicate whose results (Alex's, Barb's, Carl's, or Donna's) you used for the comparison and explain your answer.

	Sample Size	Sample Mean	Sample SD
Earl 1	10	29.0	6.0
Earl 2	10	31.0	6.0
Fred 1	75	28.0	6.0
Fred 2	75	32.0	6.0
Grace 1	10	20.0	6.0
Grace 2	10	40.0	6.0
Harry 1	10	28.0	9.0
Harry 2	10	32.0	9.0
Ida 1	10	28.0	1.5
Ida 2	10	32.0	1.5

Activity 22-7: Hypothetical Commuting Times

7-13, 22-2, 22-6, 22-7

Reconsider the previous activity. Consider four more hypothetical commuters who are trying out two different routes for driving to work (Jacques, Katrina, Liam, and Manuel). The following boxplots present distributions of their sample driving times. Suppose each person does a two-sample t-test of whether one route tends to be faster than the other. Arrange these four people in order from the one who has the largest p-value to the one who has the smallest p-value. Explain your choices. *Hint:* Note the different sample sizes indicated with the boxplots.

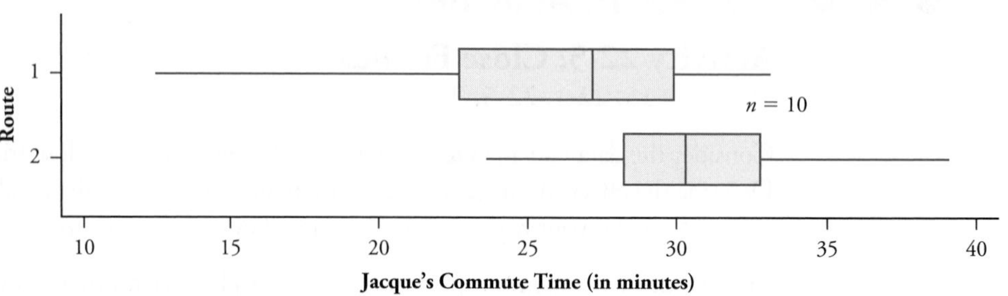

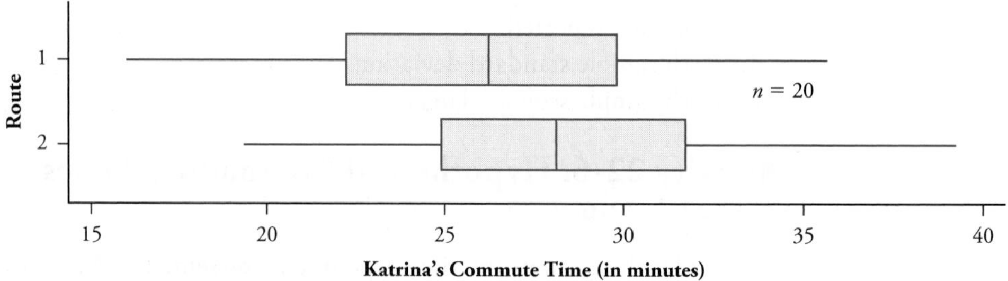

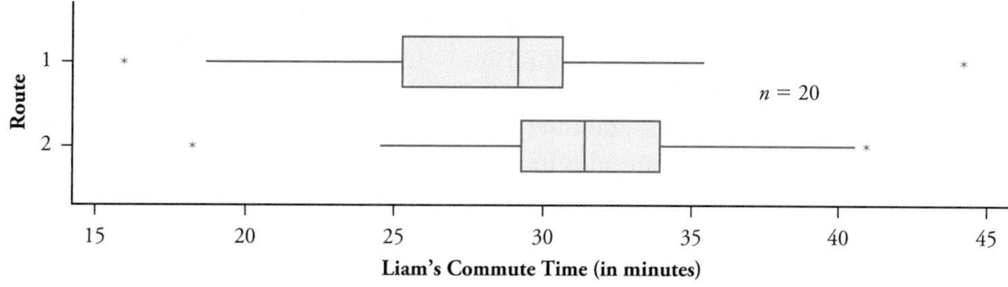

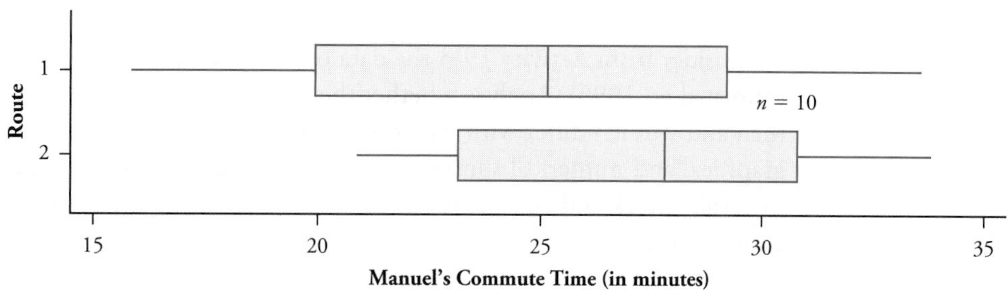

Activity 22-8: Nicotine Lozenge

1-16, 2-18, 5-6, 9-21, 19-11, 20-15, 20-19, 21-6, **22-8**

Recall the study from Activity 5-6 on whether a nicotine lozenge can help a smoker to quit. The research article reports on many background variables, such as age, weight, gender, number of cigarettes smoked, and whether the person made a previous attempt to quit smoking (Shiffman et al., 2002). Suppose the researchers want to compare the distributions of the background variables between the two treatment groups (nicotine lozenge or placebo lozenge).

a. For each of the five variables listed, indicate whether it calls for a comparison of *means* or a comparison of *proportions*.
b. Would the researchers hope to reject the null hypotheses or fail to reject the null hypotheses in these tests? Explain.

Activity 22-9: Got a Tip?

1-10, 5-28, 14-12, 15-17, 22-4, **22-9**, 25-3

Researchers investigated whether providing a fancy, foil-wrapped piece of chocolate with the dinner bill would lead to higher tips than not providing such a treat (Strohmetz et al., 2002). Ninety-two dinner parties at a restaurant in Ithaca, New York, were randomly assigned to receive such a piece of chocolate or not with their dinner bill. Of the 46 parties who received the chocolate, the average tip (as a percentage of the bill) was 17.84%, with a standard deviation of 3.06%. Of the 46 parties who did not receive the chocolate, the average tip (as a percentage of the bill) was 15.06%, with a standard deviation of 1.89%.

a. Identify the explanatory and response variables.
b. Explain why, in this case, looking at the percentage of the bill is a more useful measure than looking at the exact tip amount (as the waitress in Activity 22-4 did).
c. Is this an observational study or an experiment? Explain.
d. Do you have enough information to check whether the technical conditions for the two-sample t-test and confidence interval are satisfied? Explain.

e. Conduct the appropriate test of the researchers' conjecture. Report all aspects of the test, including your test decision at the $\alpha = .05$ level, and summarize your conclusion.

f. Produce and interpret a 95% confidence interval for estimating how much the chocolate adds to the tip percentage on average.

g. Summarize your conclusions. Be sure to address issues of genealizability and causation.

Activity 22-10: Body Temperatures

12-1, 12-19, 15-3, 15-18, 15-19, 19-3, 19-7, 20-11, **22-10,** 23-3

Reconsider from Activity 19-3 the data on body temperatures for 130 healthy adults (Shoemaker, 1996). Analyze whether the sample data provide strong evidence that men and women differ with regard to mean body temperature (BODYTEMPS). Include graphical and numerical summaries as well as a significance test and confidence interval. Check and comment on technical conditions, and summarize your conclusions.

Activity 22-11: Ideal Age

In September 2003, The Harris Poll asked a random sample of 2306 adult Americans, "If you could stop time and live forever in good health at a particular age, at what age would you like to live?" The average response from men was 39 years, and the average response from women was 43 years.

a. State the null and alternative hypotheses for testing whether this difference (between men's and women's average responses) is statistically significant.

b. What additional information do you need to conduct this test?

Suppose there were 1153 men and 1153 women in the sample. Suppose further that the standard deviation of the responses for each gender is 25 years.

c. Calculate the test statistic and p-value. What conclusion would you draw?

d. Determine and interpret a 99% confidence interval for the difference in population means.

e. Is the confidence interval consistent with the test result? Explain.

f. Do you suspect the standard deviations of the age responses might be greater than 25 years, or do you think 25 years is a reasonable upper bound for how high the standard deviation might be? Explain.

Activity 22-12: Editorial Styles

22-12, 22-13

USA Today is known as the "nation's newspaper." It strives to reach a very broad readership and is, therefore, reputed to be written at a fairly low readability level. On the other hand, the *Washington Post* generally has a reputation as a more serious newspaper aiming for a more intellectual readership. To assess whether quantitative data can reveal any evidence to support this reputation, consider the following data on sentence lengths (measured by number of words in a sentence) for the lead editorials from January 22, 2007, in *USA Today* and in the *Washington Post:*

USA Today	21	26	32	31	36	28	22	29	12	27	4	27	23	22	22	38
	17	20	12	21	15	39	29	18	18	24	15	32	7	11	30	
Washington Post	25	19	19	16	32	14	16	12	16	56	23					
	21	23	17	50	16	38	22	11	41	14						

The data are also stored in the grouped file EDITORIALS07 (with the lists USA and WAPO) .

a. Produce five-number summaries and boxplots to compare the distributions of sentence lengths between the two newspapers. Comment on key features of the distributions and on any similarities and differences that are revealed.

b. Are the technical conditions for a two-sample *t*-test satisfied? Explain.

c. Conduct a two-sample *t*-test to assess whether the sample data support the contention that the sentences in *USA Today* are shorter than those in the *Washington Post.* Report all aspects of the test and summarize your conclusion. Be sure to specify the populations to which you would be willing to generalize your results.

d. Do you agree that sentence length is a reasonable measure of a newspaper's readability level? Suggest another quantitative variable that you could have measured to compare the editorial styles of these two newspapers.

Activity 22-13: Editorial Styles

22-12, **22-13**

Reconsider the previous activity. Remove from the analysis all sentences identified by the 1.5 × IQR rule (Topic 10) as outliers (for each sample). Repeat your analysis from the previous activity with these outliers removed (EDITORIALS07). Comment on how your analysis changes, and explain why the changes make sense.

Activity 22-14: Children's Television Viewing

1-15, 19-16, 20-4, **22-14,** 22-15

Recall from Activity 1-15 the study in which one group of third- and fourth-graders was randomly assigned to receive an 18-lesson, 6-month classroom curriculum intended to reduce their use of television, videotapes, and video games (Robinson, 1999). The other group of children received their usual elementary school curriculum. All children were asked to report how many hours per week they spent on these activities, both before the curriculum intervention and afterward.

a. If the randomization achieved its goal, would there be a significant difference between the two groups prior to the curriculum intervention? Explain.

The following summary statistics pertain to the reports of television watching, in hours per week, *prior* to the intervention:

Baseline	Sample Size	Sample Mean	Sample SD
Control Group	103	15.46	15.02
Intervention Group	95	15.35	13.17

b. State, in words and in symbols, the null and alternative hypotheses for testing whether the two groups are indeed similar in television viewing habits prior to the intervention.

c. Use the summarized sample information to calculate the test statistic and p-value.
d. Would you reject the null hypothesis (that the population means are equal prior to the intervention) at the $\alpha = .05$ level?

The following summary statistics pertain to the reports of television watching at the *conclusion* of the study:

Follow-up	Sample Size	Sample Mean	Sample SD
Control Group	103	14.46	13.82
Intervention Group	95	8.80	10.41

e. Conduct a two-sample t-test of whether the mean number of hours of television viewing per week is higher in the control group than in the intervention group at the conclusion of the study. State the hypotheses, and report the test statistic and p-value. Also indicate whether you would reject the null hypothesis (of no difference) at the .05 level.
f. Even though you do not have the raw data, explain how the summary statistics reveal that the distributions of television viewing hours must be non-normal.
g. Explain why the non-normality of these distributions does not hinder the validity of using this test procedure.
h. Write a paragraph or two summarizing your findings about the comparison of television viewing habits between these two groups before and after the study. Would you conclude the curriculum intervention succeeded in reducing television viewing? Also refer to the role of randomization in addressing this question.

Activity 22-15: Children's Television Viewing

1-15, 19-16, 20-4, 22-14, 22-15

Reconsider the previous activity. In addition to asking the children in the obesity study about their television viewing habits, researchers also asked them to report how many hours they spent per week watching videotapes and playing video games. Summary statistics for the *baseline* comparisons are reported here:

Baseline: Videotapes	Sample Size	Sample Mean	Sample SD
Control Group	103	5.52	10.44
Intervention Group	95	4.74	6.57

Baseline: Video Games	Sample Size	Sample Mean	Sample SD
Control Group	103	3.85	9.17
Intervention Group	95	2.57	5.10

a. Conduct appropriate tests of whether the control and intervention group means differ significantly on either of these variables. Report the test statistics and p-values, and write a few sentences detailing your findings.

Summary statistics for the *follow-up* comparisons on these variables are reported here:

Follow-up: Videotapes	Sample Size	Sample Mean	Sample SD
Control Group	103	5.21	8.41
Intervention Group	95	3.46	4.86

Follow-up: Video Games	Sample Size	Sample Mean	Sample SD
Control Group	103	4.24	10.00
Intervention Group	95	1.32	2.72

b. Repeat part a for these follow-up data.

Activity 22-16: Classroom Attention

Researchers in a 1979 study recorded the lengths of individual instructional time (in seconds) that second-grade instructors spent with their students (Leinhardt, Seewald, and Engel, 1979). They compared these times between girls and boys in the subjects of reading and mathematics. Numerical summaries of their results appear here:

	Sample Size	Sample Mean (reading)	Sample SD (reading)	Sample Mean (mathematics)	Sample SD (mathematics)
Boys	372	35.90	18.46	38.77	18.93
Girls	354	37.81	18.64	29.55	16.59

a. Comment on whether the technical conditions necessary for the validity of the two-sample t-procedures seem to be satisfied.
b. Conduct appropriate tests of significance to determine whether the sample mean instructional times differ significantly between boys and girls in either subject.
c. Produce confidence intervals (one for reading, one for mathematics) for the difference in population means between boys and girls.
d. Write a paragraph or two summarizing your conclusions.

Activity 22-17: UFO Sighters' Personalities

20-8, **22-17**

Reconsider the study about UFO sighters described in Activity 20-8 (Spanos et al., 1993). The group of 25 people who claimed to have had an intense experience with a UFO had a mean IQ of 101.6, and the standard deviation of these IQs was 8.9. A control group of 53 community members who had not reported UFO experiences had a mean IQ of 100.6 with a standard deviation of 12.3.

a. Is this an observational study or a controlled experiment? Explain.
b. Identify the explanatory and response variables. Also identify each as quantitative or categorical.

c. Use this sample information to test whether the mean IQs of community members and UFO sighters differ significantly. Show the details of the test and write a one-sentence conclusion.
d. Find a confidence interval for the difference in population means, and explain what the interval reveals.
e. Given the type of study involved, even if the sample data revealed a significantly higher mean IQ for the control group, would you be able to draw any causal conclusion that seeing a UFO affects intelligence? Explain.

Activity 22-18: Tennis Simulations

7-22, 8-21, 9-16, **22-18**

Refer again to the data in Activity 8-21 on page 160 concerning simulation results of game lengths for three different scoring systems for tennis (TENNISSIM).

a. Use your calculator to analyze whether the sample data suggest the mean game length with conventional scoring differs significantly from the mean game length with no-ad scoring and, if so, by about how much. Write a paragraph describing your findings.
b. Repeat part a, comparing no-ad scoring with handicap scoring.
c. Do the inference techniques you employed in parts a and b say anything about whether the *variability* of games' lengths differs among the scoring methods? Do the techniques reveal anything about whether the *shapes* of the distributions differ? Why might these be interesting considerations? Explain.

Activity 22-19: Ice Cream Servings

5-9, **22-19,** 22-20

Recall from Activity 5-9 the study about how much ice cream people scoop into their bowls (Wansink et al., 2006). Researchers randomly assigned some people to use a 17-ounce bowl and some people to use a 34-ounce bowl. Sample results for the actual volumes taken (in ounces, only considering results for those using a small spoon) are summarized in the following table:

	Sample Size	Sample Mean	Sample SD
17-ounce Bowl	20	4.38	2.05
34-ounce Bowl	17	5.81	2.26

a. Is this an experiment or an observational study? Explain.
b. Do you have enough information to check whether the technical conditions of a two-sample *t*-test are satisfied? Explain.
c. State the appropriate hypotheses for conducting a one-sided two-sample *t*-test to compare the results between the two groups. Justify the direction of the alternative hypothesis you choose.
d. Conduct this test. Report the test statistic and *p*-value.
e. Assuming that the technical conditions are met, is the difference in volumes served between the two bowl sizes statistically significant at the .05 level? How about at the .01 level?
f. Can you draw a cause-and-effect conclusion between the bowl size and volume of ice cream served? Explain.

Activity 22-20: Ice Cream Servings
5-9, 22-19, **22-20**

Reconsider the previous activity. Researchers also randomly assigned some people to use a 2-ounce spoon and others to use a 3-ounce spoon. Sample results for the actual volumes served (in ounces, again only considering those using a small bowl) are summarized in the following table:

	Sample Size	Sample Mean	Sample SD
2-ounce Spoon	20	4.38	2.05
3-ounce Spoon	26	5.07	1.84

Repeat parts c–f of the previous activity for the question of whether *spoon* size affects volume of ice cream served.

Activity 22-21: Natural Selection
10-1, 10-6, 10-7, 12-20, **22-21,** 23-3

Recall from Activities 10-1, 10-6, and 10-7 the study of natural selection in which Bumpus compared physical measurements of sparrows that survived a harsh winter storm with sparrows that perished (BUMPUS).

a. Analyze the *total length* measurements for evidence of a significant difference between the two groups of sparrows. Include graphical and numerical summaries as well as a significance test and confidence interval. Check technical conditions, and write a paragraph summarizing your conclusions. Be sure to address issues of generalizability and causation.
b. Repeat, for the *weight* variable.
c. Repeat, for one of these other variables of your choosing: *alar extent, length of head and beak, humerus bone length, femur bone length, tibiotarsus bone length, skull width,* or *keel of sternum.*

Activity 22-22: Close Friends
19-17, 19-18, 22-1, 22-5, **22-22**

Reconsider the "close friends" data from Activity 22-1.

a. Conduct a test of whether men and women differ with regard to the proportion who respond with zero names to the survey question asking who they talk to about important matters. Report all aspects of the test, and summarize your conclusion. *Hint:* First, consider whether you are conducting a test to compare *means* or *proportions.*
b. Repeat part a, but for responses with six names rather than zero names.

Activity 22-23: Hypothetical SAT Coaching
22-23, 22-24

Suppose 5000 students are randomly assigned to take an SAT coaching course or not, with the following results concerning improvements in their SAT scores:

	Sample Size	Sample Mean	Sample SD
Coaching Group	2500	46.2	14.4
Control Group	2500	44.4	15.3

a. Conduct a test of whether the sample data provide evidence that SAT coaching is helpful. (Feel free to use your calculator.) State the hypotheses, and report the *p*-value. Draw a conclusion in the context of this study.

b. Produce a 99% confidence interval for the difference in population mean improvements between the two groups. Also interpret this interval.

c. Do the sample data provide *very* strong evidence that SAT coaching is helpful? Explain whether the *p*-value or the confidence interval helps you answer this question.

d. Do the sample data provide strong evidence that SAT coaching is *very* helpful? Explain whether the *p*-value or the confidence interval helps you answer this question.

Activity 22-24: Hypothetical SAT Coaching

22-23, **22-24**

Reconsider the previous activity.

a. Describe what a Type I error would mean in this context.

b. Describe what a Type II error would mean in this context.

Activity 22-25: Hypothetical ATM Withdrawals

9-24, 19-21, **22-25**

Reconsider the hypothetical ATM withdrawals that you analyzed in Activities 9-24 on page 186 and 19-21 on page 398 (HYPOATM).

a. Choose any pair of machines and calculate a 90% confidence interval for estimating the difference in population mean withdrawal amounts between the two machines.

b. Does this interval include the value zero?

c. Would this interval be any different if you had chosen a different pair of machines? Explain.

d. Are these three machines identical in their distributions of withdrawal amounts? Explain. Make sure your discussion relates to what you learned in parts b and c.

Activity 22-26: Sporting Examples

2-6, 3-4, 8-14, 10-11, **22-26**

Recall from Activity 2-6 on page 23 that a statistics professor compared the performance of two sections of students: one that used exclusively sports-themed examples and the other that used a typical variety of examples (Lock, 2006). Also recall that students signed up in advance for one section or the other and that the sports-themed section met at an earlier time than the traditional section. The data are stored in the grouped file SPORTSEXAMPLES.

a. Conduct a test of whether the *total points* variable differs significantly between the two sections. Report all aspects of the test, including your test decision at the

$\alpha = .10$ level and summarizing your conclusion. Be sure to include a check of technical conditions.

b. Determine and interpret a 90% confidence interval for the difference in population means (with the *total points* variable) between the two groups.

c. Does the way this study was conducted enable you to draw a cause-and-effect conclusion between the sports-themed examples and either better or worse performance in the course? Explain.

d. Does the way this study was conducted enable you to generalize your conclusions to all students who study statistics? Explain.

Activity 22-27: Your Choice

6-30, 12-22, 19-22, 21-28, **22-27,** 26-23, 27-22

Write a paragraph detailing a real situation about which you would be interested in performing a test of significance to compare two *means*. Describe precisely the context involved, and explain whether the study would be a controlled experiment or an observational study. Also identify very carefully the observational units and variables involved. Finally, suggest how you might go about collecting the sample data.

TOPIC

23 • • • Analyzing Paired Data

"Melts in your mouth, not in your hands" is a famous advertising slogan. How long does it take a chocolate chip to melt in your mouth? Does a peanut butter chip melt any faster or slower than a chocolate chip? How could you design a randomized experiment to find out? And how would you analyze the resulting data? Is there a better study design for answering this question? Does the study design affect how you analyze the results? The answer to these last two questions is yes, as you will see in this topic.

Overview

In the previous topic, you learned two-sample *t*-procedures that enable you to analyze the results of randomized experiments and independent random samples to determine whether the difference in means between two groups for a quantitative variable is statistically significant and to estimate the size of the difference. In this topic, you will study *paired* data, where the two-sample *t*-procedures are not appropriate. You will learn how to apply paired *t*-procedures and see the advantages of collecting data with a matched-pairs design.

Preliminaries

1. Guess a typical age difference between a husband and wife.

2. Which would you expect to have more variability: individual ages of married people or differences in ages within a husband/wife couple?

3. Would you expect a chocolate chip to melt faster, slower, or at the same speed as a peanut butter chip?

4. Your instructor will randomly assign you to place either a chocolate or a peanut butter chip on your tongue and hold it against the roof of your mouth. Record how long it takes (in seconds) for the chip to melt completely—without any "encouragement" on your part. Record this value.

5. Repeat this measurement using the other kind of chip, again determining how long it takes the chip to melt in your mouth.

6. Record the melting-time data for yourself and your classmates. Keep track of which kind of chip goes with which time, and also keep track of which kind of chip was tested first by each student.

In-Class Activities

Activity 23-1: Marriage Ages

8-17, 9-6, 16-19, 17-22, **23-1,** 23-12, 26-4, 29-17, 29-18

Reconsider the data presented in Activity 9-6 concerning the ages (in years) at marriage for a sample of 24 couples who obtained their marriage licenses in Cumberland County, Pennsylvania, in 1993. The data are recorded in the following table and stored in the grouped data file MARRIAGEAGES, containing the lists **HSBND, WIFE,** and **DFFR. DFFR** contains the differences in ages. Also following are the graphical displays and summary statistics:

Couple #	Husband's Age	Wife 's Age	Couple #	Husband's Age	Wife's Age	Couple #	Husband's Age	Wife's Age
1	25	22	9	31	30	17	26	27
2	25	32	10	54	44	18	31	36
3	51	50	11	23	23	19	26	24
4	25	25	12	34	39	20	62	60
5	38	33	13	25	24	21	29	26
6	30	27	14	23	22	22	31	23
7	60	45	15	19	16	23	29	28
8	54	47	16	71	73	24	35	36

	Sample Size	Sample Mean	Sample SD
Husband's Age	24	35.71	14.56
Wife's Age	24	33.83	13.56

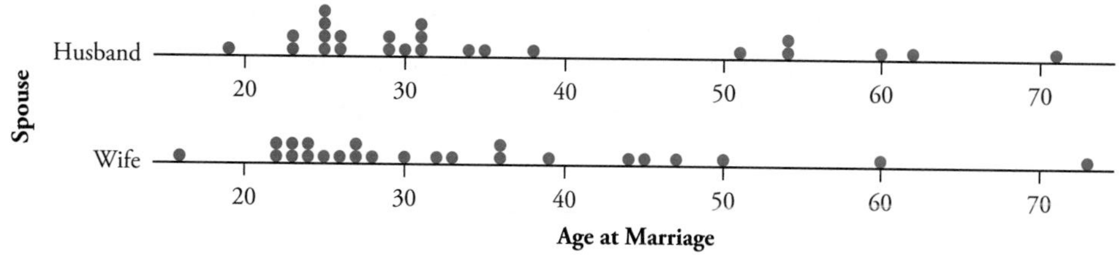

a. Use these summary statistics to conduct a two-sample t-test of whether these sample data provide evidence that the population mean age of husbands exceeds that of wives. Report the hypotheses, test statistic, and p-value. Also state your test decision at the $\alpha = .05$ level and summarize your conclusion. (Feel free to use your calculator.)

b. Use a two-sample interval to estimate the difference in population mean ages with 90% confidence. (Feel free to use your calculator.)

Unfortunately, this analysis is completely inappropriate for these data. The two-sample t-procedure is not valid because these samples are not independent. This analysis would have been appropriate only if the researcher had gathered ages for one sample of 24 husbands and then independently gathered ages for a completely different sample of 24 wives.

c. Consider the following graph that displays the husband's age and wife's age for each couple. Notice that the graph includes a "$y = x$" line showing where the husband and wife have the same age. Comment on whether most of the values are above or below this line, and indicate whether there is a clear tendency for husbands to be older than their wives.

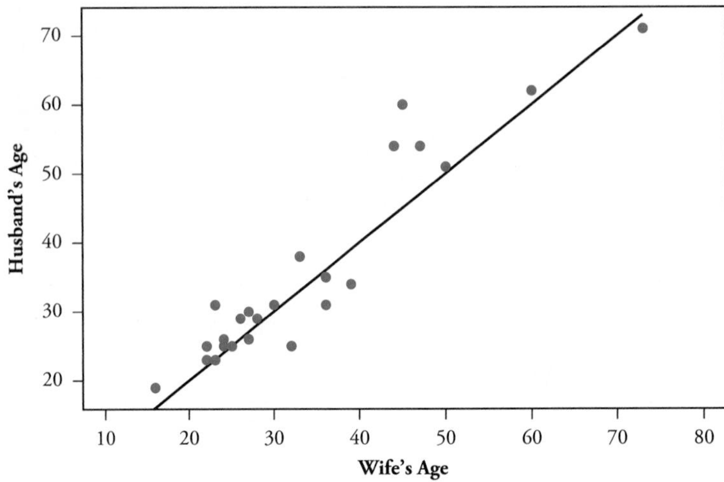

d. Does this graph reveal a random scatter, or do the husband's and wife's ages appear to be related? In particular, do older people tend to marry older people and younger people tend to marry younger people? Explain.

The key to devising a correct analysis of these data is realizing these data are *paired* because the observational units are *couples,* not independent individuals. The appropriate analysis is, therefore, to calculate the *differences* in ages for each couple, and then apply *one*-sample *t*-procedures (from Topics 19 and 20) to those differences.

A **paired *t*-procedure** applies one-sample *t*-procedures to the *differences* within a pair. The null hypothesis is H_0: $\mu_d = 0$, and the test statistic is

$$t = \frac{\bar{x}_d}{s_d/\sqrt{n}}$$

with a *p*-value based on the *t*-distribution with $(n - 1)$ degrees of freedom, where n is the number of *pairs* in the sample. (The subscript "*d*" reminds you that you are now analyzing *differences.*) A confidence interval for the population mean difference μ_d is

$$\bar{x}_d \pm t^* \frac{s_d}{\sqrt{n}}$$

The technical conditions are the same as with a one-sample *t*-procedure, except the observational units are pairs and the data are differences (e.g., a large sample size of pairs or the population of differences follows a normal distribution).

A dotplot and summary statistics for the age differences follow:

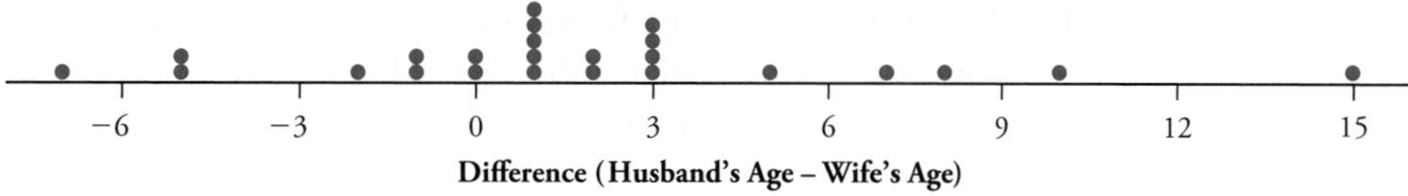

	Sample Size	Sample Mean	Sample SD
Difference	24	1.875	4.812

e. Determine a 90% confidence interval for the population mean *difference* in ages between husbands and wives. Also interpret this interval.

f. How does this interval compare to the (incorrect) interval from part b? Comment on its midpoint and width.

g. Conduct a paired t-test of whether the sample data provide strong evidence that the population mean difference exceeds zero. Report the hypotheses, test statistic, and p-value. Also state your test decision at the $\alpha = .05$ level, and summarize your conclusion. (Feel free to use your calculator.)

h. How does your conclusion from this test compare to the (incorrect) conclusion from the test in part a?

i. Check that the technical conditions for applying these t-procedures are met.

j. Explain why the paired analysis produces such a different conclusion from the independent-samples analysis.

k. Was the researcher wise for gathering paired data rather than independent-samples data to investigate this research question? In other words, was pairing helpful for estimating the population mean age difference among married couples? Explain.

Pairing is effective here because of the variation in the ages of people who apply for marriage licenses. Look at the dotplots shown before part a: the ages extend from the teens to the seventies, with substantial overlap between the husbands' ages and wives' ages. But there is a strong relationship between the ages of the husband and wife within a couple, as the graph in part c revealed. Thus, there is much less variation in the age *differences* than there is in the ages themselves. (It would be pretty surprising to find many couples with a 50-year age difference between the husband and the wife! But it would be much less surprising to find a husband and wife from two different couples with such an age difference.) Note the small standard deviation of the differences (4.812 years) compared to the standard deviations of the ages (14.56 years for husbands and 13.56 years for wives). You first encountered this reduction in variability in Activity 9-6, and now you use this to your benefit—based on the principle you discovered in Activity 20-2 that a smaller standard deviation produces a more statistically significant result, when all else (in that case, the sample size and the sample mean) remains the same. Researchers often think about increasing the sample size to decrease sampling variability, but here the researchers managed to decrease the variability in the measurements themselves with an effective study design.

Watch Out

- Remember, the first analysis you conducted in this activity, using two-sample *t*-procedures, is completely wrong. (We asked you to perform the wrong procedure so you could see how much more powerful the correct procedure is.) It's crucial that you determine which *t*-procedure (paired or unpaired) is appropriate when you are comparing two groups on a quantitative response.
- Determining which *t*-procedure to use depends entirely on how the data are collected. If the sampling or experimental design is paired, then use paired-*t* procedures. But if the samples are drawn independently for the two groups, or if randomization is used to assign subjects to separate treatment groups, then use the two-sample *t*-procedures from Topic 22.
- To help decide whether the data are collected with a paired design, ask whether there's a link between each observation in one group with a specific observation in the other group. In this study, the link between the groups is each husband in one group is married to one specific wife in the other group. In a before/after study, the link between the groups is that the data are recorded on the same individuals.
- In a paired design, mixing up the order of values in one group would create a problem. With a nonpaired design, you can mix up the order of values in one group, or both groups, without affecting the analysis at all.
- If a study has a different number of observations between the two groups, then it can't be a paired design (unless data are not recorded for some observational units). However, even if the sample sizes are the same between the two groups, you cannot necessarily conclude the design is paired.

Activity 23-2: Chip Melting

23-2, 26-20, 29-15

Consider the chip-melting data collected in the Preliminaries section. To compare the chocolate and peanut butter melting times, you will first analyze only the data collected in Preliminaries question 4, when you were each randomly assigned to test just *one* type of chip.

a. Identify and classify the explanatory and response variables in this study.

Explanatory: Type:

Response: Type:

b. Is this an observational study or an experiment? Explain.

c. Examine and comment on graphical and numerical summaries for comparing the melting times between the two types of chips. Be sure to comment on whether one type of chip tends to melt more quickly than the other type.

d. Conduct a two-sample t-test to assess whether the data provide strong evidence that the population mean melting times differ between the two types of chips. Report all aspects of the test, including a check of technical conditions, and summarize your conclusion.

Now consider all of the data collected in Preliminaries question 5, when each student tested both types of chips.

A **completely randomized experiment** randomly assigns each subject (observational unit) to one of the treatment groups. In a **matched-pairs experiment,** similar subjects are paired up, and then one member of each pair is randomly assigned to one treatment group and the other member to the other group. Sometimes the subjects are paired with themselves, so each subject receives both treatments, with the order of treatment determined at random.

When a matched-pairs design involves a quantitative response, the data can be analyzed with paired t-procedures—if the technical conditions are satisfied.

e. Consider all the melting time data collected on yourself and your classmates in the Preliminaries. Do these data call for a two-sample t-test or a paired t-test? Explain.

f. Analyze the data, using graphical and numerical summaries as well as the appropriate t-test and confidence interval. Write a paragraph or two summarizing your analysis and conclusions.

g. How did your conclusions from these two studies compare? Explain why this result makes sense. Be sure to identify what source of variability was taken into account with this study. In other words, how did your study reduce variability in melting times so as to better compare the two types of chips?

Activity 23-3: Body Temperatures, Natural Selection, and Memorizing Letters

5-5, 7-15, 8-13, 9-19, 10-1, 10-6, 10-7, 10-9, 12-1, 12-19, 12-20, 15-3, 15-18, 15-19, 19-3, 19-7, 20-11, 22-3, 22-10, 22-21, **23-3**

a. Recall the body temperature data from Activity 22-3, where you compared the mean body temperatures for samples of 65 men and 65 women. Would it be appropriate to analyze these data with a paired *t*-test? Explain.

b. Recall from Activity 10-1 the Bumpus natural selection study (1898), comparing lengths of sparrows that survived or perished in a winter storm. Explain why it would be not only inappropriate but also downright impossible to analyze these data with a paired *t*-test.

c. Recall from Activities 5-5 and 22-3 the experiment you conducted about memorizing letters. Would it be appropriate to analyze these data with a paired *t*-test? Explain.

d. Explain how you could alter the memory experiment, so a paired *t*-test would be an appropriate analysis to perform on the data. Also explain why randomization would still be important and what you would use it for.

Activity 23-4: Alarming Wake-Up

A recent study published in the journal *Pediatrics* (Smith et al., 2006) addressed the important issue of how to awaken children during a house fire so they can escape safely. Researchers worked with a volunteer sample of 24 healthy children aged 6–12 by training them to perform a simulated self-rescue escape procedure when they heard an alarm. Researchers then measured the children's reactions to a conventional smoke alarm and to a personalized recording of the mother's voice saying the child's name and urging him or her to wake up. One response variable measured was how long the child took to escape.

a. Identify the explanatory variable in this study.

b. Describe how you would design a *completely randomized* experiment to investigate the question of whether children react more quickly to the personalized alarm than to the conventional alarm. Explain how your study would make use of randomization and what purpose the randomization would serve.

c. Write out the null and alternative hypotheses you would test once the data were collected.

d. Describe how you would design a *matched-pairs* experiment to investigate this question. Again explain how your study would make use of randomization and what purpose the randomization would serve.

e. Write out the null and alternative hypotheses you would test once the data were collected.

f. Which of these two experimental designs would you recommend to the researchers? Explain why.

Activity 23-5: Muscle Fatigue

23-5, 23-15, 26-22, 27-20

Do women or men experience fatigue more quickly while exercising? To investigate this issue, researchers recruited healthy young adults to participate in a study (Hunter et al., 2004). The researchers first measured the strength of each person by the torque exerted at the wrist during a maximal voluntary contraction. Then they matched up each of the 10 men in the study with a woman of comparable strength (within 5% of the maximal torque exerted). Next, they asked each subject to complete a prescribed series of exercises with their elbow flexor muscles and elbow extensor muscles until they lacked the strength to continue. Researchers recorded how long it took before failure at this exercise task for each subject.

a. Why do you think the researchers matched the subjects by strength?

b. Identify and classify the explanatory and response variables in this study.

c. Is this an observational study or a randomized experiment? Explain.

The resulting times until task failure (in seconds) are given here:

Pair	A	B	C	D	E	F	G	H	I	J
Man's Time Until Fatigue	314	335	379	384	392	511	566	629	697	923
Woman's Time Until Fatigue	1204	1041	3963	368	334	846	888	2446	2181	813

d. Calculate the *differences* in time until fatigue for each pair. Then create and examine a dotplot of the sample distribution of these differences. Also calculate numerical summaries and comment on your findings.

e. Are the technical conditions for a paired *t*-test satisfied? Justify your answer.

f. Conduct a paired t-test to assess whether the data provide evidence that men and women differ, on average, with regard to times until fatigue. Report all aspects of the test, including your test decision at the $\alpha = .05$ significance level.

g. Produce a 90% confidence interval for estimating the population mean difference in times until fatigue between men and women. Also interpret the interval, and comment on whether it is consistent with your test decision.

h. Summarize your conclusions from this study, particularly with regard to issues of causation and generalizability.

Solution

a. The researchers expected that strength would be related to time until fatigue; therefore, matching men and women based on strength would result in less variability in the time differences than in the individual times. This reduction in variability allows for a more powerful test, one that is better able to detect a difference in fatigue times between the genders if a difference really exists.

b. The explanatory variable is gender, which is categorical and binary. The response variable is time until fatigue, which is quantitative.

c. This is an observational study. Subjects were not assigned to a gender.

d. The differences in times until fatigue, subtracting the man's time from the woman's time in each pair, are 890, 706, 3584, –16, –58, 335, 322, 1817, 1484, –110. A dotplot follows.

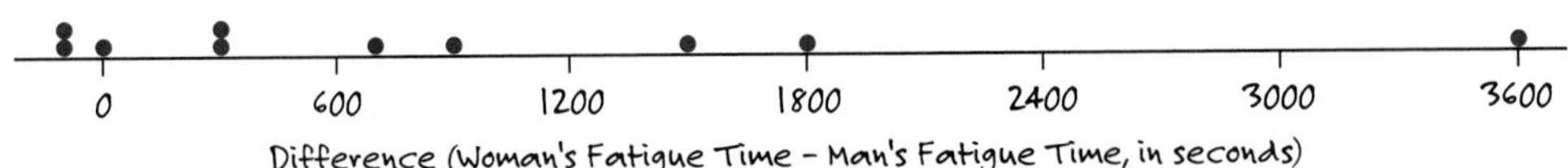

Three of these differences are negative, and seven are positive. Therefore, in most pairs, the woman lasted longer until fatigue set in than the man did. The mean of these differences is 895 seconds, so the women in the sample outlasted the men by almost 15 minutes on average. The standard deviation is 1148 seconds. The distribution of differences appears to be a bit skewed to the right.

e. Because of the small sample size ($n = 10$ pairs), the distribution of differences must be approximately normal in order for the technical conditions to be satisfied. But you have already seen that the distribution is a bit skewed to the right. This skewness is also seen in a normal probability plot, which does not look very linear:

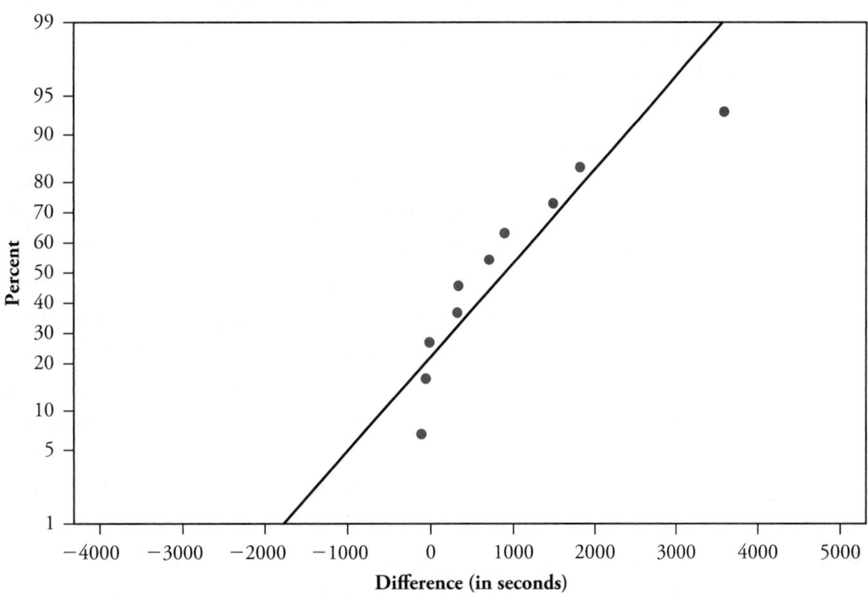

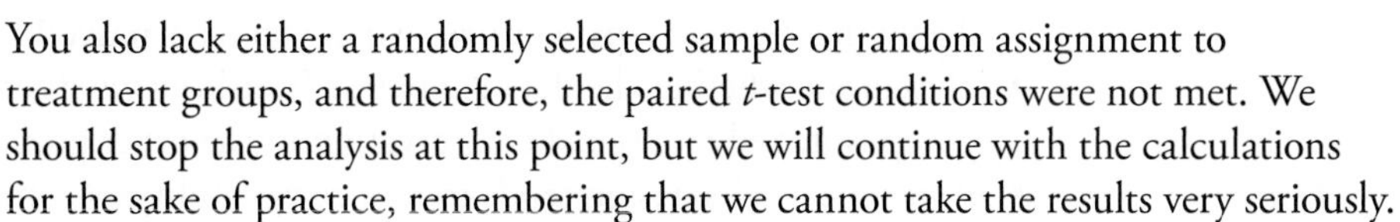

You also lack either a randomly selected sample or random assignment to treatment groups, and therefore, the paired t-test conditions were not met. We should stop the analysis at this point, but we will continue with the calculations for the sake of practice, remembering that we cannot take the results very seriously.

f. You are testing H_0: $\mu_d = 0$, vs. the alternative H_a: $\mu_d \neq 0$, where μ_d represents the population mean of the differences in times until fatigue between healthy young adult men and women. The test statistic is

$$t = \frac{895}{1148/\sqrt{10}} \approx 2.465$$

The p-value, from Table III (t-Distribution Critical Values) with 9 degrees of freedom and remembering the alternative hypothesis is two-sided, is between 2(.01) and 2(.025), which means the p-value is between .02 and .05. The calculator gives a p-value of .036. This p-value is fairly small, so you would reject the null hypothesis at the $\alpha = .05$ level. The sample data provide fairly strong evidence that the mean time until fatigue differs between men and women.

g. A 90% confidence interval for μ_d is

$$895 \pm 1.833 \frac{1148}{\sqrt{10}}$$

which is 895 ± 665.4, which gives the interval (229.6, 1560.4). This interval means you can be 90% confident that women take between 230 seconds and 1560 seconds longer, on average, to reach fatigue than men do (between 4–26 minutes).

h. This is not a randomized experiment, so even if the test were valid, you would not be able to attribute the longer time to fatigue for women to any particular cause. Moreover, because the subjects volunteered, you should be cautious about generalizing the results even to all healthy young adults in the area of the study. Finally, the technical conditions of the paired t-test were not met, because of the small sample size and skewed distribution of differences, so you cannot take the inference results very seriously. Researchers might want to repeat the study with large sample sizes.

Wrap-Up

This topic introduced you to the analysis of paired data. You learned how to recognize such data, by asking whether each observation in one group is linked to a specific observation in the other group. You also learned how to analyze such data, specifically by applying the one-sample t-procedures from Topics 19 and 20 to the *differences* for each pair.

Perhaps most importantly, you examined why the **matched-pairs experimental design** often leads to a more powerful test—one more likely to detect a statistically significant difference when a difference really exists—than a **completely randomized design.** Similarly, random sampling of pairs can be more informative than taking independent random samples from two populations. In the case of the marriage ages activity, you reduce variability by examining the age differences rather than the individual ages, because people tend to marry people of similar age. With the chip-melting experiment, you guarantee each type of chip experiences the same mouth conditions by giving each type of chip to each subject; that is, you account for the person-to-person variation in chip-melting times. Randomization still plays an important role, though, because it makes sense to randomly decide which type of chip a person uses first; otherwise, the order of use would be a potentially confounding variable; perhaps the second chip would tend to melt more quickly.

Some useful definitions to remember and habits to develop from this topic include

- Apply these paired ***t*-procedures** when you have **paired** data to analyze.
- Determine whether the data are paired by asking whether each observation in one group is linked to a specific observation in the other group.
- A matched-pairs design tends to reduce variability as compared to a completely randomized design, thereby producing a more powerful test, one better able to detect a difference between the groups if a difference really exists.
- Consider using a matched-pairs design when you design experiments where the treatments are randomly assigned within the pair or the order of the treatments is randomized (e.g., if each subject is his or her own control).

In the next unit, you will return to categorical variables, which can have more than two possible categories. You will learn a different kind of test procedure, the chi-square test.

Homework Activities

Activity 23-6: Cow Milking

Suppose you want to compare whether hand-milking or machine-milking tends to produce more milk from cows, and 20 cows are available for an experiment. Consider three different designs for that experiment:

- Design A: Randomly assign ten of the cows to be milked by hand and ten to be milked by machine.
- Design B: Use both milking methods for each of the 20 cows. For each cow, randomly decide which method to use on the first day of the study and then use the other method on the second day.
- Design C: Examine records of how much milk the cows have produced in the past, and order them from most to least productive. For the top two milk producers, randomly assign one to hand-milking and the other to machine-milking. Do the same for the next two and the next two and so on.

For each of these designs, indicate whether it calls for a *matched-pairs* analysis or a *completely randomized* analysis.

Activity 23-7: Car Ages

Suppose you want to investigate whether students and faculty at your school differ with regard to the average age of the cars they drive. You take a random sample of 50 students who drive to campus and a random sample of 25 faculty members who drive to campus, and then determine how old their cars are.

a. Does this study call for a matched-pairs analysis or an independent-samples analysis? Explain.
b. Would your answer to part a change if you sampled 25 students and 25 faculty members? Explain.

Activity 23-8: Freshman Fifteen

Suppose you want to study whether the campus legend about the "freshman fifteen" has any basis in fact: Do freshmen tend to gain fifteen pounds, on average, during their first term in college? Suppose you take a random sample of 50 freshmen at your college, weigh them at the beginning of the term, and reweigh them at the end of the term. Would a paired *t*-test be appropriate to perform on these data? Explain.

Activity 23-9: Sickle Cell Anemia and Child Development

Sickle cell disease is an inherited red-blood-cell disorder that primarily affects people of African descent. Some people carry the sickle cell trait without having the disease. An early sickle cell study investigated whether children who carry the sickle cell trait tend to have slower physical and cognitive development than similar children who do not carry the sickle cell trait (Kramer, Rooks, and Pearson, 1978). The researchers studied 50 children with the sickle cell trait and 50 without the trait. Their first step was to match at birth each child in the sickle cell group with a specific child in the control group, based on variables such as gender, birth weight, and socioeconomic status. Then, between the ages of three and five years, several measurements of physical and cognitive growth were recorded for each child. The researchers found no statistically significant differences between the groups on any of the response variables.

a. Is this an observational study or an experiment? Explain.
b. Is this a matched-pairs study or not? Explain.
c. Based on the conclusion of no statistically significant differences, what can you say about the *p*-values in this study? Explain.

Activity 23-10: Running to Home

Suppose you want to investigate whether a baseball or softball player can run from second base to home plate more quickly by taking a wide angle or a narrow angle around third base.

a. Describe in detail how you could implement a matched-pairs design to collect data on this question.
b. Would randomization play a useful role in a matched-pairs experiment on this issue? Explain.
c. Explain why it's reasonable to expect the matched-pairs design to reduce variability in this context. In other words, what type of variability is accounted for or at least reduced?

Activity 23-11: Friendly Observers

5-27, 21-1, **23-11**

Recall from Activities 5-27 and 21-1 the psychology experiment about whether subjects perform worse if they are watched by an observer with a vested interest in their performance (Butler and Baumeister, 1998). Describe how you could revise that experiment to employ a matched-pairs design. Also describe the advantages of redoing that experiment with a matched-pairs design.

Activity 23-12: Marriage Ages

8-17, 9-6, 16-19, 17-22, 23-1, **23-12**, 26-4, 29-17, 29-18

The sample of 24 marriage ages you analyzed in Activity 23-1 is actually a subsample from a larger sample of 100 marriages for which data are stored in the file MARRIAGEAGES100 (with the lists **HUSB** and **WIFE**). Use your calculator's **2-SampTTest** (press STAT and arrow over to highlight the **TESTS** menu) to reproduce your analysis of Activity 23-1, using all 100 couples in the larger sample.
(*TI hint:* Rather than finding the mean and standard deviation of the two lists, select **Data** for **Inpt** and insert **HUSB** as **L1** and **WIFE** as **L2**). Write a paragraph reporting on your findings about whether husbands tend to be older than their wives. Also comment on how your findings differ from those found when analyzing only 24 couples' ages.

Activity 23-13: Catnip Aggression

23-13, 23-14

A student who did volunteer work at an animal shelter wanted to see whether cats really respond aggressively to catnip (Jovan, 2000). Using a sample of 15 cats, she recorded how many "negative interactions" (such as hissing or clawing) each cat engaged in during a fifteen-minute period prior to being given a teaspoon of catnip and also during the fifteen-minute period following. Her conjecture was the number of negative interactions would increase after the introduction of the catnip. Her data follow:

Cat Name	Negative Interactions Before Catnip	Negative Interactions After Catnip
Amelia	0	0
Bathsheba	3	6
Boris	3	4
Frank	0	1
Jupiter	0	0
Lupine	4	5
Madonna	1	3
Michelangelo	2	1
Oregano	3	5
Phantom	5	7
Posh	1	0
Sawyer	0	1
Scary	3	5
Slater	0	2
Tucker	2	2

a. Explain how you know these data came from a matched-pairs design.
b. List the technical conditions required for the appropriate t-test and t-interval to be valid in this situation. Investigate and comment on whether the technical conditions are satisfied.
c. Conduct the appropriate t-test of whether these sample data provide evidence that cats' negative interactions differ before and after receiving catnip. Report the hypotheses, test statistic, and p-value. (Feel free to use your calculator.) Summarize your conclusion, including a test decision at the $\alpha = .05$ level.
d. Produce a 90% confidence interval for the population mean difference in negative interactions before and after catnip. Also interpret the interval.

Activity 23-14: Catnip Aggression

23-13, **23-14**

Reconsider the catnip data from the previous activity.

a. Even though it's not an appropriate analysis, perform a two-sample t-test (rather than the appropriate paired t-test) on the data. Report the hypotheses, test statistic, and p-value. Summarize the conclusion you would draw if this were the correct analysis.
b. How has the p-value changed from its value in the paired test?
c. Does this analysis indicate the pairing was useful in this study? Explain.

Activity 23-15: Muscle Fatigue

23-5, **23-15**, 26-22, 27-20

Recall the data from Activity 23-5, where researchers studied whether women or men experience fatigue more quickly while exercising.

a. Even though it's not an appropriate analysis, perform a two-sample t-test (rather than the appropriate paired t-test) on the data. Report the hypotheses, test statistic, and p-value. Summarize the conclusion you would draw if this were the correct analysis.
b. How has the p-value changed from its value in the paired test?
c. Does this analysis indicate the pairing was useful in this study? Explain.

Activity 23-16: Maternal Oxygenation

23-16, 26-19

During labor, mothers whose fetuses show abnormal heart rate patterns are often administered oxygen, in the hope of increasing the percentage of oxygen in the blood of the fetus. A recent study recorded this fetal oxygen saturation for 24 mothers whose fetuses showed abnormal heart patterns (Haydon et al., 2006). These measurements were taken three times: first, with room air as a baseline; after administering 40% oxygen to the mother; and finally, after administering 100% oxygen to the mother. The data (percentage of oxygen in the blood of the fetus) are given in the following table.

Mother	A	B	C	D	E	F	G	H	I	J	K	L
Baseline	56	53	38	39	35	55	53	47	55	30	53	42
40% Oxygen	63	48	44	44	35	64	56	54	52	39	59	43
100% Oxygen	57	40	44	52	50	47	58	60	55	48	52	57

(*continued*)

Mother	M	N	O	P	Q	R	S	T	U	V	W	X
Baseline	34	38	40	40	51	50	39	28	53	34	35	47
40% Oxygen	47	51	46	49	53	42	40	43	47	38	47	58
100% Oxygen	50	45	46	49	61	46	37	45	47	55	54	45

a. Consider testing whether fetus oxygen levels were significantly higher, on average, after the mother was administered 40% oxygen than at baseline. Would a paired or unpaired test be appropriate? Explain.

b. Examine and comment on whether the technical conditions for this test are satisfied. (The data are stored in the grouped file `MATERNALOXYGENATION`, containing lists **BASE**, **FRTY**, and **HNDRD**, respectively.)

c. Conduct this test (whether the technical conditions are satisfied). Report all aspects, including a test decision at the $\alpha = .10$ level, and summarize your conclusion.

d. Now consider whether fetus oxygen levels were significantly higher, on average, after the mother was administered 100% oxygen as opposed to 40% oxygen. Repeat parts a, b, and c for this question.

Activity 23-17: Mice Cooling

Medical examiners can use the temperature of a dead body at a murder scene to estimate the time of death. But can a clever murderer disguise the time of death by reheating the victim's body? A scientist actually investigated this issue on mice. Hart (1951) used 10 mice as the experimental units. He killed each mouse and then measured the cooling constant of its body. Then he reheated the mouse's body and measured its cooling constant in that reheated state. The results are shown in the following table and stored in the grouped file `COOLMICE`, containing lists **FRESH** and **HEAT**, respectively:

Mouse	A	B	C	D	E	F	G	H	I	J
Freshly Killed	573	482	377	390	535	414	438	410	418	368
Reheated	481	343	383	380	454	425	393	435	422	346

Mouse	K	L	M	N	O	P	Q	R	S
Freshly Killed	445	383	391	410	433	405	340	328	400
Reheated	443	342	378	402	400	360	373	373	412

a. Explain why these data call for a matched-pairs analysis.

b. Produce and comment on relevant graphical displays and numerical summaries for investigating the question of whether cooling constants for reheated mice are similar to those of freshly killed mice.

c. Conduct the appropriate test of whether the data suggest a significant difference in average cooling constants between freshly killed and reheated mice. Include a check of technical conditions, and summarize your conclusion.

d. Construct a 95% confidence interval for estimating the population mean difference in cooling constants. Also interpret this interval.

Activity 23-18: Clean Country Air

Is country air better to breathe than city air? One way to address this question is by measuring how quickly a person's lungs clear out unhealthy particles. Researchers in a 1973 study found seven pairs of identical twins, where one twin lived in the country and the other lived in a city (Camner and Phillipson, 1973). They asked each twin to inhale an aerosol of radioactive Teflon particles and then measured the percentage of radioactivity remaining after one hour. Results are in the following table:

Twin Pair	A	B	C	D	E	F	G
Rural Environment	10.1	51.8	33.5	32.8	69.0	38.8	54.6
Urban Environment	28.1	36.2	40.7	38.8	71.0	47.0	57.0

Analyze these data, using both descriptive and inferential techniques, to address the question of whether country air is better for a person's lungs and breathing than city air. Write a paragraph or two summarizing your findings and conclusions.

Activity 23-19: Exam Score Improvements

23-19, 23-20, 23-21, 27-6

The following data are scores on the first and second exams for a sample of students in an introductory statistics course. The data are stored in the grouped file EXAMSCORES with the lists **EXAM1**, **EXAM2**, and **ID**. *Note:* The * denotes a missing value; that student did not take the second exam. This student is not included in the **EXAM1** and **EXAM2** lists.

Student	Exam 1	Exam 2	Improvement	Student	Exam 1	Exam 2	Improvement
1	98	80		13	91	92	
2	76	71		14	83	80	
3	90	82		15	83	84	
4	95	68		16	93	96	
5	97	96		17	96	90	
6	89	93		18	98	97	
7	77	50		19	84	95	
8	94	64		20	76	67	
9	88	*	*	21	97	77	
10	95	84		22	72	56	
11	87	76		23	80	78	
12	84	69		24	74	64	

Disregard for now the student with the missing value; consider this a sample of 23 students.

a. Construct a dotplot of the *improvements* in scores from exam 1 to exam 2 (those who scored lower on exam 2 than on exam 1 have negative improvements).
b. Identify the student (by number) with the greatest improvement from exam 1 to exam 2, and also report by how many points he or she improved.
c. Which student had the largest decline from exam 1 to exam 2, and how large was that decline?
d. What proportion of these students scored higher on exam 1 than on exam 2?
e. Use your calculator to compute the sample mean and sample standard deviation of the improvements.
f. Perform the test of whether the mean improvement *differs* significantly from zero. State the null and alternative hypotheses, and record the test statistic and p-value.
g. Based on the significance test conducted in part f, would you reject the null hypothesis at the $\alpha = .10$ significance level? Are the sample data statistically significant at the .05 level? How about at the .01 level?
h. Determine and interpret a 95% confidence interval for the mean of the population of *improvements.*
i. Write a few sentences commenting on the question of whether there seems to be a significant difference in scores between the two exams. Be sure to address whether the technical conditions for these inference procedures are met.

Activity 23-20: Exam Score Improvements

23-19, **23-20,** 23-21, 27-6

Reconsider the previous activity and the data stored in the file EXAMSCORES.

a. Now treat the missing value as a score of **0** on the second exam. Recalculate the p-value of the test from part f and the confidence interval in part h. *TI hint:* To edit one or more lists, you can use the **SetUpEditor** feature located by pressing [STAT]. Select **SetUpEditor** and then enter as many lists as you need, separated by commas. Press the [ENTER] key and then press [STAT] and [1] to select **EDIT** to access the List editor. You can delete entries by pressing the [DEL] key or insert new entries by pressing [2ND] [INS] within this editor.
b. Comment on whether disregarding the missing value or treating it as a zero makes much difference in this analysis.

Activity 23-21: Exam Score Improvements

23-19, 23-20, **23-21,** 27-6

Reconsider the previous two activities.

a. Explain how you would carry out a significance test to investigate whether the proportion of test takers who improved is significantly less than one-half. *Hint:* Define the new population parameter, state the new null and alternative hypotheses, and select the appropriate test procedure.
b. Carry out the test described in part a. In addition to stating your conclusion, comment on the validity of the technical conditions.

Activity 23-22: Comparison Shopping

23-22, 23-23, 26-21

The following price data were collected by Cal Poly students over a two-day period at two grocery stores, Luckys and Vons, in the same town (the data are also stored in the file SHOPPING, containing the lists LUCKY and VON).

Product	Luckys (in $)	Vons (in $)	Product	Luckys (in $)	Vons (in $)
Baker's angel flake coconut	1.60	1.50	Hershey's semi-sweet choc. chips (12 oz)	2.19	2.68
C&H granulated sugar (10 lb)	4.75	3.99	Iceberg lettuce	0.89	.99
Campbell's Tomato Soup	0.75	0.60	Jiffy Crunch Peanut Butter (18 oz)	2.49	2.42
Carnation evaporated milk (5 oz)	0.48	0.49	Kikkoman soy sauce	1.29	1.49
Carrots (2 lb bag)	1.86	2.09	Kraft Minute Rice (28 oz)	3.49	2.98
Crest regular toothpaste (6.4 oz)	1.99	2.85	Kraft Parmesan Cheese	3.99	2.29
Dannon yogurt, strawberry	0.85	0.84	Lender's raisin bagels (6 pack, fresh)	1.99	3.86
Ding Dongs (12 pack)	3.79	3.79	Milk (1/2 gallon)	2.07	2.15
Dr. Pepper (2 liter)	0.79	0.59	Minute Maid frozen OJ (8.54 oz)	1.89	1.99
Dreyer's Vanilla Frozen Yogurt	4.99	4.99	Navel oranges (price/oz)	6.18	4.36
Excedrin (40 tablets)	6.99	6.99	Northern bath tissue (12 rolls)	4.01	3.99
Franco-American Spaghetti O's	1.29	1.39	Pepperidge Farm Brussels Cookies	2.00	1.98
Gladlock Zipper Sandwich bags	3.29	3.29	PopTarts, strawberry (8-pack)	1.67	2.03
Gold Medal flour (5 lb, regular)	1.62	1.69	Special K	5.29	5.39
Green beans, cut (8 oz)	0.49	0.53	Tide (50 oz)	4.69	4.75
Ground round (22% fat, price/lb)	3.29	3.29	Wishbone Italian dressing (16 fl oz)	2.99	2.89
Heinz ketchup (28 oz)	2.49	2.18	Wonder bread, wheat	2.49	2.99

a. Explain how you know that these data are from a matched-pairs design. Also discuss the advantages of pairing the data in this manner.

b. Calculate the differences in the prices of these items between the two stores.

c. Examine visual displays and numerical summaries of these price differences. Write a few sentences commenting on the question of whether one store seems to have lower prices than the other store.

d. Conduct the appropriate test of significance to assess whether the sample data provide significant evidence that the prices tend to differ between these two stores. Report your hypotheses, sampling distribution specified by the null hypothesis, test statistic, and p-value. Also look into and comment on whether the technical conditions seem to be satisfied. Write a few sentences summarizing your conclusion.

e. Determine a 95% confidence interval for the mean price difference between the two stores. What does this interval tell you about how much you can expect to save?

f. If you had subtracted in the opposite order to calculate the differences in part b, what would have changed about the test and interval? Describe specifically how the values would have changed.

g. Remove the most extreme outlier and repeat this analysis. Comment on the degree to which your conclusions change, depending on whether you include the suspicious value. Also, can you hypothesize a cause for the unusual behavior of the outlier?

Activity 23-23: Comparison Shopping

23-22, **23-23**, 26-21

Reconsider the previous activity and the data stored in the file SHOPPING.

a. Suppose for each of the grocery items you indicate whether it cost more at Luckys, at Vons, or the same at both stores. Would this be a categorical or a quantitative variable? Explain.

b. Determine how many of these items cost more at Luckys, how many cost more at Vons, and how many cost the same at both stores.

c. Ignoring the items that cost the same at both stores, how many items remain? What proportion of them cost more at Luckys?

d. Consider testing the null hypothesis that half of all items cost more at Luckys vs. the alternative hypothesis that the proportion of all items that cost more at Luckys is not one-half. Clearly identify the population and parameter of interest in this test.

e. Carry out the test described in part d. Write a paragraph describing your conclusion, including details of your calculations and a check of the technical conditions.

f. Determine a 96% confidence interval for the proportion of all items that cost more at Luckys.

UNIT

6

Inferences with Categorical Data

TOPIC

24 • • • Goodness-of-Fit Tests

How can statistics be useful in catching people who try to cheat on their taxes by making up the numbers they report? Well, it turns out the leading (leftmost) digits of many numbers that occur in the world follow a predictable pattern, so auditors can test for whether the numbers on a tax return follow that pattern or not. To do this yourself, you need a new test procedure, one that allows you to assess whether sample data conform to a hypothesized model. That's what you will learn in this topic.

Overview

In this topic, you return to studying *categorical* variables. For the first time, you will learn an inference technique that applies to a categorical variable with *more than two* categories. One of the most famous and widely used procedures in all of statistics is the chi-square goodness-of-fit test. This procedure assesses how closely sample results conform to a hypothesized model about the proportional breakdown of the various categories.

Preliminaries

1. What day of the week were you born on? *Hint:* If you do not know, you might consult an online yearly calendar such as the one at `www.timeanddate.com`.

2. Do you suspect that any one day of the week is more or less likely to be a person's birthday than any other day of the week?

3. Guess which digit (1–9) is *most* likely to be the leading (leftmost) digit for numbers appearing in an almanac.

4. Guess which digit (1–9) is *least* likely to be the leading (leftmost) digit for numbers appearing in an almanac.

In-Class Activities

Activity 24-1: Birthdays of the Week

24-1, 24-2, 24-3

Are people equally likely to be born on any of the seven days of the week? Or are some days more likely to be a person's birthday than other days? To investigate this question, days of birth were recorded for the 147 "noted writers of the present" listed in *The World Almanac and Book of Facts 2000.* The counts for the seven days of the week are given here:

Mon	Tues	Wed	Thu	Fri	Sat	Sun
17	26	22	23	19	15	25

a. Identify the observational units and variable here. Is the variable categorical or quantitative? If it is categorical, is it also binary?

Observational units:

Variable: Type:

b. Construct a bar graph of these data. Comment on what it reveals about whether the seven days of the week are equally likely to be a person's birthday.

c. Let π_M represent the proportion of *all* people who were born on a Monday, or equivalently, the probability that a randomly selected person was born on a Monday. Similarly define π_{Tu}, π_W, . . . , π_{Su}. Are these parameters or statistics? Explain.

d. The null hypothesis says that the seven days of the week are *equally likely* to be a person's birthday. In that case, what are the values of π_M, π_{Tu}, . . . , π_{Su}?

The test procedure you are about to learn is called a **chi-square goodness-of-fit test.** It applies to a categorical variable, which need not be binary. The null hypothesis asserts specific values for the population proportion in each category. The alternative hypothesis simply says that at least one of the population proportions is not as specified in the null hypothesis.

As always, the test statistic measures how far the *observed* sample results deviate from what is expected if the null hypothesis is correct. With a chi-square test, you construct the test statistic by comparing the observed sample counts in each category to the *expected counts* under the null hypothesis.

e. Intuitively, what value would make sense for the expected count of Monday birthdays in this study (with a sample size of 147), under the null hypothesis that one-seventh of all birthdays occur on Mondays? Explain.

You calculate the **expected count** for a particular category by multiplying the sample size by the hypothesized population proportion for that category:

$$E_i = n \times \pi_{i0}$$

Note: The subscript 0 on π_{i0} emphasizes that this is the value conjectured for π_i in the null hypothesis.

f. Calculate the expected counts for each of the seven days. Record them in the middle row of the following table. *Hint:* You should not need to do seven different multiplications in this case.

	Mon	Tues	Wed	Thu	Fri	Sat	Sun	Total
Observed Count (O)	17	26	22	23	19	15	25	147
Expected Count (E)								147
$\frac{(O-E)^2}{E}$								

Now the question is how to construct a test statistic to measure how far the observed counts fall from the expected counts. If you simply subtract the expected counts from the observed counts, some will be positive and some negative, and the effects of their deviations will cancel each other out rather than accumulate. Instead, you will *square* the differences between the observed and expected counts. The problem with squaring is a squared difference of four is large if you're talking about small numbers (in the single digits or teens), but rather small if you're talking about big numbers (in the hundreds or thousands). So, you will divide this squared difference by the expected count to "standardize" the squared deviation. Then you add those values together across all the categories to calculate the test statistic. In symbols, the test statistic is

$$X^2 = \sum \frac{(O_i - E_i)^2}{E_i}$$

where the symbol Σ says to sum the standardized squared deviations over all categories.

g. For each of the seven days of the week, calculate $\frac{(O-E)^2}{E}$. Record these values in the bottom row of the preceding table.

Monday: Tuesday: Wednesday:

Thursday: Friday: Saturday:

Sunday:

h. Now add those seven $\frac{(O-E)^2}{E}$ values to calculate the test statistic.

i. What kind of values (e.g., large or small) of the test statistic constitute evidence against the null hypothesis that the seven days of the week are equally likely to be a person's birthday? Do you think the value calculated in part h provides convincing evidence? If you are not sure, what additional information do you need? Explain your reasoning.

As always, your next step is to calculate the *p*-value. The *p*-value tells you the probability of getting sample data at least as far from the hypothesized proportions as these data are, by random chance if the null hypothesis is true. So again, a small *p*-value indicates the sample data is unlikely to have occurred by chance alone if the null hypothesis is true, providing evidence in favor of the alternative.

In this case, you do not use the normal or *t*-distribution to calculate the *p*-value. When the null hypothesis is true, the test statistic follows a **chi-square distribution.** (Activity 24-12 asks you to use simulation to approximate the sampling distribution of the chi-square test statistic.) The degrees of freedom is equal to the number of categories minus one. Critical values for the chi-square distribution have been tabulated in Table IV (Chi-Square Distribution Critical Values) in the back of the book.

You read the chi-square table in the same manner as you read the *t*-table. Because large values of the test statistic provide evidence against the null hypothesis, the *p*-value for the chi-square test is the probability of exceeding the value of the test statistic. When the null hypothesis is true, the test statistic averages out to equal the degrees of freedom, so a test statistic greater than the degrees of freedom provides at least a little evidence against the null hypothesis. When the test statistic is large enough to produce a small *p*-value, then the sample data provide strong evidence against the null hypothesis. A sketch of a chi-square distribution with 6 degrees of freedom follows:

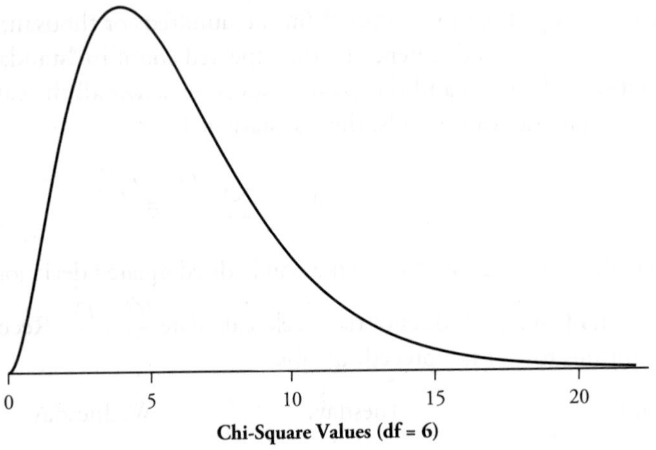

j. Calculate the p-value for this test as accurately as possible, using Table IV. Also shade the region whose area corresponds to this p-value on the preceding sketch.

k. Write a sentence interpreting what this p-value means in the context of this study about birthdays of the week.

l. Based on this p-value, what test decision (reject or fail to reject H_0) would you reach at the $\alpha = .10$ level? How about at the $\alpha = .05$ and $\alpha = .01$ levels?

$\alpha = .10$ level: $\alpha = .05$ level: $\alpha = .01$ level:

As with all inference procedures, certain technical conditions must be satisfied for this chi-square procedure to provide accurate p-values. In addition to requiring a random sample from the population of interest, all *expected* counts need to be at least five. When this condition is not met, one option is to combine similar categories together to force all expected counts to be at least five.

m. Is the expected value condition satisfied for this birthday study? What about the random sampling condition? If not, would you be comfortable in generalizing the results to a larger population anyway? Explain.

n. Summarize your conclusion about whether these sample data provide evidence against the null hypothesis that any of the seven days of the week are equally likely to be a person's birthday.

Watch Out

- Be aware the subscripts here mean something different than they did when comparing a binary response between two groups. When we compared π_{nicotine} to π_{placebo}, we meant the proportion of successes *in the nicotine lozenge group* and the proportion of successes *in the placebo group.* In that case, the subscript indicated the group of interest. But with $\pi_M, \pi_{Tu}, \ldots, \pi_{Su}$, the subscripts indicate the different categories of the variable of interest, and the symbols denote the population proportion who fall into each of those categories. These π values must sum to 1 when added across all possible categories.
- The alternative hypothesis simply says that *at least one* of the population proportions is not as specified in the null hypothesis. (Because the proportions must sum to one across all categories, this also means at least *two* population proportions are not as specified.) It does *not* say, however, that *all* of the population proportions have been misspecified.
- In fact, there are no one-sided (directional) alternative hypotheses with a chi-square test, because the alternative hypothesis includes all the different ways for the null hypothesis to be false. However, in calculating the p-value, you only want to know whether the calculated test statistic is too large to have plausibly

24

happened by chance alone, so always look at the area to the right of the calculated test statistic.

- Be sure to look at the *expected* counts, not the *observed* counts, for checking the technical condition.
- Now, it is even more important to be careful about rounding errors in intermediate calculations.
- It's especially important never to *accept* the null hypothesis. Even if the test leads to a large p-value, all you can say is that the sample data are not inconsistent with the hypothesized model. In this case, the sample data are consistent with the hypothesis that the seven birthdays are equally likely. But the sample data would also be consistent with many other hypotheses about the seven birthdays. A careful conclusion says that you fail to reject H_0—that the sample data do not provide any reason to doubt the seven birthdays are equally likely. But do not go so far as to say you firmly believe the seven birthdays to be equally likely.
- There is no confidence interval counterpart to this goodness-of-fit test, although you can always find a one-sample confidence interval to estimate the population proportion in one particular category.

Chi-square test of goodness-of-fit:
H_0: The population proportions are as hypothesized for all categories:

$$(\pi_1 = \pi_{10}, \pi_2 = \pi_{20}, \ldots, \pi_m = \pi_{m0})$$

H_a: At least one population proportion differs from the others:

$$(\text{At least one } \pi_i \neq \pi_{i0})$$

Test statistic:

$$X^2 = \sum \frac{(O_i - E_i)^2}{E_i}$$

where $E_i = n \times \pi_{k0}$ (*sample size* times *hypothesized proportion for that category*).
p-value $= \Pr(\chi_k^2 \geq X^2)$, where χ_k^2 represents a chi-square distribution with degrees of freedom $k = m - 1$ (*number of categories* minus 1).
The **technical conditions** necessary for this test to be valid are

- The observations arise from a random sample of the population.
- The expected counts are at least five for each category.

Activity 24-2: Birthdays of the Week

24-1, **24-2,** 24-3

Reconsider the previous activity. Now let's use the same sample data to test a different hypothesis about birthdays. Suppose you suspect that weekend days are less likely to be birthdays, perhaps because doctors want the weekend off and so do not schedule Caesarean deliveries for weekends. Let's test whether the data provide evidence against the hypothesis that weekend days are half as likely as other days to be someone's birthday and that all weekdays are equally likely. The probabilities still need to add up to one, so you are now testing H_0: $\pi_M = 1/6$, $\pi_{Tu} = 1/6$, $\pi_W = 1/6$, $\pi_{Th} = 1/6$, $\pi_F = 1/6$, $\pi_{Sa} = 1/12$, $\pi_{Su} = 1/12$.

a. Calculate the expected counts under this null hypothesis. Record them in the middle row of the following table. *Hint:* These expected counts will not be the same for every day this time. Round them to three decimal places, *not* to integers.

	Mon	Tues	Wed	Thu	Fri	Sat	Sun	Total
Observed Count	17	26	22	23	19	15	25	147
Expected Count								147
$\frac{(O-E)^2}{E}$								

b. Record each category's contribution to the test statistic in the bottom row of the table. Then add those contributions to calculate the value of the test statistic.

c. Use Table IV to determine the p-value of this test as accurately as possible.

d. What test decision would you make at the $\alpha = .01$ level?

e. Which category (day of the week) has the greatest contribution to the test statistic? (In other words, which day has the greatest $\frac{(O-E)^2}{E}$ value?) Is the observed count higher or lower than the expected count for that day? Explain what this result reveals.

f. Summarize the conclusion you would draw about the proposed hypothesis, also explaining the reasoning process behind your conclusion.

These last two questions reveal that when a chi-square test gives a significant result, you can identify the category that differs most substantially from the hypothesized model by finding the greatest value of $\frac{(O-E)^2}{E}$.

Watch Out

- Although a common null hypothesis is that all categories are equally likely, other null hypotheses (i.e., specifying different values for the category proportions) can also be tested (as long as the hypothesized category proportions sum to 1).
- Expected counts need not be integers. Do not round them to integers. As you learned in Topic 11 when you first studied probabilities and expected values, expected counts are often not literally "expected" at all.

24

Activity 24-3: Birthdays of the Week

24-1, 24-2, **24-3**

Reconsider Activity 24-1, with the original hypothesis that any of the seven days of the week are *equally likely* to be a person's birthday. But now suppose the sample size is ten times larger and the sample results turn out proportionally identical to those in the actual study. In other words, suppose the observed counts turn out to be those given here:

	Mon	Tues	Wed	Thu	Fri	Sat	Sun	Total
Observed Count	170	260	220	230	190	150	250	1470

a. Before doing any calculations, do you expect the *p*-value for this test to be less than before (in Activity 24-1), greater than before, or the same as before? Explain your reasoning.

b. How do the sample proportions for each day compare to those with the actual data? How would the bar graph compare? *Hint:* You need not do any calculations; simply think about how the data have been altered.

c. Use the following information to calculate the test statistic for testing the "equally likely" hypothesis. *Hint:* Do not waste time recalculating any values already provided.

	Mon	Tues	Wed	Thu	Fri	Sat	Sun	Total
Observed Count	170	260	220	230	190	150	250	1470
Expected Count	210	210	210	210	210	210	210	1470
$\frac{(O-E)^2}{E}$	7.619	11.905	0.476	1.905	1.905	17.143	7.619	

Test statistic:

d. Determine the *p*-value of the test, as accurately as you can. Also summarize the conclusion you would draw from this test.

e. How do this *p*-value and conclusion compare to those from Activity 24-1? Explain why this makes sense.

This activity should remind you that *sample size* plays a crucial role in significance tests. The larger a sample size, the more statistically significant the result (if all else remains proportionally the same). But the *practical* significance of the result might be questionable with a very large sample size.

Activity 24-4: Kissing Couples

13-3, 13-4, 16-6, 17-12, **24-4,** 24-13

Recall from Activities 13-3, 16-6, and 17-2 the study in which researchers observed a sample of 124 kissing couples and found that 80 of them leaned their heads to the right (Güntürkün, 2003). Suppose the researchers want to test the hypothesis that 3/4 of the population lean to the right and 1/4 to the left.

a. Identify the observational units and variable in this study. Is the variable categorical or quantitative? If it is categorical, is it also binary?

Observational units:

Variable: Type:

b. Does the one-sample z-test from Topic 17 apply to these data? (Do not worry about the random sampling issue for now.)

Even though it's not necessary to apply a chi-square goodness-of-fit test to a *binary* categorical variable (because the one-sample z-test works fine), you can apply it.

c. State the null and alternative hypotheses for testing the hypothesis that 3/4 of the population lean to the right and 1/4 to the left when kissing.

d. Compute the expected counts (for both the "lean right" and "lean left" categories) for testing this hypothesis. Then calculate the test statistic and determine the p-value as accurately as possible with your calculator. To calculate the p-value, you can use your calculator's χ^2 **GOF-Test,** which can be found by pressing STAT and arrowing over to the TESTS menu (if you do not have this feature, you will need to download the program CHIGOF). Before using the χ^2 **GOF-Test,** you need to create two lists, one with the observed values and one with the expected values. Also report the degrees of freedom.

e. Comment on whether the technical conditions for this test are satisfied.

f. Summarize the conclusion you would draw from this test.

In testing H_0: $\pi_{right} = .75$ vs. H_a: $\pi_{right} \neq .75$, the z-test statistic turns out to equal

$$z = \frac{.645 - .75}{\sqrt{\frac{.75(1 - .75)}{124}}} \approx -2.70$$

and the two-sided p-value is .0069.

g. Comment on how the value of the z-test statistic compares to the value of the chi-square test statistic. *Hint:* Try to find a mathematical relationship between the two test statistics.

h. Your calculator computes the p-value for the chi-square test to be .007. How do the p-values compare between the two tests?

You should have found that these two tests produce identical p-values and conclusions. The relationship between the test statistics is that the chi-square test statistic is the square of the one-sample z-test statistic: $X^2 = z^2$.

Watch Out
This equivalence only holds when the one-sample z-test has a two-sided alternative. If the z-test has a one-sided alternative, then its p-value will be half as large as the chi-square p-value (as long as the sample result is in the direction conjectured by the alternative).

Activity 24-5: Leading Digits

24-5, 24-9, 24-19

The leading digit of a number is simply its first (leftmost) digit. A mathematician named Benford conjectured that with many real datasets, 1 is the most common leading digit, followed by 2, 3, and so on. In fact, Benford developed a formula that he used to predict the proportion of data values that have a certain leading digit. His formula says the proportion of data values having a leading digit of i is $\log_{10}\left(1 + \frac{1}{i}\right)$:

Leading Digit	1	2	3	4	5	6	7	8	9	Total
Benford's Probability	.301	.176	.125	.097	.079	.067	.058	.051	.046	1.000

Let's investigate how well the Benford model fits a sample of real data from *The World Almanac and Book of Facts 2006*. You will analyze the leading digits of the populations of the 194 countries in the world. For example, China is the most populous nation with 1,306,313,812 people, so its leading digit is 1. Kazakhstan (made famous by the *Borat* movie) has a much smaller population of 15,185,844 people, but its leading digit is also 1. The following table shows how many countries have each leading digit:

Leading Digit	1	2	3	4	5	6	7	8	9	Total
Number of Countries	55	36	22	23	15	14	12	12	5	194

a. Display these data with a bar graph, using sample proportions (rather than observed counts) for the vertical axis. Comment on whether the bar graph appears to be consistent with the probabilities from Benford's model.

b. State the null and alternative hypotheses for testing whether Benford's probabilities accurately describe the population proportions in the almanac.

c. Calculate the expected counts, and comment on whether the technical conditions for the chi-square test are satisfied.

d. Calculate the test statistic and *p*-value.

e. What test decision would you make at any reasonable significance level?

f. Summarize your conclusion and explain the reasoning process that supports it.

g. Explain how this activity suggests a way to check tax returns for fraud.

Solution

a. A bar graph follows:

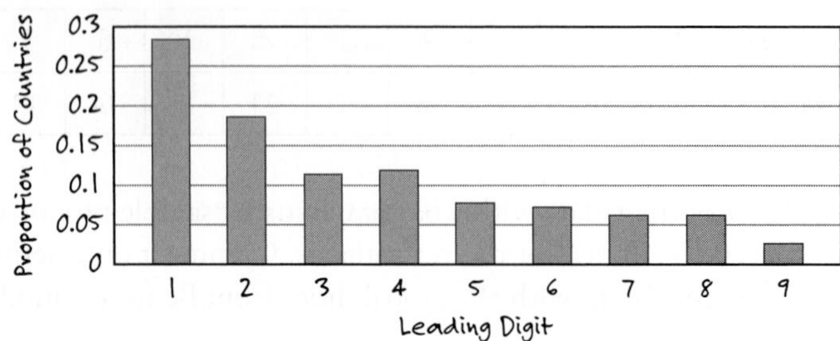

This graph does appear to be consistent with the Benford probabilities, insofar as the most common leading digit in this sample is 1 by a wide margin, followed by 2, and then gradually tapering off from there. *Note:* Even though the leading digits are numbers, you can consider this variable to be categorical, because the number simply separates the countries into groups. That's why a bar graph is appropriate rather than a histogram, and also why a chi-square goodness-of-fit test is appropriate.

b. The null hypothesis is H_0: $\pi_1 = .301$, $\pi_2 = .176$, $\pi_3 = .125$, $\pi_4 = .097$, $\pi_5 = .079$, $\pi_6 = 0.067$, $\pi_7 = .058$, $\pi_8 = .051$, and $\pi_9 = .046$. In other words, the null hypothesis says the Benford probability model is correct for describing country populations. The alternative hypothesis simply says this model is not correct, meaning at least one of the hypothesized Benford probabilities is wrong. In symbols, you can write H_a: at least one π_i differs from its hypothesized value.

c. The expected count in the "1" category is 194(.301) = 58.394 countries. This is fairly close to the observed count of 55 countries with a leading digit of 1. Calculating the expected counts for the other categories in the same way produces these results:

Leading Digit	Observed Count	Expected Count
1	55	58.394
2	36	34.144
3	22	24.250
4	23	18.818
5	15	15.326
6	14	12.998
7	12	11.252
8	12	9.894
9	5	8.924
Total	194	194.000

All of these expected counts are greater than five, so that technical condition is satisfied. If you consider the population to be all numbers appearing in the almanac, you do not literally have a random sample of numbers from the almanac, but these are likely to be representative of population values.

d. The contribution to the test statistic from the "1" category is

$$\frac{(55 - 58.394)^2}{58.394} \approx 0.197$$

Calculating the other contributions in the same way produces these results:

Leading Digit	Observed Count	Expected Count	$\frac{(O - E)^2}{E}$
1	55	58.394	0.197
2	36	34.144	0.100
3	22	24.250	0.209
4	23	18.818	0.929
5	15	15.326	0.007
6	14	12.998	0.077
7	12	11.252	0.050
8	12	9.894	0.448
9	5	8.924	1.725
Total	194	194.000	3.742

The test statistic equals 3.742, the sum of the nine values in the last column. (You might get a slightly different, more accurate answer if you carry more than three decimal places in your intermediate calculations.) Looking at Table IV with 8 degrees of freedom (one less than the nine categories), this test statistic is way off the chart to the left. Therefore, the p-value is much greater than .2. Your calculator computes the p-value to be about .88, as seen in this graph (created on a computer for better resolution):

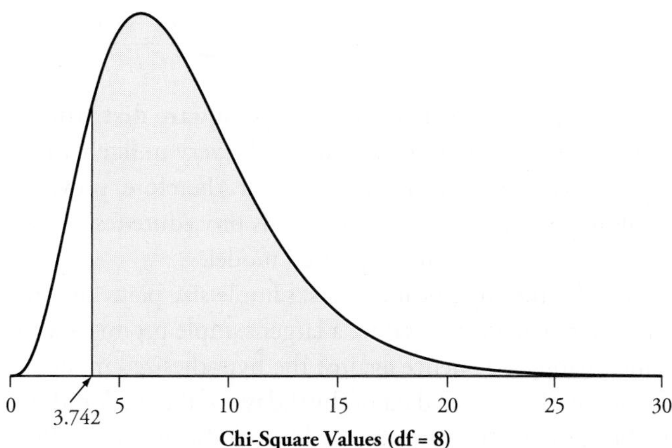

e. This is a very large p-value, much greater than all common significance levels, so you fail to reject the null hypothesis that Benford's probabilities adequately model these data.

f. The large p-value (based on the small value of the chi-square test statistic) from this chi-square analysis reveals the sample data are extremely consistent with the

Benford probabilities. If Benford's model were correct, it would not be the least bit surprising to observe sample data as found with these nations' populations. In other words, the sample data provide no reason to doubt that Benford's model describes the distribution of leading digits in the almanac.

g. An accountant or IRS agent could apply a chi-square test to see how well the numbers in a tax return follow Benford's probabilities. If the p-value turns out to be very small, that suggests the leading digits of numbers on that tax return differ substantially from what Benford's probabilities predict. This certainly does not prove the tax return is fraudulent, but it might suggest the numbers were made up and not legitimate. A tax return whose leading digits do not follow Benford's probabilities might be worth a closer look.

Watch Out

- Always remember this chi-square goodness-of-fit test applies only to a *categorical* variable. In this case, you are treating the *leading digit* variable as categorical, even though it involves a number, because you are only interested in the numbers as categories, not as numbers to add or average. But do not attempt to perform this test on a quantitative variable, such as sleep time, where you are interested in estimating the average value of the variable.
- The exception to the previous point is that you can always convert a quantitative variable into a categorical one, for example by classifying sleep time into several categories. You might ask respondents to classify less than six hours of sleep as "little," six to eight hours as "moderate," and more than eight hours as "a lot."

Wrap-Up

This topic has introduced you to the **chi-square goodness-of-fit test.** This procedure applies to data on a single categorical variable, especially a variable with more than two categories. The basic idea of this test procedure is to compare the *observed* counts in the various categories with the *expected* counts, under the hypothesis that the population proportions/probabilities being tested are correct. These are compared through the **chi-square test statistic**

$$X^2 = \sum \frac{(O-E)^2}{E}$$

and the p-value is calculated from the **chi-square distribution.** As always, a small p-value indicates the sample data would be very unlikely to occur by chance alone if the hypothesized proportions were correct, therefore, providing strong evidence against that null hypothesis. In other words, this procedure tests how well the observed counts conform with (i.e., *fit*) a hypothesized model.

As with other significance tests, sample size plays an important role in a chi-square test. If all else remains the same, a larger sample produces a smaller p-value and, therefore, stronger evidence against the hypothesized model than a smaller sample does. For example, the sample data on birthdays of the week did not provide much reason to doubt the seven days are equally likely to be birthdays. But with a larger sample size, the same observed sample proportions would have differed enough to reject the hypothesis of equal likeliness.

You also reanalyzed the data from the kissing study to learn that in the case of a *binary* categorical variable, this chi-square goodness-of-fit test is equivalent to a two-sided one-sample z-test. The p-values are identical, and the chi-square test statistic equals the square of the z-test statistic.

Some useful definitions to remember and habits to develop from this topic include

- Apply this chi-square goodness-of-fit test when you have a *single* categorical variable with two or more categories. In the case of two categories, you may use the one-sample *z*-test instead.
- It is always a good idea to start with graphical and numerical summaries before proceeding to apply an inference procedure. In this case, a bar graph is the appropriate graphical display.
- When you reject the hypothesized model, identify the category with the largest value of $\frac{(O-E)^2}{E}$ as the category with the greatest discrepancy between observed and expected counts.
- The chi-square goodness-of-fit test is one of the few significance test procedures that does not have an analogous confidence interval procedure. You can always find a confidence interval for the population proportion in one particular category, however.
- Always remember to check technical conditions. For this test, check that the expected counts exceed five in all categories. Also ask about whether the sample was randomly selected before generalizing to a larger population.
- Also, remember to never *accept* a null hypothesis, even if the sample data are shown to be quite consistent with that hypothesis.

In the next topic, you will learn a different chi-square test, one that applies to a *pair* of categorical variables. In other words, this test can be used to analyze data presented in two-way tables. You will find the strategy of comparing observed counts to expected counts is the same as with a goodness-of-fit test, and the calculation of the chi-square test statistic is also the same. But the hypotheses to be tested will be different, so the method for calculating expected counts will be different as well.

Homework Activities

Activity 24-6: Mendel's Peas

24-6, 24-7

Chi-square tests have been useful in genetics for comparing how well observed data conform to the predictions of genetic theories. In one of Gregor Mendel's early studies (1865), he crossed two hybrid pea plants and recorded the form (round or wrinkled) and albumen color (yellow or green) of the fertilized seeds. His theory predicted these should appear in the following proportions:

Round, Yellow	Wrinkled, Yellow	Round, Green	Wrinkled, Green
9/16	3/16	3/16	1/16

The observed counts in the 556 fertilized seeds (Mendel, 1865) follow:

Round, Yellow	Wrinkled, Yellow	Round, Green	Wrinkled, Green
315	101	108	32

Conduct a chi-square test to assess how well the observed counts fit Mendel's predictions. Report all aspects of the test. Also write a brief report summarizing your conclusion and explaining the reasoning process.

Activity 24-7: Mendel's Peas

24-6, **24-7**

Reconsider the previous activity. The following analyses and interpretations of those data all contain an error. Explain what is wrong in each case. *Hint:* No cases contain an arithmetic error, so do not bother to check the arithmetic.

a. The expected counts are all 556/4 = 139. The test statistic is

$$\frac{(315-139)^2}{139}+\frac{(101-139)^2}{139}+\frac{(108-139)^2}{139}+\frac{(32-139)^2}{139}\approx 322.52$$

b. The expected counts are 312.75, 104.25, 104.25, and 34.75. The test statistic is

$$\frac{(315-312.75)^2}{312.75}+\frac{(101-104.25)^2}{104.25}+\frac{(108-104.25)^2}{104.25}+\frac{(32-34.75)^2}{34.75}\approx 0.47$$

The p-value is much greater than .2, so you reject the null hypothesis that Mendel's theory is correct.

c. The expected counts are 312.75, 104.25, 104.25, and 34.75. The test statistic is

$$\frac{(315-312.75)^2}{312.75}+\frac{(101-104.25)^2}{104.25}+\frac{(108-104.25)^2}{104.25}+\frac{(32-34.75)^2}{34.75}\approx 0.47$$

Because this is greater than .05, you fail to reject the null hypothesis that Mendel's theory is correct.

d. The expected counts are 312.75, 104.25, 104.25, and 34.75. The test statistic is

$$\frac{(315-312.75)^2}{315}+\frac{(101-104.25)^2}{101}+\frac{(108-104.25)^2}{108}+\frac{(32-34.75)^2}{32}\approx 0.49$$

Note: The error in this case has only a small effect on the value of the test statistic.

e. The expected counts are 313, 104, 104, 35. The test statistic is

$$\frac{(315-313)^2}{313}+\frac{(101-104)^2}{104}+\frac{(108-104)^2}{104}+\frac{(32-35)^2}{35}\approx 0.51$$

Note: Again, the error in this case has only a small effect on the value of the test statistic.

Activity 24-8: Poohsticks

Medical researchers in England conjectured that transient ischemic attacks (strokes lasting only a few minutes) might result from many emboli released into the internal carotid artery reaching the same destination in the blood system (Knight, 2004). To investigate this theory, they conducted a study based on a game played by Winnie-the-Pooh and friends! They dropped sticks (actually, they used fir cones) from a bridge into a stream and then watched the cones' progress downstream to see where they eventually came to rest. They selected six possible resting points (which they labeled A through F) and recorded how many cones came to rest at each point (Knight, 2004). Their results follow:

Resting Point	A	B	C	D	E	F	Total
Number of Cones Coming to Rest	9	0	5	3	31	23	71

Conduct a chi-square test of whether the data provide strong evidence that these six resting points are not equally likely. Report the hypotheses, test statistic, and p-value. Also check the technical conditions, and summarize your conclusion.

Activity 24-9: Leading Digits

24-5, **24-9**, 24-19

Reconsider Activity 24-5. Consider testing how well the Benford probabilities model the leading digits of the populations of the 58 counties in California, as reported in *The World Almanac and Book of Facts 2006*. The counts for these leading digits follow:

Leading Digit	1	2	3	4	5	6	7	8	9	Total
Number of Counties	19	11	4	8	2	4	3	4	3	58

a. Display these data with a bar graph, and describe the distribution of leading digits.
b. Calculate the expected counts.
c. Are all the expected counts equal to five or more?
d. In order to make all the expected counts equal five or more, combine the "5" and "6" categories, and then combine the "7," "8," and "9" categories. Report the observed counts for this new table:

Leading Digit	1	2	3	4	5 or 6	7, 8, or 9	Total
Number of Counties							58

e. Apply a chi-square goodness-of-fit test of the Benford probabilities to this new table. Report all aspects of the test, including your test decision at the $\alpha = .10$ level. Also summarize your conclusion in context. *Hint:* Be sure to use a degrees of freedom appropriate for the new table in part d.

Activity 24-10: Political Viewpoints

17-13, **24-10**

The 2004 General Social Survey asked a random sample of adult Americans to classify their political viewpoint as liberal, moderate, or conservative. The results are summarized in the table:

Political Viewpoint	Liberal	Moderate	Conservative	Total
Number of Respondents	319	497	493	1309

a. Calculate the proportion of respondents in each of the three categories.
b. Are these values (your answers to part a) parameters or statistics? Explain.
c. Display these data with a bar graph. Does it appear the population might be equally likely to classify themselves into these three political categories? Explain.
d. Conduct a chi-square goodness-of-fit test to investigate whether the population is equally likely to classify themselves into these three political categories. Report all aspects of the test, and summarize your conclusions.

Activity 24-11: Wayward Golf Balls

A naïve teacher of statistics bought a house on a golf course, discovering too late that the property was situated in a prime spot for poor golfers to lose their golf balls. As golf balls are marked with identifying numbers, typically the digits 1–4, he wondered whether they were equally likely to be any of the digits 1–4. He tallied the identification numbers on the first 500 golf balls that he collected as follows:

Identification Number	1	2	3	4	Other	Total
Number of Golf Balls	137	138	107	104	14	500

a. Create a bar graph to display these results. Describe the distribution of ball identification numbers.

Now discard the 14 golf balls in the "other" category from the rest of your analysis, so the sample size becomes 486 rather than 500.

b. Calculate the expected counts under the "equally likely" hypothesis for a sample of 486 golf balls.
c. Calculate the test statistic.
d. Determine the p-value as accurately as you can.
e. Would you reject the "equally likely" hypothesis at the $\alpha = .10$ level? What about at the .05 and .01 levels?
f. Suppose there had been 10 times as many golf balls in his yard, with the same proportional breakdown of identification numbers. How would you expect the test statistic and p-value to change in this case? How would your conclusion change? Explain why your answer makes sense.
g. Make the change in part f, and recalculate the test statistic and p-value. Do you need to rethink your answer to part f?

Activity 24-12: Halloween Treats

In an effort to curb the obesity problem among America's youth, researchers investigated whether trick-or-treaters might be equally likely to choose a toy as opposed to candy when presented with a choice between the two (Schwartz, Chen, and Brownell, 2003). Seven households in five suburban Connecticut neighborhoods offered two bowls to trick-or-treaters aged 3 to 14. One bowl contained lollipops or fruit candy, and the other contained small toys such as plastic bugs that glow in the dark. Of the 283 children who were simultaneously offered the two bowls, 148 chose candy and 135 chose toys.

a. Test the "equally likely" hypothesis with a chi-square test. Report all aspects of the test, including a check of technical conditions. Summarize the conclusion you would draw, in this context, at the $\alpha = .20$ significance level.

b. Use these sample data to produce an 80% confidence interval for the population proportion of trick-or-treaters who would choose the toy in such a situation. Also interpret this interval.
c. Does this interval include the value .5? Explain how this is consistent with your test decision in part a.
d. To what population would you feel comfortable generalizing the results of this study? Explain.

Activity 24-13: Kissing Couples

13-3, 13-4, 16-6, 17-12, 24-4, **24-13**

Reconsider the kissing study, most recently analyzed in Activity 24-4.

a. Conduct a chi-square goodness-of-fit test of whether the population proportions of kissing couples who lean to the right and left are both 1/2.
b. Conduct a chi-square goodness-of-fit test of whether the population proportions of kissing couples who lean to the right and left are 2/3 and 1/3, respectively.

Activity 24-14: Candy Colors

1-13, 2-19, 13-1, 13-2, 15-7, 15-8, 16-4, 16-22, **24-14,** 24-15

A class of statistics students examined a sample of 541 Reese's Pieces candies and found 273 were orange, 154 were brown, and 114 were yellow.

a. Display these data with a bar graph, and comment on the distribution of color.
b. Suppose you conduct a goodness-of-fit test of whether the three candy colors are equally likely. Without doing any calculations, would you expect the *p*-value to be small? Explain, based on your intuition alone without doing any formal calculations.
c. Conduct a goodness-of-fit test of whether the distribution of colors in the population of all Reese's Pieces is 50% orange, 25% brown, and 25% yellow. Report all aspects of the test, including a check of technical conditions, and summarize your conclusion.
d. Is there more than one (null) hypothesis about the color distribution that would not be rejected at the .05 significance level? Explain, without performing any calculations.

Activity 24-15: Candy Colors

1-13, 2-19, 13-1, 13-2, 15-7, 15-8, 16-4, 16-22, 24-14, **24-15**

Reconsider the previous activity. Suppose the distribution of colors in the population of all Reese's Pieces really is 50% orange, 25% brown, and 25% yellow. Suppose further you take a random sample of 50 candies and test whether the three colors are equally likely.

a. Are you guaranteed to reject the "equally likely" hypothesis? Explain.
b. If you do not reject the null hypothesis, have you made a correct decision, a Type I error, or a Type II error? Explain.
c. Suppose your friend takes a random sample of 200 candies, four times as many as in your sample. Who is more likely to commit a Type II error: you or your friend, or are you both equally likely to make a Type II error? Explain.

Activity 24-16: Flat Tires

17-1, 17-2, 17-4, 17-10, 17-16, 18-18, **24-16**

Reconsider your class data from the Preliminaries questions in Topic 17, about which tire you would select if you had to make up an answer for which tire was flat.

a. Display your class sample data for the four tire choices with a bar graph, and describe what it reveals about your class's choices.
b. Conduct a goodness-of-fit test of whether your class sample data suggest the four tires are not equally likely to be chosen. Be sure to include a check of technical conditions, and summarize your conclusion.
c. Repeat part b for the hypothesis that the right front tire has probability .4 of being chosen, and each of the other three tires has probability .2.

Activity 24-17: Lifetime Achievements

6-3, 6-6, **24-17**

Reconsider the data collected on yourself and your classmates in the Preliminaries questions for Topic 6, about which lifetime achievement you would select if given a choice: an Academy Award, a Nobel Prize, or an Olympic Medal. Test whether your sample data provide evidence that these three achievements are not equally likely to be chosen in the population of all students at your school. Report all aspects of the test, and summarize your conclusions.

Activity 24-18: Calling "Heads" or "Tails"

13-10, 17-14, 17-15, **24-18**

Reconsider Activities 17-14 and 17-15, where you analyzed data collected on yourself and your classmates on the question of whether you would call "heads" or "tails" if you were asked to predict the outcome of a coin flip. Recall that the author of *Statistics You Can't Trust,* Steve Campbell, claimed that 70% of all people would call "heads" in this situation.

a. Use your class data to conduct a chi-square test of this claim. Report the hypotheses, expected values, test statistic, and p-value. Also include a check of technical conditions.
b. What test decision would you reach at the $\alpha = .10$ significance level? Also summarize your conclusion in context.
c. Suppose the sample size had been twice as large and the sample proportions of heads and tails had remained the same. Repeat parts a and b for this situation.
d. Comment specifically on how the expected values, test statistic, and p-value changed when you doubled the sample size.

Activity 24-19: Leading Digits

24-5, 24-9, **24-19**

Reconsider Activity 24-5. Find some data, involving at least 100 observations, and record the distribution of leading digits. (You might look in an almanac or on the Internet.) Analyze the data with a bar graph and a chi-square test of the Benford probabilities. Report the details of your analysis and summarize your conclusions. *Hint:* Be sure to describe the study's context and the source of your data.

TOPIC 25 Inference for Two-Way Tables

One of the questions on the General Social Survey asks about the general level of happiness that Americans feel about their lives. Has the distribution of responses about happiness level changed much over the past few generations? Do the happiness level responses differ significantly among people with different political viewpoints? Which groups tend to feel more, or less, happiness? You will investigate these issues, and others that can be analyzed with the same statistical technique, in this topic.

Overview

You will continue to study techniques for analyzing categorical variables in this topic. In Topic 21, you learned inference techniques for comparing two groups on a binary categorical response. In this topic, you will learn a technique that is more general in two respects: First, this technique lets you compare results across more than two groups. Second, the response variable can have more than two categories. The technique you will learn is a chi-square test for two-way tables. You first studied such tables in Topic 6 with descriptive methods such as segmented bar graphs and conditional proportions; now you will study an inferential method for generalizing to a larger population or drawing cause-and-effect conclusions. You will find this type of chi-square test has much in common with, though it also differs from, a chi-square goodness-of-fit test. Again, you will see the importance of considering how the data were collected before determining the scope of conclusions that can be drawn.

Preliminaries

1. Would you rate your general level of happiness as very happy, pretty happy, or not too happy?

2. Would you expect that Americans' levels of happiness have increased over the past few generations, decreased, or stayed about the same?

3. Which political group (liberals, moderates, or conservatives) would you expect to express the most happiness about their lives?

4. Which political group (liberals, moderates, or conservatives) would you expect to express the least happiness about their lives?

In Class Activities

Activity 25-1: Pursuit of Happiness

2-16, 3-25, 13-17, **25-1,** 25-2, 25-4

Have Americans become more happy or less happy over the past few generations? Or has their happiness level remained fairly constant? The General Social Survey (GSS) interviews a random sample of adult Americans every two years, and one of the questions asks respondents to rate their general happiness level. Sample results for 1972, 1988, and 2004 are summarized in the following 2 × 3 table:

	1972	1988	2004	Total
Very Happy	486	498	419	1403
Less Than Very Happy	1120	968	918	3006
Total	1606	1466	1337	4409

a. Identify the observational units in this study.

b. Identify and classify the explanatory and response variables in this study.

Explanatory: Type:

Response: Type:

c. Is this an observational study or an experiment? Explain.

d. What kind of graphical display is appropriate for these data?

e. Comment on what the following graph reveals about the sample proportions of Americans who considered themselves very happy for each of these years.

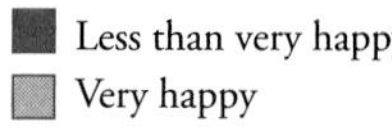

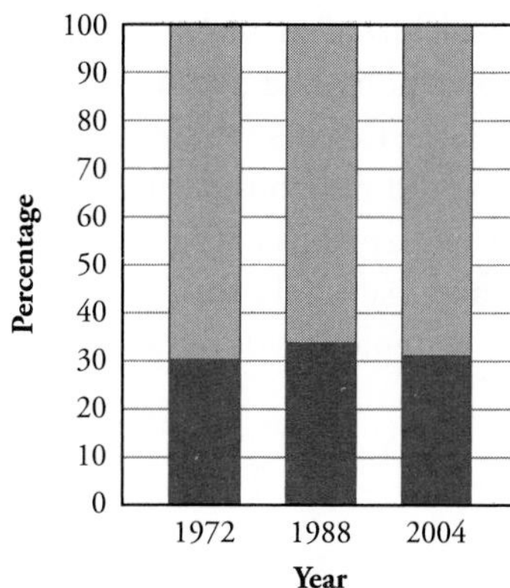

f. Explain why the two-sample *z*-test from Topic 21 is not appropriate for testing whether the population proportions who considered themselves very happy differed significantly over these three years.

In asking whether these sample data provide evidence that the proportions of Americans who identified themselves as very happy differed among the three *populations* for these years, the issue of *sampling variability* arises yet again. Even if the *population* proportions were identical for those three years, you would expect the *sample* proportions to differ somewhat based solely on random sampling. As you have done with all of the other inference techniques you have studied, you have to ask how likely it is to have observed such extreme sample data if, in fact, the "very happy" proportions were the same for all three populations (years). However, it's a little harder to quantify this extremeness now that you are comparing more than two groups. To quantify this, you will adopt a strategy similar to the goodness-of-fit test: you will compare the *observed* counts in the table with the counts *expected* under the hypothesis of equal population proportions/distributions.

g. Use appropriate symbols to state the null hypothesis that the population proportion of adult Americans who considered themselves to be very happy was the same for these three years.

h. For the three years combined, what proportion of respondents were very happy?

i. If this same proportion (your answer to h) of the 1606 respondents in the year 1972 had been very happy, how many people would this represent? Record your answer with two decimal places.

j. Repeat part i first for the year 1988 and then 2004.

1988: 2004:

You have calculated the *expected counts* under the null hypothesis that the population proportion of adult Americans who considered themselves to be very happy was the same for these three years (and consequently also for the population proportions who did not consider themselves very happy). A more general technique for calculating the expected count of the cell in row i and column j of a two-way table is to take the marginal total for row i times the marginal total for column j divided by the grand total (sample size of the study). The notation for this technique is

$$E_{ij} = \frac{R_i C_j}{n} = \frac{(\text{row total for row } i) \times (\text{column total for column } j)}{(\text{grand total for entire table})}$$

k. Use this more general formula to calculate the expected count of "less than very happy" people in the year 1988. Record this count in parentheses in the bottom middle cell of the following table:

	1972	1988	2004	Total
Very Happy	486 (511.05)	498 (466.50)	419 (425.45)	1403
Less Than Very Happy	1120 (1094.95)	968 ( )	918 (911.55)	3006
Total	1606	1466	1337	4409

Now the task is again to use a *test statistic* to measure how far the observed counts fall from the expected counts. To do this, you will use the same calculation as with the goodness-of-fit test:

$$X^2 = \sum_{i,j} \frac{(O_{ij} - E_{ij})^2}{E_{ij}} = \sum_{\text{all cells}} \frac{(\text{observed} - \text{expected})^2}{\text{expected}}$$

l. Calculate the value of $\frac{(O - E)^2}{E}$ for the "less than very happy" people in 1988 (i.e., for the bottom middle cell of the table).

m. The other contributions to the test statistic calculation are provided here. Insert your answer to part l and sum these to calculate the value of the test statistic:

$$X^2 = 1.228 + 2.127 + 0.098$$
$$+ 0.573 + \_\_\_\_ + 0.046$$
$$=$$

n. What kind of values (e.g., large or small) of the test statistic constitute evidence against the null hypothesis that the three populations have the same proportions of very happy Americans? Explain.

Once again, it turns out that under the null hypothesis of equal population proportions, the test statistic X^2 has a *chi-square* distribution. The degrees of freedom is now equal to $(r - 1)(c - 1)$, where r is the number of rows and c is the number of columns. Because large values of the test statistic provide evidence against the null hypothesis, the *p*-value for the chi-square test is the probability of exceeding the value of the test statistic by chance alone. As always, the *smaller* the *p*-value, the less likely the observed data would have occurred by sampling variability. Thus, the *smaller* the *p*-value, the stronger the evidence that there *is* a difference in the population proportions among the populations.

o. Use Table IV to determine (as accurately as possible) the *p*-value for these sample data.

p. Would these sample data have been very unlikely to occur by chance alone if the population proportions of very happy people had been identical for these three years? What test decision would you reach at the $\alpha = .05$ significance level?

q. Summarize the conclusion you draw from this study.

25

This test is called a **chi-square test for homogeneity of proportions.** As with the two-sample *z*- and *t*-tests in Topics 21 and 22, this technique applies for two scenarios: independent random sampling from several populations (as in this activity), or random assignment to several treatment groups (as in Activity 25-3). However, you do not have to limit yourself to binary response variables, but can instead compare the entire distribution across the groups. As always, the scope of conclusions (with regard to generalizability and causation) depends on which of these scenarios (or both) describes how the study was conducted.

Chi-square test of equal proportions in a two-way table:

H_0: The population distributions/proportions are the same for all populations.

H_a: The set of proportions for at least one population differs from the others.

Test statistic:

$$X^2 = \sum_{\text{all cells}} \frac{(O_{ij} - E_{ij})^2}{E_{ij}}$$

where $E_{ij} = \frac{R_i C_j}{n}$ (*row total* times *column total* divided by *grand total*).
p-value $= \Pr(\chi_k^2 \geq X^2)$, where χ_k^2 represents a chi-square distribution with degrees of freedom $k = (r - 1)(c - 1)$.

The **technical conditions** necessary for the test to be valid are

- The observations arise from independent random samples from the populations or random assignment to treatment groups.
- The *expected* counts are at least five for each cell of the table.

Watch Out

- Notice that although this chi-square test has many similarities with the test from the previous topic, they are not the same. This chi-square test applies when you have *two* categorical variables (in this case, *year of survey* and *self-reported happiness level*). In other words, this test applies when the data can be organized in a two-way table, and your goal is to compare proportions across groups. On the other hand, the chi-square goodness-of-fit test (from Topic 24) applies when you have only *one* categorical variable and want to know whether its distribution conforms to a set of prespecified proportions/probabilities.
- You do not need to specify numerical values for the population proportion with this test, which is another contrast to the chi-square goodness-of-fit test.
- Because so many calculations are involved with this test, be very careful with rounding errors in intermediate calculations. For this reason, we often use the graphing calculator to perform these calculations.

Activity 25-2: Pursuit of Happiness

2-16, 3-25, 13-17, 25-1, **25-2,** 25-4

Reconsider the previous activity. When the General Social Survey asked about respondents' general feelings of happiness, they actually presented three options for responses, rather than simply two: very happy, pretty happy, not too happy. The full table for the same three years, along with a segmented bar graph, follows:

	1972	1988	2004	Total
Very Happy	486	498	419	1403
Pretty Happy	855	832	738	2425
Not Too Happy	265	136	180	581
Total	1606	1466	1337	4409

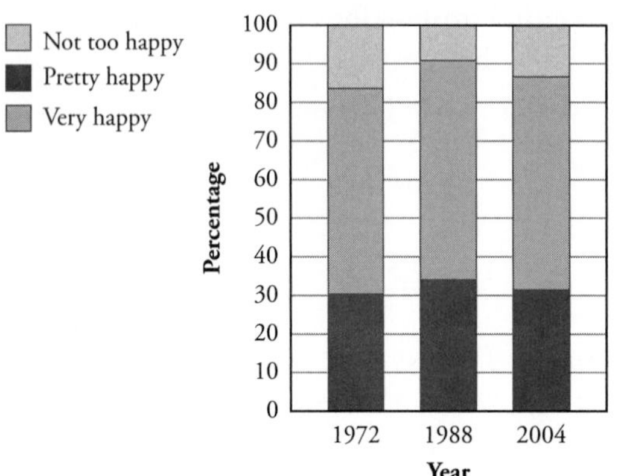

a. State the hypotheses (in words) to be tested with a chi-square test on the data in this table.

b. Use your calculator to compute the expected values, test statistic, and p-value. Report the test statistic and p-value. To do this, you can use the **χ^2-Test.** Before using the **χ^2-Test,** however, you need to set up a matrix containing the observed counts:

1. Press [2ND] [MATRIX].
2. Arrow over to the EDIT menu and press [ENTER] to select matrix [A].
3. The table at the beginning of this activity has three categories and three years, so your matrix needs three rows and three columns (i.e., the dimensions should be 3 × 3).
4. Now enter the numbers from the table into the matrix using the arrow keys to move the cursor. Your screen should look like the one shown here:

```
MATRIX[A] 3 ×3
[ 486    498    419  ]
[ 855    832    738  ]
[ 265    136    180  ]
```

5. Once matrix [A] is completed, press [STAT] and arrow over to the TESTS menu and select **χ^2-Test.**
6. Select **Calculate** and then press [ENTER]. The expected values are automatically entered into the matrix [B].

Test statistic: p-value:

c. Do the sample data provide evidence that the distributions of self-reported happiness levels differ among these three years, at the .05 significance level?

When the chi-square test result turns out to be significant, you can get a sense for where the primary differences are by examining the individual cells' contributions to the test statistic.

d. Identify which two table cells are most responsible for the large value of the test statistic. You can either do this by hand or use the CELL program on your calculator to get the cell contributions. Then state whether the observed count is higher than expected or lower than expected for these cells.

e. Summarize what your analysis reveals about whether the distributions of happiness levels differ among the years 1972, 1988, and 2004.

Activity 25-3: Got a Tip?

1-10, 5-28, 14-12, 15-17, 22-4, 22-9, **25-3**

Can telling a joke affect whether a customer in a coffee bar leaves a tip for the waiter? A recent study investigated this question at a coffee bar at a famous resort on the west coast of France (Gueguen, 2002). The waiter randomly assigned coffee-ordering customers into one of three groups: one group received a card telling a joke with the bill, another group received a card containing an advertisement for a local restaurant, and a third group received no card at all. Results are summarized in the following 2×3 table:

	Joke Card	Advertisement Card	No Card	Total
Left a Tip	30	14	16	60
Did Not Leave a Tip	42	60	49	151
Total	72	74	65	211

a. Identify the observational units in this study.

b. Identify and classify the explanatory and response variables in this study.

Explanatory: Type:

Response: Type:

c. Is this an observational study or an experiment? Explain.

d. Does this study involve random sampling from populations, or random assignment to groups, or both, or neither?

e. Use your calculator to conduct a chi-square test on these data. Report the hypotheses, test statistic, and *p*-value. Also indicate your test decision at the .10 significance level.

f. Would you conclude from this study that the joke card *caused* customers to be more likely to leave a tip? Explain why this conclusion is valid or not.

g. Would you generalize your conclusion to all waiters and waitresses in all food and drink establishments? Explain why or why not.

Activity 25-4: Pursuit of Happiness

2-16, 3-25, 13-17, 25-1, 25-2, **25-4**

Reconsider the General Social Survey discussed in Activities 25-1 and 25-2. But this time, consider only the year 2004 with a different variable: the respondent's political inclination, classified as liberal, moderate, or conservative. The sample results are summarized in the table:

	Liberal	Moderate	Conservative
Very Happy	90	128	187
Pretty Happy	177	291	258
Not Too Happy	51	76	48

a. Identify and classify the explanatory and response variables in this table.

Explanatory: Type:

Response: Type:

b. Explain how the data collection in this case differs from Activities 25-1 and 25-2.

These data fit neither of the two scenarios specified for the chi-square homogeneity of proportions test. The respondents were certainly not assigned to political groups by randomization. They were randomly selected, but the researchers did not take one random sample of liberals, an independent random sample of moderates, and an independent random sample of conservatives. Rather, the researchers took *one* random sample of people and *then* classified them on *both* of these variables (political inclination and happiness level).

It turns out the same chi-square test applies to two-way tables where the data are collected by one random sample from a population. The difference is that the null hypothesis being tested is that in the population the two variables are *independent,* and the alternative hypothesis is that there is a *relationship* between the variables. (Recall from Topic 6 that two categorical variables are independent if the conditional distribution of the response is the same for all category groups of the explanatory variables.)

c. Use your calculator to conduct the chi-square test of whether happiness level is independent of political viewpoint. Report all aspects of the test, and summarize your conclusions. *Hint:* If the test indicates strong evidence of a relationship between the variables, examine the table cells that contribute most to the value of the test statistic in order to describe the relationship (as in part d of Activity 25-2). A segmented bar graph can also be helpful for describing the relationship between the two variables.

Watch Out
Even if the test gives a very small *p*-value, indicating very strong evidence of a relationship between the variables, you cannot legitimately conclude this relationship is causal *unless* the observational units were randomly assigned to the groups. In this case, the sample data provide very strong evidence of a relationship between reported happiness level and political self-classification, with a higher proportion of conservatives and a smaller proportion of liberals being very happy. But you cannot conclude that a person's political viewpoint causes them to be happy or unhappy because this is an observational study, and therefore, many other variables also differ among these political groups.

Chi-square test of independence in a two-way table:
H_0: The two categorical variables are independent (not related) in the population.
H_a: There is a relationship between the two categorical variables in the population.

Test statistic:

$$X^2 = \sum_{\text{all cells}} \frac{(O_{ij} - E_{ij})^2}{E_{ij}}$$

where $E_{ij} = \frac{R_i C_j}{n}$ (*row total* times *column total* divided by *grand total*).
p-value $= \Pr(\chi_k^2 \geq X^2)$, where χ_k^2 represents a chi-square distribution with degrees of freedom $k = (r - 1)(c - 1)$.

The **technical conditions** necessary for the test to be valid are

- The observations arise from a random sample from the population.
- The *expected* counts are at least five for each cell of the table.

Activity 25-5: Pets as Family

A 2006 report from the Pew Research Center examined how close people felt to their pets. Researchers interviewed a random sample of adult Americans in late 2005. Those with a pet dog or cat were asked whether they felt "close" or "distant" to their pets. Of the 1181 respondents with a dog, 94% answered that they felt close to their pets. Of the 687 with a cat, 84% responded that they felt close to their pets.

You will test whether these sample data provide evidence of a significant association between which pet an adult American owns (dog or cat) and whether the person says that he or she feels close to his or her pet.

a. First, organize the data into a 2 × 2 table, with the explanatory variable in columns and the response variable in rows. Then conduct a chi-square test. (Feel free to use your calculator.) Report the hypotheses, test statistic, and *p*-value. Also check technical conditions and summarize your conclusion.

b. Now apply a two-sample z-test (as in Topic 21) to compare the proportions of pet owners who feel close to their pets between dog owners and cat owners. Report the hypotheses, test statistic, and p-value, and summarize your conclusion. Also comment on whether the technical conditions of this procedure appear to be met.

c. How do the p-values and conclusions compare between your analyses of parts a and b?

d. How do the test statistics compare between your analyses of parts a and b? *Hint:* Try to find a mathematical relationship between them, as in part g of Activity 24-4.

You should have found the chi-square test and two-sample z-test give identical p-values and conclusions, as long as the alternative hypothesis for the z-test is two-sided. The test statistics are not identical but are closely related; the chi-square test statistic is the square of the z-test statistic: $X^2 = z^2$.

Watch Out

- You can only apply a two-sample z-test with a 2 × 2 table (i.e., when the explanatory and response variables are both binary). But the chi-square test applies to any size of table.
- With a 2 × 2 table, a z-test is usually preferred over a chi-square test because the z-test allows more readily for a one-sided alternative and also because the z-procedure includes a confidence interval to estimate the magnitude of the difference in population proportions between the two groups.

Activity 25-6: Newspaper Reading

25-6, 25-17, 25-18

The following table summarizes data from the random sample of adult Americans who participated in the 2004 General Social Survey, with regard to the question of how often people read a newspaper:

	Male	Female
Every Day	191	167
Few Times a Week	89	133
Once a Week	53	81
Less Than Once a Week	56	65
Never	33	38

a. Does this study fall into any of the three scenarios for a chi-square two-way table test: independent random sampling from multiple populations, random assignment to treatment groups, or random sampling from one population with two categorical variables? If so, which one? Explain.

b. Use descriptive methods (graphical and numerical, but not inferential) to analyze these data for evidence that men and women differ with regard to how often they read a newspaper. Summarize your preliminary conclusions before you conduct a formal test.

c. Analyze these data with a chi-square test. Report all aspects of the test, including a check of technical conditions and a test decision at the .05 significance level, and summarize your conclusion.

Solution

a. The General Social Survey took *one* random sample, and this table summarizes responses on two categorical variables (*gender* and *newspaper reading frequency*), so the third scenario applies.

b. The following segmented bar graph displays the distributions (conditional proportions) of newspaper reading frequency for each gender:

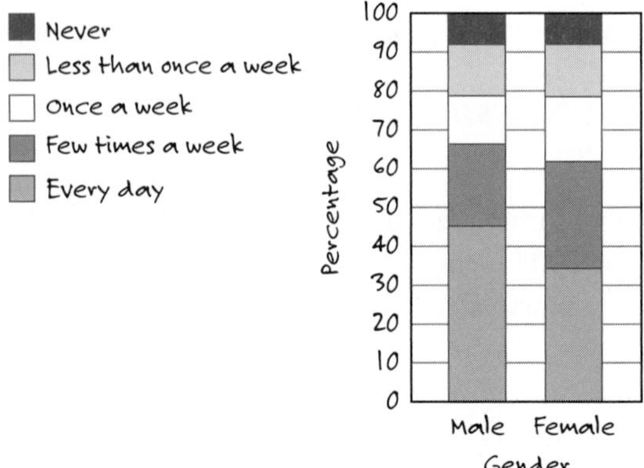

This graph reveals that a higher proportion of men than women read the newspaper every day (191/422 = .453 for men vs. 167/484 = .345 for women). A higher proportion of women than men read the paper a few times a week (.211 for men vs. .275 for women) and once a week (.126 for men vs. .167 for women).

c. The null hypothesis asserts newspaper reading frequency is independent of gender. In other words, the null hypothesis says the proportional distribution of the various reading frequency categories is identical for men and women in the population of all American adults. The alternative hypothesis says newspaper reading frequency is related to gender, which means these population distributions are not the same for men and women.

The expected counts appear in parentheses in the table:

	Male	Female	Total
Every Day	191 (166.75)	167 (191.25)	358
Few Times a Week	89 (103.40)	133 (118.60)	222
Once a Week	53 (62.42)	81 (71.58)	134
Less Than Once a Week	56 (56.36)	65 (64.64)	121
Never	33 (33.07)	38 (37.93)	71
Total	422	484	906

For an example of one of these calculations, the expected count for the (every day, male) cell is found by taking (358)(422)/906 = 166.75.

All of these expected counts are larger than five. The subjects were randomly selected from the population of adult Americans, so the technical conditions are satisfied.

The chi-square test statistic, computed as $\Sigma \frac{(O - E)^2}{E}$, turns out to be

$$\begin{aligned} X^2 &= 3.526 + 3.075 + 2.006 + 1.749 + 1.420 \\ &\quad + 1.238 + 0.002 + 0.002 + 0.000 + 0.000 \\ &= 13.020 \end{aligned}$$

Comparing this value to a chi-square distribution with (5 − 1)(2 − 1) = 4 degrees of freedom reveals that the *p*-value is slightly greater than .01. This is less than .05, so the test decision is to reject the null hypothesis at the .05 significance level. The data provide fairly strong evidence that newspaper reading frequency is related to gender.

In particular, the table cells in the top row contribute the most to the test statistic calculation. This finding indicates more men than expected read the newspaper every day and fewer women than expected read the newspaper every day. The sample proportions of men and women who read the newspaper every day are 191/422 ≈ .453 for men and 167/484 ≈ .345 for women.

Wrap-Up...........

In this topic, you revisited the issue of analyzing two-way tables, as you first did in Topic 6. You have expanded your knowledge beyond using segmented bar graphs and conditional proportions to describe such tables. You learned to apply the **chi-square test for two-way tables** to such data.

You saw similarities between the chi-square test for goodness-of-fit and the chi-square test for two-way tables. Both tests involve calculating expected counts and comparing observed to expected counts through the chi-square statistic, calculated in the same manner as $X^2 = \Sigma \frac{(O - E)^2}{E}$. The primary difference between these procedures and the chi-square goodness of fit test is that chi-square tests for two-way tables apply to studies with *two* categorical variables, not just one.

You investigated three scenarios to which this test applies:

1. When independent random samples are drawn from two or more populations with a categorical response variable. In this scenario, the test compares proportions across several groups, such as comparing the proportions of people who classify themselves as very happy across several different years. The results can be generalized to the larger populations.
2. When subjects are randomly assigned to two or more treatment groups. In this scenario, the test again compares proportions across several groups, such as comparing the proportions of customers who leave a tip, depending on the type of card presented with the bill. Cause-and-effect conclusions can be drawn, for example, determining that offering a joke card with the bill causes an increase in the proportion of customers who leave a tip.
3. When one random sample is selected from a population and two categorical variables are recorded for each observational unit. In this scenario, the test assesses whether those two variables are *independent,* for example, whether happiness level is independent of political viewpoint. The results can be generalized to the larger population.

In all three of these scenarios, the mechanics of the calculations are identical. What changes are the way the hypotheses are described and the scope of conclusions that can be drawn.

Some useful definitions to remember and habits to develop from this topic include

- Apply the chi-square goodness-of-fit test when you have *two* categorical variables, possibly having several categories each.
- Remember to begin a thorough analysis of a two-way table with a segmented bar graph.
- Be sure to identify which of the three scenarios applies and state your hypotheses and conclusions accordingly.
- If you find a significant result, look for the greatest contributions to the chi-square statistic in order to identify the strongest reason for the difference or relationship.
- As always, be sure to summarize your conclusions in context, and be careful to state variables as variables. In particular, be careful when describing your analysis in terms of either "differences between groups" or "relationships between variables."

Whereas this topic has considered studies where both variables are categorical, the next unit presents studies where both the explanatory and response variables are quantitative. You will first study graphical methods for displaying such data and numerical measures for summarizing such data. Then you will study a mathematical model for describing the relationship between the variables. Finally, you will learn an inference technique for generalizing findings beyond the sample data.

Homework Activities

Activity 25-7: Government Spending

6-1, **25-7,** 25-8, 25-24

Recall from Activity 6-1 that the General Social Survey asks people how they feel about government spending on the *environment.* The following table organizes the sample results from 2004 according to the respondents' political inclination:

	Liberal	Moderate	Conservative	Total
Too Little	127	158	113	398
About Right	27	80	91	198
Too Much	1	17	32	50
Total	155	255	236	646

a. Display these data with a segmented bar graph, and comment on what the graph reveals about the sample.

b. Does the observed count of one liberal who thinks the government spends too much render the chi-square test invalid for these data? Explain.

c. Would a chi-square test in this setting be a test of equal proportions or of independence? Explain.

d. Conduct the appropriate chi-square test on these data. Report the hypotheses, test statistic, and p-value. Report the test decision at the $\alpha = .025$ level. Also summarize your conclusion in this context.

e. If your test result is statistically significant, identify the 2 or 3 cells with the greatest discrepancies between observed and expected counts, as measured by $\frac{(O-E)^2}{E}$. Also comment on whether the observed counts are higher or lower than expected for these cells, and summarize what that result reveals.

Activity 25-8: Government Spending

6-1, 25-7, **25-8,** 25-24

Repeat parts a and d of the previous activity for the data on opinions about federal government spending on the *space program*. The data are organized in this table:

	Liberal	Moderate	Conservative	Total
Too Little	23	25	26	74
About Right	75	109	109	293
Too Much	53	120	92	265
Total	151	254	227	632

Activity 25-9: Baldness and Heart Disease

6-23, **25-9**

Recall from Activity 6-23 the study that classified samples of males, one group suffering from heart disease and one group not, according to their degree of baldness (Lasko et al., 2003). The resulting table is reproduced here:

	None	Little	Some	Much	Extreme
Heart Disease	251	165	195	50	2
Control	331	221	185	34	1

a. Is this an observational study or an experiment? Explain.
b. If the table were to be analyzed as is, then one of the technical conditions for the validity of the chi-square test would be violated. Explain why.
c. Combine the "much" and "extreme" categories into a "much or more" category and produce the resulting 2 × 4 table.
d. Conduct a chi-square test on this 2 × 4 table. Report the hypotheses (in context), test statistic, and *p*-value.
e. Write a few sentences summarizing your findings about a possible relationship between baldness and heart disease. Be sure to address the issue of causation.

Activity 25-10: Removing Gingivitis

Researchers in a dental study investigated the effectiveness of mouthwash for removing gingivitis from tooth surfaces of male patients (Donner and Banting, 1989). Subjects were given a baseline examination and a professional teeth-cleaning. Those with gingivitis on the facial surfaces of lower anterior teeth were randomly assigned to one of four treatment groups. These groups differed in how much active ingredient was in a mouthwash that subjects were asked to use on a daily basis. Three months later the subjects were examined again for the presence of gingivitis on these teeth. The results are summarized as follows:

Treatment Group (amount of active ingredient in mouthwash)	None (placebo)	Low	Intermediate	High
Number of Gingivitis Surfaces Presented	352	190	309	171
Proportion of Gingivitis Surfaces Cured	.222	.290	.311	.427

a. Which data collection scenario was used in this study: independent random sampling from multiple populations, random assignment to treatment groups, or random sampling from one population? Explain.
b. Identify the observational units in this study.
c. Identify the explanatory and response variables in this study.
d. Convert the information provided into a two-way table of *counts*.
e. Display the data with an appropriate graph, and comment on what the graph reveals.
f. Conduct a test of whether the data suggest the population proportions of success (no gingivitis) differ among the four treatment groups. Include all aspects of the test, including a check of technical conditions. Also summarize your conclusions.

Activity 25-11: Asleep at the Wheel

25-11, 25-12, 25-13

Researchers in New Zealand investigated whether sleepiness is related to car crashes (Connor et al., 2002). They took a sample of drivers who had been in a car crash and compared it to a control sample of drivers who had not been in a car crash. They interviewed all of these drivers about how much sleep they got in the day and week preceding the crash (or non-crash, for the control group). One variable considered was the Stanford sleepiness scale, which classified a person's sleepiness into three categories. Sample results are summarized in the table:

	Crash Group	Control Group
Most Alert	175	322
Moderately Alert	272	256
Least Alert	63	8

a. Is this an observational study or an experiment? Explain.
b. Display the results with an appropriate graph, and comment on what the graph reveals.
c. Conduct a test of whether the proportional breakdown of sleepiness categories differs significantly between the crash group and the control group. Report all aspects of the test, and state your test decision at the $\alpha = .05$ level.
d. Summarize your conclusion about whether the data suggest a relationship between sleepiness and car crashes. Be sure to comment on whether a cause-and-effect conclusion is warranted.

Activity 25-12: Asleep at the Wheel

25-11, **25-12,** 25-13

Reconsider the study described in the previous activity. Researchers also recorded the measured alcohol level (in mg/100 ml) for each driver. Results are summarized in the following table:

	Crash Group	Control Group
0	263	474
1–50	32	9
>50	118	3

Analyze these data with appropriate graphical, numerical, and inferential techniques. Write a paragraph or two describing your analysis and summarizing your conclusions.

Activity 25-13: Asleep at the Wheel

25-11, 25-12, **25-13**

Reconsider the study described in the previous two activities. Researchers also recorded the educational level for each driver. Results are summarized in the following table:

	Crash Group	Control Group
Less Than Three Years of High School	251	157
At Least Three Years of High School, But No Further Study	137	154
Further Study (beyond high school)	178	276

Analyze these data with appropriate graphical, numerical, and inferential techniques. Write a paragraph or two describing your analysis and summarizing your conclusions.

Activity 25-14: Children's Television Advertisements

6-11, **25-14**

Recall from Activity 6-11 that researchers analyzed food advertisements shown during children's television programming on three networks (Black Entertainment Television, Warner Brothers, and the Disney Channel) in July 2005 (Outley and Taddese, 2006). They classified each commercial according to the type of product advertised as well as the network on which it appeared. Their data are summarized in the following two-way table:

	BET	WB	Disney	Total
Fast Food	61	32	0	93
Drinks	66	9	5	80
Snacks	3	0	2	5
Cereal	15	16	4	35
Candy	17	26	0	43
Total	162	83	11	256

a. Show that the technical conditions of the chi-square test are not satisfied with the Disney Channel included in the analysis.
b. Which of the food categories is problematic for satisfying the technical conditions?
c. Create a new two-way table that excludes the Disney Channel and combines the "snacks" and "candy" categories.
d. Apply a chi-square test to the table in part c. Report the hypotheses, test statistic and *p*-value. Also summarize your conclusion in this context.

Activity 25-15: Weighty Feelings

6-13, **25-15**, 26-13

Recall from Activity 6-13 that one of the questions in the 2003–2004 NHANES (National Health and Nutrition Examination Survey) asked respondents whether they feel their current weight is about right, overweight, or underweight. Responses by gender are summarized in the following table:

	Female	Male	Total
Underweight	116	274	390
About Right	1175	1469	2644
Overweight	1730	1112	2842
Total	3021	2855	5876

a. Which data collection scenario was used in this study: independent random sampling from multiple populations, random assignment to treatment groups, or random sampling from one population? Explain.

b. Create a segmented bar graph to compare the distribution of weight self-images between men and women. Comment on what the graph reveals.
c. Conduct a chi-square test of whether these sample data suggest the distribution of weight self-images is the same between males and females. Report all aspects of the test, and summarize your conclusions.

Activity 25-16: Hand Washing

2-2, 21-16, **25-16**

Recall from Activity 2-2 that researchers for the American Society for Microbiology and the Soap and Detergent Association monitored the behavior of more than 6300 users of public restrooms. They observed people in public venues in four locations (Turner Field in Atlanta, Museum of Science and Industry and Shedd Aquarium in Chicago, Grand Central Station and Penn Station in New York City, and Ferry Terminal Farmers Market in San Francisco). The results by city are summarized in the table:

	Atlanta	Chicago	New York City	San Francisco
Washed Hands	1175	1329	1170	1521
Did Not Wash Hands	413	180	333	215

a. Identify the explanatory and response variables in this table.
b. What are the dimensions of this table? (In other words, this is a ? × ? table.)
c. For the four locations combined, what proportion of restroom users washed their hands?
d. Show how to use your answer to part c to calculate expected counts for the table, under the null hypothesis that the population proportions who wash their hands are the same in all four locations.
e. Calculate the chi-square test statistic and *p*-value. State the test decision at the $\alpha = .01$ level, and summarize your conclusion.
f. Which cells of the table have the largest contributions to the test statistic calculations? Explain what that reveals.

Activity 25-17: Newspaper Reading

25-6, **25-17**, 25-18

Recall in Activity 25-6 you used sample data from the 2004 General Social Survey to compare the newspaper reading frequencies of men and women. Conduct a similar analysis to compare the newspaper reading frequencies of people with different political inclinations. The following table displays the data:

	Liberal	Moderate	Conservative
Every Day	94	126	137
Few Times a Week	59	82	80
Once a Week	29	50	49
Less Than Once a Week	23	44	45
Never	15	21	29

Write a paragraph or two describing your analysis and summarizing your conclusions.

Activity 25-18: Newspaper Reading

25-6, 25-17, **25-18**

Reconsider the previous activity. The General Social Survey has asked the question about newspaper reading frequency for many years. The following table summarizes responses in the 1980s, 1990s, and between 2000–2004:

	1980s	1990s	2000–2004
Every Day	5399	4109	1425
Few Times a Week	2243	2055	900
Once a Week	1247	1241	520
Less Than Once a Week	790	895	519
Never	576	505	323

Analyze these data to address the question of whether newspaper reading frequency of adult Americans has changed over the past few decades and, if so, how. Use appropriate graphical, numerical, and inferential techniques. Write a paragraph or two describing your analysis and summarizing your conclusions.

Activity 25-19: Suitability for Politics

6-8, 6-9, 21-15, **25-19,** 25-20

Recall from Activity 6-8 that another question asked on the 2004 General Social Survey was whether the respondent generally agreed or disagreed with the assertion that "men are better suited emotionally for politics than women." The following two-way tables summarize the results, one by the respondent's gender and one by his/her political viewpoint:

	Liberal	Moderate	Conservative
Agree	40	68	96
Disagree	168	225	215

	Male	Female
Agree	109	103
Disagree	276	346

a. Is the appropriate chi-square test for these data about equal proportions or independence? Explain.

b. Conduct a chi-square test of whether the sample data provide strong evidence of a relationship between *political inclination* and *reaction to the statement.* Report all aspects of the test, and summarize your conclusion.

c. Repeat part b for the issue of whether there is a relationship between *gender* and *reaction to the statement.*

d. Identify which p-value is smaller and comment on what that reveals.

Activity 25-20: Suitability for Politics

6-8, 6-9, 21-15, 25-19, **25-20**

Reconsider the data from Activity 6-9, which considered whether reactions to the assertion that "men are better suited emotionally for politics than women" have changed over the past few decades. The following two-way table summarizes the responses over the years:

	1970s	1980s	1990s	2000s
Agree	2398	2563	1909	802
Disagree	2651	4597	6427	2609

Analyze these data with appropriate graphical, numerical, and inferential techniques. Write a paragraph or two describing your analysis and summarizing your conclusions.

Activity 25-21: Cold Attitudes

5-26, **25-21**

Reconsider Activity 5-26, which described a study in which researchers investigated whether people who tend to think positive thoughts catch a cold less often than those who tend to think negative thoughts (Cohen et al., 2003). The scientists recruited over 300 initially healthy volunteers, and they first interviewed them over two weeks to gauge their emotional state, eventually assigning them to one of three categories (high, medium, low) for positive emotions and similarly into three categories for negative emotions. Then the researchers injected rhinovirus, the germ that causes colds, into each subject's nose. The subjects were then monitored for the development of cold-like symptoms.

a. Is this an observational study or an experiment? *Hint:* Ask yourself whether the explanatory variable was assigned or observed by the researchers.

There were 112 subjects in the low category, 111 in the medium category, and 111 in the high category for positive emotions. It turned out that 33% of subjects in the low category developed a cold, compared to 26% in the middle category and 19% in the high category.

b. Organize this information into a two-way table of counts. *Hint:* Report integer values for these observed counts, based on the approximate percentages in the previous paragraph.

c. Conduct a chi-square test of whether the proportions of subjects who developed a cold differ significantly (at the $\alpha = .075$ level) among the three groups. Report all aspects of the test, including a check of technical conditions. Also summarize your conclusion, being sure to address the issue of cause-and-effect.

d. Repeat parts b and c for the negative emotion categories. There were 112 subjects in the low category, 111 in the medium category, and 111 in the high category for negative emotions. It turned out that 28% of subjects in the low category developed a cold, compared to 25% in the middle category and 26% in the high category.

Activity 25-22: Preventing Breast Cancer

6-14, 6-15, 21-3, 21-18, **25-22**

Reconsider the Study of Tamoxifen and Raloxifene (STAR) described in Activity 6-14 and analyzed in Activity 21-3. Recall the initial results released in April 2006 revealed that 163 of 9726 women in the tamoxifen group had developed invasive breast cancer, compared to 167 of the 9745 women in the raloxifene group.

a. Organize this information into a 2 $\times$ 2 table. Be sure to put the explanatory variable in columns.
b. Calculate the expected counts, under the null hypothesis that the population proportions of women who would develop breast cancer are the same with both treatments.
c. Conduct a chi-square test on these data. Report the hypotheses, test statistic, and p-value. Summarize your conclusion.
d. Comment on how the chi-square test statistic and p-value compare to the test statistic and p-value from the two-sample z-test in Activity 21-3 on page 427.

Activity 25-23: A Nurse Accused

1-6, 3-20, 6-10, **25-23**

Reconsider the murder case against nurse Kristen Gilbert, as described in Activity 1-6 and analyzed in Activity 6-10 (Cobb and Gerlach, 2006). Hospital records for an 18-month period indicated that of 257 8-hour shifts during which Gilbert worked, a patient died during 40 of those shifts. But of 1384 8-hour shifts during which Gilbert did not work, a patient died during only 34 of those shifts.

a. Conduct a chi-square test of whether the difference in death rates between the two groups (Gilbert shifts and non-Gilbert shifts) is statistically significant at the $\alpha = .0001$ level.
b. Write a sentence or two interpreting what the p-value means in this context.
c. Despite the very small p-value, could a defense attorney reasonably argue that the higher rate of deaths on Gilbert's shifts is due to a confounding variable? Explain.
d. Despite the very small p-value, could a defense attorney reasonably argue that the higher rate of deaths on Gilbert's shifts is due to random chance? Explain.

Activity 25-24: Government Spending

6-1, 25-7, 25-8, **25-24**

Reconsider the data from Activity 25-8 about spending on the *space program,* organized in the following table:

	Liberal	Moderate	Conservative	Total
Too Little	23	25	26	74
About Right	75	109	109	293
Too Much	53	120	92	265
Total	151	254	227	632

This activity asks you to approximate the sampling distribution of the chi-square statistic under the null hypothesis of independence between these two variables (*political inclination* and *opinion about government spending on the space program*). To do this, you will keep the marginal totals fixed and randomly assign, for example, the 74 "too little" responses among the 151 liberals, 254 moderates, and 227 conservatives.

a. Use your calculator's **randInt** feature:

1. Press [MATH] and arrow over to the **PRB** menu to carry out this randomization once.
2. Let 1, 2, and 3 represent Liberal, Moderate, and Conservative, respectively.
3. For the first row, enter **randInt(1,3,74)** to create a list of length 74 with the integers 1, 2, and 3.
4. Store this as a named list.
5. Next, enter the frequencies of these integers in the first row of the following table. The easiest way to get the frequencies is to create a histogram of the data you just generated.
6. Setting the window correctly is extremely important to get the correct values. Let **Xmin=1, Xmax=4, Xscl=1, Ymin=0, Ymax=40, Yscl=1,** and **Xres=1** (do not use [ZOOM] [9]).
7. Once you have created the histogram, press the [TRACE] key to obtain the frequencies.
8. Repeat these directions for the second row.
9. For the third row, subtract to guarantee the totals in the bottom row. Record the simulation results in the following table:

	Liberal	Moderate	Conservative	Total
Too Little				74
About Right				293
Too Much				265
Total	151	254	227	632

b. Use your calculator to calculate the value of the chi-square test statistic for this simulated table.

c. Is this simulated test statistic greater than or equal to the test statistic value from the actual sample data?

25

UNIT

7

Relationships in Data

TOPIC

26 • • • Graphical Displays of Association

Is the selling price of a house related to its size? Is a state's birth rate associated with its death rate? Is there a relationship between the fuel efficiency rating of a car and its weight? Is the life expectancy of a country's people connected in some way with how many television sets the country has? You will begin to learn statistical techniques for addressing such questions, and many more, in this topic.

Overview

Of all of the scenarios that you have studied thus far, you have yet to study relationships between two *quantitative* variables. This scenario is the subject of the final unit, which will again progress from graphical displays to numerical summaries and then proceed to mathematical models and statistical inference. In this topic, you will begin the process by investigating the concept of association and exploring the use of graphical displays called scatterplots to display the association between quantitative variables. You will also study some modifications of this basic graph, such as labeled scatterplots, and learn how to discern additional information from them.

Preliminaries

1. Do you expect that bigger houses tend to cost more than smaller houses?

2. Do you think that if one house is larger than another, then it must always cost more than the smaller house?

3. Do you expect that heavier cars tend to have better fuel efficiency (miles per gallon ratings) than smaller cars or to have worse fuel efficiency, or do you expect no relationship between a car's weight and its fuel efficiency?

4. Take a guess as to the number of televisions per thousand people in the United States; do the same for China and for Haiti.

   United States:

China:

Haiti:

5. Do you expect that people in countries with more televisions per thousand people tend to have longer life expectancies or shorter life expectancies than people in countries with fewer televisions, or do you suspect there is no relationship between number of televisions in a country and the life expectancy of its people?

6. Measure the length of your right foot. Also measure the handspan (distance from tip of thumb to tip of little finger, when they are extended as far as possible) of your right hand.

7. Record the foot lengths and handspans for yourself and your classmates. Also record each student's height and gender.

In-Class Activities

Activity 26-1: House Prices

8-10, **26-1,** 27-5, 28-2, 28-12, 28-13, 29-3

The following table reports the price and size (in square feet) for a sample of houses in Arroyo Grande, California. These data were obtained from the Web site `zillow.com` on February 7, 2007, for a random sample of houses listed on that site as recently sold.

Address	Price ($)	Size (in sq ft)	Address	Price ($)	Size (in sq ft)
2130 Beach St.	311,000	460	1030 Sycamore Dr.	490,000	1664
2545 Lancaster Dr.	344,720	1030	620 Eman Ct.	492,000	1160
415 Golden West Pl.	359,500	883	529 Adler St.	500,000	1545
990 Fair Oaks Ave.	414,000	728	646 Cerro Vista Cir.	510,000	1567
845 Pearl Dr.	459,000	1242	926 Sycamore Dr.	520,000	1176
1115 Rogers Ct.	470,000	1499	227 S. Alpine St.	541,000	1120
579 Halcyon Rd.	470,000	1419	654 Woodland Ct.	567,500	1549
1285 Poplar St.	470,000	952	2230 Paso Robles St.	575,000	1540
1080 Fair Oaks Ave.	474,000	1014	2461 Ocean St.	580,000	1755
690 Garfield Pl.	475,000	1615	833 Creekside Dr.	625,000	1844

a. What are the observational units?

b. How many variables are reported in the table for each observational unit? What type (categorical or quantitative) is each variable?

c. Do the data suggest that bigger houses tend to cost more than smaller ones? *Hint:* Notice the houses are presented in order by price, from least to most expensive.

The simplest graph for displaying two quantitative variables simultaneously is a **scatterplot,** which uses a vertical axis for one of the variables and a horizontal axis for the other variable. For each observational pair, a dot is placed at the intersection of its two values. The convention is to place the response variable on the vertical axis and the explanatory variable on the horizontal axis.

Consider the following scatterplot of *price* vs. *size* (our convention is to say *y* vs. *x*, with the first variable (*y*) being on the vertical axis).

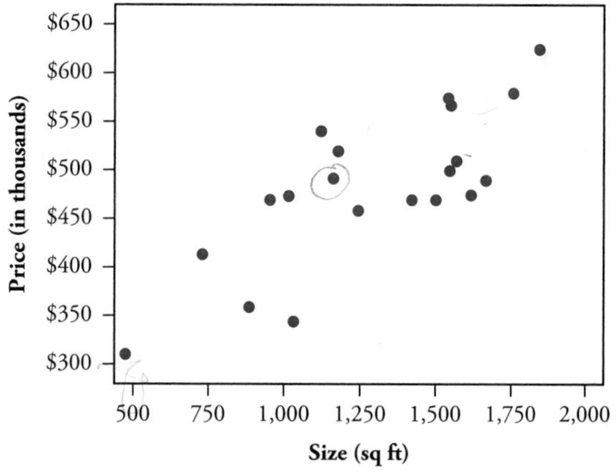

26

d. What is the address of the house represented by the point in the bottom left of the graph? How about the house represented by the top-right point?

Bottom left: Top right:

e. Circle the point corresponding to the house at 845 Pearl Drive (1242 square feet, $459,000).

f. Does the scatterplot reveal any relationship between house sizes and their prices? In other words, does knowing a house's size provide any useful information about its price? Write a sentence describing the relationship between the two variables.

Two variables display an **association** (or relationship) if knowing the value of one variable is useful (to some degree) in predicting the value of the other variable.

Three aspects of the association between quantitative variables are direction, strength, and form. **Direction** refers to whether greater values of one variable tend to occur with greater values of the other variable (*positive* association) or with smaller values of the other variable (*negative* association). The **strength** of the association indicates how closely the observations follow the relationship between the variables. In other words, the strength of the association reflects how accurately you could predict the value of one variable based on the value of the other variable. The **form** of the association can be *linear,* or it can follow some more complicated pattern.

g. Is house size positively or negatively associated with price? Would you describe the association as strong, moderate, or weak? Is the association between house *size* and *price* roughly linear?

Direction:

Strength:

Form:

h. Find an example of a *pair* of houses, where one house is larger than the other but costs less. Provide their addresses here and circle the pair of points on the previous scatterplot.

The concept of association is another example of a statistical tendency. It is *not always* the case that a larger house costs more, but larger houses certainly *tend* to cost more.

Watch Out

- When it is appropriate to consider one variable the explanatory variable and the other the response variable, always put the explanatory variable on the horizontal (x-) axis and the response variable on the vertical (y-) axis.
- If a scatterplot is described as "y vs. x" (e.g., *weight* vs. *height*), this indicates that the first variable should be placed on the vertical axis and the second variable on the horizontal axis.
- When constructing a scatterplot, be careful not to lose the bivariate aspect of the data. For example, make sure the right price stays with the right size house; don't mix up houses.

Activity 26-2: Birth and Death Rates

The following scatterplot displays the birth rates and death rates (per 1000 residents) for the 50 states as of 1997, according to *The World Almanac and Book of Facts 1999:*

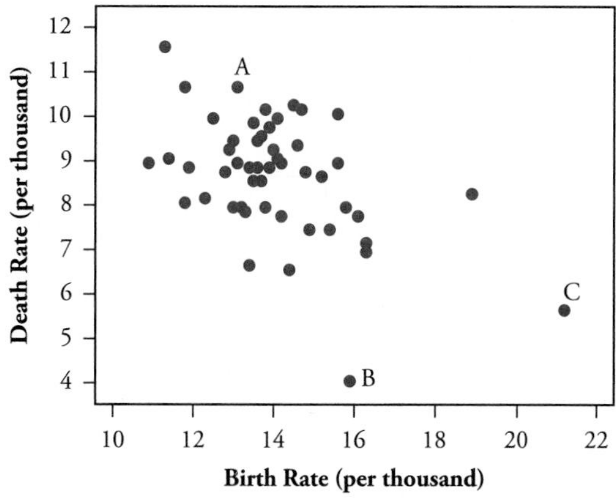

a. Would you say that a state's birth rate is related to (associated with) its death rate? If so, describe the direction, strength, and form of the relationship.

b. Notice three states have been identified by the letters *A*, *B*, and *C* in the graph. One of these states is Alaska, one is Florida, and the third is Utah. Think about context to make a reasonable guess as to which is which, and explain your reasoning.

A: Reasoning:

B: Reasoning:

C: Reasoning:

This activity should remind you of the importance of always thinking about context and looking for explanations for unusual observations.

Activity 26-3: Car Data

26-3, 27-1, 28-10

The following nine scatterplots pertain to models of cars described in *Consumer Reports' 1999 New Car Buying Guide.* The eight variables represented here are

- City MPG (in miles per gallon) rating
- Highway MPG (in miles per gallon) rating
- Weight (in pounds)
- Percentage of weight in the front half of the car
- Time to accelerate (in seconds) from 0 to 60 miles per hour
- Time to cover 1/4 mile (in seconds)
- Fuel capacity (in gallons)
- Page number of the magazine on which the car was described

Evaluate the direction and strength of the association between the variables in each graph. Do this by arranging the associations revealed in the scatterplots from those that reveal the most strongly negative association between the variables, to those that reveal virtually no association, to those that reveal the most strongly positive association. Arrange them by letter in the following table. (Because you will use each letter only once, you should probably look at all nine plots first.)

	Negative				None	Positive			
	Strongest			Weakest		Weakest			Strongest
Letter of Scatterplot	D	G	A	H	C	E	I	F	B

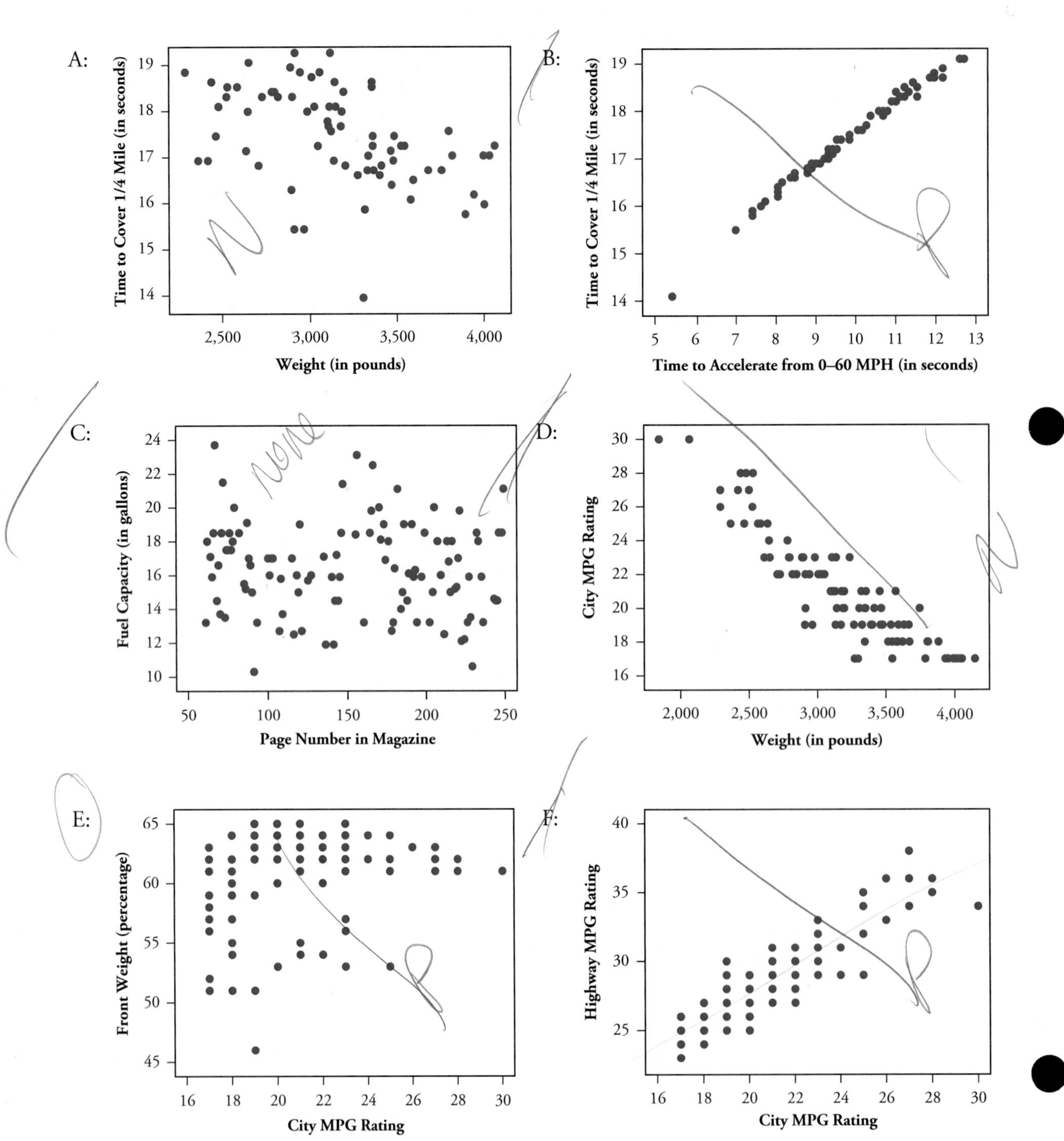

G:

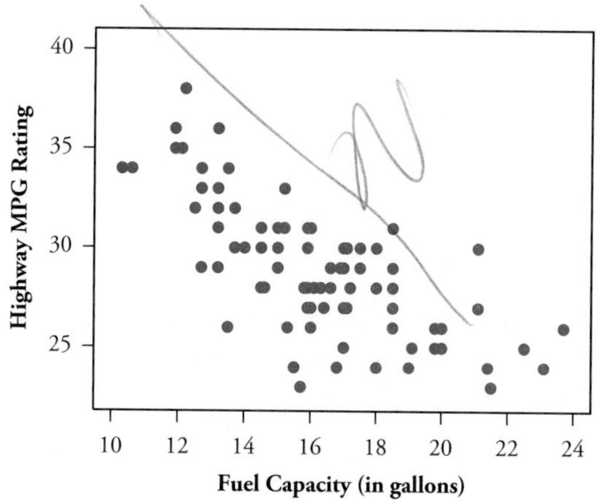

Highway MPG Rating

Fuel Capacity (in gallons)

H:

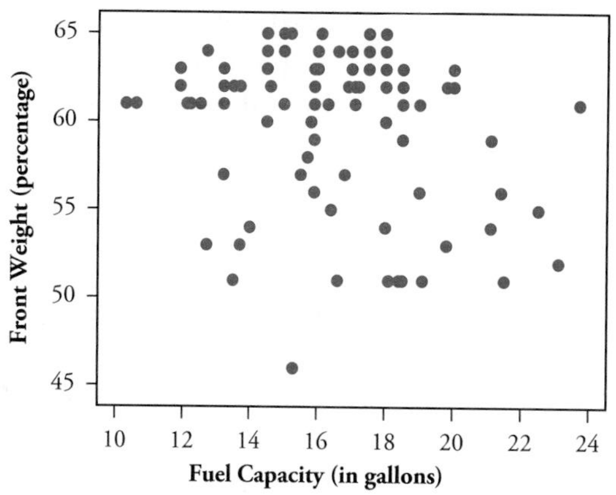

Front Weight (percentage)

Fuel Capacity (in gallons)

I:

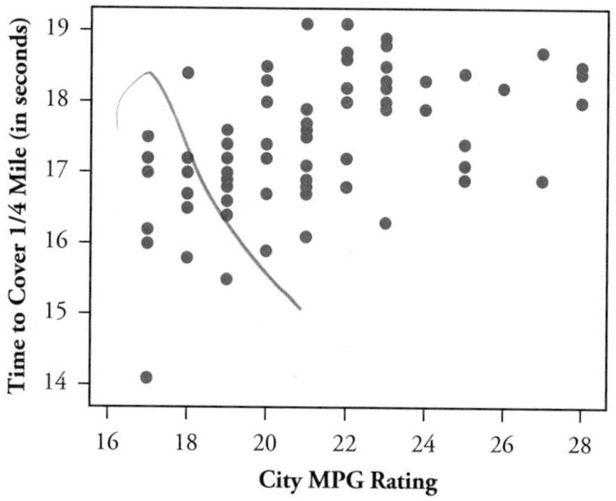

Time to Cover 1/4 Mile (in seconds)

City MPG Rating

Activity 26-4: Marriage Ages

8-17, 9-6, 16-19, 17-22, 23-1, 23-12, **26-4,** 29-17, 29-18

Refer to the data from Activity 23-1 on page 469 concerning the ages of 24 couples applying for marriage licenses. The following scatterplot displays the relationship between husband's age and wife's age. The line drawn on the graph is the $y = x$ line, where the husband's age equals the wife's age.

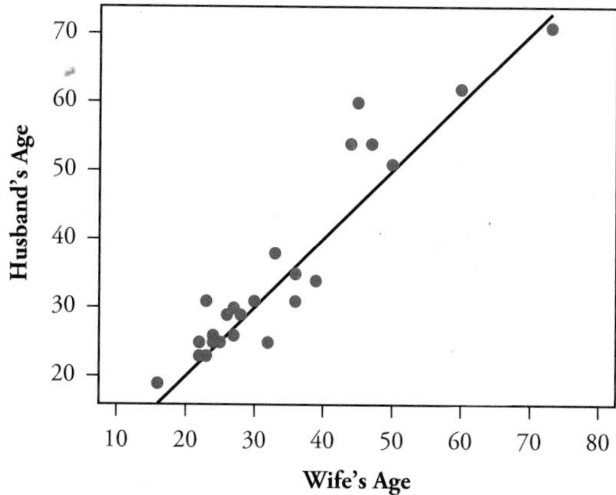

a. Based on the scatterplot, does there seem to be an association between the husband's age and the wife's age? Describe the association's direction, strength, and form.

b. Refer to the original listing of the data to determine how many of the 24 couples' ages fall exactly on the line. In other words, how many couples listed the same age for both husband and wife on their marriage license?

c. Again looking back at the data, for how many couples is the husband older than the wife? Do these couples fall above or below the line drawn in the scatterplot?

d. For how many couples is the husband younger than the wife? Do these couples fall above or below the line drawn in the scatterplot?

e. Summarize what you can learn about the ages of marrying couples by noting that the majority of couples produce points that fall above the $y = x$ line.

This activity illustrates that when working with *paired* data, including a $y = x$ line on a scatterplot can provide valuable information about whether one member of the pair (e.g., the husband) tends to have a greater value for the variable than the other member of the pair (e.g., the wife).

Activity 26-5: Heights, Handspans, and Foot Lengths

26-5, 28-1, 29-12, 29-13, 29-14

Consider the class data collected in the Preliminaries section about heights, handspans, and foot lengths. Such data might be useful for trying to predict the height of a criminal based on a handprint or footprint left at the scene of a crime.

a. Use your graphing calculator to create a scatterplot of *height* vs. *handspan*. To do this, enter the data into two lists, **HGHT** and **SPAN**. Press [2ND] [**STAT PLOTS**], select **Plot1,** and set it up so it is similar to this:

Press [ZOOM] and then [9] to display the graph (remember to turn off the other plots). Do these variables appear to be associated? If so, describe the association.

A *categorical* variable can be incorporated into a scatterplot by constructing a **labeled scatterplot,** which assigns different labels to the dots based on the category of the observational unit. For example, you might indicate observations coming from males with the label *M* and from females with the label *F.*

b. Use your graphing calculator to create a labeled scatterplot of *height* vs. *handspan,* using different labels for men and women. There are two ways you can do this exercise:

- You can divide these data between men and women (four lists total: two for men and two for women) and then use **Plot1** and **Plot2** simultaneously employing different symbols.
- Your other option is to create a new list (call it **GNDR** for gender) and press [0] if the corresponding data is from a male student and [1] if the data is from a female student (your calculator only accepts numeric values in lists). Download the program named LBLSCAT and select **SPAN** as the **X-LIST**, **HGHT** as the **Y-LIST**, and **GNDR** as the **CATEGORICAL VARIABLE LIST**. You will need to type in names for the different categories, e.g., you could use **M** for male and **F** for female. Use the graph to comment on the three questions posed next:

Do men and women tend to differ with regard to handspan? How can you tell from the graph?

Do men and women tend to differ with regard to height? How can you tell from the graph?

Does the *association* between handspan and height differ between men and women? How can you tell from the graph?

c. Conduct a similar analysis of *height* vs. *foot length,* describing the relationship between them and comparing that relationship between the two sexes.

Activity 26-6: Televisions and Life Expectancy

26-6, 27-3, 28-19

The following table provides information on life expectancy and number of televisions per thousand people in a sample of 22 countries, as reported by *The World Almanac and Book of Facts 2006:*

Country	Life Expectancy	Televisions per 1000 People	Country	Life Expectancy	Televisions per 1000 People
Angola	38.45	15	Mexico	75.25	272
Australia	80.45	716	Morocco	70.75	165
Cambodia	59.00	9	Pakistan	63.00	105
Canada	80.15	709	Russia	67.30	421
China	72.40	291	South Africa	43.30	138
Egypt	71.05	170	Sri Lanka	73.25	102
France	79.70	620	Uganda	51.60	28
Haiti	52.95	5	United Kingdom	78.45	661
Iraq	68.75	82	United States	77.80	844
Japan	81.25	719	Vietnam	70.70	184
Madagascar	57.00	23	Yemen	61.80	286

a. Which of the countries listed has the fewest televisions per thousand people? Which country has the most televisions? Record those countries and numbers.

Fewest: Most:

b. Use your graphing calculator to produce a scatterplot of *life expectancy* vs. *televisions per thousand people.* The data are stored in the grouped file TVLIFE, containing the lists **LFEXP** and **PERTV,** respectively. Does there appear to be an association between the two variables? If so, describe its direction, strength, and form.

both increase

c. Because the association between the variables is so strong, you might conclude that simply sending televisions to the countries with lower life expectancies would cause their inhabitants to live longer. Comment on this argument.

d. If two variables are strongly associated, does it follow there must be a cause-and-effect relationship between them? Explain.

This example illustrates again the very important distinction between *association* and *causation*. Two variables might be strongly associated without having a cause-and-effect relationship between them. Often with observational studies, both variables are related to a third (*confounding*) variable.

e. In the case of life expectancy and television sets, suggest a confounding variable that is associated both with a country's life expectancy and with the prevalence of televisions in that country. *Hint:* Be sure to express this as a variable.

Activity 26-7: Airline Maintenance

The March 2007 issue of *Consumer Reports* presents data about airline maintenance. The following table presents the percentage of an airline's maintenance outsourced to other companies and the percentage of the airline's delays caused by the airline:

Airline	Outsourced (%)	Delays (%)	Airline	Outsourced (%)	Delays (%)
Alaska	92	42	Southwest	68	20
Hawaiian	80	70	AirTran	66	14
US Airways	77	24	Frontier	65	31
Northwest	76	43	United	63	27
America West	76	39	Delta	48	26
Continental	69	20	American	46	26
JetBlue	68	18	ATA	18	19

a. What are the observational units in this table?

b. Produce a scatterplot of these data, using *percentage of delays* as the response variable.

c. Describe the association between these variables. *Hint:* Comment on direction, strength, and form.

d. Is it reasonable to conclude there is a cause-and-effect relationship between these variables, based on these data? Explain.

Solution

a. The observational units are the airlines.

b. The scatterplot follows.

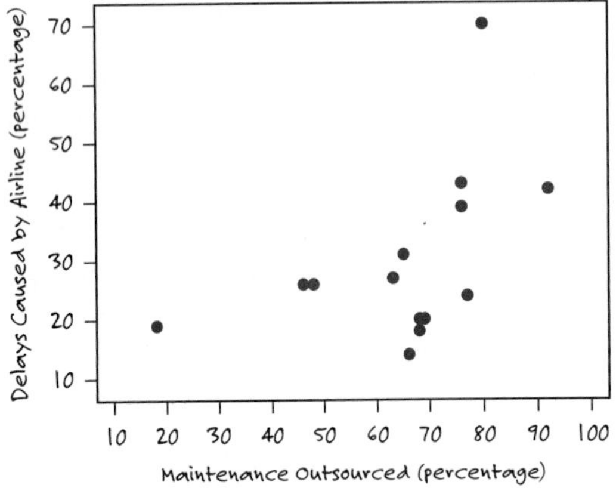

c. The scatterplot reveals a fairly strong, positive association between these variables. Airlines that outsource a higher percentage of their maintenance tend to have a greater percentage of delays caused by the airline, as compared to airlines that outsource a smaller percentage of their maintenance. The form of the association is clearly nonlinear, as the relationship appears to follow a curved pattern.

d. No, you cannot conclude from these data that outsourcing maintenance *causes* airlines to experience more delays. This is an observational study, not an experiment. Many other variables could explain the observed association between outsourcing and delays. For example, perhaps less financially successful airlines tend to outsource more of their maintenance and also to have more delays.

Watch Out

- Remember to express comments about association as a tendency and not as a hard-and-fast rule. Airlines with a higher percentage of outsourced maintenance *tend* to have a higher percentage of delays caused by the airline, but there are exceptions to this tendency.
- Do not assume all relationships are linear or that a nonlinear relationship must mean a weak association. In this example, the relationship is clearly curved and not linear, yet the association between variables is fairly strong. This strength indicates that you could make fairly good predictions about an airline's percentage of delays if you knew its outsource percentage.

- Never forget you cannot draw a cause-and-effect conclusion from an observed association, even a very strong association, unless the data are collected from a randomized experiment.

Wrap-Up...........

This topic introduced you to an issue that occupies a prominent role in statistical practice: exploring relationships between quantitative variables. You discovered the concept of **association** and learned how to construct and interpret **scatterplots** as visual displays of association. You also saw examples of linear relationships with house prices and of curved relationships with the airline maintenance data. You learned about positive association (e.g., marriage ages) and negative association (e.g., birth and death rates). You also encountered strong, moderate, and weak associations with the many variables recorded on new car models.

You learned that information from a categorical variable can be included in a **labeled scatterplot,** for example, when you included gender in the scatterplot of *height* vs. *handspan.* With paired data such as the marriage ages, including a $y = x$ line on the scatterplot can be informative.

Finally, your analysis of the data on life expectancy of people in different countries versus the number of televisions owned per 1000 people in these countries served as a reminder that observing an association between two variables does not necessarily establish a cause-and-effect relationship between those variables. People in countries with more televisions per thousand people do tend to have longer life expectancies than people in countries with fewer televisions, but the explanation for this association probably has more to do with the overall quality of life and technological sophistication of the country rather than any beneficial effect of televisions themselves.

Some useful definitions to remember and habits to develop from this topic include

- Always plot the response variable on the vertical (y-) axis and the explanatory variable on the horizontal (x-) axis.
- Because scatterplots display two variables at once, it is even more important to label axes clearly.
- When describing the association revealed in a scatterplot, comment on **form** (linear or curved), **direction** (positive or negative), and **strength** (strong, moderate, or weak). Also investigate and comment on any *unusual* observations that the plot reveals.
- To include information on a categorical variable, create a labeled scatterplot by using different labels for the various categories.
- With paired data, drawing a $y = x$ line on the scatterplot can help you compare the values for the two members of the pair to determine whether there is a tendency for one member of the pair to have larger values than the other. To create the $y = x$ line on your graphing calculator, press the [Y=] key and then press the [X,T,Θ,n] key. You should see a Y1 = X line on your screen.
- Remember an observed association between two variables does not necessarily imply a cause-and-effect relationship exists between them.

Just as you moved from graphical displays to numerical summaries when describing distributions of data in Unit 2, the next topic considers a numerical measure of the degree of association between two quantitative variables: the *correlation coefficient.*

26

Homework Activities

Activity 26-8: Miscellany

For each of the following pairs of variables, indicate what you would expect for the direction (positive, negative, or none at all) and strength (none, weak, moderate, or strong) of the association between them. Also provide a brief explanation for each answer.

- **a.** SAT score and college GPA for college students
- **b.** Distance from the equator and average January temperature for U.S. cities
- **c.** Lifetime and weekly cigarette consumption
- **d.** Serving weight and calories of fast-food sandwiches
- **e.** Airfare and distance to destination
- **f.** Number of letters in a person's last name and combined number of Scrabble points those letters are worth
- **g.** Average driving distance and average score per round among professional golfers *Hint:* Low scores are better than high scores in golf.
- **h.** Number of miles driven and asking price for used Honda Civics listed for sale on the Internet.
- **i.** Distance from the sun and diameter of planets in our solar system
- **j.** Distance from sun and number of (earth) days to revolve around sun among planets in our solar system
- **k.** Price of a textbook and number of pages in text
- **l.** Number of classes missed and overall course score in a class of statistics students
- **m.** Expected lifetime and gestation period among different kinds of mammals

Activity 26-9: *Challenger* Disaster

26-9, 27-17

The following data were obtained from 23 space shuttle launches prior to the *Challenger* disaster in January 1986 (Dalal, Folkes, and Hoadley, 1989). Each of four joints in the shuttle's solid rocket motor is sealed by two O-ring seals. After each launch, the reusable rocket motors were recovered from the ocean. This table lists the number of O-ring seals showing evidence of thermal distress and the outside temperature (in degrees Fahrenheit) at launch for each of the 23 flights. (The data are stored in the grouped file CHALLENGER with lists **ORING** and **TEMP,** respectively.)

Flight Date	O-ring Failures	Temperature	Flight Date	O-ring Failures	Temperature
4/12/81	0	66	11/28/83	0	70
11/12/81	1	70	2/3/84	1	57
3/22/82	0	69	4/6/84	1	63
11/11/82	0	68	8/30/84	1	70
4/4/83	0	67	10/5/84	0	78
6/18/83	0	72	11/8/84	0	67
8/30/83	0	73	1/24/85	3	53

Flight Date	O-ring Failures	Temperature	Flight Date	O-ring Failures	Temperature
4/12/85	0	67	10/3/85	0	79
4/29/85	0	75	10/30/85	2	75
6/17/85	0	70	11/26/85	0	76
7/29/85	0	81	1/12/86	1	58
8/27/85	0	76			

a. Produce a scatterplot of *O-ring failures* vs. *outside temperature.*

b. Write a few sentences commenting on whether the scatterplot reveals any association between O-ring failures and outside temperature.

c. The forecasted low temperature for the morning of the fateful launch was 31 degrees. What does the scatterplot reveal about the likelihood of O-ring failure at such a temperature?

d. Now eliminate those flights that had no O-ring failures. Produce a scatterplot of *O-ring failures* vs. *outside temperature* for the remaining seven flights. Does *this* scatterplot reveal much association between O-ring failures and outside temperature? If so, does it make the case as strongly as the previous scatterplot did? Explain.

e. Some NASA officials argued that flights on which no O-ring failures occurred provided no information about the issue. They, therefore, examined only the second scatterplot and concluded there was little evidence of an association between temperature and O-ring failures. Comment on the wisdom of this approach and specifically on the claim that flights on which no O-ring failures occurred provided no information.

Activity 26-10: Broadway Shows

2-14, **26-10,** 26-11, 27-15

Recall from Activity 2-14 the data on Broadway shows collected for the week of June 19–25, 2006. The following scatterplots present pairs of variables from gross receipts, attendance, seating capacity, and top ticket price.

A:

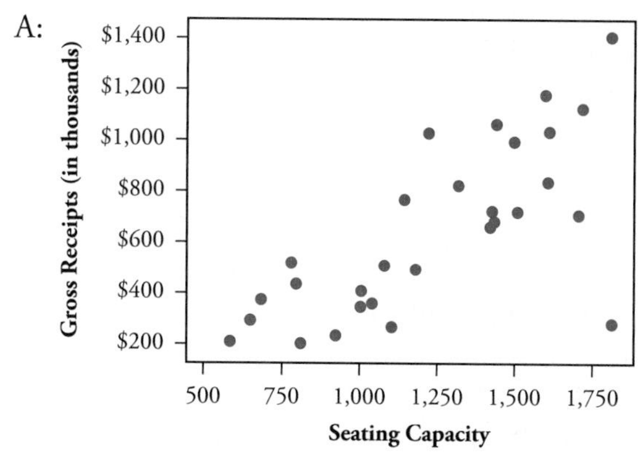

B:

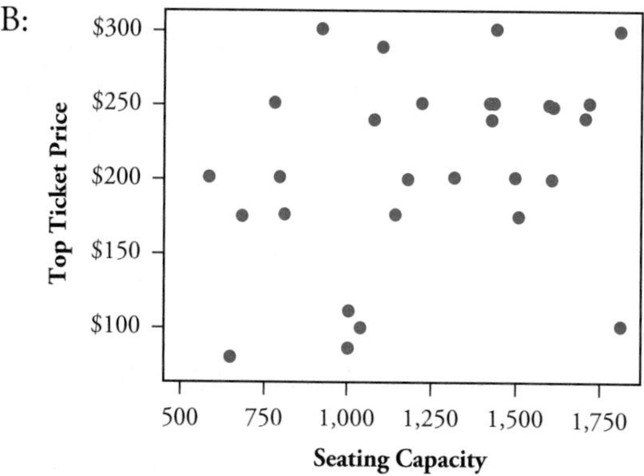

C:

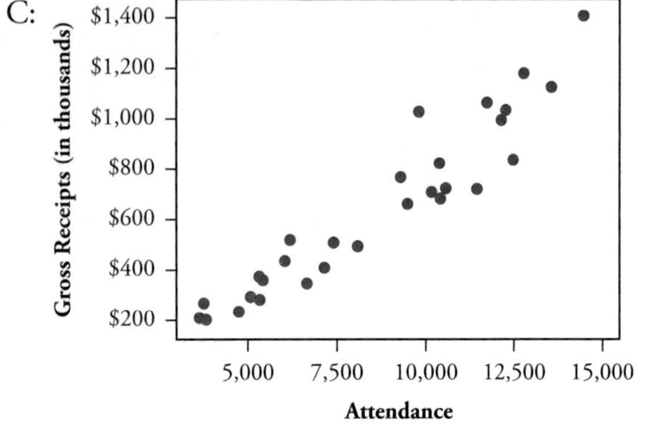

D:

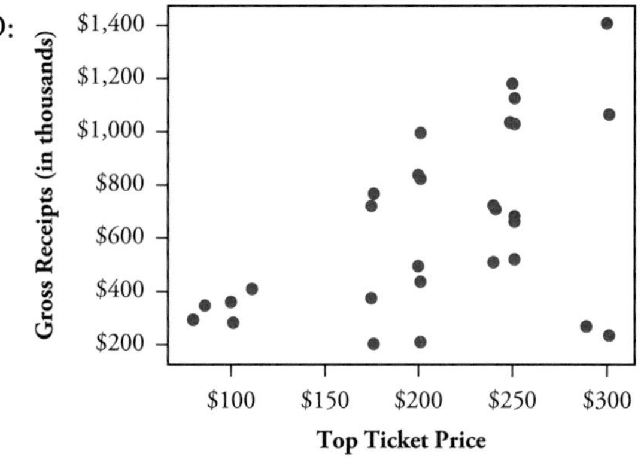

E:

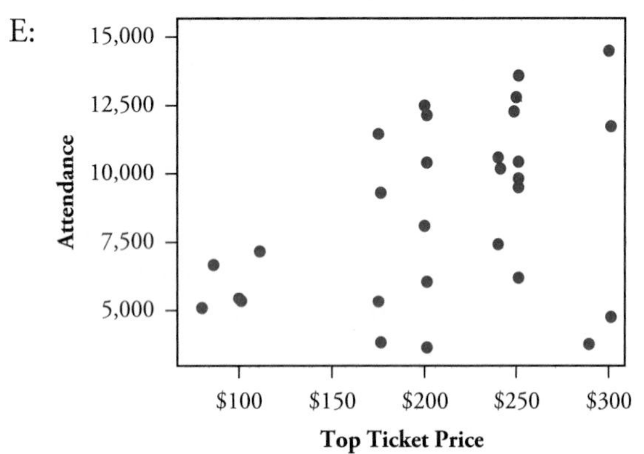

F:

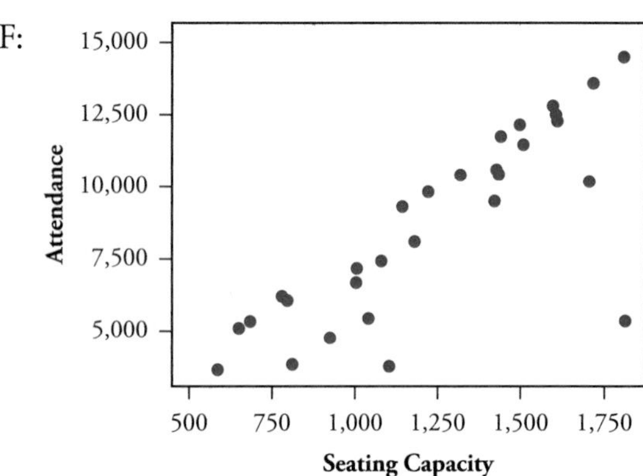

a. In which direction is the association between the variables in all of these scatterplots?

b. Arrange these scatterplots in order (by letter) from weakest to strongest association.

Activity 26-11: Broadway Shows

2-14, 26-10, **26-11,** 27-15

Reconsider the data from Activity 2-14 and the previous activity concerning Broadway shows. (The data are stored in the grouped file BROADWAY06). Notice there are two different types of shows listed: plays and musicals. The lists are named **GROSS** (gross in \$), **PTNTL** (potential in \$), **AVGPD** (average paid in \$), **TOPTK**, (top ticket in \$), **ATTND** (attendance), **CPCTY** (capacity), **PRCAP** (percent capacity), and **TYPE** (**0** for musical and **1** for play).

a. Produce a labeled scatterplot of *gross receipts* vs. *percentage capacity,* using different labels to distinguish plays from musicals. See Activity 26-5 part b for instructions on creating the scatterplot on your graphing calculator.

b. Ignoring for the moment the distinction between plays and musicals, comment on the association between receipts and percentage capacity.

c. Now comment on differences in the relationship between receipts and percentage capacity based on whether the show is a play or a musical. For shows with similar percentage capacities, does one type of show tend to take in more money than the other type of show? Explain.

Activity 26-12: Monthly Temperatures

9-20, **26-12,** 27-12

Recall from Activity 9-20 the following table reporting the average monthly temperatures for San Francisco, California, and for Raleigh, North Carolina:

	Jan	Feb	Mar	Apr	May	Jun	Jul	Aug	Sep	Oct	Nov	Dec
Raleigh	39	42	50	59	67	74	78	77	71	60	51	43
San Francisco	49	52	53	56	58	62	63	64	65	61	55	49

Below are two scatterplots of these data, labeled A and B, with a "$y = x$" line drawn on plot B:

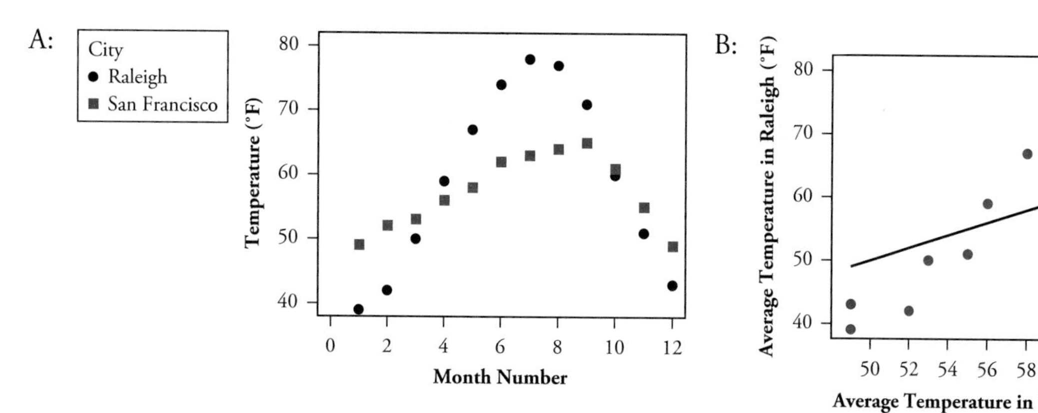

a. In how many of the 12 months is Raleigh's monthly temperature higher than San Francisco's? Explain how you could use either graph to discern this.

b. In which of the twelve months is Raleigh's monthly temperature higher than San Francisco's? With which graph can you discern this? Explain.

c. How would you characterize the strength of the association between the two cities' monthly temperatures? With which graph is it easier to judge this association? Explain.

Activity 26-13: Weighty Feelings

6-13, 25-15, **26-13**

Recall from Activity 6-13 that one of the questions in the 2003–2004 NHANES (National Health and Nutrition Examination Survey) asked respondents whether they feel their current weight is about right, overweight, or underweight. The following labeled scatterplot displays *self-reported weight* vs. *self-reported height,* using different labels (circle, diamond, square) for the three categories of self-opinion about weight:

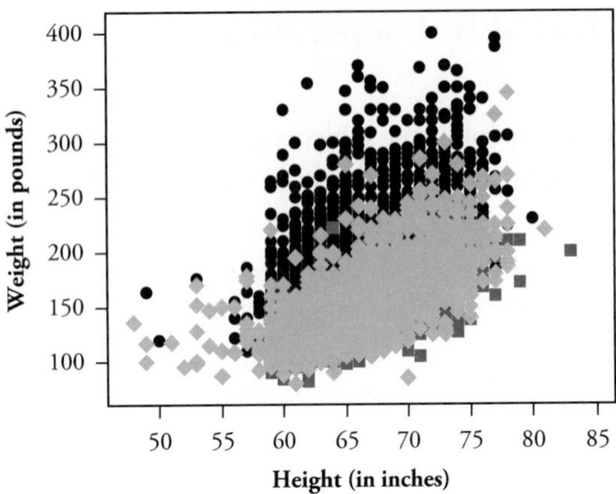

a. Which label (circle, diamond, square) goes with which category of self-opinion about weight: about right, overweight, or underweight? Explain your reasoning.

b. Write a paragraph describing what this labeled scatterplot reveals about the relationship among weight, height, and self-opinion about weight.

Activity 26-14: Fast-Food Sandwiches

26-14, 26-15, 28-16

The fast-food chain Arby's lists nutritional information for all of its sandwiches on its Web site. These data (as of June 2006) are in the grouped file ARBYS06, and data for the first 5 (of 42) sandwiches appear in the following table:

Sandwich	Serving Weight (g)	Calories	Calories from Fat	Fat Total (g)	Cholesterol (mg)	Sodium (mg)	Protein (g)	Meat
Roast Turkey Reuben Sandwich	309	611	270	30	94	1429	44	turkey
Roast Turkey & Swiss Sandwich	359	725	270	30	91	1788	45	turkey
Roast Ham & Swiss Sandwich	359	705	279	31	63	2103	36	ham
Roast Beef & Swiss Sandwich	264	777	372	41	89	1743	37	beef
Corned Beef Reuben Sandwich	309	606	293	33	83	1849	34	beef

a. Use your graphing calculator to create a scatterplot of *calories* vs. *serving weight.* Describe the association between these variables revealed by the scatterplot. The lists are named SWGHT (serving weight), CAL (calories), CAFT (calories from fat), TFAT (total fat), CHLST (cholesterol), SDM (sodium), PROTN (protein), and MEAT (where **0** is for poultry, **1** is for pork and beef, and **2** is for other).

b. Now use your graphing calculator to produce a labeled scatterplot of *calories* vs. *serving weights,* using different labels for the various types of meat. Comment on any tendencies you observe with regard to type of meat. In particular, does one type of meat tend to have more or fewer calories than other meat types of similar serving size? Explain.

Activity 26-15: Fast-Food Sandwiches

26-14, **26-15,** 28-16

Consider again the data on Arby's sandwiches from the previous activity (ARBYS06). Use your graphing calculator to examine labeled scatterplots for other pairs of variables, still using meat as the categorical variable. Select one labeled scatterplot that you find interesting, and describe what it reveals. In particular, comment on any differences you observe among the different types of meat.

Activity 26-16: College Alumni Donations

26-16, 27-16

For the graduating classes of 1961–1998 at Harvey Mudd College, the following table lists the percentage of alumni who made a financial contribution to the college during 1998–1999. Also listed is the average amount of a contribution among those who made a contribution. (The data are stored in the file HMCDONORS with lists **YEAR** (class year), **PERGV** (giving %), **AVGFT** (average gift), and **PAVGF** (previous average gift).)

Year	Alumni (%)	Avg Gift ($)	Year	Donors (%)	Avg Gift ($)	Year	Donors (%)	Avg Gift ($)	Year	Donors (%)	Avg Gift ($)
1961	67	773	1971	65	228	1981	45	339	1991	50	111
1962	81	12187	1972	60	349	1982	40	339	1992	44	142
1963	64	2951	1973	50	399	1983	51	239	1993	30	83
1964	51	2453	1974	63	2567	1984	36	255	1994	32	80
1965	58	2476	1975	56	406	1985	48	172	1995	51	58
1966	64	723	1976	47	444	1986	41	157	1996	44	66
1967	60	614	1977	54	344	1987	50	178	1997	32	124
1968	61	298	1978	53	295	1988	40	234	1998	35	236
1969	54	4945	1979	49	3106	1989	43	156			
1970	61	402	1980	48	201	1990	46	133			

a. Ignore the class year information and use your graphing calculator to analyze the distribution of donor percentages. Write a few sentences commenting on key features of this distribution.

b. Continue to ignore the class year information and use your graphing calculator to analyze the distribution of average gifts. Write a few sentences commenting on key features of this distribution.

c. Now use your graphing calculator to produce a scatterplot of *donor percentage* vs. *class year.* Comment on any patterns and any unusual observations revealed in the scatterplot.

d. Use your graphing calculator to produce a scatterplot of *average gift* vs. *class year.* Comment on any patterns and any unusual observations revealed in the scatterplot.

e. Does one class stand out as deviating from the pattern established by the majority of classes in either scatterplot? If so, identify the class and comment on what makes it unusual.

Activity 26-17: Peanut Butter

The September 1990 issue of *Consumer Reports* rated 37 varieties of peanut butter. Each variety was given an overall sensory quality rating, based on taste tests by a trained sensory panel. Also listed was the cost (in cents per three tablespoons, based on the average price paid by *Consumer Reports* shoppers), and sodium content (per three tablespoons, in milligrams) of each product. Finally, each variety was classified as creamy (cr) or chunky (ch), natural (n) or regular (r), and salted (s) or unsalted (u). The data are in the grouped file PEANUTBUTTER with lists COST, SODIU (sodium), QUAL (quality), CRCH (cr/ch), RN (r/n), and SU (s/u), respectively. (*TI note:* Some grouped files containing multiple lists may have been sent to the calculator's archive. If the TI-Connect software does this, you will need to unarchive the grouped file on your calculator first). The first few rows are repeated here:

Brand	Cost (in cents)	Sodium (in mg)	Quality	cr/ch	r/n	s/u
Jif	22	220	76	cr	r	s
Smucker's Natural	27	15	71	cr	n	u
Deaf Smith Arrowhead Mills	32	0	69	cr	n	u
Adams 100% Natural	26	0	60	cr	n	u
Adams	26	168	60	cr	n	s

a. What are the observational units?
b. Classify each variable as categorical (also binary) or quantitative.
c. Select any *pair* of quantitative variables you would like to examine. Use your graphing calculator to produce a scatterplot of these variables, and write a paragraph commenting on the relationship between them. *Hint:* Be sure to make clear which pair of variables you are analyzing, both in the axes labels and in your paragraph.
d. Now select *one* of the three categorical variables, and use your graphing calculator to produce a labeled scatterplot of your two variables from part a, using this new variable for the labels. Comment on how the relationship between the two quantitative variables differs (if at all) for the two categories of your third variable.

Activity 26-18: Digital Cameras

10-5, **26-18,** 27-21

Recall from Activity 10-5 that the July 2006 issue of *Consumer Reports* rated 78 brands of digital cameras. The following labeled scatterplot displays *rating score* vs. *price,* with labels indicating the type of camera (compact, subcompact, advanced compact, and super-zoom).

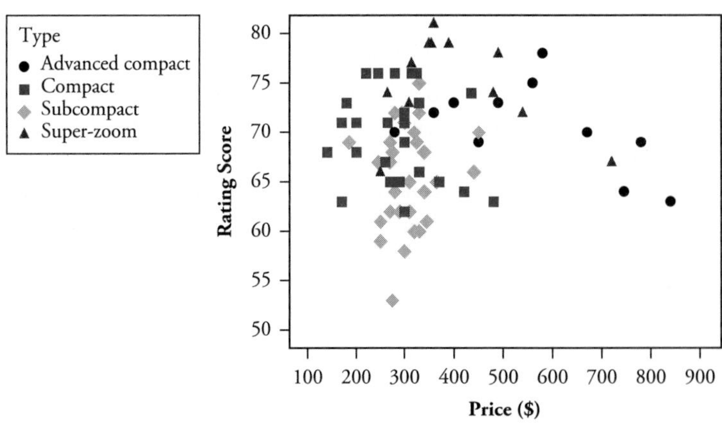

a. Does the scatterplot suggest more expensive cameras tend to have higher ratings? Describe the association between rating score and price. *Hint:* Ignore type of camera for now. Comment on direction, strength, form, and unusual observations.

b. Now consider type of camera. Summarize what the scatterplot reveals about the relationship between rating score and price for the different types of cameras.

Activity 26-19: Maternal Oxygenation

23-16, **26-19**

Recall from Activity 23-16 the study that investigated whether administering oxygen to mothers in labor increased the percentage of oxygen in the blood of a fetus exhibiting abnormal heart patterns. The following scatterplot displays the percentage of oxygen in the blood of the fetus, first with room air as a baseline and then after administering 40% oxygen to the mother:

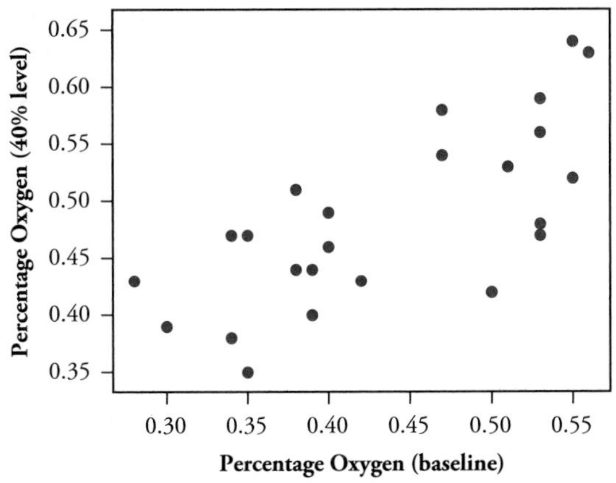

a. Does this scatterplot indicate fetuses tend to have a higher percentage of oxygen after the mother is administered 40% oxygen than at the baseline measurement? Explain how you can tell. *Hint:* You might want to draw a $y = x$ line on the scatterplot.

b. The journal article (Haydon et al., 2006) claimed fetuses with the lowest initial oxygen levels appeared to increase their oxygen percentages the most. Does the scatterplot support this conclusion? Explain.

Activity 26-20: Chip Melting

23-2, **26-20**, 29-15

Reconsider the data collected in Topic 23 and analyzed in Activity 23-2 on the time it takes a chocolate or peanut butter chip to melt in a person's mouth.

a. Produce a scatterplot with the chocolate-chip melting time on one axis and the peanut-butter-chip melting time on the other axis. Do the data suggest an association between these variables? Explain.

b. Draw a $y = x$ line on the scatterplot. Do most of the points fall on the same side (above or below) of the line? If so, or even if not, explain what this result reveals.

Activity 26-21: Comparison Shopping

23-22, 23-23, **26-21**

Reconsider the data from Activity 23-22 on page 486, which compared prices of many items at two grocery stores (SHOPPING).

a. Identify the observational units and the two quantitative variables recorded about the observational units in this study.
b. Use your graphing calculator to examine a scatterplot of the *prices at Luckys* vs. *prices at Vons.* Write a paragraph describing the relationship between the prices at the two stores. Also address the issue of where the points tend to fall compared to the $y = x$ line.
c. Are there any products for which the scatterplot makes you suspicious of the prices recorded? Explain.

Activity 26-22: Muscle Fatigue

23-5, 23-15, **26-22,** 27-20

Recall from Activity 23-5 the matched-pairs study comparing time until muscle fatigue between men and women. At the beginning of the study, researchers matched men and women based on strength. Study results, measured in times until task failure (in seconds), are reproduced here:

Pair	A	B	C	D	E	F	G	H	I	J
Man's Time Until Fatigue	314	335	379	384	392	511	566	629	697	923
Woman's Time Until Fatigue	1204	1041	3963	368	334	846	888	2446	2181	813

a. Produce a scatterplot of these data, and draw a $y = x$ line. *Hint:* Clearly label which variable you put on which axis.
b. What does the $y = x$ line reveal about whether men or women tend to last longer before muscle fatigue? Explain.
c. Does this scatterplot reveal much association between the variables? Does this association indicate that men and women of similar strength tend to have similar times until muscle fatigue? Explain.

Activity 26-23: Your Choice

6-30, 12-22, 19-22, 21-28, 22-27, **26-23,** 27-22

Describe two different *pairs* of quantitative variables whose relationship you might be interested in studying. Be very specific in describing these variables, and identify the *observational units* as well as the variables.

TOPIC

27 Correlation Coefficient

In 1970, the United States Selective Service instituted a draft to decide which young men would be forced to join the armed forces. Wanting to be completely fair, they used a random lottery process that assigned draft numbers to birthdays: those born on days with low draft numbers were drafted. But was the lottery process carried out in a fair, truly random manner? In this topic, you will learn a new technique for analyzing such data and answering this question.

Overview

In the previous topic, you saw how scatterplots provide useful visual information about the relationship between two quantitative variables. Rather than relying on visual impressions alone, however, it is also handy to have a *numerical* measure of the strength of association between two variables—just as you made use of numerical summaries for various aspects of a single variable's distribution. This topic introduces you to such a measure and asks you to investigate some of its properties. This measure, one of the most famous in statistics, is the correlation coefficient.

Preliminaries

1. If every student in a class scores exactly ten points lower on the second exam than on the first exam, does that indicate a positive association, negative association, or no association between the two exam scores?

2. If every student in a class scores exactly half as many points on the second exam than on the first exam, does that indicate a positive association, negative association, or no association between the two exam scores?

3. Would you consider the strength of the association to be higher in situation 1 than in situation 2, or higher in situation 2 than in situation 1, or equally strong in both situations?

4. In a perfectly fair, random lottery, would you expect to see much of a relationship between draft number and sequential date (where January 1 is coded as 1 and December 31 as 366) of the birthday? Explain.

In-Class Activities

Activity 27-1: Car Data

26-3, **27-1,** 28-10

Recall from Activity 26-3 on page 541 the nine scatterplots related to car data.

The following table arranges those scatterplots according to the direction and strength of association revealed in them:

	Negative				None	Positive			
	Strongest			Weakest		Weakest			Strongest
Letter of Scatterplot	D	G	A	H	C	E	I	F	B
Correlation Coefficient									

The **correlation coefficient,** denoted by r, is a number that measures the degree to which two quantitative variables are associated.

The calculation of r is very tedious to do by hand, so you will begin by letting your calculator compute correlation coefficients while you explore their properties. Download the grouped file CARS99, which contains the lists **PNUM** (page number), **CMPG** (city mpg), **HMPG** (highway mpg), **FCAP** (fuel capacity), **WGT** (weight), **FWGT** (front weight), **ACC30** (acceleration 0–30 mph), **ACC60** (acceleration 0–60 mph), **MILE** (time to travel ¼ mile), and **TYPE** (**0** for family car, **1** for small car, **2** for luxury car, **3** for large car, **4** for sports car, and **5** for upscale car). (*TI note:* This grouped file may be sent to the calculator's archive.) Next, download the program CORR.

a. Use the CORR program on your calculator to compute the value of the correlation coefficient between *time to travel 1/4 mile* and *weight*. Record this value in the preceding table in the column corresponding to scatterplot A.

b. Now use your calculator to compute the value of the correlation coefficient for the other eight scatterplots. Record these in the table below the appropriate letters.

c. Based on these results, what do you suspect is the largest value that a correlation coefficient can assume? What do you suspect is the smallest value?

Largest: Smallest:

d. Under what circumstances do you think the correlation coefficient assumes its largest or smallest value? *Hint:* Consider what would have to be true of the scatterplot.

e. How does the value of the correlation relate to the *direction* of the association?

f. How does the value of the correlation relate to the *strength* of the association?

These examples should convince you that a correlation coefficient has to be between +1 and −1, and it equals one of those values when the observations form a perfectly straight line. The sign of the correlation coefficient reflects the direction of the association (e.g., positive values of r correspond to a positive linear association). The magnitude of the correlation coefficient indicates the strength of the association, with values closer to +1 or −1 signifying a stronger linear association.

Activity 27-2: Governors' Salaries

The following table reports governors' salaries for the fifty states (as of the year 2005), along with the median housing prices for the states. These data are stored in the grouped file GOVERNORS05 with lists **MEDHP** (median housing price), **MEDHV** (median housing value), and **GSLRY** (governor's salary).

State	Governor's Salary	Median Housing Price	State	Governor's Salary	Median Housing Price
Alabama	$96,361	$85,100	Louisiana	$95,000	$85,000
Alaska	$85,776	$144,200	Maine	$70,000	$98,700
Arizona	$95,000	$121,300	Maryland	$145,000	$146,000
Arkansas	$77,028	$72,800	Massachusetts	$135,000	$185,700
California	$175,000	$211,500	Michigan	$177,000	$115,600
Colorado	$90,000	$166,600	Minnesota	$120,303	$122,400
Connecticut	$150,000	$166,900	Mississippi	$122,160	$71,400
Delaware	$114,000	$130,400	Missouri	$120,087	$89,900
Florida	$129,060	$105,500	Montana	$96,462	$99,500
Georgia	$128,903	$111,200	Nebraska	$85,000	$88,000
Hawaii	$94,780	$272,700	Nevada	$117,000	$142,000
Idaho	$98,500	$106,300	New Hampshire	$104,758	$133,300
Illinois	$150,691	$130,800	New Jersey	$175,000	$170,800
Indiana	$95,000	$94,300	New Mexico	$110,000	$108,100
Iowa	$107,482	$82,500	New York	$179,000	$148,700
Kansas	$103,813	$83,500	North Carolina	$123,819	$108,300
Kentucky	$112,705	$86,700	North Dakota	$88,926	$74,400

(continued)

27

State	Governor's Salary	Median Housing Price	State	Governor's Salary	Median Housing Price
Ohio	$132,292	$103,700	Texas	$115,345	$82,500
Oklahoma	$117,571	$70,700	Utah	$104,600	$146,100
Oregon	$93,600	$152,100	Vermont	$168,466	$111,500
Pennsylvania	$144,416	$97,000	Virginia	$175,000	$125,400
Rhode Island	$105,194	$133,000	Washington	$148,035	$168,300
South Carolina	$106,078	$94,900	West Virginia	$95,000	$72,800
South Dakota	$103,222	$79,600	Wisconsin	$131,768	$112,200
Tennessee	$85,000	$93,000	Wyoming	$105,000	$96,600

a. What are the observational units for these data?

b. Use your calculator to produce a scatterplot of *governor's salary* vs. *median housing price.* Describe the association (direction, strength, and form) between these two variables.

c. Based on this scatterplot, guess the value of the correlation coefficient between *governor's salary* and *median housing price.*

d. Use the CORR program on your calculator to compute the value of this correlation. Record this value, and comment on the accuracy of your guess.

e. Does one of the states appear to be unusual in the scatterplot? Which state? Describe what is unusual about it, as compared to the other states.

f. Suppose Hawaii gives its governor a $100,000 raise. Make this change in the data. Then reproduce the scatterplot and recalculate the value of the correlation coefficient. Has the correlation coefficient changed much?

g. Repeat part f, after giving the governor of Hawaii an additional $100,000 raise.

h. Now suppose Hawaii decides to make its governorship an unpaid position. Change the governor of Hawaii's salary to **$0.** Then reproduce the scatterplot and recalculate the value of the correlation coefficient. Has the correlation coefficient changed much?

i. Based on these calculations, would you say the correlation coefficient is a *resistant* measure of association? Explain.

j. Why do you suspect government agencies tend to report the *median* rather than the *mean* housing price?

Activity 27-3: Televisions and Life Expectancy

26-6, **27-3,** 28-19

Reconsider the data from Activity 26-6 about life expectancy and number of televisions per thousand people in a sample of 22 countries (TVLIFE). A scatterplot is reproduced here.

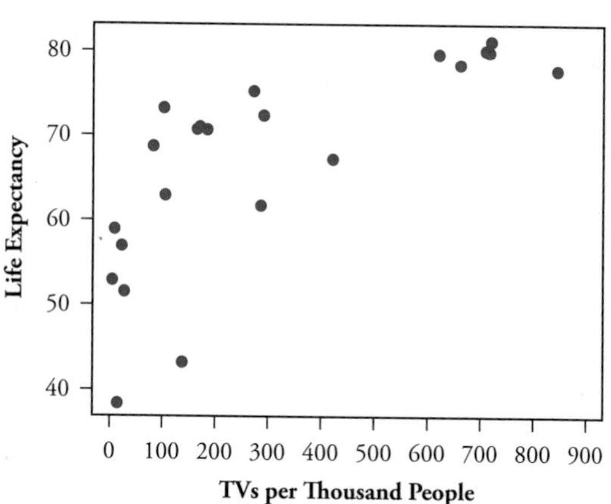

a. Describe the direction and strength of the association between life expectancy and number of televisions per thousand people in these countries. Also comment on whether this association follows a linear form.

b. Based on this scatterplot, guess the value of the correlation coefficient between *life expectancy* and *televisions per thousand people* in these countries.

c. Use the CORR program on your calculator to compute this correlation coefficient. How accurate is your guess?

d. Would you say the value of the correlation coefficient is fairly high, even though the association between the variables is not linear?

e. Does the fairly high value of the correlation coefficient provide evidence of a cause-and-effect relationship between number of televisions and life expectancy? Explain.

Watch Out

- Correlation measures the degree of *linear* association between two quantitative variables. But even when two variables display a nonlinear relationship, the correlation between them still might be quite high when there is a strong increasing or decreasing trend. With these data, the relationship is clearly curved and not linear, and yet the correlation is still fairly high. Do not assume from a high correlation coefficient that the relationship between the variables must be only linear. Always look at a scatterplot, in conjunction with the correlation coefficient, to assess the form (linear or not) of the association.
- No matter how close a correlation coefficient is to ± 1, and no matter how strong the association between two variables, a cause-and-effect conclusion cannot necessarily be drawn from observational data. There are far more plausible explanations for why countries with lots of televisions per thousand people tend to have long life expectancies. For example, the technological sophistication of the country is related to both number of televisions and life expectancy.

Activity 27-4: Guess the Correlation

This activity will give you practice at judging the value of a correlation coefficient by examining a scatterplot.

a. Open the applet `Guess the Correlation`. Keep **15** for the Number of Points, and click New Sample. The applet will generate some "pseudo-random data" and produce a scatterplot.

Guess the Correlation

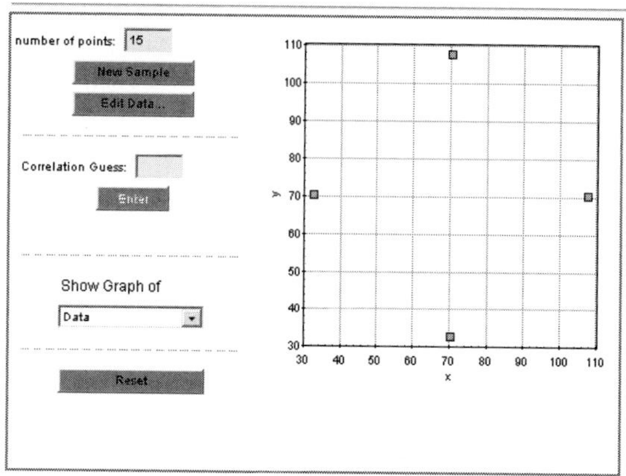

Based solely on the scatterplot, guess the value of the correlation coefficient. Enter your guess in the Correlation Guess field in the applet, and click Enter. The applet then reports the actual value of the correlation coefficient. Record your guess and the actual value in the first empty column of the following table:

Repetition Number	1	2	3	4	5	6	7	8	9	10
Your Guess										
Actual Value										

b. Click New Sample to generate another scatterplot of pseudo-random data. Enter your guess for the value of the correlation coefficient in the applet. Then record your guess and the actual value of the correlation coefficient in the preceding table. Repeat for a total of ten repetitions.

c. After the ten repetitions, guess the value of the correlation coefficient between *your guesses* for *r* and the *actual values* of *r*.

d. From the applet's pull-down menu below Show Graph Of, select Guess vs. Actual. The applet will create the scatterplot of your ten guesses and the corresponding actual correlation coefficients and will also report the correlation coefficient between your guesses and the actual values. Record this correlation coefficient. Does the value surprise you?

e. Use the applet to examine a scatterplot of *your errors* vs. *the actual values.* Is there evidence you are better at guessing certain correlation coefficient values than other values? Explain.

f. Use the applet to examine a scatterplot of *your errors* vs. *the repetition (trial) number.* Is there evidence your guesses were more accurate or less accurate as you went along? Explain.

g. Suppose all of your guesses had been too high by exactly 0.1, what would the correlation coefficient between your guesses and the actual values be? *Hint:* Think about what the scatterplot would look like.

h. Repeat part g, supposing your guesses had all been too low by exactly 0.5.

i. If the correlation coefficient between your guesses and the actual values is 1.0, does this mean you guessed perfectly every time? What does this value reveal about the utility of the correlation coefficient as a measure of your guessing prowess? Explain.

Activity 27-5: House Prices

8-10, 26-1, **27-5,** 28-2, 28-12, 28-13, 29-3

Reconsider the data on house prices from Activity 26-1. The mean house price is \$482,386, and the standard deviation is \$79,801.5. The mean house size is 1288.1 square feet, and the standard deviation is 369.191 square feet.

You can gain some insight into how the correlation coefficient r measures association by examining the formula for its calculation:

$$r = \frac{1}{n-1}\sum_{i=1}^{n}\left(\frac{x_i - \bar{x}}{s_x}\right)\left(\frac{y_i - \bar{y}}{s_y}\right)$$

where x_i denotes the ith observation of one variable, y_i the ith observation of the other variable, $\bar{x}$ and $\bar{y}$ the respective sample means, s_x and s_y the respective sample standard deviations, and n the sample size. This formula says to standardize each x and y value into its z-score, multiply these z-scores together for each observational unit, add those results, and finally divide the sum by one less than the sample size.

The following table begins the process of calculating the correlation between house price and size by calculating the houses' z-scores for price and size and then multiplying the results.

Address	Price (\$)	Price z-score	Size	Size z-score	Product of z-scores
2130 Beach St.	311,000		460	−2.243	
2545 Lancaster Dr.	344,720	−1.725	1030	−0.699	1.206
415 Golden West Pl.	359,500	−1.54	883	−1.097	1.69
990 Fair Oaks Ave.	414,000	−0.857	728	−1.517	1.3
845 Pearl Dr.	459,000	−0.293	1242	−0.125	0.037

Address	Price ($)	Price *z*-score	Size	Size *z*-score	Product of *z*-scores
1115 Rogers Ct.	470,000	−0.155	1499	0.571	−0.089
579 Halcyon Rd.	470,000	−0.155	1419	0.355	−0.055
1285 Poplar St.	470,000	−0.155	952	−0.91	0.141
1080 Fair Oaks Ave.	474,000	−0.105	1014	−0.742	0.078
690 Garfield Pl.	475,000	−0.093	1615	0.885	−0.082
1030 Sycamore Dr.	490,000	0.095	1664	1.018	0.097
620 Eman Ct.	492,000	0.12	1160	−0.347	−0.042
529 Adler St.	500,000	0.221	1545	0.696	0.154
646 Cerro Vista Cir.	510,000	0.346	1567	0.755	0.261
926 Sycamore Dr.	520,000	0.471	1176	−0.304	−0.143
227 S. Alpine St.	541,000	0.734	1120	−0.455	−0.334
654 Woodland Ct.	567,500	1.067	1549	0.707	0.754
2230 Paso Robles St.	575,000	1.161	1540	0.682	0.792
2461 Ocean St.	580,000	1.223	1755		
833 Creekside Dr.	625,000	1.787	1844	1.506	2.691

a. Calculate the *z*-score for the price of 2130 Beach St. and for the size of 2461 Ocean St. Then calculate the product of the *z*-scores for these two houses. Show your calculations below and record the results in the table.

b. The sum of the products turns out to equal 14.819. Use this information, and the fact that there are 20 houses in this sample, to determine the value of the correlation coefficient between *house price* and *size*.

c. What do you notice about the *size z*-score for most of the houses with negative *price z*-scores? Explain how the signs of these *z*-scores result from the strong positive association between house *price* and *size*.

d. Download the grouped file HOUSEPRICESAG, containing the lists **HPRC** (price), **SIZE**, **BATH** (number of baths), and **BEDRM** (number of bedrooms), and confirm your calculation in part b by using your calculator to compute the value of the correlation coefficient between *house price* and *size*.

27

Activity 27-6: Exam Score Improvements

23-19, 23-20, 23-21, **27-6**

Reconsider the data on exam scores from Activity 23-19 (EXAMSCORES).

a. Use your calculator to produce a scatterplot of *exam 2 score* vs. *exam 1 score.* Comment on the direction, strength, and form of the association revealed.

b. Use your calculator to compute the correlation coefficient between *exam 1 score* and *exam 2 score.*

c. Now suppose each student scores 10 points lower on exam 1 than he/she actually did. How would you expect this result to affect the value of the correlation coefficient between *exam 2 score* and *exam 1 score?* Explain.

d. Use your calculator to make this change (subtract 10 points from everyone's score on exam 1.) Reproduce the scatterplot of *exam 2 score* vs. *new exam 1 score* and recalculate the correlation coefficient. How did the correlation value change?

e. Now suppose each student scores twice as many points on exam 2 as he/she actually did. How would you expect this result to affect the value of the correlation coefficient between *exam 2 score* and *exam 1 score?* Explain.

f. Use your calculator to make this change: double everyone's score on exam 2. Reproduce the scatterplot of this *new exam 2 score* vs. *new exam 1 score,* and recalculate the correlation. How did the correlation value change?

These questions demonstrate another property of the correlation coefficient: It does not change if the scale of measurement is altered by adding a constant or multiplying by a constant.

g. Now consider a different (hypothetical) class of students. Suppose each student scores exactly 10 points higher on exam 2 than he/she does on exam 1. What do you think the value of the correlation coefficient would be between *exam 1 score* and *exam 2 score?* Explain your reasoning. *Hint:* Consider what the scatterplot would look like.

h. Make up some hypothetical bivariate data with the property described in part g. *Hint:* Choose any values at all for the exam 1 scores, and then make sure each exam 2 score is 10 points higher. Do this for at least 5 hypothetical students. Then use your calculator to produce a scatterplot and calculate the correlation coefficient. Does this confirm the value you expected in part g, or do you need to revise your thinking?

i. Now suppose each student scores exactly twice as many points on exam 2 than he/she does on exam 1. What do you think the value of the correlation coefficient would be between *exam 1 score* and *exam 2 score?* Explain your reasoning. *Hint:* Consider what the scatterplot would look like.

j. Make up some hypothetical bivariate data with the property described in part i. Then use your calculator to produce a scatterplot and calculate the correlation. Does this confirm the value you expected in part i, or do you need to revise your thinking?

Watch Out

- A correlation coefficient is a number! In fact, it is a number between −1 and 1, inclusive. Although this might seem obvious by now, many students say "the same" and do not give a number in response to question g.
- The slope, or steepness, of the points in a scatterplot is unrelated to the value of the correlation coefficient. If the points fall on a perfectly straight line with a positive slope, then the correlation coefficient equals 1.0 whether that slope is very steep or not steep at all. What matters for the magnitude of the correlation is how closely the points concentrate around a line, not the steepness of a line.

Activity 27-7: Draft Lottery

10-18, **27-7,** 29-9

Recall from Activity 10-18 that in 1970 the United States Selective Service conducted a lottery to decide which young men would be drafted into the armed forces (Fienberg, 1971). Each of the 366 birthdays of the year was assigned a draft number. Young men born on days assigned low draft numbers were drafted. The grouped file DRAFTLOTTERY lists the draft number assigned to each birthday. The "sequential date" column lists the birthday as a number from 1–366 (January 1 is coded as 1 and December 31 as 366). The lists in the grouped file are SEQ70 (sequential date for 1970), DRF70 (draft numbers for 1970), SEQ71 (sequential date for 1971), and DRF71 (draft numbers for 1971). *Note:* The lottery numbers assigned by month are also included in lists within this grouped file.

a. What draft number was assigned to your birthday?

b. In a perfectly fair, random lottery, what should the correlation coefficient between *draft number* and *sequential date of the birthday* equal? Explain.

c. Use your calculator to produce a scatterplot of *draft number* vs. *sequential date of the birthday.* Based on the scatterplot, guess the value of the correlation coefficient. Explain the reasoning behind your guess.

d. Use your calculator to compute the value of the correlation coefficient. Does its value surprise you? If so, look back at the scatterplot to see whether, in hindsight, its value makes sense. Summarize what the value of this correlation coefficient reveals about how the draft numbers were distributed across birthdays throughout the year.

e. Data for 1971 are also stored in the grouped file DRAFTLOTTERY. Examine a scatterplot, and calculate the correlation coefficient between draft number and sequential date for that year's lottery. Comment on your findings.

Solution

a. Answers will vary.

b. With a perfectly fair, random lottery, there should be no association between *draft number* and *sequential date for the birthday.* In other words, these variables should be independent, so the correlation coefficient would equal zero. With an actual lottery, you would not expect the correlation coefficient to equal exactly zero, but it should be close to zero.

c. The scatterplot is shown here.

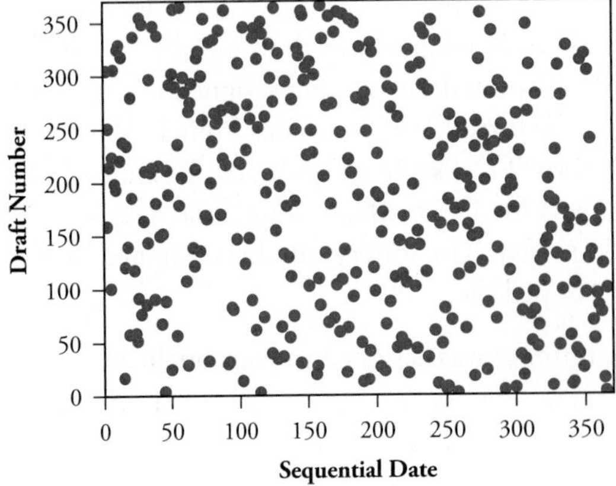

It's hard to see a relationship between the variables in this scatterplot, so a reasonable guess for the value of the correlation coefficient would be close to zero.

d. Your calculator reveals the correlation coefficient to equal −.226. This indicates a weak negative association between draft number and sequential date. Although not large, this correlation value is farther from zero than most people expect. Looking at the scatterplot more closely, you can see there are few points in the top right and bottom left of the graph. This result suggests few birthdays late in the year were assigned high draft numbers, and few birthdays early in the year were assigned low draft numbers, which means young men born late in the year were at a disadvantage and had a better chance of getting a low draft number. Birthdays late in the year were not mixed as thoroughly as those earlier in the year, so they tended to be selected early in the process and thereby assigned a low draft number.

e. The scatterplot for the 1971 draft lottery data is shown here.

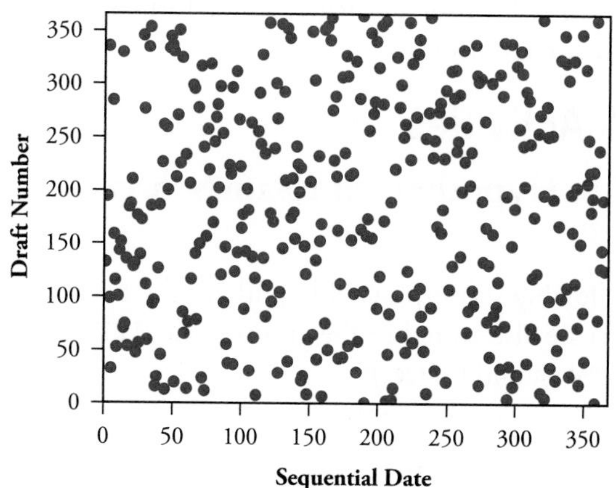

The correlation coefficient is .014, which is very close to 0. This value indicates there is no evidence of association between draft number and sequential date, suggesting the lottery process was fair and random in 1971. The mixing mechanism was greatly improved after the anomaly with the 1970 results was spotted.

27

Wrap-Up...........

In this topic, you discovered the correlation coefficient as a measure of the linear relationship between two variables. Analyzing pairs of variables for the house data, you discovered some of the properties of this measure. For example, a correlation value has to be between −1 and +1, inclusive. The sign of the correlation coefficient reflects the direction of the association. The magnitude of the correlation coefficient reflects the strength of the association, with correlation coefficients close to −1 or +1 indicating very strong association, and correlation coefficients close to 0 reflecting very weak linear association. But also keep in mind that you discovered the correlation coefficient is not *resistant* to outliers, as altering simply one state's value for governor's salary changed the value of the correlation considerably. It is important to always accompany your interpretation of the correlation coefficient with a scatterplot.

You also learned how to calculate a correlation coefficient based on *z*-scores and gained practice judging the value of a correlation based on a scatterplot. Finally, with the data on televisions and life expectancy, you saw again that you should not infer a causal relationship between variables based on a high correlation.

Some useful definitions to remember and habits to develop from this topic include

- The **correlation coefficient** is a number that measures the direction and strength of linear association between two quantitative variables.
- The correlation coefficient is not resistant to outliers. One very unusual point can produce a large correlation coefficient even when most of the data reveals no pattern, or a small correlation coefficient when most of the data follows a clear linear pattern.
- Always examine a scatterplot in addition to calculating a correlation coefficient. A clear nonlinear relationship can have a small (close to zero) correlation, and a correlation can be close to –1 or +1, even if the relationship follows a curve or other nonlinear pattern.
- Never forget a large correlation coefficient between two variables does not necessarily establish a cause-and-effect relationship between those variables.

The next topic will expand your understanding of relationships between variables by introducing you to least squares regression, a formal mathematical model often used for describing such relationships.

Homework Activities

Activity 27-8: Hypothetical Exam Scores

7-12, 8-22, 8-23, 9-22, 10-22, **27-8**

Consider the following scatterplots of hypothetical scores on two exams for Class A and Class B (the data are also stored in the grouped file HYPOEXAMS, which contains lists EXM1A, EXM1B, EXM2B, and EXM2C.):

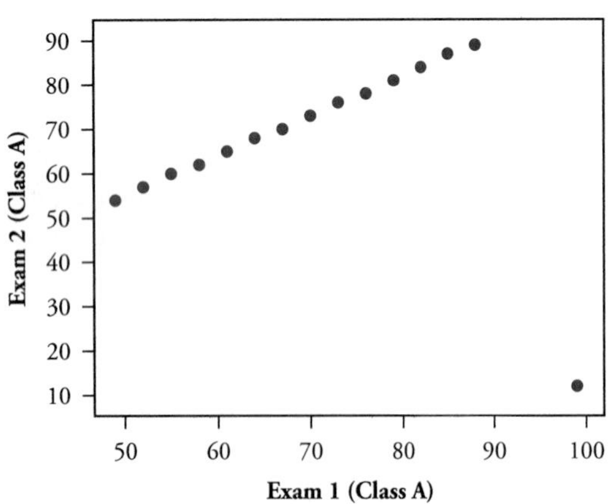

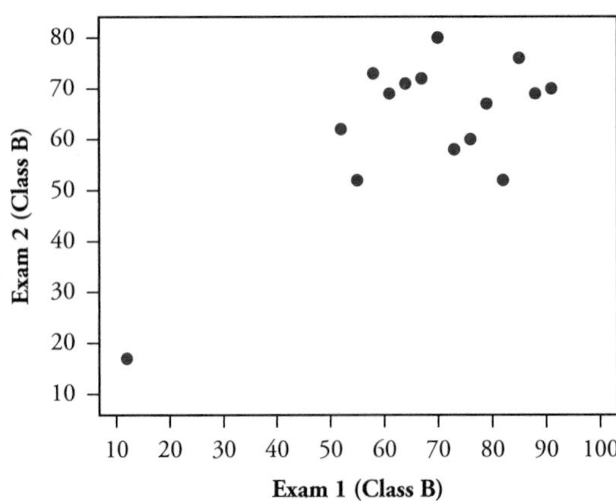

a. In class A, do most of the exam scores follow a linear pattern? Are there any exceptions?

b. In class B, are most of the exam scores scattered haphazardly with no apparent pattern? Are there any exceptions?

c. Use your calculator to compute the correlation coefficient between *exam 1 score* and *exam 2 score* for each of these classes. Are you surprised at either of the values? Explain.

d. Describe how these scatterplots pertain to the issue of resistance of the correlation coefficient.

Now consider the following scatterplot of exam data for class C:

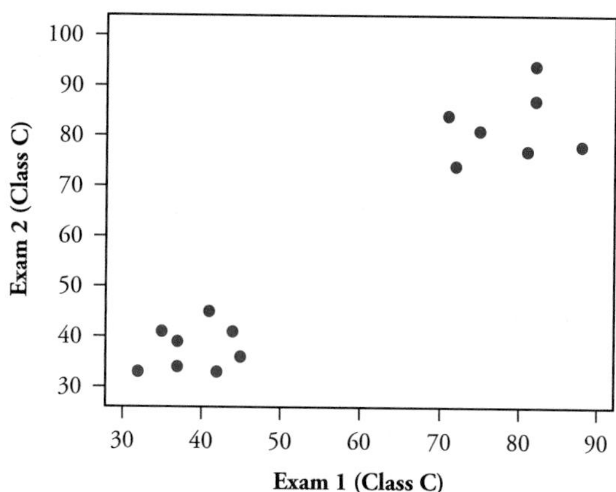

e. Describe what the scatterplot reveals about the relationship between exam scores in class C.

f. Use your calculator to compute the correlation coefficient between exam scores in class C. Is its value higher than you expected? Explain what this example reveals about correlation.

Activity 27-9: Proximity to the Teacher

1-11, 5-23, 27-9, 29-6

Recall from Activity 5-23 the idea of studying whether students who sit closer to the teacher tend to have higher quiz scores than students who sit farther away from the teacher. Suppose you measure distance from the teacher and average quiz score for a group of students. Explain how you know each of the following statements is in error:

a. The correlation between *distance* and *quiz average* is −1.8.

b. The correlation between *distance* and *quiz average* is −.8, and the correlation between *quiz average* and *distance* is −.4.

c. The correlation is −.8, so there is no association between *distance* and *quiz average*.

d. The correlation between *quiz average* and *gender* is −.8.

e. The correlation between *distance* and *quiz average* is −.8, so students who sit farther away tend to score higher.

f. The correlation between *distance* and *quiz average* is −.8, so sitting closer to the teacher must cause students to score higher on quizzes.

Activity 27-10: Monopoly

27-10, 27-11

The following table presents data on properties in the board game Monopoly. Listed are the property's position on the game board, purchase price, rental charge, rental charge with one house, and rental charge with one hotel (the data are also stored in the grouped file MONOPOLY, containing lists HOTEL, HOUSE, POSIT (position), PRICE, and RENT, respectively):

27

Property	Pos	Price ($)	Rent ($)	House ($)	Hotel ($)	Property	Pos ($)	Price ($)	Rent ($)	House ($)	Hotel ($)
Mediterranean	1	60	2	10	250	Kentucky	21	220	18	90	1050
Baltic	3	60	4	20	450	Indiana	23	220	18	90	1050
Oriental	6	100	6	30	550	Illinois	24	240	20	100	1100
Vermont	8	100	6	30	550	Atlantic	26	260	22	110	1150
Connecticut	9	120	8	40	600	Ventnor	27	260	22	110	1150
States	11	140	10	50	750	Marvin Gardens	29	280	24	120	1200
St. Charles Place	13	140	10	50	750	Pacific	31	300	26	130	1275
Virginia	14	160	12	60	900	North Carolina	32	300	26	130	1275
St. James Place	16	180	14	70	950	Pennsylvania	34	320	28	150	1400
Tennessee	18	180	14	70	950	Park Place	37	350	35	175	1500
New York	19	200	16	80	1000	Boardwalk	39	400	40	200	2000

a. Suppose you are interested in predicting the rent of a property based on its price. Use your calculator to produce a scatterplot of *rent* vs. *price.* Based on the scatterplot, guess the value of the correlation coefficient between *rent* and *price.*

b. Use your calculator to compute the correlation coefficient in part a. How close was your guess to the actual value?

c. Change Boardwalk's price and rent values to those listed in the following table. In each instance (trial), produce a scatterplot and guess the value of the correlation for the revised data. Then calculate the actual value of the correlation coefficient. Record your guesses and the actual values.

Trial	1 ($)	2 ($)	3 ($)	4 ($)	5 ($)	6 ($)	7 ($)
Boardwalk Price	400	400	400	400	100	1	1
Boardwalk Rent	40	100	1	1000	40	40	100

d. Summarize what this activity reveals about whether the correlation coefficient is a resistant measure of association. Explain.

Activity 27-11: Monopoly

27-10, **27-11**

Reconsider the Monopoly data from the previous activity.

a. What are the observational units for these data?

b. How many variables are presented for each observational unit?

c. Select one pair of variables, other than the *price* vs. *rent* variables analyzed in the previous activity. Produce a scatterplot and comment on the association revealed.

d. Use your calculator to compute the correlation coefficient for your pair of variables from part c.
e. Would you expect the correlation coefficient from part d to increase or decrease if Boardwalk were removed from the analysis? Explain your reasoning.
f. Remove Boardwalk, and recalculate the value of the correlation coefficient. Was your expectation in part e correct? If not, explain the flaw in your reasoning.

Activity 27-12: Monthly Temperatures

9-20, 26-12, **27-12**

Reconsider Activities 9-20 and 26-12 and the data on average monthly temperatures in Raleigh, North Carolina, in degrees Fahrenheit:

	Jan	Feb	Mar	Apr	May	Jun	Jul	Aug	Sep	Oct	Nov	Dec
Avg Temp	39	42	50	59	67	74	78	77	71	60	51	43

The following scatterplot displays Raleigh's average monthly temperature vs. the month number (January = 1, February = 2, and so on):

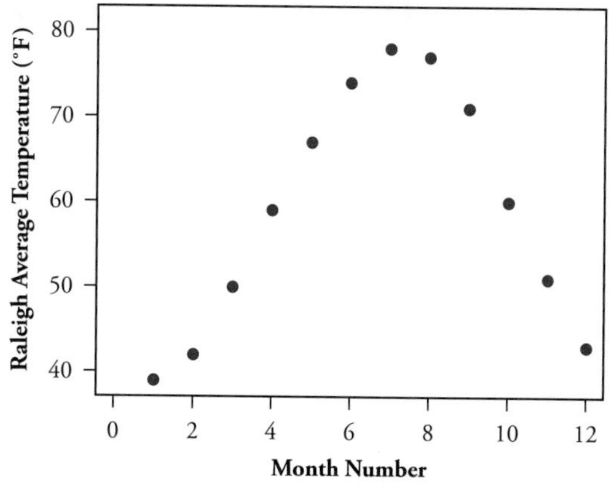

a. Does there appear to be any relationship between *temperature* and *month* in Raleigh? If so, describe the relationship.
b. Use your calculator to compute the correlation coefficient between these variables. Does this correlation value seem to indicate a strong or a weak relationship?
c. Explain why the correlation is so close to 0 even though the scatterplot reveals a clear relationship between *temperature* and *month.*

Activity 27-13: Planetary Measurements

8-12, 10-20, 19-20, **27-13,** 28-20, 28-21

Reconsider the data from Activity 8-12 on planetary measurements. The following scatterplot displays the period of revolution around the sun (in earth days) vs. the distance from the sun (in millions of miles).

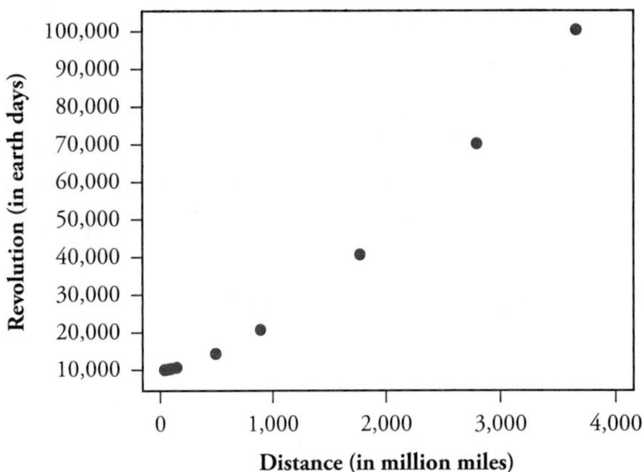

a. Describe the association between these variables as revealed in the scatterplot.

b. Would a straight line appear to be a reasonable summary of the relationship between *revolution* and *distance?* Explain.

c. The correlation coefficient between *revolution* and *distance* turns out to equal .989. This value is very close to 1. Does this value mean a straight line is the best model for a reasonable summary of the relationship between *revolution* and *distance?* Explain.

Activity 27-14: Ice Cream, Drownings, and Fire Damage

a. Suppose a beach community keeps track of the amount of ice cream sold in a given month and the number of drownings that occur in that month. Would you expect to find a negative correlation, a positive correlation, or a correlation close to zero? Explain your reasoning.

b. If the community in part a were to find a strong positive correlation between ice cream sales and drownings, would that mean ice cream *causes* drowning? If not, suggest an alternative explanation (i.e., a *confounding* variable) for the strong association.

c. Explain why you would expect to find a positive correlation between the number of fire engines that respond to a fire and the amount of damage done in the fire. Does this imply the damage would be less extensive if fewer fire engines were dispatched? Explain.

Activity 27-15: Broadway Shows

2-14, 26-10, 26-11, **27-15**

Reconsider Activity 26-10 on page 551, where you looked at scatterplots for variables pertaining to Broadway shows (see the data stored in the file `BROADWAY06`, containing the lists **ATTND, ACGPD, CPCTY, GROSS, PRCAP, PTNTL, TOPTK,** and **TYPE,** respectively.).

a. Use your calculator to compute the correlation coefficient between the following pairs of variables:

A: *gross* and *capacity*
B: *top price* and *capacity*
C: *gross* and *attendance*
D: *gross* and *top price*
E: *attendance* and *top price*
F: *attendance* and *capacity*

b. Arrange these in order from the smallest correlation to the largest correlation. Does this ordering agree with the ordering based on scatterplots alone in Activity 26-10 on page 551?

Activity 27-16: College Alumni Donations

26-16, **27-16**

Reconsider the data from Activity 26-16 on page 555, concerning alumni donations at Harvey Mudd College during 1998–1999 (HMCDONORS). The following scatterplots display average gift vs. percentage giving per class and also average gift vs. previous year's average gift per class.

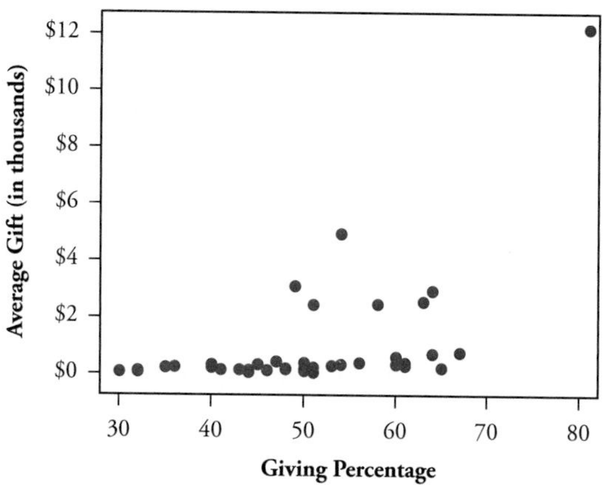

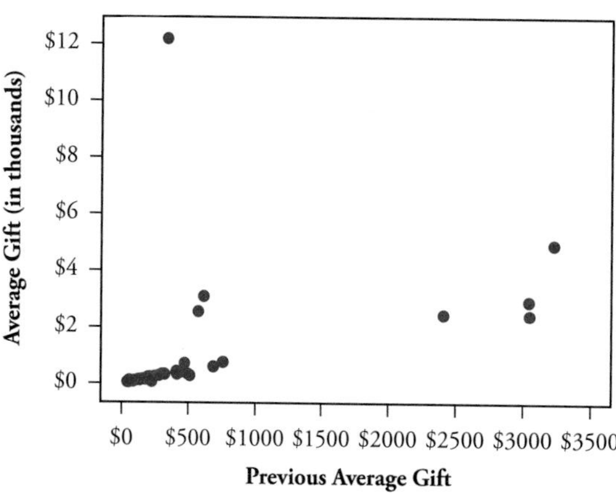

a. Use your calculator to re-create these scatterplots. Identify which graduating class is an outlier in its average gift.

b. Calculate the correlation coefficient for both of these pairs of variables.

c. Based on the scatterplot, do you think the correlation between *average gift* and *donor percentage* would increase, decrease, or not change much if the outlier were removed? Explain your reasoning.

d. Based on the scatterplot, do you think the correlation between *average gift* and *previous average gift* would increase, decrease, or not change much if the outlier were removed? Explain your reasoning.

e. Remove the outlier and recalculate both correlation coefficients. How did they change? Were your predictions in parts c and d correct?

Activity 27-17: *Challenger* Disaster

26-9, **27-17**

Recall from Activity 26-9 on page 550 the data on O-ring failures and outside temperatures from 23 space shuttle launches prior to the fatal *Challenger* mission (Dalal, Folkes, and Hoadley, 1989). Use your calculator to determine the value of the correlation coefficient between temperature and number of O-ring failures (CHALLENGER). Then exclude the flights in which zero O-rings failed and recalculate the correlation. Explain why these correlation values turn out to be so different.

Activity 27-18: Solitaire

11-22, 11-23, 15-4, 15-14, 21-20, **27-18**

The following table reports data on 25 winning games of computer solitaire. The "Losses" column gives the number of losing games that preceded the win; the "Time" column is the time in seconds to complete the game; and the "Points" column is the number of points that the computer listed as the game score (see the grouped data file SOLITAIRE, containing the lists **WIN**, **LOSS**, **TIME**, and **POINT**, respectively).

27

Win #	Losses	Time	Points	Win #	Losses	Time	Points	Win #	Losses	Time	Points
1	2	188	4349	10	6	241	3447	18	33	192	4267
2	3	193	4276	11	11	194	4257	19	7	185	4384
3	13	175	4636	12	4	150	5300	20	9	131	5979
4	11	218	3858	13	4	157	5110	21	11	142	5582
5	5	178	4541	14	11	183	4444	22	6	182	4444
6	4	254	3340	15	0	161	5008	23	4	152	5240
7	1	192	4292	16	7	175	4651	24	1	183	4404
8	9	181	4534	17	21	144	5502	25	4	160	5023
9	5	174	4666								

Use your calculator to produce scatterplots and calculate correlation coefficients between all *three pairs* of these variables. Write a paragraph reporting your findings about possible relationships among these variables.

Activity 27-19: Climatic Conditions

The following data from the *Statistical Abstract of the United States 1992* pertain to a number of climatic variables for a sample of 25 American cities. These variables measure long-term averages of

- January high temperature (in degrees Fahrenheit)
- January low temperature
- July high temperature
- July low temperature
- Annual precipitation (in inches)
- Days of measurable precipitation per year
- Annual snow accumulation
- Percentage sunshine

City	Jan High	Jan Low	July High	July Low	Precip	Days Precip	Snow	Sun
Atlanta	50.4	31.5	88	69.5	50.77	115	2	61
Baltimore	40.2	23.4	87.2	66.8	40.76	113	21.3	57
Boston	35.7	21.6	81.8	65.1	41.51	126	40.7	58
Chicago	29	12.9	83.7	62.6	35.82	126	38.7	55
Cleveland	31.9	17.6	82.4	61.4	36.63	156	54.3	49
Dallas	54.1	32.7	96.5	74.1	33.7	78	2.9	64
Denver	43.2	16.1	88.2	58.6	15.4	89	59.8	70
Detroit	30.3	15.6	83.3	61.3	32.62	135	41.5	53
Houston	61	39.7	92.7	72.4	46.07	104	0.4	56
Kansas City	34.7	16.7	88.7	68.2	37.62	104	20	62
Los Angeles	65.7	47.8	75.3	62.8	12.01	35	0	73

City	Jan High	Jan Low	July High	July Low	Precip	Days Precip	Snow	Sun
Miami	75.2	59.2	89	76.2	55.91	129	0	73
Minneapolis	20.7	2.8	84	63.1	28.32	114	49.2	58
Nashville	45.9	26.5	89.5	68.9	47.3	119	10.6	56
New Orleans	60.8	41.8	90.6	73.1	61.88	114	0.2	60
New York	37.6	25.3	85.2	68.4	47.25	121	28.4	58
Philadelphia	37.9	22.8	82.6	67.2	41.41	117	21.3	56
Phoenix	65.9	41.2	105.9	81	7.66	36	0	86
Pittsburgh	33.7	18.5	82.6	61.6	36.85	154	42.8	46
St. Louis	37.7	20.8	89.3	70.4	37.51	111	19.9	57
Salt Lake City	36.4	19.3	92.2	63.7	16.18	90	57.8	66
San Diego	65.9	48.9	76.2	65.7	9.9	42	0	68
San Francisco	55.6	41.8	71.6	65.7	19.7	62	0	66
Seattle	45	35.2	75.2	55.2	37.19	156	12.3	46
Washington	42.3	26.8	88.5	71.4	38.63	112	17.1	56

Use your calculator to compute the correlation coefficient between all pairs of these eight variables; the data are stored in the grouped file `CLIMATE`, containing lists **JANHI, JANLO, JULHI, JULLO, PRECD, PRECI, SNOW,** and **SUN,** respectively. *Hint:* There are a total of 28 such pairs of variables. It's probably easiest to record the correlation values in a table similar to the one shown here:

	Jan High	Jan Low	July High	July Low	Precip	Days Precip	Snow	Sun
Jan High	XXX							
Jan Low	XXX	XXX						
July High	XXX	XXX	XXX					
July Low	XXX	XXX	XXX	XXX				
Precip	XXX	XXX	XXX	XXX	XXX			
Days Precip	XXX	XXX	XXX	XXX	XXX	XXX		
Snow	XXX	XXX	XXX	XXX	XXX	XXX	XXX	
Sun	XXX	XXX	XXX	XXX	XXX	XXX	XXX	XXX

a. Which pair of variables has the *strongest* (either positive or negative) linear association? What is the value of the correlation between those variables?

b. Which pair of variables has the *weakest* (either positive or negative) linear association? What is the value of the correlation between those variables?

c. Suppose you want to predict the annual snowfall for an American city and you are allowed to look at that city's averages for these other variables. Which variable would be *most* useful to you? Which variable would be *least* useful?

d. Suppose you want to predict the average July high temperature for an American city and you are allowed to look at that city's averages for these other variables. Which variable would be *most* useful to you? Which variable would be *least* useful?

e. Use your calculator to explore the relationship between *annual snowfall* and *annual precipitation* more closely. Produce and comment on a scatterplot of these two variables.

Activity 27-20: Muscle Fatigue

23-5, 23-15, 26-22, **27-20**

Reconsider the matched-pairs study comparing muscle fatigue between men and women from Activity 23-5 (Hunter et al., 2004). In Activity 26-22, you analyzed a scatterplot of time until muscle fatigue for men and women.

a. Calculate the correlation coefficient between time until muscle fatigue for men and time until muscle fatigue for women.

b. Comment on what this correlation coefficient suggests about whether men and women of similar strength tend to have similar times until muscle fatigue.

Activity 27-21: Digital Cameras

10-5, 26-18, **27-21**

Recall from Activities 10-5 and 26-18 the data on ratings and prices of 78 brands of digital cameras (`DIGITALCAMERAS`. See Activity 10-5 for more details).

a. Use your calculator to produce a scatterplot of *rating score* vs. *price.* Then calculate the value of the correlation coefficient. Comment on what this value indicates about the direction and strength of the association between these variables.

b. Now consider the four types of cameras (compact, subcompact, advanced compact, super-zoom) separately. Repeat part a for each type of camera.

Activity 27-22: Your Choice

6-30, 12-22, 19-22, 21-28, 22-27, 26-23, **27-22**

Think of a situation in which you would expect two quantitative variables to have a high (either positive or negative) correlation, even though no cause-and-effect relationship exists between them. Describe these variables and include an alternative explanation (other than cause-and-effect) for the strong association between them.

TOPIC

28 • • • Least Squares Regression

Much of statistics involves making predictions. If you find a footprint at the scene of a crime, can its length help you predict the height of the person who left the print? Can you predict the sale price of a house based on its square footage, and how much does each extra square foot add to the predicted price of the house? On average, how much can you expect each additional page of a textbook to add to its price, and what proportion of the variability in textbook prices is explained by knowing the number of pages, as opposed to other factors or random variation? You will investigate all of these questions in this topic.

Overview

In the two previous topics, you studied scatterplots as visual displays of the relationship between two quantitative variables and the correlation coefficient as a numerical measure of the linear association between those variables. With this topic, you will investigate the widely used technique called least squares regression. This technique provides a simple mathematical model for describing the relationship between two quantitative variables, enabling you to make predictions about one variable's value from the other's value.

Preliminaries

1. Which would you expect to be a better predictor of the price of a textbook: number of pages or year of publication?

2. Guess how much (on average) each additional page of a textbook adds to its price.

3. Would you guess that knowing the number of pages in a textbook explains about 50% of the variability in prices, about 65% of this variability, or about 80% of this variability?

4. Do you think that larger animals need to reach higher speeds before they break into a trot, as compared to smaller animals, or do you think larger animals can trot at lower speeds, or do you think an animal's size has little to do with the speed at which it can break into a trot?

In-Class Activities

Activity 28-1: Heights, Handspans, and Foot Lengths

26-5, **28-1,** 29-12, 29-13, 29-14

Consider again the issue of predicting a person's height from his/her foot length, as you might need to do if you discovered a footprint at the scene of a crime.

a. Open the `Least Squares Regression` applet. You will find a scatterplot of *height (in inches)* vs. *foot length (in centimeters)* for a sample of 20 students. Describe the direction, strength, and form of the association revealed in this scatterplot. *Hint:* Be sure to comment on whether the relationship between height and foot length appears to be linear.

Least Squares Regression

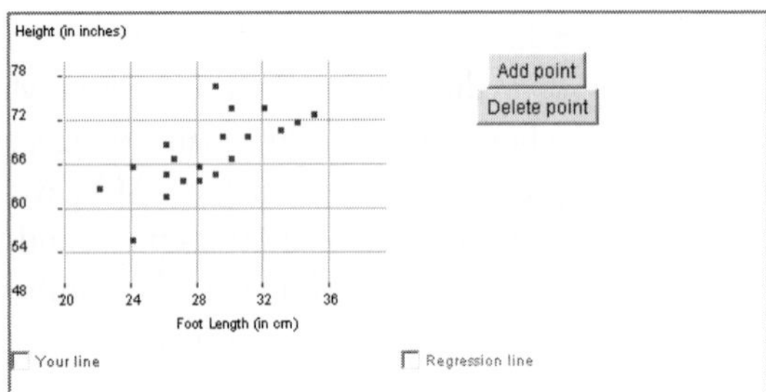

b. Click Your line to add a blue line to the scatterplot. After clicking Move line, you can place the mouse over one end of the line and drag to change the slope of the line. You can also use the mouse to move the green dot up and down vertically to change the intercept of the line. Move the line until you believe it "best" summarizes the relationship between height and foot length for these data. Write down the resulting equation for this line.

The equation of a generic line can be written as $\hat{y} = a + bx$, where y denotes the response variable and x denotes the explanatory variable (also called the **predictor variable**). In this case, x represents foot length and y represents height, and it is good form to use variable names in the equation. The caret on the y (read as "y-hat") indicates that its values are predicted, not actual, heights. The symbol a represents the value of the **y-intercept** of the line, and b represents the value of the **slope** of the line.

Watch Out

"Least squares line" and "regression line" are used interchangeably, so do not be confused by the use of one phrase rather than the other phrase. You might also see the terms "explanatory variable" and "predictor variable" used interchangeably.

c. Do you think that everyone in the class obtained this same line?

d. Suggest a criterion for deciding which line "best" summarizes the relationship.

One way to measure the "fit" of a line is to calculate the *residuals* for all of the observational units.

> A **residual** is the difference between the observed y value and the y value predicted by your line for the corresponding x value. In other words, the residual is the vertical distance from an observation to the regression line.

e. Click Show Residuals to represent visually these residuals for your line on the scatterplot. The applet also reports the sum of the absolute residuals (SAE) under your equation. Record this SAE value for your line. Did anyone in class have a smaller SAE value?

A more common criterion for determining the "best" line is to look at the **sum of squared residuals (SSE).**

f. Click Show Squared Residuals to represent them visually and to determine the SSE value for your line. Record this value.

g. Continue to adjust your line until you think the sum of the squared residuals (SSE) is as small as possible. Report your new equation and new SSE value. Did anyone in class have a smaller SSE value?

The line that achieves the exact minimum value of the sum of the squared residuals is called the **least squares line,** or the **regression line.**

h. Now click Regression Line to determine and display the equation for the line that actually does minimize (as shown using some calculus) the sum of the squared residuals. Record its equation and SSE value.

Regression line:

SSE (least-squares):

One of the primary uses of regression is *prediction.* You can use the regression line to predict the value of the y-variable for a given value of the x-variable simply by plugging that value of x into the equation of the regression line. This process is equivalent to finding the y-value of the point on the regression line corresponding to the x-value of interest.

i. Use the least squares regression line to predict the height of someone whose foot length is 28 cm. Does this prediction seem reasonable, based on the scatterplot?

Watch Out

Remember to provide measurement units when reporting predictions. In other words, be clear that the predicted height is in inches, not centimeters or any other units.

j. Use the least squares regression line to predict the height of someone whose foot length is 29 cm.

k. By how much do these predictions differ? Does this number look familiar? Explain.

The **slope coefficient** of a least squares regression line is interpreted as the predicted change in the response (y) variable associated with a one-unit increase in the explanatory (x) variable.

l. What height would the least squares regression line predict for a person with a 0 cm foot length? Does this make any sense? Explain.

The **intercept coefficient** of a least squares regression line is interpreted as the predicted value of the response (y) variable when the explanatory (x) variable has a value of zero. In many contexts (such as predicting height from foot length), this prediction makes no sense.

m. What height would the least squares regression line predict for a person with a 45 cm foot length? Would you consider this prediction as reliable as the one for a person with a 28 cm foot length? Explain.

Extrapolation means trying to predict the response variable for values of the explanatory variable beyond those contained in the data. When you have no information about the behavior of the data outside the values contained in your dataset (e.g., you have no reason to believe the relationship between height and foot length remains roughly linear beyond these values), extrapolation is not advisable.

n. Uncheck the Your line box to remove it from the display. Click the point for one of the students with a foot length between 28 and 32 cm. Drag this point up and down, changing the student's height without altering his/her foot length. Does the least squares regression line change much as you change this student's height?

o. Repeat the previous question, first using the student with the shortest foot length and then using the student with the longest foot length. Is it even possible to change the *height* value enough to make the slope negative? Do these points seem to have more or less *influence* on the least squares regression line than the point near the middle of the foot lengths from part n? Explain.

An observation is considered **influential** if removing it from the dataset substantially changes the least squares regression equation. Typically, observations that have extreme explanatory (x) variable values (far below or far above the sample mean $\bar{x}$) have more potential to be influential.

p. Now reload the applet and click the Your line box to redisplay the blue line. Notice this line is flat at the mean of the height ($\bar{y}$) values. Click the Show Squared Residuals box to determine the SSE if you were to use $\bar{y}$ as your predicted value for every value of foot length (x), as if you knew nothing about the foot lengths. Record this SSE value.

SSE($\bar{y}$):

q. Recall the SSE value for the least squares regression line. Determine the *percentage change* in the SSE between the $\bar{y}$ line and the least squares line. *Hint:* Calculate 100% × [*SSE*($\bar{y}$) − *SSE*(*least squares*)/*SSE*($\bar{y}$)].

The percentage change value from part q indicates the reduction in the prediction errors from using the least squares line instead of the $\bar{y}$ line. This value shows how much you gain by having access to the x-values in predicting the y-values. This percentage change value is referred to as the **coefficient of determination** and is interpreted as the percentage of variability in the response variable that is explained by the least squares regression line with the explanatory variable.

This coefficient provides you with a measure of the accuracy of your predictions and is most useful for comparing different models (e.g., different choices of explanatory variable). The coefficient of determination is equal to the square of the correlation coefficient, so it is denoted by r^2.

Watch Out

- Always use good statistical notation when writing the equation of a least squares line. The preferred form for this equation ($\widehat{height} = 38.302 + 1.033\,footlength$) differs in three ways from what you probably learned in algebra class ($y = mx + b$):
  - It uses variable names instead of generic y and x labels.
  - It puts a "hat" on the y variable ($\hat{y}$) to indicate the line gives a *prediction.*
  - It lists the intercept first and slope second, rather than the familiar $y = mx + b$ form.

- When checking for influential observations, do not simply assume the observations with extreme x values are influential. Those particular observations have more *potential* to be influential, but they might not prove to be influential on the least squares line if their y values are consistent with the pattern of the other observations. To check for influence, see whether deleting the point in question changes the least squares line substantially.
- It is very easy to be a bit sloppy, and therefore incorrect, when interpreting r^2. It does not represent the proportion of points that fall on the line, or the proportion of the y-variable that is explained by the x-variable. Rather, r^2 is the proportion of the variability in the y-variable that is explained by the least squares line with the x-variable. Of course, when writing your interpretation in a given context, use the variable names rather than generic x and y labels.
- Also exert care when interpreting the slope. Again, always put your comments in context, and instead of saying "the change in y," be sure to say "the average change in y" or "the predicted change in y," because you are working with a *model* and not an exact relationship in your data. Also try not to make the relationship sound causal (e.g., do not say that "increasing the foot size by 1 cm increases the height. . .") with observational studies.

Activity 28-2: House Prices

8-10, 26-1, 27-5, **28-2,** 28-12, 28-13, 29-3

Reconsider the data from Activity 26-1 on prices and sizes of a sample of houses that sold in the year 2006 in Arroyo Grande, California (HOUSEPRICESAG). The scatterplot is reproduced here, with the least squares line superimposed:

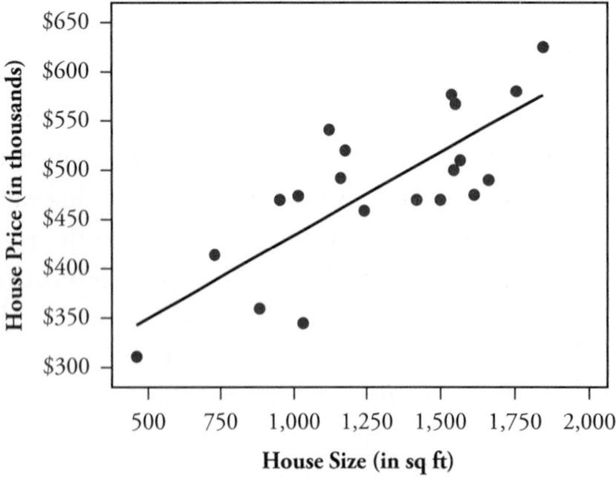

a. Does the least squares line appear to provide a reasonable model for predicting the price of a house based on its size?

You have not yet considered how to calculate the slope and intercept coefficients of the least squares line. Let the equation of a generic least squares line be $\hat{y} = a + bx$. The most convenient expressions for calculating the intercept and slope coefficients for the least squares line involve the means and standard deviations of the two variables, along with the correlation coefficient between them. It turns out the slope can be calculated from $b = r\frac{s_y}{s_x}$. The intercept coefficient can then be calculated from $a = \bar{y} - b\bar{x}$.

Watch Out
Notice that the word "coefficient" appears often here. Be especially careful not to confuse a slope coefficient with a correlation coefficient.

b. Use the following summary statistics to calculate the slope and intercept coefficients of the least squares line for predicting house price from house size. Then report the equation of the least squares line. *Note:* It's important to keep track of which is the explanatory variable (x) and which is the response variable (y).

	Mean	Standard Deviation	Correlation
House Price ($)	482,386	79,802	.780
House Size (sq ft)	1288.1	369.2	

Slope b =

Intercept a =

Least squares line equation:

c. Use your calculator to confirm your calculation of the least squares line. *Hint:* Do not be surprised if the coefficients differ a bit based on rounding discrepancies.

1. Press the [STAT] key, and then select **LinReg(a+bx)** from the CALC menu. Be careful when making your selection! The calculator has two formulas that are very similar—both **LinReg(a+bx)** and **LinReg(ax+b).**
2. Enter the list names after **LinReg(a+bx),** separated by a comma (press the [,] key) and followed by a function name. Press [VARS] and arrow over to select the Y-VARS menu and then down to select **Function.** (*TI hint:* If you omit the function name, then you will obtain the coefficients for the least squares line, but the least squares line will not be stored as a function in the Y= menu.) For this activity, your Home screen should look like this:

```
LinReg(a+bx) ʟSI
ZE,ʟPRICE,Y1
```

3. Press [ENTER].

d. Use this least squares line to predict the price of the house at 845 Pearl Drive, which has a size of 1242 square feet. *Hint:* When you finish the calculation, look at the scatterplot and line above to make sure your answer is reasonable. The easiest way to use your least squares line to make predictions is to use the function that you used to store the least squares line in the Y = menu (in the screen shown following part b we used **Y1**).

1. Select the function by pressing VARS, and arrowing down to select **Function** from the Y-VARS menu.
2. Use function notation to obtain your prediction. For example, if you use **Y1** as your function, then your Home screen will look similar to this one:

3. Press ENTER.

e. The actual (observed) price for the house at 845 Pearl Drive was $459,000. Was your prediction in part c too high or too low? By how much?

A common theme in statistical modeling is to think of each data point as being composed of two parts: the part that is explained by the model (often called the *fit*) and the "leftover" part (often called the *residual*) that is the result either of chance variation or of variables you have not yet considered or measured.

In the context of least squares regression, the **fitted value** for an observation is simply the y-value that the regression line would predict for the x-value of that observation (i.e., the fitted value is $\hat{y}$). The **residual** is the difference between the actual y-value and the fitted value $\hat{y}$ (*residual* = *actual* – *fitted*), so the residual measures the vertical distance from the observed y-value to the regression line.

f. Identify the fitted value and residual for the house at 845 Pearl Drive. *Hint:* You do not have to do any calculations beyond what you've done in parts c and d, but be careful with the sign of the residual.

Fitted value: Residual:

g. Consider the houses whose points lie above the line. What can you say about the residual values for those houses?

h. Based on the scatterplot alone (with the least squares line drawn on it), circle the point corresponding to the house with the greatest negative residual. Also identify the address for this house, from the list in Activity 26-1 on page 538.

i. Write a sentence interpreting the value of the *slope* coefficient in this context.

j. How much does the predicted price of a house increase with each additional 100 square feet in size?

k. Write a sentence interpreting the value of the *intercept* coefficient. Does this value make sense in this context? Explain.

l. What proportion of the variability in house prices is explained by the least squares line with size? Explain how to calculate this value. *Hint:* Make use of the correlation coefficient given in part b.

m. Would it be reasonable to use this least squares line to predict the price of a 3500 square foot house? Explain.

n. Would it be reasonable to use this least squares line to predict the price of a 2000 square foot house in Canton, New York (a small town in the far northern part of New York state)? Explain.

Watch Out

- Be sure to subtract in the correct order (*observed* minus *predicted*) when calculating a residual. (Remember that points *above* the line have *positive* residuals.)
- Never take a prediction very seriously if it results from extrapolating well beyond the actual data.
- Remember not to generalize from the sample data to a larger population unless the sample was drawn randomly or you have some other reason to believe the sample is representative of the population.

28

Activity 28-3: Animal Trotting Speeds

Heglund and Taylor (1988) describe a study in which they trained animals to use a treadmill in order to study how speed, stride, and gait compare across different kinds of animals. One part of their study recorded the speeds at which animals progressed from walking to trotting, looking for a relationship with the size of the animal as measured by body mass. The data, with body mass measured in kilograms and speed in meters per second, follow and are also stored in the grouped file TROTSPEEDS with lists **MASS** (body mass) and **TROT** (trot speed):

Animal	Body Mass	Trot Speed	Animal	Body Mass	Trot Speed
White mouse	0.029	0.19	Suni	3.5	1.15
Ground squirrel	0.193	0.67	Dog (2)	3.89	1.39
White rat	0.362	0.11	Dik-dik	4.35	0.99
Dog (1)	0.96	0.84	Dog (3)	9.21	1.15

(*continued*)

Animal	Body Mass	Trot Speed	Animal	Body Mass	Trot Speed
Gazelle	11.2	1.53	Waterbuck	114	1.92
African goat	20	1.53	Pony (2)	140	1.69
Fat-tailed sheep	23	1.41	Zebu cattle	160	2.08
Dog (4)	25	1.25	Pony (3)	170	1.81
Wildebeest	98	1.89	Donkey	170	1.8
Pony (1)	110	1.64	Eland	213	1.75

The following scatterplot displays *trot speed* vs. *body mass*. The least squares line (*predicated trot speed* = 0.9894 + 0.005484 *body mass*; $r^2 = .535$) has been added:

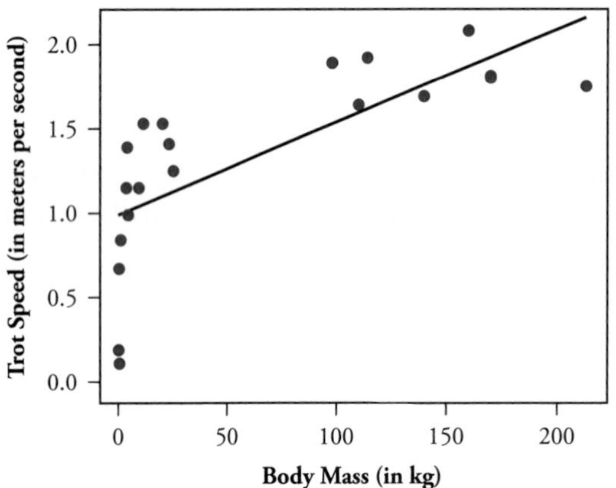

a. Describe the association between *trot speed* and *body mass* revealed in this scatterplot. Does it appear larger animals tend to break into a trot at higher speeds than smaller animals? *Hint:* Remember to comment on the direction, strength, and form of the association.

b. Does the least squares line appear to provide a reasonable model for summarizing the relationship between *trot speed* and *body mass?* Explain.

c. Looking at the graph, what do you notice about the residuals for the smallest animals (those with the smallest body mass values, say, less than 1 kilogram)? What do you notice about the residuals for the medium-sized animals (with body masses between, say, 1 and 120 kilograms)? *Hint:* Look for whether the residuals tend to be positive or negative.

Small animals: Medium-sized animals:

With these data, the poor fit of the least squares line is fairly obvious from the scatterplot. In less clear-cut cases, you would examine a residual plot to help decide whether a linear model adequately describes the relationship in the data.

A **residual plot** is a scatterplot of the residuals vs. either the explanatory variable or the fitted values. When a straight line is a reasonable model, the residual plot should reveal a seemingly random scattering of points. When a nonlinear model fits the data better, the residual plot reveals a pattern of some kind.

d. Use your calculator to produce a residual plot of *residuals* vs. *body mass* for these data. Whenever you use the **LinReg(a+bx)** command, the residuals are automatically placed in a list named **RESID.** Does the residual plot reveal a pattern, suggesting a linear model is not appropriate?

When a straight line is not the best mathematical model for a relationship, you can often *transform* one or both variables to make the association more linear.

A **transformation** is a mathematical function applied to a variable, re-expressing that variable on a different scale.

Common transformations include logarithm, square root, and other powers. Often trial and error is needed to select the transformation that establishes a linear relationship.

e. Use your calculator to take the logarithm (base 10) of the *body mass* variable. Then produce and examine a scatterplot of *trot speed* vs. log(*body mass*). (*TI hint*: Remember that you can apply the LOG function to an entire list, and then save this list with a new name.) Does the association appear to be fairly linear now?

f. Use your calculator to determine the least squares line for predicting *trot speed* from log(*body mass*). Also produce a scatterplot with this line drawn on it. Record the equation of this line and the value of r^2.

g. Use your calculator to produce a residual plot for the least squares line in part f. Does this plot reveal a random scattering of points, indicating the linear model is reasonable for the transformed data?

h. Use this transformation model (from part f) to predict the trot speed for an animal with a body mass of 10 kilograms. Then do the same for an animal with a body mass of 100 kilograms. *Hint:* Remember to take the logarithm (base 10) of the body mass as your first step.

10 kilograms:

100 kilograms:

i. Calculate the difference between these predictions. Does this value look familiar? Explain.

Watch Out

- When making a prediction with a model based on a transformed variable, it is easy to forget to apply the transformation before plugging into the equation of the line. One good reminder is to use the transformation as the name for the new variable, such as log(*body mass*).
- The value of r^2 alone is not sufficient for assessing the fit of the linear model. Even a nonlinear relationship could have a large value of r^2. Examine the original scatterplot and a residual plot to help decide whether a linear model is appropriate.

Activity 28-4: Textbook Prices

28-4, 28-5, 29-5

Two undergraduate students at Cal Poly investigated factors that might be related to the price of college textbooks (Shaffer and Kaplan, 2006). They took a random sample of 30 textbooks from the campus bookstore and recorded the price, number of pages, and year of publication for each book. The data are stored in the grouped file TEXTBOOKPRICES with the lists TXPRC (price), PAGES, and TXYR (year), respectively.

a. Which of these variables is the most reasonable to consider the response variable? Explain.

b. Use your calculator to examine scatterplots of *price* vs. *pages* and *price* vs. *year*. Comment on what these graphs reveal about the association between *price* and these other variables.

c. Which of the two explanatory variables appears to be a better predictor of *price?*

d. Use your calculator to determine the least squares line for predicting a textbook's price based on its number of pages. Report the equation of this regression line, and draw the line on the scatterplot. *Hint:* Be sure to use good statistical notation in reporting this equation.

e. Use this least squares line to predict the price of a 500-page textbook.

f. Identify and interpret the value of the slope coefficient, in the context of this study.

g. Determine and interpret the value of r^2, in the context of this study.

h. Create and comment on a residual plot for this least squares line.

i. Use your calculator to determine the least squares line for predicting a textbook's price based on its year of publication. Report the equation of this regression line and the value of r^2.

j. Create a residual plot for this least squares line. Comment on what it reveals about how well the linear model summarizes the relationship between *price* and *year.*

Solution

a. It seems reasonable to regard *price* as the response variable, because it is natural to take an interest in predicting a textbook's price from other variables.

b. These scatterplots follow:

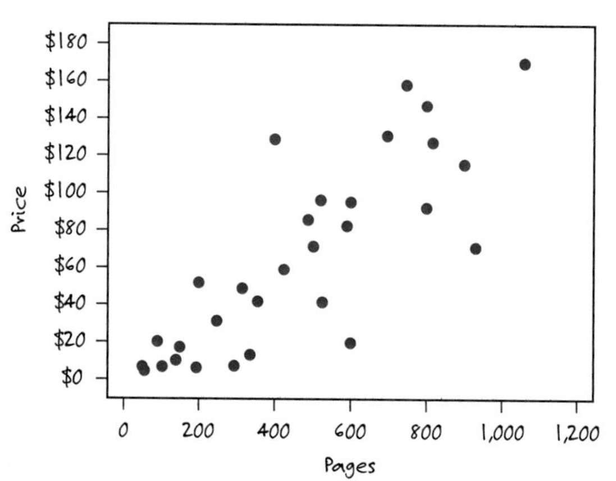

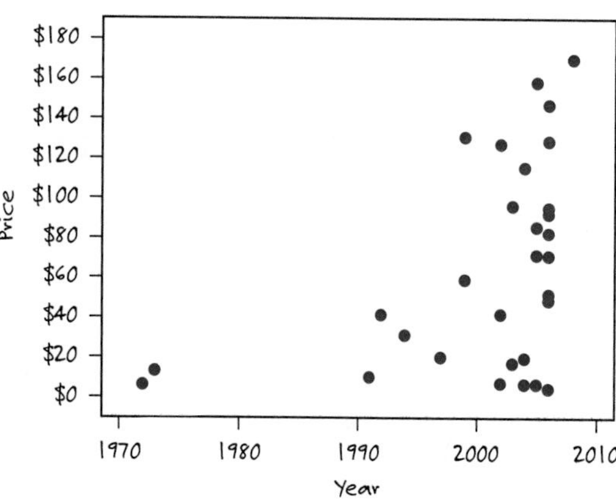

The scatterplot of *price* vs. *pages* reveals a fairly strong, positive, linear association. The scatterplot of *price* vs. *year* also indicates a positive association, but the association is much weaker and not very linear. Two unusual textbooks from the early 1970s, with very low prices, appear to be outliers, because they differ substantially from the pattern of the other textbooks, and potentially influential observations.

c. Number of pages appears to be a much better predictor of price than year. The relationship is much stronger and also more linear.

d. The equation of this least squares line is *predicted price* = −3.42 + 0.147 *pages.* It is shown on the scatterplot here.

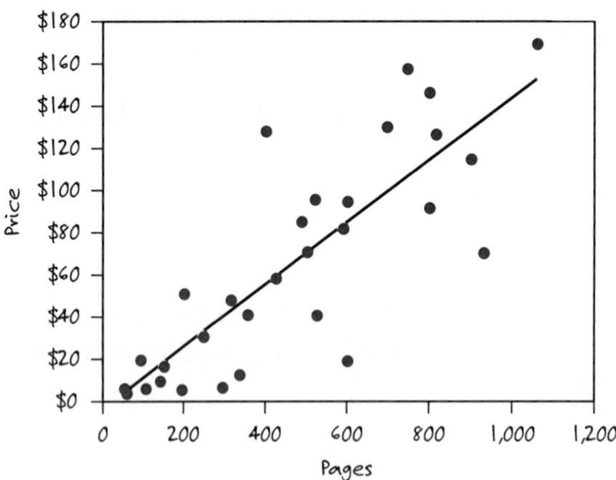

e. The predicted price for a 500-page textbook is *predicted price* = −3.42 + 0.147(500) = 70.08 dollars.

f. The slope coefficient is 0.147 dollars/page, which indicates the predicted price of a textbook increases by 0.147 dollars (about 15 cents) for each additional page.

g. The value of the correlation coefficient between price and pages is $r = .823$, so $r^2 = (.823)^2 = .677$. This coefficient says that 67.7% of the variability in textbook prices is explained by the least squares line with number of pages. The other 32.3% is explained by other factors, which could include random variation.

h. A plot of residuals vs. pages follows.

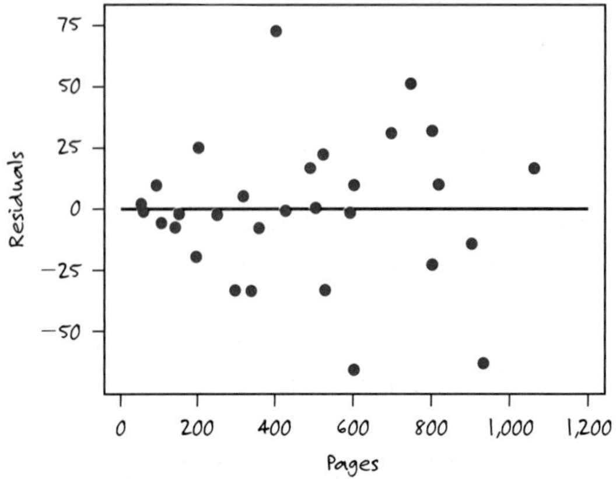

This residual plot reveals no obvious pattern, which suggests the least squares line is a reasonable model for the relationship between *price* and *pages.*

i. The equation of this least squares line is *predicted price* = −4969 + 2.516 *year,* with $r^2 = .186$. The least squares line is shown on the scatterplot here.

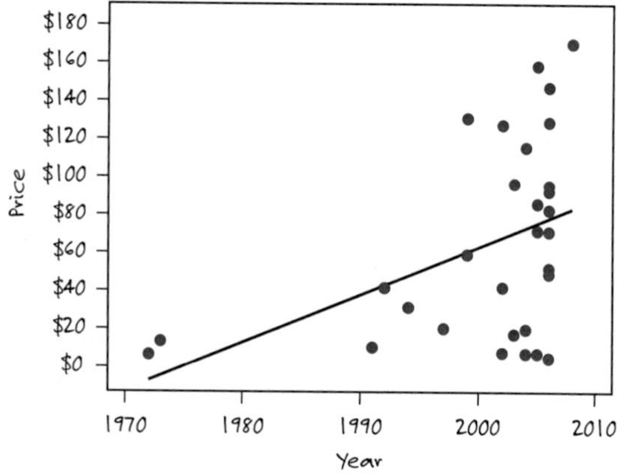

j. The residual plot is shown here.

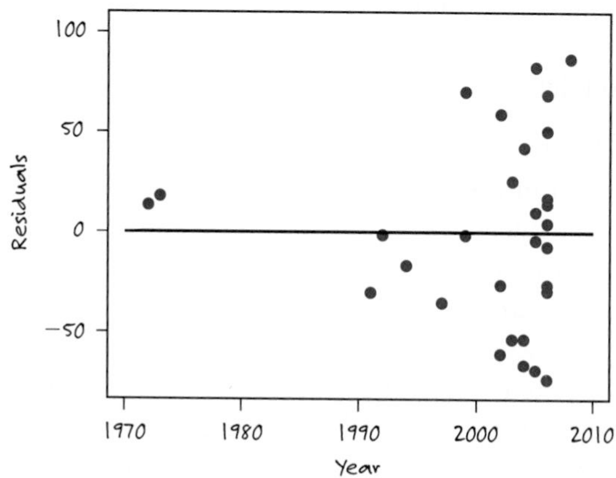

This plot reveals that the middle years (1990–2000) tend to have negative residuals. This pattern suggests a linear model is not very appropriate for predicting price from year of publication. A transformation might lead to a more appropriate linear model.

Watch Out

- As we cautioned earlier, be sure to keep the roles of the variables straight (which is the explanatory variable and which is the response variable). This distinction matters a lot to the calculation of the least squares line.
- Also as we cautioned earlier, remember to refer to context in your conclusions and mention the appropriate units of measurement (dollars for the response variable, in this case) as well.
- Always start by looking at visual displays, even when the goal is a more sophisticated analysis, such as linear regression.
- Again as we mentioned earlier, be very careful when interpreting the slope coefficient and r^2 values. Seemingly small changes in wording can alter the meaning substantially.

Wrap-Up

This topic led you to the study of a mathematical model for describing the relationship between two quantitative variables. In studying least squares regression, you encountered a variety of related terms and concepts. Chief among these is the idea of minimizing squared vertical deviations from the line as a criterion for deciding which line "best" fits a set of data. You used least squares regression for prediction and learned not to do extrapolation or overgeneralization, for example, with predicting house prices. You also learned to interpret the slope coefficient as the predicted (or expected or average) change in the response variable for a one-unit increase in the explanatory variable. For example, the slope in the house price example indicates the predicted increase in price for each additional square foot.

You studied the concepts of fitted values and residuals and the interpretation of r^2 (called the coefficient of determination) as the proportion of variability in the response variable explained by the regression line with the explanatory variable. You also encountered the idea of an influential observation, which refers to an observation whose removal has a big impact on the least squares line.

You saw that a linear model is not appropriate for all sets of bivariate data. For example, the data on trotting speed and body mass displayed a strong positive association, but clearly a nonlinear relationship. Further evidence of this nonlinearity was provided by a residual plot. You used a transformation to achieve a linear relationship, allowing you to fit a least squares line to the transformed data and then to make predictions.

Some useful definitions to remember and habits to develop from this topic include

- The terms **least squares line** and **regression line** are used interchangeably to refer to the line that minimizes the sum of squared vertical deviations between the points and the line.
- The vertical distances between the observed y-values and the predicted y-values are called **residuals;** those predicted y-values are called **fitted values.**
- The **slope coefficient** can be interpreted as the predicted change in the response variable associated with a one-unit increase in the x-variable.
- The **intercept coefficient** can be interpreted as the predicted value of the response variable when the explanatory variable has a value of zero. In many cases, this interpretation is not sensible for the context at hand.
- Be careful not to take seriously predictions that involve **extrapolation,** which refers to making predictions for x-values beyond the range of x-values in the data used to generate the least squares line.
- An **influential observation** is one whose removal has a great impact on the least squares line. Observations with extremely large or small x-values often have more potential to be influential. Always check for whether such observations have an undue influence on the least squares line.
- The **coefficient of determination,** equal to the square of the correlation coefficient (r^2), indicates the proportion of variability in the response variable that is explained by the least squares line with the explanatory variable.
- Always examine scatterplots and **residual plots** to see whether a linear model provides a reasonable summary of the relationship between the two variables.
- When a linear model is not appropriate, try a **transformation** of one or both variables in an effort to achieve linearity.

In the next topic, you will move from a *descriptive* regression analysis to an *inferential* analysis. You will conduct significance tests and form confidence intervals for population regression parameters. The most important test you will study assesses whether the sample data provide evidence of an association between two quantitative variables in the population.

Homework Activities

Activity 28-5: Textbook Prices

28-4, **28-5,** 29-5

Reconsider the analysis of textbook prices from Activity 28-4. Identify what is wrong with each of the following statements.

a. The slope coefficient is 0.147, meaning all books cost exactly 0.147 (dollars) times the number of pages.
b. The slope coefficient is 0.147, meaning the predicted price of a book is 0.147 (dollars) times the number of pages.
c. The slope coefficient is 0.147, meaning the predicted price of a one-page book is 0.147 dollars.
d. The slope coefficient is 0.147, meaning the number of pages is predicted to increase by 0.147 for each additional dollar the book costs.
e. The coefficient of determination is $r^2 = .677$, meaning the least squares line correctly predicts the price for 67.7% of the textbooks in the sample.
f. The coefficient of determination is $r^2 = .677$, meaning 67.7% of the textbooks fall close to the least squares line.
g. The coefficient of determination is $r^2 = .677$, meaning 67.7% of the textbooks contain pages.
h. The coefficient of determination is $r^2 = .677$, meaning 67.7% of the textbook prices are explained by the least squares line with number of pages.
i. The coefficient of determination is $r^2 = .677$, meaning 67.7% of the variability in textbooks is explained by the least squares line with number of pages. *Hint:* Read this carefully; something important is missing.

Activity 28-6: Airfares

28-6, 28-7, 28-8, 28-9, 29-16

The following table reports the distance (in miles) as well as the cheapest available airfare (in dollars) for twelve destinations from Baltimore, Maryland, as reported in the *Harrisburg Patriot-News* on January 8, 1995:

Destination	Distance (in miles)	Airfare ($)	Destination	Distance (in miles)	Airfare ($)
Atlanta	576	178	Miami	946	198
Boston	370	138	New Orleans	998	188
Chicago	612	94	New York	189	98
Dallas	1216	278	Orlando	787	179
Detroit	409	158	Pittsburgh	210	138
Denver	1502	258	St. Louis	737	98

a. What are the observational units for these data?
b. Produce a scatterplot with *distance* as the explanatory variable and *airfare* as the response variable. Comment on the direction, strength, and form of the association.

c. Use the following summary statistics to determine the least squares line for predicting the *airfare* to a destination based on its *distance:*

	Mean	SD	Correlation
Distance	712.667 miles	402.686 miles	.795
Airfare	166.917 dollars	59.454 dollars	

d. Use the least squares line to predict the airfare to a destination that is 750 miles away.
e. Use the least squares line to predict the airfare to a destination that is 7500 miles away. Explain why it's not advisable to take this prediction very seriously.
f. Report the value of the slope coefficient, and write a sentence interpreting it in this context.
g. By how much does the least squares line predict the airfare to increase (on average) for each additional 100 miles flown? Explain your answer.
h. What proportion of the variability in these airfares is explained by the least squares line with distance? Explain how you determine this value.

Activity 28-7: Airfares

28-6, **28-7,** 28-8, 28-9, 29-16

Reconsider the airfare data from the previous activity.

a. Calculate the fitted value and residual for Atlanta. The data are stored in the file `AIRFARES`, containing lists **AIRF** and **DIST.**
b. Which of the twelve destinations has the greatest residual (in absolute value)? Also report the value of that residual.
c. Use your calculator to produce a plot of *residuals* vs. *distance.* Does this residual plot reveal an obvious pattern that questions the adequacy of the linear model?

Activity 28-8: Airfares

28-6, 28-7, **28-8,** 28-9, 29-16

Consider again the airfare data from the previous two activities (`AIRFARES`).

a. Which destination has the most potential to be influential? Explain. *Hint:* Base your answer on the *distances.*
b. Remove that destination (your answer to part a) from the analysis, and use your calculator to recalculate the least squares line. Report its equation as well as the value of r^2. Have these changed much?
c. Based on the analysis in part b, would you conclude this destination (your answer to part a) is, in fact, influential? Explain.
d. Suppose Denver's airfare is reduced to $5. Make this change, and recalculate the least squares line. Report its equation as well as the value of r^2. Have these changed much?
e. Return Denver to its actual value. Now suppose Orlando's airfare is doubled. Make this change, and recalculate the least squares line. Report the least squares equation as well as the value of r^2. Have these changed much?

Activity 28-9: Airfares

28-6, 28-7, 28-8, **28-9,** 29-16

Recall that the slope b and intercept a of a least squares line can be calculated from summary statistics by $b = r\frac{s_y}{s_x}$ and $a = \bar{y} - b\bar{x}$. Also refer to the summary statistics given in part c of Activity 28-6 for the airfare data. Use these equations to determine how the slope and intercept would change in each of the following scenarios:

a. Suppose $500 is added to each airfare.
b. Suppose each airfare is doubled.
c. Suppose each distance is cut in half.
d. Suppose 1000 miles is added to each distance.

Activity 28-10: Car Data

26-3, 27-1, **28-10**

Reconsider the data from Activity 26-3, but focus now on predicting the city miles per gallon rating of a car from its weight. The means and standard deviations of these variables and the correlation between them are reported in the following table:

	Mean	SD	Correlation
Weight (pounds)	3185.5	494.5	−.907
City MPG	20.962	3.165	

a. Use these statistics to determine (by hand) the coefficients of the least squares line for predicting a car's miles per gallon rating from its weight. Report the equation of this line.
b. The MPG rating for the Audi TT was not provided. Use the regression line to predict the city MPG rating for this car, whose weight is 2655 pounds.
c. By how many miles per gallon does the least squares line predict a car's fuel efficiency to drop (on average) for each additional 100 pounds of weight? *Hint:* Use the slope coefficient to answer this question.
d. What proportion of the variability in cars' miles per gallon ratings is explained by the least squares line with weight?

Activity 28-11: Electricity Bills

The following table and the grouped data file `ELECTRICBILL`, containing lists AVTMP and BILL, respectively, list the average temperature (in degrees Fahrenheit) for 28 months and the electricity bill for each month:

Month	Avg Temp	Bill	Month	Avg Temp	Bill
April 1991	51	$41.69	September 1991	74	$37.88
May 1991	61	$42.64	October 1991	59	$35.94
June 1991	74	$36.62	November 1991	48	$39.34
July 1991	77	$40.70	December 1991	44	$49.66
August 1991	78	$38.49	January 1992	34	$55.49

(continued)

February 1992	32	$47.81	November 1992	45	$43.82
March 1992	41	$44.43	December 1992	39	$44.41
April 1992	43	$48.87	January 1993	35	$46.24
May 1992	57	$39.48	February 1993	*	*
June 1992	66	$40.89	March 1993	30	$50.80
July 1992	72	$40.89	April 1993	49	$47.64
August 1992	72	$41.39	May 1993	*	*
September 1992	70	$38.31	June 1993	68	$38.70
October 1992	*	*	July 1993	78	$47.47

a. Before you examine the relationship between *average temperature* and *electric bill,* examine the distribution of electric bill charges themselves. Create a dotplot of the electric bill charges, and write a few sentences describing the distribution of electric bill charges.

b. Produce a scatterplot of *electric bill* vs. *average temperature.* Does the scatterplot reveal a positive association, a negative association, or not much association at all between these variables? If there is an association between the variables, how strong is it?

These temperatures have a mean of 55.88 degrees and a standard deviation of 16.21 degrees. The electric bills have a mean of $43.18 and a standard deviation of $4.99. The correlation coefficient between average temperature and electric bill is –.695.

c. Use this information to determine the equation of the least squares (regression) line for predicting the *electric bill* from the *average temperature.*

d. Interpret the slope coefficient in the context of these data.

e. Use this least squares line to determine the fitted value and residual for March 1992.

f. Without doing any calculations, identify the month with the greatest fitted value. Explain your answer.

g. What proportion of the variability in electric bills is explained by the regression line with average temperature?

Activity 28-12: House Prices

8-10, 26-1, 27-5, 28-2, **28-12,** 28-13, 29-3

Reconsider your regression analysis of the data on prices and sizes of a sample of houses (HOUSEPRICESAG) in Activity 28-2.

a. Examine the scatterplot in Activity 28-2 on page 586 and the original data in Activity 26-1 on page 538. Identify (by its address) the house with the most unusual size, either unusually small or unusually large.

b. Remove this house from the analysis, and use your calculator to recalculate the least squares line for predicting *house price* from *house size.* Report the equation of this line, as well as the value of r^2.

c. Use the least squares line from part b to predict the price of a 1242 square foot house, and compare your answer to part d in Activity 28-2.

d. Have the line, value of r^2, and prediction changed substantially from your earlier (complete) analysis? In other words, does the house you identified in part a have much influence on the line? Explain.

Activity 28-13: House Prices

8-10, 26-1, 27-5, 28-2, 28-12, **28-13,** 29-3

Consider again the previous activity and the goal of predicting the price of a house from its size. The file HOUSEPRICESAG also contains data on the number of bedrooms and the number of bathrooms in the house.

a. First, consider number of bedrooms on its own. Produce and examine graphical displays and numerical summaries of the distribution of number of bedrooms. Write a few sentences describing the distribution.
b. Now consider *number of bedrooms* as an explanatory variable for predicting *house price.* Produce and comment on a scatterplot of *house price* vs. *number of bedrooms,* and also calculate and report the correlation coefficient.
c. Determine the least squares line for predicting the price of a house based on its number of bedrooms. Report the equation of this line and also the value of r^2.
d. Examine residual plots to investigate whether a linear model is appropriate here. Summarize your conclusions.
e. Repeat parts a–d with an analysis of the number of bathrooms.
f. Based on your regression analyses, which of the three explanatory variables (*house size, number of bedrooms, number of bathrooms*) appears to be the best for predicting the price of a house? Explain how you are making your decision.

Activity 28-14: Honda Prices

7-10, 10-19, 12-21, **28-14,** 28-15, 29-10, 29-11

Recall from Activities 10-19 and 12-21 the sample of 116 Honda Civics for sale on the Internet in July 2006. The response variable to be predicted is the car's *asking price,* and potential explanatory variables are *mileage* and *year of manufacture.* For this activity, consider *mileage* as the explanatory variable.

a. Produce a scatterplot of *price* vs. *mileage* (HONDAPRICES). Describe the association, including any unusual cases, that the scatterplot reveals.
b. Determine the least squares line for predicting *price* from *mileage.* Draw the line on the scatterplot.
c. Interpret the value of the slope coefficient in this context.
d. Calculate and interpret the value of r^2 in this context.
e. Use the least squares line to predict the price for a Honda Civic with 50,000 miles. Then do the same for a Honda Civic with 150,000 miles.
f. Identify the outlier in the mileage variable. Then investigate whether this car is influential by removing it and recalculating the least squares line and r^2. Do these change much, indicating this car is influential?
g. With the potentially influential car removed, examine a residual plot and comment on whether it displays a pattern. Summarize what this means about whether a straight line is the best model for describing the relationship between *price* and *mileage.*

Activity 28-15: Honda Prices

7-10, 10-19, 12-21, 28-14, **28-15,** 29-10, 29-11

Reconsider the previous activity about predicting the price of a Honda Civic for sale on the Internet. Redo the previous activity with *year of manufacture* as the explanatory variable (HONDAPRICES). In part e, make predictions for a 1998 Honda Civic and for a 2003 Honda Civic.

Activity 28-16: Fast-Food Sandwiches

26-14, 26-15, **28-16**

Recall the data on Arby's fast-food sandwiches from Activity 26-14 (ARBYS06). Suppose you want to predict the calories from fat in a sandwich based on its total calories.

a. Produce a scatterplot appropriate for this goal, and describe the association between *calories from fat* and *total calories.*

b. Determine the equation of the appropriate least squares line. Draw this line on the scatterplot, and also report the value of r^2.

c. Analyze a residual plot to see whether the least squares line is a reasonable model for these data.

d. Interpret the value of the slope coefficient in this context.

e. Interpret the value of r^2 in this context.

f. Suppose Arby's introduces a new sandwich that is 750 calories. Use your least squares line to predict the number of calories from fat in this sandwich.

Activity 28-17: Box Office Blockbusters

28-17, 28-18

How well can you predict the amount of money a movie will make at the box office after the first weekend of its release? To investigate this question, consider the data stored in the grouped file BOXOFFICE05 with lists **GROSS, SCRN, OPEN,** and **OSCRN,** respectively, which pertain to the top 100 movies (by gross box office income) in 2005. The first few rows of the file are reproduced here:

Title	Rank	Gross (millions)	Screens	Opening Gross (millions)	Opening Screens
Star Wars: Episode III—Revenge of the Sith	1	$380.271	3663	$108.436	3661
The Chronicles of Narnia: The Lion, the Witch, and the Wardrobe	2	$291.711	3853	$65.556	3616
Harry Potter and the Goblet of Fire	3	$290.013	3858	$102.686	3858
War of the Worlds	4	$234.280	3910	$64.879	3908
King Kong	5	$218.080	3627	$50.130	3568

For now consider only the *opening weekend gross income* as a predictor of *overall gross income.* Conduct a full regression analysis, including graphical displays and residual plots. Write a paragraph or two summarizing your recommendation for how to predict *overall gross income* from *opening weekend gross income.*

Activity 28-18: Box Office Blockbusters

28-17, **28-18**

Refer to the previous activity. Now consider the other two explanatory variables for predicting the overall gross income: *total number of screens on which the movie appeared* and *number of screens for the opening weekend.* Conduct a full regression analysis, including graphical displays and residual plots, for these two explanatory variables (BOXOFFICE05). Select one of the variables as a better predictor of overall gross income than the other, and write a paragraph or two summarizing your analysis. Also comment on whether either of these variables is a better predictor of overall gross income than the opening weekend's gross income.

Activity 28-19: Televisions and Life Expectancy

26-6, 27-3, **28-19**

Reconsider the data from Activity 26-6 about life expectancy and number of televisions per thousand people in a sample of 22 countries.

a. Use your calculator to produce a scatterplot of *life expectancy* vs. *number of televisions per thousand people,* with the least squares line added. Report the equation of this line and the value of r^2. Also examine a residual plot. Summarize whether the least squares line provides a reasonable summary of the relationship between these variables.

b. Transform the explanatory variable by taking the log (base 10) of the number of televisions per thousand people. Repeat part a, using this transformed explanatory variable.

c. Now transform the explanatory variable by taking the square root of the number of televisions per thousand people. Repeat part a, using this newly transformed explanatory variable.

d. Is the least squares model more appropriate with the transformed variables than with the original data? If so, for which of the two transformations do you consider the least squares model more appropriate? Explain the evidence behind your conclusions.

Activity 28-20: Planetary Measurements

8-12, 10-20, 19-20, 27-13, **28-20,** 28-21

Reconsider the data on planetary measurements from Activity 8-12 (PLANETS).

a. Produce a scatterplot displaying the relationship between a planet's distance from the sun and its position number. Would a least squares line be a good fit for this relationship? Explain.

b. Regardless of your answer to part a, determine and report the least squares line for predicting *distance* from *position number.* Produce and examine a residual plot, and comment on whether the residuals seem to be scattered randomly or to follow a pattern. Summarize whether a least squares line is a reasonable model for this relationship.

c. Use your calculator to create two new variables: *square root of distance* and *logarithm of distance.* Examine scatterplots of each of these variables vs. *position.* Which transformation (square root or logarithm) seems to produce a more linear relationship?

d. For whichever transformation you select in part b as more appropriate, use your calculator to determine the least squares line for predicting that transformation of distance from position. Record this equation, indicating which transformation you are using.
e. Report the value of r^2 for this regression equation.
f. Look at a residual plot and comment on whether the residuals seem to be scattered randomly or to follow a pattern.

Activity 28-21: Planetary Measurements

8-12, 10-20, 19-20, 27-13, 28-20, **28-21**

Reconsider the planetary data from Activity 8-12 (PLANETS). Consider again the problem of using a planet's distance from the sun to predict the period of its revolution around the sun, as you considered in Activity 27-13.

a. Produce a scatterplot of *period of revolution* vs. *distance*. Does there appear to be a strong relationship between the two? Does it appear to be linear?
b. Create a new variable that equals the square of distance. Produce a scatterplot of *revolution* vs. this new variable. Comment on whether the relationship now appears to be linear.
c. Create new variables by trying other powers of distance (e.g., square root, cube) until you find one such that the relationship between period of revolution and this transformation of distance appears to be quite linear. Report the power that appears to produce the most linear relationship. *Hint:* Do not limit yourself only to integers for powers.
d. Use your calculator to determine the least squares line for predicting revolution from your "best" power transformation of distance. Record the equation of this least squares line for predicting period of revolution.
e. Examine a residual plot for this regression model and comment on what it reveals about the fit of the model.

Note: If you linearize this relationship successfully, you produce empirical verification of Kepler's third law of planetary motion.

Activity 28-22: Gestation and Longevity

28-22, 28-23

The following table lists the average gestation period (in days) and longevity (in years) for a sample of animals, as reported in *The World Almanac and Book of Facts 2006.*

Animal	Gestation (in days)	Longevity (in years)	Animal	Gestation (in days)	Longevity (in years)
Baboon	187	20	Camel	406	12
Bear, black	219	18	Cat	63	12
Bear, grizzly	225	25	Chimpanzee	230	20
Bear, polar	240	20	Chipmunk	31	6
Beaver	105	5	Cow	284	15
Buffalo	285	15	Deer	201	8

(continued)

Animal	Gestation (in days)	Longevity (in years)	Animal	Gestation (in days)	Longevity (in years)
Dog	61	12	Monkey	164	15
Donkey	365	12	Moose	240	12
Elephant	645	40	Mouse	21	3
Elk	250	15	Opossum	13	1
Fox	52	7	Pig	112	10
Giraffe	457	10	Puma	90	12
Goat	151	8	Rabbit	31	5
Gorilla	258	20	Rhinoceros	450	15
Guinea pig	48	4	Sea lion	350	12
Hippopotamus	238	41	Sheep	154	12
Horse	330	20	Squirrel	44	10
Kangaroo	36	7	Tiger	105	16
Leopard	98	12	Wolf	63	5
Lion	100	15	Zebra	365	15

a. Use your calculator to produce a scatterplot of *gestation* vs. *longevity.* (The data are stored in the grouped file GESTATION06, containing lists **GEST** and **LONG.**) Also calculate the correlation coefficient. Comment on what the scatterplot and correlation coefficient reveal.

b. Use your calculator to determine the least squares line for predicting an animal's gestation period from its longevity. Report its equation and the value of r^2.

c. Use the least squares line to calculate the fitted value and residual for a baboon. *Hint:* First, find the baboon's longevity and gestation period in the data table.

d. Use your calculator to create a residual plot of the animals' *residual value* vs. *longevity.* Is there any relationship between residuals and longevities? Explain in a sentence or two what this relationship signifies about the accuracy of predictions for animals with long vs. short lifetimes.

e. Which animal is clearly an outlier both in longevity and in gestation period? Determine its residual value. Does it have the greatest residual (in absolute value) of any animal?

f. Which animal has the greatest (in absolute value) residual? Is its gestation period longer or shorter than expected for an animal with its longevity?

g. Remove the giraffe from the analysis. Reproduce the scatterplot and determine the least squares line with the giraffe omitted. Report the equation of the line and value of r^2. Have these changed considerably from the original analysis that included the giraffe in the dataset?

h. Return the giraffe's values to the analysis, and then remove the elephant. Again reproduce the scatterplot and determine the least squares line with the elephant omitted. Have these changed considerably from the original analysis?

i. In which case (giraffe or elephant), did the removal of one animal affect the regression line more? In other words, which animal has more *influence* on the least squares line?

Activity 28-23: Gestation and Longevity

28-22, **28-23**

Reconsider the data from the previous activity concerning the relationship between an animal's longevity and its gestation period.

a. Use the original least squares line you found in the previous activity to predict the gestation period of a human being, assuming a longevity of 75 years. Show the details of your calculation.

b. Do you accept this prediction from part a as being reasonable? If not, explain why the least squares line does not produce a reasonable prediction in this case.

Activity 28-24: Residual Plots

Below are four scatterplots with regression lines drawn in:

1. City MPG rating vs. weight ($r^2 = .823$)

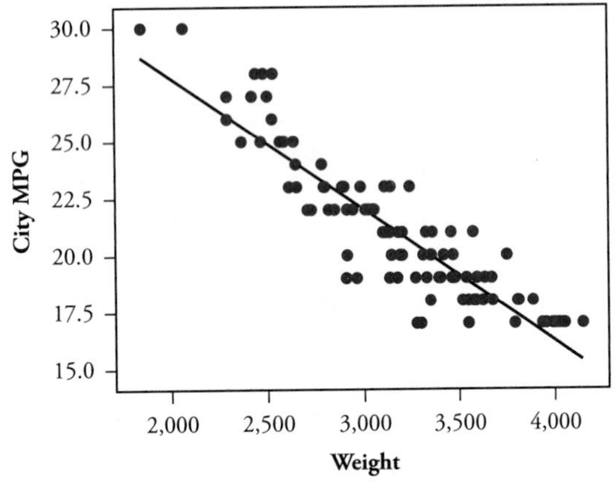

2. Distance from sun vs. position ($r^2 = .828$)

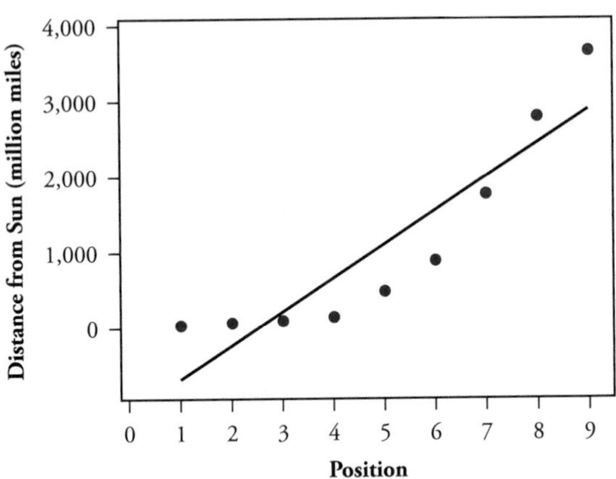

3. Rent vs. price for Monopoly properties ($r^2 = .988$)

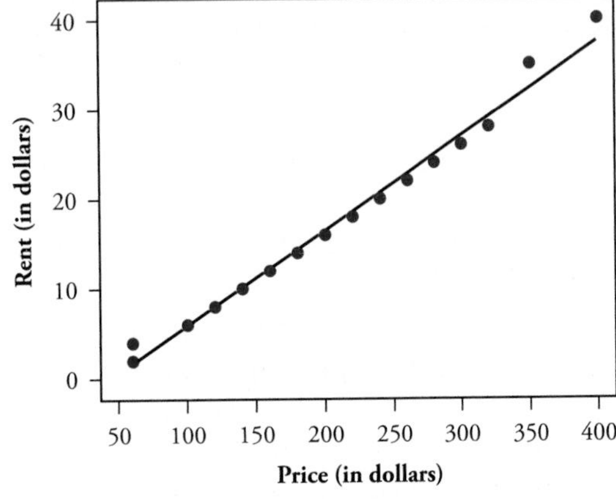

4. Airfare vs. distance ($r^2 = .632$)

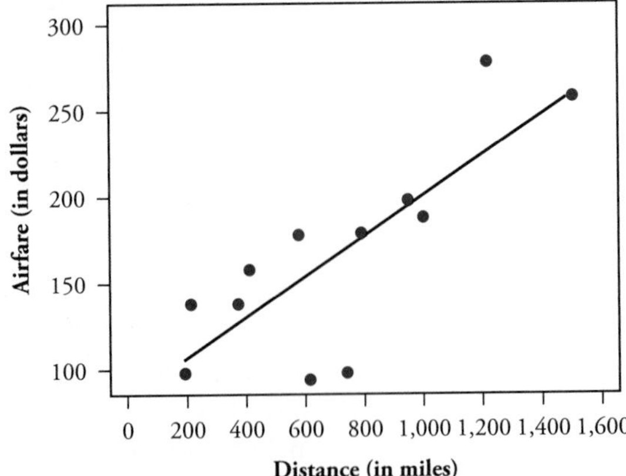

a. For each of these datasets, a scatterplot of the residuals from the regression line vs. the explanatory or predictor variable is shown next. Your task is to match up each residual plot with its regression scatterplot.

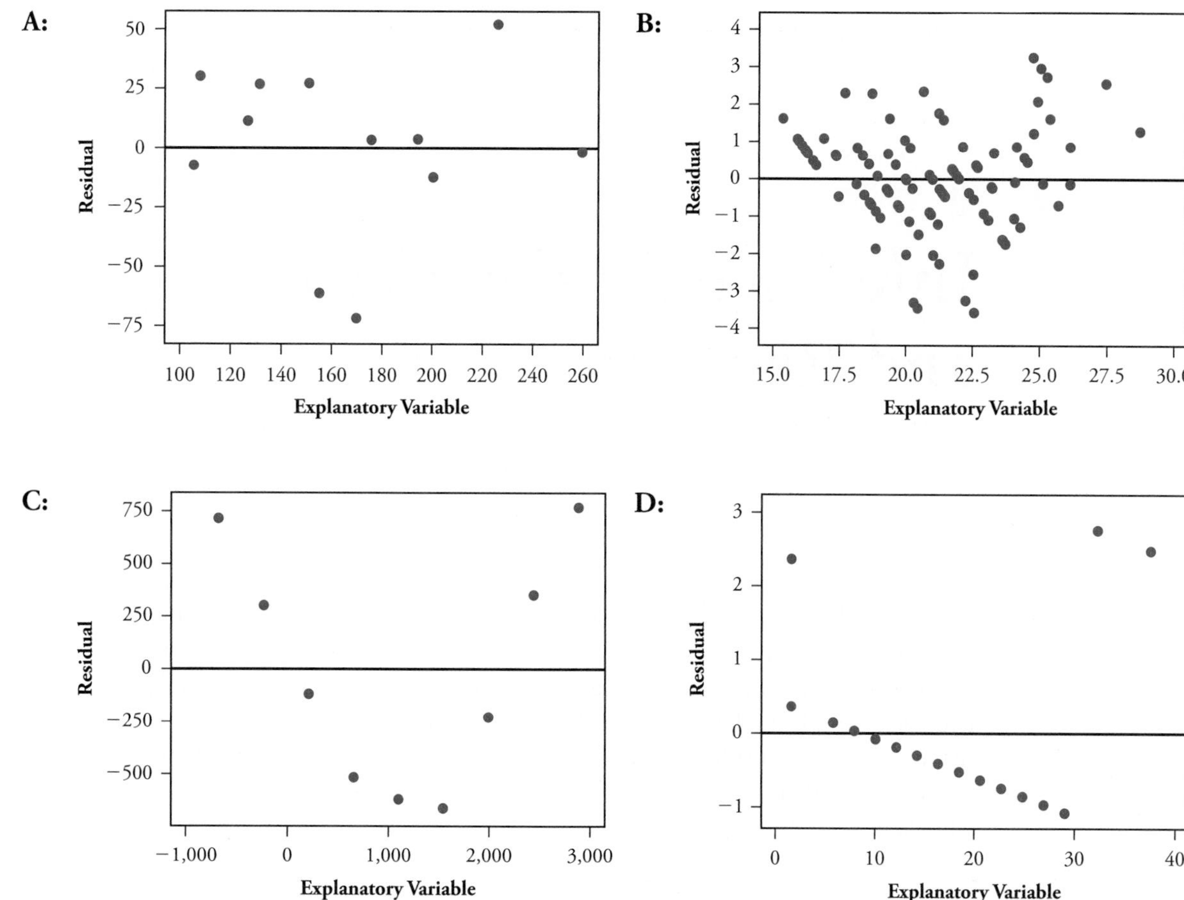

b. Now match up the data with the following descriptions of the residual plots:

- The residuals are randomly scattered.
- The residuals appear in negatively sloping bands.
- The residuals show a distinct curved pattern.
- The residuals show a clear linear pattern with three clear outliers.

c. Look back at the original scatterplots with the regression lines drawn in. In which two plots, do the lines summarize the relationship in the data about as well as possible? In which two plots do the lines fail to capture important aspects of the relationship? Explain.

d. Do the scatterplots for which the least squares line summarizes the data as well as possible correspond to the greatest values of r^2? Explain.

Activity 28-25: Wrongful Conclusions

8-6, 16-21, 17-8, **28-25**

It can be shown that the sum of the residuals from a least squares line always equals zero.

a. Does it follow from this fact that the *mean* of the residuals must equal zero? Explain.

b. Does it follow from this fact that the *median* of the residuals must equal zero? Explain.

c. Does the observation with the greatest value of the predictor variable necessarily have the greatest fitted value? Explain.

d. Does the observation with the greatest value of the response variable necessarily have the greatest residual? Explain.

e. Is it possible for an observation to have the greatest residual if its value of the predictor variable equals the mean of that variable? Explain.

TOPIC

29 • • •

Inference for Correlation and Regression

By how much do additional pages in a college textbook increase the price? You partially addressed this question in the previous topic, but now you will return to the issue of inference, where your goal is to infer from the sample to the larger population from which the sample was drawn. Does a large sample correlation necessarily imply a relationship in the population as well? You will investigate this question by considering whether scientists have discovered a biochemical measurement that can predict the level of romantic feelings experienced by someone who claims to be in love.

Overview

In earlier topics of this unit, you studied graphical and numerical techniques for analyzing the relationship between two quantitative variables. In the previous topic, you used least squares regression lines to describe such a relationship. This topic leads you from a descriptive analysis to an inferential analysis, where the goal is to decide how strongly the sample data provide evidence of a relationship in the population from which the sample was selected. You will learn how to conduct significance tests and produce confidence intervals about a population slope coefficient. You will also study a significance test procedure related to a population correlation coefficient. Once again, you will discover the concept of a sampling distribution underlying these inference procedures. You will also see, yet again, that the general structure, reasoning, and interpretation of confidence intervals and significance tests remain unchanged.

Preliminaries

1. Do you think there is a positive association between the amount of time a student claims to spend studying per week and his or her grade point average?

2. Guess the value of the correlation coefficient between students' self-reported studying times and grade point averages.

In-Class Activities

Activity 29-1: Studying and Grades

29-1, 29-2, 29-7, 29-8

Students conducting a project at the University of the Pacific (UOP) investigated whether students who claim to study more tend to have higher grade point averages. They surveyed 80 of their fellow students, with self-reported responses for study hours per week and grade point average (GPA) appearing in the following table and in the grouped data file UOPGPA, containing the lists **GPA, HOURS,** and **ID,** respectively:

ID	Hours	GPA	ID	Hours	GPA	ID	Hours	GPA	ID	Hours	GPA
1	0.5	1.90	21	2.0	2.94	41	4.0	3.30	61	2.0	3.60
2	1.0	2.30	22	5.5	2.94	42	5.0	3.30	62	4.0	3.60
3	4.0	2.30	23	3.0	3.00	43	7.0	3.40	63	2.0	3.67
4	5.0	2.40	24	5.0	3.00	44	6.0	3.40	64	2.0	3.70
5	1.5	2.50	25	2.0	3.00	45	4.5	3.40	65	5.0	3.70
6	4.0	2.50	26	5.0	3.00	46	7.0	3.40	66	3.0	3.71
7	3.0	2.50	27	2.0	3.00	47	5.0	3.40	67	4.5	3.74
8	6.0	2.50	28	4.5	3.00	48	7.0	3.44	68	1.5	3.75
9	2.0	2.50	29	3.0	3.00	49	4.0	3.45	69	3.0	3.75
10	1.5	2.70	30	3.5	3.00	50	3.5	3.48	70	8.0	3.83
11	2.0	2.70	31	7.0	3.00	51	8.0	3.49	71	7.0	3.87
12	5.0	2.77	32	4.0	3.00	52	6.0	3.50	72	8.0	3.90
13	2.0	2.78	33	4.0	3.05	53	5.0	3.50	73	2.0	3.90
14	4.0	2.80	34	3.5	3.10	54	3.0	3.50	74	4.0	3.92
15	2.0	2.80	35	1.0	3.20	55	3.0	3.50	75	5.0	3.92
16	4.0	2.80	36	2.0	3.20	56	4.0	3.50	76	4.0	3.94
17	4.0	2.80	37	3.5	3.20	57	3.0	3.50	77	4.0	3.96
18	3.0	2.80	38	4.0	3.20	58	2.0	3.50	78	6.0	3.98
19	0.25	2.83	39	3.0	3.25	59	4.0	3.55	79	6.0	3.98
20	5.0	2.90	40	3.5	3.30	60	8.0	3.58	80	3.5	4.00

a. Identify the observational units and variables in this study. Classify each variable as explanatory or response and as categorical or quantitative.

Observational units:

Explanatory variable: Type:

Response variable: Type:

b. Use your calculator to construct a scatterplot of *GPA* vs. *study hours*. Describe any association revealed by the scatterplot. Is the association what you expected? Explain.

c. Use your calculator to determine the equation of the least squares line for predicting grade point average from study hours per week. Record the equation, being sure to write it in the context of these two variables.

d. Report and interpret the value of r^2 for this least squares line.

e. By how many grade points does this equation predict the GPA to rise or fall for each additional hour of study? Explain where this number appears in the least squares equation.

f. Is this least squares line based on a population or a sample? Would you refer to its coefficients as parameters or statistics? Explain.

g. If the student researchers were to take another sample of 80 students from the population, would they be likely to get the exact same regression equation from the new sample? Explain.

These regression coefficients are *statistics.* As with all statistics (such as a sample proportion or a sample mean), sample regression coefficients are subject to *sampling variability.* The sample regression line varies from sample to sample if many different random samples are selected from the population. In order to draw an inference from the student researchers' sample to the larger population of *all* UOP students, you need to study this sampling distribution of the sample regression slope. The slope is typically of more interest than the intercept because a non-zero slope indicates there *is* a relationship between the variables.

To investigate how often a sample regression line at least this extreme would occur by chance, if in fact there were no association between study hours and GPA in the population, you will simulate taking many random samples from a hypothetical population of 1000 students for whom there is no association between study hours and GPA.

h. Open the applet `Sampling Regression Lines`. You will see a scatterplot of 1000 points, for which the individual distributions of hours studied and of GPA in the hypothetical population are very similar to those in this sample, but the population has been constructed so there is no association between study hours and GPA, as the null hypothesis asserts. Leave the population parameters set with a slope of 0 (indicating no association between the variables in the population), an intercept of 3.24 (which is the sample mean GPA), an x mean of 3.928 (the sample mean study hours), an x standard deviation of 1.84, and a sigma of 0.45 (which is roughly the standard deviation of the GPAs for a fixed value of study hours).

Sampling Regression Lines

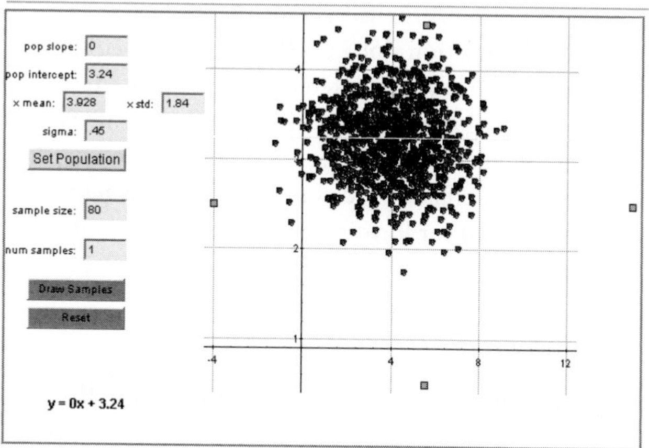

Set the sample size to **80** and the number of samples to **1.** Click Draw Samples. Record the equation of the resulting sample least squares line. Is this line the same as the least squares line in part c, based on the actual sample data?

i. Click Draw Samples two more times. Record the sample least squares line each time. Did you get a different sample line every time?

j. Now ask for **100** samples (for 80 students in each sample). Describe the pattern of variation in the sample least squares lines as they scroll by.

k. A separate window should appear, showing a dotplot of the 100 sample slope coefficients and of the 100 sample intercept coefficients. Describe the distribution of the slope coefficients. *Hint:* Remember to discuss shape, center, and spread of this sampling distribution. Also report the mean and standard deviation of these slope coefficients.

l. How many of the 100 simulated sample slopes are as large as (or larger than) the slope coefficient from the actual sample (0.0894)? What does this result suggest about the size of the p-value for testing whether the population slope is zero against the alternative hypothesis that it is positive? What conclusion would you draw about those hypotheses? Explain.

The distribution in part k approximates the sampling distribution of the sample slope coefficient. As you have seen with many other sampling distributions, the shape is approximately normal, and it is centered at the value of the population parameter (in this case, the population slope coefficient, which equals zero). This sampling distribution forms the basis for inference (significance test and confidence interval) procedures concerning the population slope coefficient. We will denote the sample slope b and the population slope β.

Details of the inference procedures for testing whether the population slope β differs from zero and for forming a confidence interval to estimate the population slope β follow.

Inference procedures for the population slope β:
Confidence interval: $b \pm t^* \text{ SE}(b)$, where t^* is the appropriate critical value with $n - 2$ degrees of freedom.
Null hypothesis: H_0: $\beta = 0$ (indicating no relationship between the variables)
Alternative hypothesis:

i. H_a: $\beta < 0$ *or*
ii. H_a: $\beta > 0$ *or*
iii. H_a: $\beta \neq 0$

Test statistic:

$$t = \frac{b}{\text{SE }(b)}$$

where SE(b) denotes the standard error of the sample slope coefficient b. You will learn how to use your calculator to perform these procedures. This standard error measures the variability in sample slope coefficients that would result from repeatedly taking random samples from the population.

Technical conditions will be described in the next activity.

p-value:

i. $\Pr(T_{n-2} \leq t)$ *or*
ii. $\Pr(T_{n-2} \geq t)$ *or*
iii. $2 \Pr(T_{n-2} \geq |t|)$

The structure, reasoning, and interpretation of this test are the same as with the other tests you have studied. In particular, the confidence interval is formed by taking the *sample statistic* plus or minus *a critical value* multiplied by *the standard error of the statistic.* For the significance test, the test statistic is equal to the *sample statistic* minus *the hypothesized value* (zero in this case), divided by *the standard error.* As always, the p-value measures how likely such an extreme sample (or more extreme sample) would be if the null hypothesis were true, so a small p-value provides evidence against the null hypothesis. What's different here is that the parameter of interest is a population *slope* coefficient, not a mean or proportion. Also note the degrees of freedom for this *t*-procedure is two less than the sample size.

m. State the null and alternative hypotheses (in symbols and in words) for testing the student researchers' conjecture that students who study more tend to have higher grade point averages. *Hint:* Does the conjecture call for a one-sided or a two-sided alternative?

n. Use your calculator and the actual sample data observed to determine the standard error of the slope coefficient, denoted by SE(*b*). To do this, use the **LinRegTTest** feature:

1. Press [STAT] and arrow right to select the **TESTS** menu and then select **LinRegTTest** (you will need scroll up or down the screen to locate this feature). *TI hint:* Place **Y1** (located by pressing [VARS] and selecting **Function** from the **Y-VARS** menu) at the **RegEQ** prompt and set the frequency to **1.** Your Home screen should look like this:

```
LinRegTTest
 Xlist:HOURS
 Ylist:GPA
 Freq:1
 β & ρ:≠0 <0 >0
 RegEQ:Y1
 Calculate
```

2. To calculate SE(*b*), divide **b** by **t.** You can find both b and t by pressing [VARS] and arrowing down to select **Statistics** and then the **EQ** and **TEST** menus, respectively, or you can download and use the STDER program on your calculator. Is this reasonably close to the standard deviation of your 100 simulated sample slope coefficients?

o. Use the standard error from part n to calculate (by hand) the test statistic for testing whether the population slope differs from zero.

p. Use the *t*-Distribution of Critical Values table (Table III) and your test statistic in part o to determine (as best you can) the *p*-value of the test. Is this *p*-value consistent with your simulation results? Explain.

q. Does the *p*-value suggest the association between *GPA* and *study hours* found in the sample of 80 students would be very unlikely if, in fact, no association existed in the population? Explain.

r. Use the sample slope and its standard error to form a 95% confidence interval for the population slope coefficient. *TI hint:* You can use your calculator's **LinRegTInt** command to find this interval. The setup for **LinRegTInt** is similar to the setup for the **LinRegTTest** command. See part n.

s. Interpret this interval, remembering what a slope coefficient represents and to relate your interpretation to the context.

t. Write a few sentences summarizing your conclusion about the evidence found in the students' sample concerning an association between *GPA* and *study hours.*

Watch Out

Please note that SE(b) is simply our notation for the standard error of the sample slope. The parentheses do not mean the standard error is to be multiplied by the slope. The (b) is only there to remind you that this value represents the standard error of the sample *slope* and not the standard error of some other statistic.

Activity 29-2: Studying and Grades

29-1, **29-2,** 29-7, 29-8

Reconsider the research question and sample data from the previous activity.

The **technical conditions** required for the validity of these inference procedures regarding the slope coefficient are as follows:

1. The data arise from a simple random sample from the population or a randomized experiment.
2. The two variables are linearly related.
3. At every given x-value, the distribution of the y-values in the population is normal.
4. The standard deviations of these normal distributions (at each x-value) are the same.

You can check these last three conditions by examining residual plots:

- Check condition 2 by seeing whether a scatterplot of residuals vs. *x*-values reveals a pattern (e.g., curvature) when plotted against the *x*-values. A clear pattern indicates the condition is not met.
- Check condition 3 by seeing whether the residuals roughly follow a normal-shaped distribution (with a histogram and/or normal probability plot).
- Check condition 4 by seeing whether the scatterplot of *residuals* vs. *x-values* displays roughly the same amount of variability at all *x*-values (i.e., equal spread across the graph).

The following questions will help you see what these conditions mean.

a. Produce a histogram and normal probability plot of the residuals (whenever you use the **LinRegTTest** command, the residuals are automatically calculated and stored in the list named **RESID**). Describe the distribution of the residuals. Do these plots reveal any marked features suggesting nonnormality (condition 3)?

b. Produce a scatterplot of the *residuals* vs. *study hours.* Describe that plot. Does it reveal a strong pattern (such as curvature), suggesting a linear model is not appropriate (condition 2)? Does the plot suggest the variability of the residuals differs substantially at various *x*-values (study hours) (condition 4)?

Activity 29-3: House Prices

8-10, 26-1, 27-5, 28-2, 28-12, 28-13, **29-3**

Reconsider the data from Activities 26-1 and 28-2 on the relationship between house size and price (HOUSEPRICESAG).

a. Use your calculator to produce regression output for conducting inference about the population slope. Report the equation of the sample least squares line, and also report the standard error of the slope coefficient.

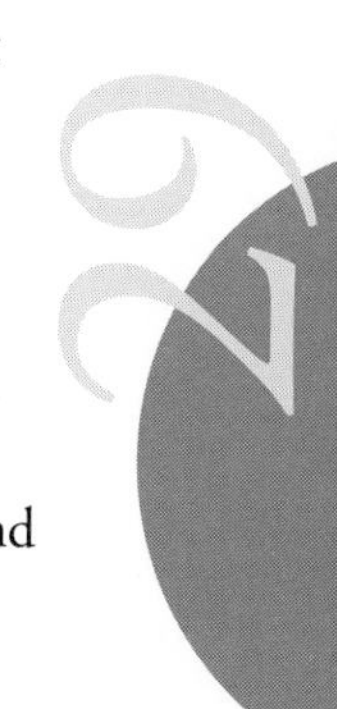

b. Conduct a test of whether the sample data provide evidence that *house price* is positively associated with *house size* in the population. Report all aspects of the test, including an examination of residual plots to check technical conditions, and summarize your conclusion.

c. Estimate the population slope coefficient with a 90% confidence interval. Interpret this interval. Also state whether the value 0 is in this interval, and comment on how this result relates to your test result in part b.

Activity 29-4: Plasma and Romance

In a recent study, researchers investigated possible biochemical mechanisms that could be involved in the early stages of romantic love (Emanuele et al., 2006). They measured plasma level of neurotrophins for a sample of 58 subjects who had recently fallen in love. They also asked each subject to rate his or her level of passionate love feelings on a numerical scale. The correlation coefficient between these variables is .347, and a scatterplot follows.

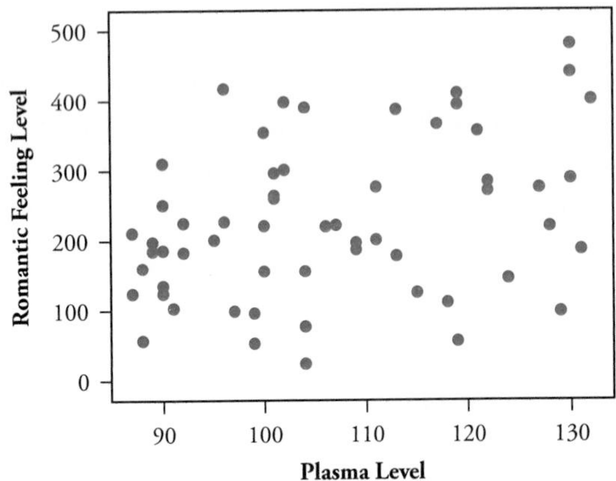

a. Comment on what the scatterplot and correlation coefficient reveal about a possible association between plasma levels and passionate feelings.

b. Is the correlation coefficient in part a, a parameter or a statistic? What symbol represents it?

c. Is it possible to obtain such a large correlation by random chance, even if the correlation between these variables equals zero in the population?

An alternative to conducting a significance test regarding the population slope coefficient is to conduct such a test about the population *correlation* coefficient. These tests turn out to be equivalent, but an advantage of the correlation test is that only the sample correlation and sample size are needed. Details of the test procedure follow, where ρ (the Greek letter rho) denotes the population correlation coefficient.

Test of significance for population correlation coefficient ρ:
Null hypothesis: H_0: $\rho = 0$
Alternative hypothesis:

i. H_a: $\rho < 0$ *or*
ii. H_a: $\rho > 0$ *or*
iii. H_a: $\rho \neq 0$

Test statistic:

$$t = \frac{r\sqrt{n-2}}{\sqrt{1-r^2}}$$

p-value:

i. $\Pr(T_{n-2} \leq t)$ *or*
ii. $\Pr(T_{n-2} \geq t)$ *or*
iii. $2\Pr(T_{n-2} \geq |t|)$

Technical conditions:

1. The data comprise a simple random sample from the population.
2. Both variables are normally distributed.

d. State the null and alternative hypotheses, in symbols and in words, for testing the researchers' claim that plasma levels are positively correlated with strength of romantic feeling.

e. Use the sample size ($n = 58$) and sample correlation coefficient ($r = .347$) to calculate the test statistic.

f. Determine the *p*-value as accurately as possible.

g. Summarize your conclusion from this test, and explain the reasoning process by which your conclusion follows from the sample data. Also comment on (and justify) whether you consider the technical conditions for this procedure have been met.

Activity 29-5: Textbook Prices

28-4, 28-5, **29-5**

Recall from Activity 28-4 that Shaffer and Kaplan took a random sample of 30 textbooks from the Cal Poly Bookstore in November 2006 (`TEXTBOOKPRICES`). In that activity, you determined and analyzed a least squares line for predicting the price of a textbook based on its number of pages.

a. Reproduce a scatterplot of *price* vs. *pages,* including the least squares line. Also report the equation of the line and the value of r^2, along with an interpretation of r^2.

b. Report the value of the slope coefficient, along with the symbol to represent it, and its standard error. Provide an interpretation of both numbers.

c. Identify the *population* of interest in this study.

d. Conduct a significance test of whether the sample data provide strong evidence (at the $\alpha = .01$ significance level) that there is a positive relationship in the population between a textbook's price and its number of pages. Report all aspects of the test, and summarize your conclusion.

e. Determine a 90% confidence interval for estimating the population slope coefficient. Also interpret this interval.

f. How would you expect a 99% confidence interval for the population slope to differ? Without doing the calculation, comment on how (if at all) its midpoint and width would change.

g. Comment on whether the technical conditions for these inference procedures appear to be satisfied. Include an analysis of appropriate residual plots.

Solution

a. The scatterplot with least squares line follows.

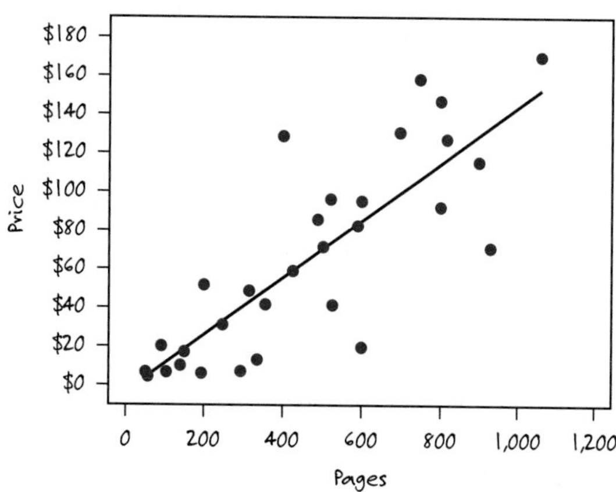

The equation of this line is *predicted price* = −3.42 + 0.147 *pages*. The value of r^2 is .677, indicating that 67.7% of the variability in textbook prices can be explained by the number of pages in the texts.

b. The slope coefficient is $b = 0.147$, and a quick calculation reports the standard error to be 0.019. The slope indicates that the predicted price increases by $0.147 on average for each additional page in the book. The standard error measures the variability in the sample slopes from repeated random samples of 30 textbooks from this population.

c. The population of interest in this study is all textbooks in the Cal Poly Bookstore in November 2006.

d. Let β represent the slope of the least squares line for predicting price from pages in the population. Then the null hypothesis is H_0: $\beta = 0$, meaning there is *no* association between textbook price and number of pages in the population. The alternative is H_a: $\beta > 0$, meaning there is a positive association between these variables in the population. The test statistic is

$$t = \frac{b}{SE(b)} = \frac{0.147}{0.019} \approx 7.74$$

(Without rounding, your calculator reports the test statistic to be 7.65.) Comparing this to the t-distribution (Table III) with $30 - 2 = 28$ degrees of freedom reveals that the p-value is much less than .0005. Such a small p-value provides extremely strong evidence of a positive relationship between a textbook's price and its number of pages in the population of all textbooks in that bookstore at that time.

29

e. The t^* critical value for 90% confidence with 28 degrees of freedom is 1.701 (from Table III). A 90% confidence interval for the population slope β is $0.147 \pm 1.701(0.019)$, which is 0.147 ± 0.032, which is the interval from 0.115 through 0.179. You can be 95% confident that in the population of all textbooks in that bookstore, the predicted price of a textbook increases between 11.5 and 17.9 cents for each additional page.

f. A 99% confidence interval for the population slope β has the same midpoint, namely the sample slope 0.147. But the 99% interval is wider than the 90% interval, in order to achieve higher confidence of capturing the population slope coefficient.

g. First, Shaffer and Kaplan did take a random sample of textbooks, so that condition is satisfied as long as you restrict your population to the Cal Poly Bookstore in November 2006. To check the normality condition, consider the following histogram and normal probability plot of the residuals.

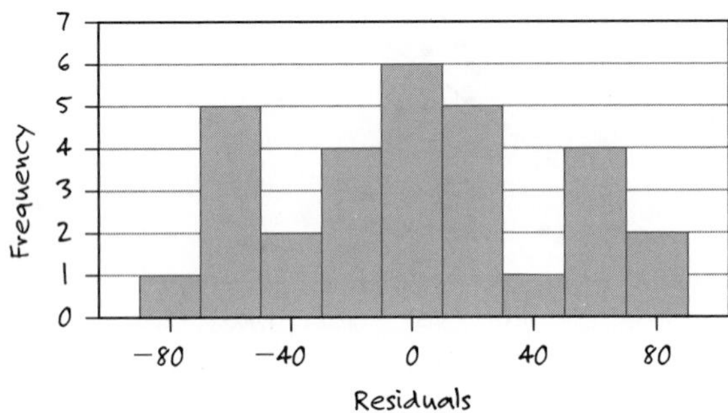

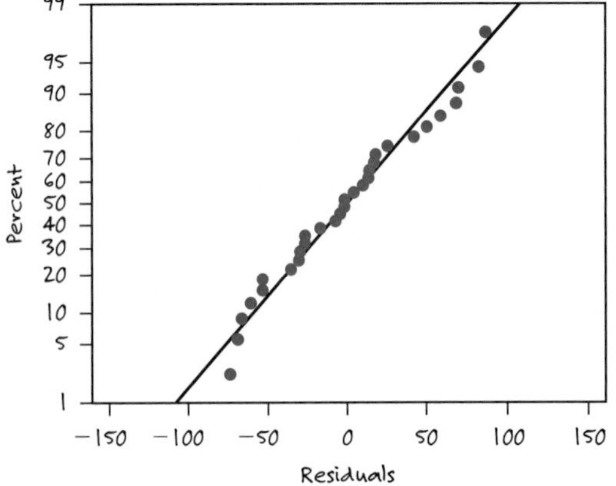

Both plots indicate the distribution of the residuals is approximately normal, so that condition is satisfied. For the conditions regarding linearity and equal variability, consider a plot of *residuals* vs. *pages:*

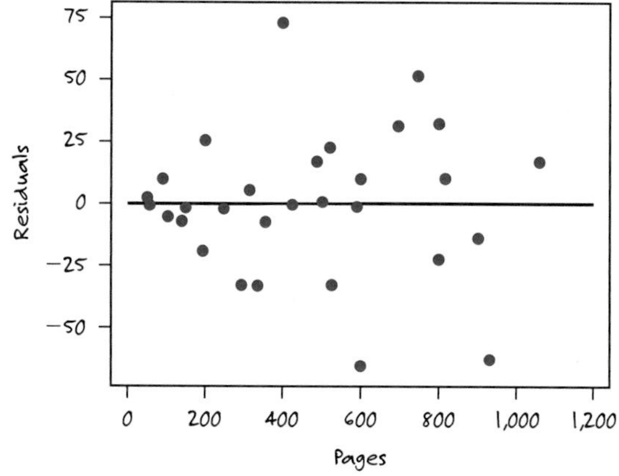

There is no obvious pattern to the residuals in this graph, so the linearity condition is met. The variability in residuals appears to be similar across all values of number of pages, although there might be a bit more variability in the residuals for larger numbers of pages. All technical conditions are met, so the significance test and confidence interval are valid.

Watch Out

- Begin every regression analysis with a scatterplot of the response variable vs. explanatory variable.
- Because there are so many components to regression analysis, it can be easy to forget to check the technical conditions. Remember to perform this check.
- Always consider what the parameter is whenever you conduct a significance test or produce a confidence interval. In this regression setting, the parameter is more complicated than a simple mean or proportion.
- Notice the degrees of freedom for the *t*-distribution here is $n - 2$, rather than $n - 1$ as with inference for a population mean.
- When reading regression output from a software or calculator program, be sure to find the output for the *slope* rather than the intercept to use the procedures introduced in this topic. The slope coefficient, not the intercept, concerns how the variables are related to each other.
- When reading regression output from a software or calculator program, be aware the *p*-value reported is usually for a two-sided test. If your test is one-sided (as in this example), remember to divide the *p*-value by two, as long as the sign of the sample slope is in the direction specified by the alternative hypothesis.
- Once again, remember to summarize your conclusions in the context of the research question; do not simply talk about the population slope. In this case, refer to the population slope for predicting a textbook's price from its number of pages. Better yet, interpret the population slope as the predicted increase in price for each additional page.

Wrap-Up...........

This topic expanded your knowledge of linear regression analysis to include *inference* procedures that complement the descriptive methods studied earlier. These procedures enable you to judge the strength of evidence that a sample provides for a relationship in the larger population. You explored the sampling distribution of sample slope coefficients, in the context of a study about whether study time is associated with grade point average. You found this sampling distribution has much in common with other sampling distributions you have studied, such as symmetry around the actual value of the population parameter.

Based on that sampling distribution, you learned how to conduct a significance test and determine a confidence interval for the population slope coefficient. In the context of a study about whether a person's level of romantic feelings is associated with a specific biochemical measurement, you saw how to conduct a test for a population correlation coefficient, based solely on the sample correlation and sample size.

You also saw that graphical methods play an important role in regression analysis, not only for descriptive purposes but also with regard to checking technical conditions of procedures.

Some useful definitions to remember and habits to develop from this topic include

- Always start a regression analysis with a scatterplot of response variable vs. explanatory variable to see whether the association is roughly linear and whether there are unusual or potentially influential observations.
- The sampling distribution of a sample slope coefficient refers to the pattern of variation in sample slopes when repeated random samples of the same size are taken from the population.
- The standard error of the sample slope is used in both the test statistic and confidence interval. This standard error measures the variability in the sample slope coefficients in repeated random samples.
- Examine residual plots to check on technical conditions for *t*-procedures regarding a population slope.
- The sample correlation and sample size are sufficient information for conducting a significance test regarding the population correlation coefficient, assuming both variables follow normal distributions.
- Apply the graphical and descriptive methods you learned earlier, in conjunction with these inferential techniques for regression.

Homework Activities

Activity 29-6: Proximity to Teacher

1-11, 5-23, 27-9, **29-6**

Reconsider the proposed study of whether students who sit closer to the teacher tend to score higher on quizzes than students who sit farther away. Suppose you measure both *distance from teacher* and *quiz average* as quantitative variables.

a. State the appropriate null and alternative hypotheses, in symbols and in words, for testing this conjecture. *Hint:* This research question calls for a one-sided alternative. Ask yourself whether the conjectured direction suggests a positive or a negative correlation between *distance from teacher* and *quiz average.*

Suppose you gather data on a random sample of students and find the correlation coefficient between *distance from teacher* and *quiz average* equals .3.

b. Is this a parameter or a statistic? Explain, and indicate the symbol used to represent it.

c. What further information do you need to conduct a significance test of whether the correlation coefficient in the population is negative?

d. Suppose the sample size is $n = 20$ students. Calculate the test statistic in this case, still supposing the sample correlation turns out to be $-.3$. Also determine the *p*-value as accurately as possible. What decision would you make at the $\alpha = .05$ level?

e. Repeat part d for a sample size of $n = 50$ students.
f. Repeat part d for a sample size of $n = 150$ students.
g. What happens to the p-value as the sample size increases? Explain why this result makes sense.

Activity 29-7: Studying and Grades

29-1, 29-2, **29-7**, 29-8

Recall that the study analyzed in Activity 29-1 found a sample correlation of $r = .343$ between grade point average and self-reported study hours per week among the 80 students sampled. Let ρ denote the correlation between these two variables in the *population* of all students at the school.

a. Conduct a test of significance of whether these sample data provide evidence at the .05 level that ρ is not zero. State the hypotheses, in words and in symbols, and show the details of your calculation for the test statistic and p-value.
b. Verify that the test statistic and p-value from part a, based on the correlation coefficient, match those found in parts o and p of Activity 29-1, based on a test for the regression slope.
c. Suppose a sample correlation coefficient of .343 were found in a study that involved a sample size of $n = 20$ students. Would you expect the p-value of this test to be greater or less than it was when $n = 80$? Explain.
d. Calculate the test statistic and p-value for the test described in part c. Is your expectation confirmed or refuted?
e. Repeat parts c and d for a sample size of $n = 200$.
f. Suppose a study involves $n = 5000$ subjects. Determine how small the sample correlation r could be and still be statistically significantly different from 0 at the $\alpha = .05$ level.

Activity 29-8: Studying and Grades

29-1, 29-2, 29-7, **29-8**

Reconsider Activity 29-1 by returning to the `Sampling Regression Lines` applet.

a. Change the sample size from 80 to **20.** Generate **100** sample regression lines. Describe the distribution of the 100 simulated sample slope coefficients. Report the mean and standard deviation of these sample slopes. Comment on how the sampling distribution of the sample slope coefficients has changed from when the sample size was 80.
b. Change the sample size back to **80,** but now change the population slope to **0.1** rather than 0. Again generate 100 sample regression lines, and then describe the distribution of the 100 simulated sample slope coefficients, and report the mean and standard deviation of these sample slopes. Comment on how the sampling distribution of the sample slope coefficients has changed from when the population slope was 0.
c. Change the population slope back to **0,** but now change the standard deviation of x-values to be **5.0** rather than 1.84. Again generate 100 sample regression lines, and then describe the distribution of the 100 simulated sample slope coefficients, and report the mean and standard deviation of these sample slopes. Comment on how the sampling distribution of the sample slope coefficients has changed.
d. Change the standard deviation of the x-values back to **1.84,** but now change σ (the standard deviation of the y-values for a given x-value) from 0.45 to **1.45.** Again generate 100 sample regression lines, and then describe the distribution of

the 100 simulated sample slope coefficients, and report the mean and standard deviation of these sample slopes. Comment on how the sampling distribution of the sample slope coefficients has changed.

Activity 29-9: Draft Lottery

10-18, 27-7, **29-9**

Reconsider your analysis of the 1970 and 1971 draft lotteries from Activity 27-7. You found that the correlation coefficient between draft number and sequential birthday for the 1970 lottery was −.226.

a. Determine the probability that a fair random lottery would produce a correlation coefficient so far from zero, simply by random chance. In other words, determine the *p*-value for testing whether the population correlation equals zero, against a two-sided alternative. Report the test statistic as well as the *p*-value. *Hint:* There were 366 numbers in this draft.

You also found that the correlation coefficient between draft number and sequential birthday for the 1971 lottery was .014.

b. Repeat part a for the 1971 lottery.
c. Summarize what these *p*-values reveal about the fairness of these two lotteries.

Activity 29-10: Honda Prices

7-10, 10-19, 12-21, 28-14, 28-15, **29-10,** 29-11

Recall from Activities 28-14 and 28-15 that data was collected in July 2006 on a sample of used Honda Civics for sale on the Internet. The goal is to predict the price of a used Honda Civic, based on either the car's year of manufacture or mileage. Consider using *mileage* as the explanatory variable (HONDAPRICES).

a. Report the equation of the least squares line for predicting *price* from *mileage.* Also identify and interpret the value of the slope coefficient.
b. Determine a 95% confidence interval for the slope coefficient of the model for predicting *price* from *mileage* in the population of all Honda Civics for sale on the Internet.
c. Does this interval include only positive values, only negative values, or some of both? Based on this interval, what can you say about the *p*-value of the corresponding test of whether the population slope coefficient differs from zero? Explain, based on the confidence interval but without conducting the test.
d. Examine residual plots to assess whether the technical conditions appear to be satisfied. Summarize your findings.
e. Remove the car that is the outlier in mileage, and repeat parts a–d.

Activity 29-11: Honda Prices

7-10, 10-19, 12-21, 28-14, 28-15, 29-10, **29-11**

Reconsider the previous activity. Repeat parts a–d of that activity, using *year of manufacture* as the explanatory variable. Also answer the following question:

e. Do you consider *mileage* or *year of manufacture* to be the better explanatory variable for predicting *price.* Justify your choice.

Activity 29-12: Heights, Handspans, and Foot Lengths

26-5, 28-1, **29-12,** 29-13, 29-14

Recall from Activity 26-5 that you collected data on heights, handspans, and foot lengths for yourself and your classmates. For this activity, consider using *handspan* to predict *height.*

a. Produce a scatterplot of *height* vs. *handspan,* and comment on the association it reveals.
b. Determine the equation of the least squares line for predicting *height* from *handspan.* Also draw this line on the scatterplot.
c. Report and interpret the value of r^2.
d. Examine residual plots to assess whether the linear model is appropriate and whether or not the technical conditions for the t-test and t-interval are satisfied. Summarize your analysis.
e. Are the coefficients in part b parameters or statistics? Explain.
f. Conduct a test of whether the slope coefficient is positive in the population of all students at your school. Report all aspects of the test, including a test decision at the .05 significance level.
g. Determine a 95% confidence interval for the population slope coefficient. Also interpret this interval.
h. Summarize your findings about whether *handspan* is a useful predictor of *height* in the population of all students at your school.

Activity 29-13: Heights, Handspans, and Foot Lengths

26-5, 28-1, 29-12, **29-13,** 29-14

Reconsider the previous activity. Repeat parts a–h, using *foot length* as the explanatory variable. Also answer the following question:

i. Do you consider *handspan* or *foot length* to be the better explanatory variable for predicting *height?* Justify your choice.

Activity 29-14: Heights, Handspans, and Foot Lengths

26-5, 28-1, 29-12, 29-13, **29-14**

Reconsider the previous two activities.

a. Now consider only predicting the height of a *male* student at your school. Conduct a full regression analysis to determine whether *handspan* or *foot length* is a better predictor of male height. Include inferential as well as descriptive components in your analysis. Write a few paragraphs summarizing your analysis and conclusions, including appropriate graphs and calculations.
b. Repeat part a for the issue of predicting the height of a *female* student at your school.

Activity 29-15: Chip Melting

23-2, 26-20, **29-15**

Reconsider the data collected in Topic 23 and analyzed in Activities 23-2 and 26-20 on the time it takes a chocolate or peanut butter chip to melt in a person's mouth. Conduct a test of whether there is a positive association between these two melting times in the population of all students at your school. Report all aspects of the test, including a check of technical conditions. Write a paragraph or two summarizing your conclusions and supporting them with appropriate graphs and calculations.

Activity 29-16: Airfares

28-6, 28-7, 28-8, 28-9, **29-16**

Consider the data from Activity 28-6 concerning airfare and distances to various destinations (AIRFARES).

a. Conduct a test of whether the sample data provide strong evidence that the population slope coefficient is positive. Report the hypotheses (in symbols and in words), test statistic, and p-value. Also write a brief conclusion.
b. Construct a 99% confidence interval for the population slope coefficient, and interpret the interval in context.
c. Does this confidence interval suggest it is plausible that the predicted price of flying to a destination increases by a quarter for each additional mile flown? How about by a dime? A nickel?
d. Examine residual plots, and comment on what they reveal concerning the technical conditions for this procedure.

Activity 29-17: Marriage Ages

8-17, 9-6, 16-19, 17-22, 23-1, 23-12, 26-4, **29-17**, 29-18

Consider again the data on marriage ages for a sample of 24 couples from Activities 23-1 and 26-4 (MARRIAGEAGES). Suppose you are interested in predicting the wife's age based on the husband's age.

a. Determine the appropriate least squares equation, and draw it on a scatterplot.
b. Test whether the slope coefficient differs significantly from zero. Report all aspects of the test, including a check of technical conditions, and summarize your conclusion.
c. Determine a 95% confidence interval for the slope coefficient in the population of all married couples in the county from which the sample was selected.
d. Does this confidence interval include the value 1? Comment on the importance of this value in this context.
e. Conduct a significance test of whether the population slope coefficient differs from the value 1. Write a paragraph showing the details of your test and summarizing your conclusion. *Hint:* Use the significance test procedure given for testing whether the slope is 0, but adjust the hypotheses accordingly and subtract 1 (rather than 0) from the sample slope coefficient in the numerator of the test statistic.

Activity 29-18: Marriage Ages

8-17, 9-6, 16-19, 17-22, 23-1, 23-12, 26-4, 29-17, **29-18**

Reconsider the previous activity (MARRIAGEAGES).

a. Which of the 24 husbands has the most potential *influence* over the regression line by virtue of his fairly extreme age? Also explain what "influence" means in this setting.
b. Suppose the 71-year-old husband had actually married a 21-year-old wife rather than a 73-year-old wife. Repeat the analysis of the previous activity in this case.
c. Comment on how much this change to one person's age alters your conclusions.

Table I: Random Digits

Line										
1	17139	27838	19139	82031	46143	93922	32001	05378	42457	94248
2	20875	29387	32582	86235	35805	66529	00886	25875	40156	92636
3	34568	95648	79767	15307	71133	15714	44142	44293	19195	30569
4	11169	41277	01417	34656	80207	33362	71878	31767	04056	52582
5	15529	30766	70264	86253	07179	24757	57502	51033	16551	66731
6	33241	87844	41420	10084	55529	68560	50069	50652	76104	42086
7	83594	48720	96632	39724	50318	91370	68016	06222	26806	86726
8	01727	52832	80950	27135	14110	92292	17049	60257	01638	04460
9	86595	21694	79570	74409	95087	75424	57042	27349	16229	06930
10	65723	85441	37191	75134	12845	67868	51500	97761	35448	56096
11	82322	37910	35485	19640	07689	31027	40657	14875	07695	92569
12	06062	40703	69318	95070	01541	52249	56515	59058	34509	35791
13	54400	22150	56558	75286	07303	40560	57856	22009	67712	19435
14	80649	90250	62962	66253	93288	01838	68388	55481	00336	19271
15	70749	78066	09117	62350	58972	80778	46458	83677	16125	89106
16	50395	30219	03068	54030	49295	48985	16247	28818	83101	18172
17	48993	89450	04987	02781	37935	76222	93595	20942	90911	57643
18	77447	34009	20728	88785	81212	08214	93926	66687	58252	18674
19	24862	18501	22362	37319	33201	88294	55814	67443	77285	36229
20	87445	26886	66782	89931	29751	08485	44910	83844	56013	26596
21	14779	15506	62210	44517	14721	99774	19102	44921	80165	03984
22	79801	42412	09555	05280	00534	10592	36738	03573	77510	66222
23	05297	26648	86929	47906	09699	36563	66286	23137	79434	51560
24	10384	40923	29328	82914	16875	15622	39567	68096	14770	89189
25	31058	85234	87674	21263	42429	99728	16261	16519	65635	59197
26	33946	72507	42425	33159	77435	31639	41819	06220	52297	04770
27	33443	10708	05353	97028	24889	93727	55007	40152	09817	02700
28	13371	02271	79272	26697	63927	01803	74220	93901	07809	69131
29	64066	18676	20977	69271	31705	38898	30547	39412	93606	89007
30	13054	79473	09691	66466	33599	89378	38296	20518	00716	56617
31	34282	50764	73730	49650	13380	49874	81896	57540	66339	76256
32	47982	45080	98353	50960	90679	61396	87926	65972	12968	90622
33	43609	07061	43180	70246	59960	36413	02223	57282	79836	45778
34	85025	13341	31367	83655	04989	88434	96865	46110	60777	54349
35	54749	23767	77894	97636	61682	77530	47755	16204	80388	66027

(*continued*)

Line										
36	44494	67165	26455	35332	42455	00522	74952	08265	53974	26516
37	24228	49167	38008	24678	74906	90101	03858	64494	01655	82336
38	28260	54215	25038	12030	55249	21816	85465	91448	06029	30113
39	80761	07480	46048	97445	80484	42483	63428	80705	19955	31948
40	35969	42174	57469	65149	75981	50368	78941	76123	94717	36337
41	56033	79072	43090	73725	35831	46077	30078	98594	09732	49898
42	06185	53730	84649	40715	25446	90966	40349	81553	36166	97485
43	96249	21285	85912	53370	13489	20214	81912	19510	03250	44944
44	60504	57017	38892	76889	05930	44207	60717	70562	74812	15672
45	27651	90466	80992	03306	75644	45917	30889	36652	75794	34054
46	64410	74350	60581	08972	26284	71779	84450	62976	36142	01946
47	95202	01751	86883	28174	38999	94490	49858	02425	40593	83850
48	78892	99777	18049	66117	78028	70955	75476	98203	01512	23591
49	12709	43956	61118	90635	09299	60528	80161	53155	90775	11483
50	62053	67152	49996	39482	68025	94097	83991	05527	48252	47713
51	19060	58444	64059	97800	16489	87323	05381	50739	06795	67614
52	93142	09643	63074	51502	27243	17788	80518	31654	37937	28985
53	53647	72844	72831	04314	34827	85166	06378	78712	57505	94858
54	74440	32656	13285	82425	29910	75178	48183	52694	82881	30463
55	00875	39143	30131	20830	55687	45897	51699	16687	14893	62227
56	92403	73577	01847	44960	01413	80761	94012	53683	01197	90235
57	73889	36517	94526	09074	79816	79249	68788	04445	12270	69892
58	73717	50576	62638	93857	05737	80588	02481	15325	64990	37749
59	42176	41558	40318	03808	77255	76219	80048	77046	06268	84857
60	03102	55200	52076	68954	36760	59891	34203	52293	14493	39429
61	43468	85530	17692	86887	43360	89660	09245	98723	49664	77925
62	27948	48317	12879	33738	08930	35252	88590	70673	53033	66886
63	56256	00738	97390	28189	50935	07846	65929	45204	96515	36730
64	73057	32866	18205	78341	16474	11094	45487	59284	89071	94574
65	04941	79671	06593	80638	10957	18261	91490	08417	39933	50915
66	33311	58411	97188	03098	06639	30810	65231	46429	95749	68249
67	96459	68595	13745	30662	02707	02666	01937	40702	75257	44362
68	29868	71432	07187	65108	35046	30404	69061	30546	87289	09953
69	27335	84350	60204	13434	07806	80533	07612	29126	58743	68703
70	13524	26818	44665	01502	56917	01550	60298	65433	08407	65654
71	30149	51222	45593	75282	11560	66997	30922	77893	21988	96037
72	67244	94022	05995	37383	42375	87856	69908	44703	92659	09319
73	96190	84676	29074	98926	60127	76094	26079	20548	22746	08079
74	64413	69462	40328	95740	80855	96865	94793	47329	73979	29327
75	90438	55215	07235	50602	69957	61746	05753	97097	44279	97163

Table II: Standard Normal Probabilities

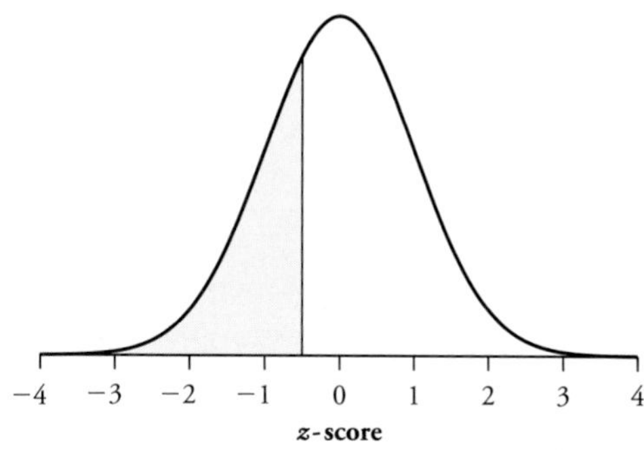

The table reports the area to the left of the value z under the standard normal curve.

	Hundredths Place									
z	.00	.01	.02	.03	.04	.05	.06	.07	.08	.09
≤ −3.5	≤ .0002									
−3.4	.0003	.0003	.0003	.0003	.0003	.0003	.0003	.0003	.0003	.0002
−3.3	.0005	.0005	.0005	.0004	.0004	.0004	.0004	.0004	.0004	.0004
−3.2	.0007	.0007	.0006	.0006	.0006	.0006	.0006	.0005	.0005	.0005
−3.1	.0010	.0009	.0009	.0009	.0008	.0008	.0008	.0008	.0007	.0007
−3.0	.0014	.0013	.0013	.0012	.0012	.0011	.0011	.0011	.0010	.0010
−2.9	.0019	.0018	.0018	.0017	.0016	.0016	.0015	.0015	.0014	.0014
−2.8	.0026	.0025	.0024	.0023	.0023	.0022	.0021	.0021	.0020	.0019
−2.7	.0035	.0034	.0033	.0032	.0031	.0030	.0029	.0028	.0027	.0026
−2.6	.0047	.0045	.0044	.0043	.0041	.0040	.0039	.0038	.0037	.0036
−2.5	.0062	.0060	.0059	.0057	.0055	.0054	.0052	.0051	.0049	.0048
−2.4	.0082	.0080	.0078	.0075	.0073	.0071	.0069	.0068	.0066	.0064
−2.3	.0107	.0104	.0102	.0099	.0096	.0094	.0091	.0089	.0087	.0084
−2.2	.0139	.0136	.0132	.0129	.0125	.0122	.0119	.0116	.0113	.0110
−2.1	.0179	.0174	.0170	.0166	.0162	.0158	.0154	.0150	.0146	.0143
−2.0	.0228	.0222	.0217	.0212	.0207	.0202	.0197	.0192	.0188	.0183
−1.9	.0287	.0281	.0274	.0268	.0262	.0256	.0250	.0244	.0239	.0233
−1.8	.0359	.0351	.0344	.0336	.0329	.0322	.0314	.0307	.0301	.0294
−1.7	.0446	.0436	.0427	.0418	.0409	.0401	.0392	.0384	.0375	.0367
−1.6	.0548	.0537	.0526	.0516	.0505	.0495	.0485	.0475	.0465	.0455
−1.5	.0668	.0655	.0643	.0630	.0618	.0606	.0594	.0582	.0571	.0559
−1.4	.0808	.0793	.0778	.0764	.0749	.0735	.0721	.0708	.0694	.0681
−1.3	.0968	.0951	.0934	.0918	.0901	.0885	.0869	.0853	.0838	.0823
−1.2	.1151	.1131	.1112	.1093	.1075	.1057	.1038	.1020	.1003	.0985
−1.1	.1357	.1335	.1314	.1292	.1271	.1251	.1230	.1210	.1190	.1170
−1.0	.1587	.1562	.1539	.1515	.1492	.1469	.1446	.1423	.1401	.1379
−0.9	.1841	.1814	.1788	.1762	.1736	.1711	.1685	.1660	.1635	.1611
−0.8	.2119	.2090	.2061	.2033	.2005	.1977	.1949	.1922	.1894	.1867
−0.7	.2420	.2389	.2358	.2327	.2297	.2266	.2236	.2207	.2177	.2148
−0.6	.2743	.2709	.2676	.2643	.2611	.2578	.2546	.2514	.2483	.2451

(continued)

z	.00	.01	.02	.03	.04	.05	.06	.07	.08	.09
−0.5	.3085	.3050	.3015	.2981	.2946	.2912	.2877	.2843	.2810	.2776
−0.4	.3446	.3409	.3372	.3336	.3300	.3264	.3228	.3192	.3156	.3121
−0.3	.3821	.3783	.3745	.3707	.3669	.3632	.3594	.3557	.3520	.3483
−0.2	.4207	.4168	.4129	.4090	.4052	.4013	.3974	.3936	.3897	.3859
−0.1	.4602	.4562	.4522	.4483	.4443	.4404	.4364	.4325	.4286	.4247
−0.0	.5000	.4960	.4920	.4880	.4840	.4801	.4761	.4721	.4681	.4641
0.0	.5000	.5040	.5080	.5120	.5160	.5199	.5239	.5279	.5319	.5359
0.1	.5398	.5438	.5478	.5517	.5557	.5596	.5636	.5675	.5714	.5753
0.2	.5793	.5832	.5871	.5910	.5948	.5987	.6026	.6064	.6103	.6141
0.3	.6179	.6217	.6255	.6293	.6331	.6368	.6406	.6443	.6480	.6517
0.4	.6554	.6591	.6628	.6664	.6700	.6736	.6772	.6808	.6844	.6879
0.5	.6915	.6950	.6985	.7019	.7054	.7088	.7123	.7157	.7190	.7224
0.6	.7257	.7291	.7324	.7357	.7389	.7422	.7454	.7486	.7517	.7549
0.7	.7580	.7611	.7642	.7673	.7704	.7734	.7764	.7794	.7823	.7852
0.8	.7881	.7910	.7939	.7967	.7995	.8023	.8051	.8079	.8106	.8133
0.9	.8159	.8186	.8212	.8238	.8264	.8289	.8315	.8340	.8365	.8389
1.0	.8413	.8438	.8461	.8485	.8508	.8531	.8554	.8577	.8599	.8621
1.1	.8643	.8665	.8686	.8708	.8729	.8749	.8770	.8790	.8810	.8830
1.2	.8849	.8869	.8888	.8907	.8925	.8944	.8962	.8980	.8997	.9015
1.3	.9032	.9049	.9066	.9082	.9099	.9115	.9131	.9147	.9162	.9177
1.4	.9192	.9207	.9222	.9236	.9251	.9265	.9279	.9292	.9306	.9319
1.5	.9332	.9345	.9357	.9370	.9382	.9394	.9406	.9418	.9429	.9441
1.6	.9452	.9463	.9474	.9484	.9495	.9505	.9515	.9525	.9535	.9545
1.7	.9554	.9564	.9573	.9582	.9591	.9599	.9608	.9616	.9625	.9633
1.8	.9641	.9649	.9656	.9664	.9671	.9678	.9686	.9693	.9699	.9706
1.9	.9713	.9719	.9726	.9732	.9738	.9744	.9750	.9756	.9761	.9767
2.0	.9773	.9778	.9783	.9788	.9793	.9798	.9803	.9808	.9812	.9817
2.1	.9821	.9826	.9830	.9834	.9838	.9842	.9846	.9850	.9854	.9857
2.2	.9861	.9864	.9868	.9871	.9875	.9878	.9881	.9884	.9887	.9890
2.3	.9893	.9896	.9898	.9901	.9904	.9906	.9909	.9911	.9913	.9916
2.4	.9918	.9920	.9922	.9925	.9927	.9929	.9931	.9932	.9934	.9936
2.5	.9938	.9940	.9941	.9943	.9945	.9946	.9948	.9949	.9951	.9952
2.6	.9953	.9955	.9956	.9957	.9959	.9960	.9961	.9962	.9963	.9964
2.7	.9965	.9966	.9967	.9968	.9969	.9970	.9971	.9972	.9973	.9974
2.8	.9974	.9975	.9976	.9977	.9977	.9978	.9979	.9979	.9980	.9981
2.9	.9981	.9982	.9983	.9983	.9984	.9984	.9985	.9985	.9986	.9986
3.0	.9987	.9987	.9987	.9988	.9988	.9989	.9989	.9989	.9990	.9990
3.1	.9990	.9991	.9991	.9991	.9992	.9992	.9992	.9992	.9993	.9993
3.2	.9993	.9993	.9994	.9994	.9994	.9994	.9994	.9995	.9995	.9995
3.3	.9995	.9995	.9996	.9996	.9996	.9996	.9996	.9996	.9996	.9997
3.4	.9997	.9997	.9997	.9997	.9997	.9997	.9997	.9997	.9997	.9998
≥ 3.5	≥ .9998									

Table III: *t*-Distribution Critical Values

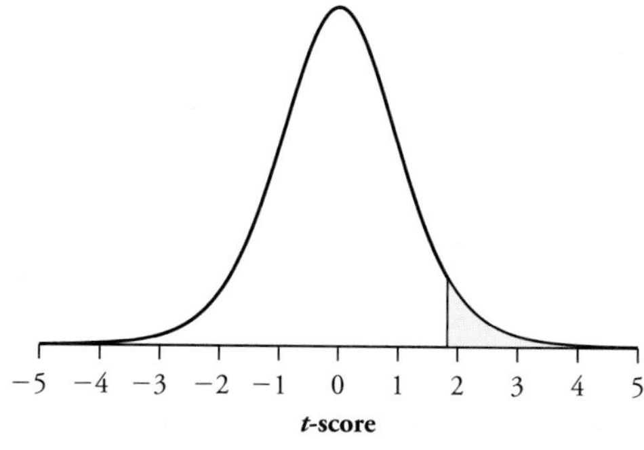

The table reports the critical value for which the area to the right is as indicated.

Area to Right	0.2	0.1	0.05	0.025	0.01	0.005	0.001	0.0005
Confidence Level	60%	80%	90%	95%	98%	99%	99.80%	99.90%
df								
1	1.376	3.078	6.314	12.706	31.821	63.657	318.317	636.607
2	1.061	1.886	2.920	4.303	6.965	9.925	22.327	31.598
3	0.978	1.638	2.353	3.182	4.541	5.841	10.215	12.924
4	0.941	1.533	2.132	2.776	3.747	4.604	7.173	8.610
5	0.920	1.476	2.015	2.571	3.365	4.032	5.893	6.869
6	0.906	1.440	1.943	2.447	3.143	3.708	5.208	5.959
7	0.896	1.415	1.895	2.365	2.998	3.500	4.785	5.408
8	0.889	1.397	1.860	2.306	2.897	3.355	4.501	5.041
9	0.883	1.383	1.833	2.262	2.821	3.250	4.297	4.781
10	0.879	1.372	1.812	2.228	2.764	3.169	4.144	4.587
11	0.876	1.363	1.796	2.201	2.718	3.106	4.025	4.437
12	0.873	1.356	1.782	2.179	2.681	3.055	3.930	4.318
13	0.870	1.350	1.771	2.160	2.650	3.012	3.852	4.221
14	0.868	1.345	1.761	2.145	2.625	2.977	3.787	4.140
15	0.866	1.341	1.753	2.131	2.602	2.947	3.733	4.073
16	0.865	1.337	1.746	2.120	2.583	2.921	3.686	4.015
17	0.863	1.333	1.740	2.110	2.567	2.898	3.646	3.965
18	0.862	1.330	1.734	2.101	2.552	2.878	3.611	3.922
19	0.861	1.328	1.729	2.093	2.539	2.861	3.579	3.883
20	0.860	1.325	1.725	2.086	2.528	2.845	3.552	3.850
21	0.859	1.323	1.721	2.080	2.518	2.831	3.527	3.819
22	0.858	1.321	1.717	2.074	2.508	2.819	3.505	3.792
23	0.858	1.319	1.714	2.069	2.500	2.807	3.485	3.768
24	0.857	1.318	1.711	2.064	2.492	2.797	3.467	3.745
25	0.856	1.316	1.708	2.060	2.485	2.787	3.450	3.725
26	0.856	1.315	1.706	2.056	2.479	2.779	3.435	3.707
27	0.855	1.314	1.703	2.052	2.473	2.771	3.421	3.690
28	0.855	1.313	1.701	2.048	2.467	2.763	3.408	3.674
29	0.854	1.311	1.699	2.045	2.462	2.756	3.396	3.659
30	0.854	1.310	1.697	2.042	2.457	2.750	3.385	3.646
31	0.853	1.309	1.696	2.040	2.453	2.744	3.375	3.633
32	0.853	1.309	1.694	2.037	2.449	2.738	3.365	3.622
33	0.853	1.308	1.692	2.035	2.445	2.733	3.356	3.611
34	0.852	1.307	1.691	2.032	2.441	2.728	3.348	3.601
35	0.852	1.306	1.690	2.030	2.438	2.724	3.340	3.591
36	0.852	1.306	1.688	2.028	2.434	2.719	3.333	3.582
37	0.851	1.305	1.687	2.026	2.431	2.715	3.326	3.574
38	0.851	1.304	1.686	2.024	2.429	2.712	3.319	3.566
39	0.851	1.304	1.685	2.023	2.426	2.708	3.313	3.558
40	0.851	1.303	1.684	2.021	2.423	2.704	3.307	3.551
50	0.849	1.299	1.676	2.009	2.403	2.678	3.261	3.496
60	0.848	1.296	1.671	2.000	2.390	2.660	3.232	3.460
80	0.846	1.292	1.664	1.990	2.374	2.639	3.195	3.416
100	0.845	1.290	1.660	1.984	2.364	2.626	3.174	3.391
500	0.842	1.283	1.648	1.965	2.334	2.586	3.107	3.310
Infinity (z)	0.842	1.282	1.645	1.960	2.326	2.576	3.090	3.291

Table IV: Chi-Square Distribution Critical Values

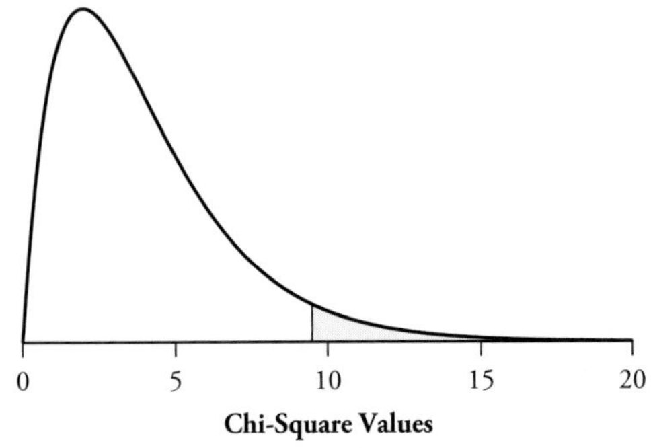

The table reports the critical value for which the area to the right is as indicated.

Area to Right	0.2	0.1	0.05	0.025	0.02	0.01	0.005	0.001	0.0005
df									
1	1.64	2.71	3.84	5.02	5.41	6.63	7.88	10.83	12.12
2	3.22	4.61	5.99	7.38	7.82	9.21	10.60	13.82	15.20
3	4.64	6.25	7.81	9.35	9.84	11.34	12.84	16.27	17.73
4	5.99	7.78	9.49	11.14	11.67	13.28	14.86	18.47	20.00
5	7.29	9.24	11.07	12.83	13.39	15.09	16.75	20.52	22.11
6	8.56	10.64	12.59	14.45	15.03	16.81	18.55	22.46	24.10
7	9.80	12.02	14.07	16.01	16.62	18.48	20.28	24.32	26.02
8	11.03	13.36	15.51	17.53	18.17	20.09	21.96	26.12	27.87
9	12.24	14.68	16.92	19.02	19.68	21.67	23.59	27.88	29.67
10	13.44	15.99	18.31	20.48	21.16	23.21	25.19	29.59	31.42
11	14.63	17.28	19.68	21.92	22.62	24.73	26.76	31.26	33.14
12	15.81	18.55	21.03	23.34	24.05	26.22	28.30	32.91	34.82
13	16.98	19.81	22.36	24.74	25.47	27.69	29.82	34.53	36.48
14	18.15	21.06	23.68	26.12	26.87	29.14	31.32	36.12	38.11
15	19.31	22.31	25.00	27.49	28.26	30.58	32.80	37.70	39.72
16	20.47	23.54	26.30	28.85	29.63	32.00	34.27	39.25	41.31
17	21.61	24.77	27.59	30.19	31.00	33.41	35.72	40.79	42.88
18	22.76	25.99	28.87	31.53	32.35	34.81	37.16	42.31	44.43
19	23.90	27.20	30.14	32.85	33.69	36.19	38.58	43.82	45.97
20	25.04	28.41	31.41	34.17	35.02	37.57	40.00	45.31	47.50
21	26.17	29.62	32.67	35.48	36.34	38.93	41.40	46.80	49.01
22	27.30	30.81	33.92	36.78	37.66	40.29	42.80	48.27	50.51
23	28.43	32.01	35.17	38.08	38.97	41.64	44.18	49.73	52.00
24	29.55	33.20	36.42	39.36	40.27	42.98	45.56	51.18	53.48
25	30.68	34.38	37.65	40.65	41.57	44.31	46.93	52.62	54.95
26	31.79	35.56	38.89	41.92	42.86	45.64	48.29	54.05	56.41
27	32.91	36.74	40.11	43.19	44.14	46.96	49.64	55.48	57.86
28	34.03	37.92	41.34	44.46	45.42	48.28	50.99	56.89	59.30
29	35.14	39.09	42.56	45.72	46.69	49.59	52.34	58.30	60.73
30	36.25	40.26	43.77	46.98	47.96	50.89	53.67	59.70	62.16
40	47.27	51.81	55.76	59.34	60.44	63.69	66.77	73.40	76.09
50	58.16	63.17	67.50	71.42	72.61	76.15	79.49	86.66	89.56
60	68.97	74.40	79.08	83.30	84.58	88.38	91.95	99.61	102.70
80	90.41	96.58	101.88	106.63	108.07	112.33	116.32	124.84	128.26
100	111.67	118.50	124.34	129.56	131.14	135.81	140.17	149.45	153.17

Appendix A: Student Glossary

This "do-it-yourself" glossary is meant to provide you with an opportunity to organize definitions of key terms in one location. Most of the following key terms appear in the textbook in bold type. You should feel free to write definitions or properties of these terms in the glossary as you encounter them. You can also record several topic numbers, activity numbers, or page numbers for each keyword, as each term may appear in more than one activity throughout the textbook.

Key Words	Page Ref.	Notes
alternative hypothesis		
anecdotal evidence		
association		
bar graph		
biased sampling		
binary		
blind		
boxplot		
categorical		
causation		
center		
Central Limit Theorem for a Sample Mean		
Central Limit Theorem for a Sample Proportion		
chi-square distribution		
chi-square goodness-of-fit test		
chi-square test of equal proportions in a two-way table		
chi-square test of independence in a two-way table		
coefficient of determination		
comparison		
completely randomized experiment		
conditional distribution		
confidence		
confidence interval		
confidence interval for a population mean μ (t-interval)		
confidence interval for a population proportion π		

(*continued*)

Key Words	Page Ref.	Notes
confidence interval for $\pi_1 - \pi_2$		
confidence interval for $\mu_1 - \mu_2$		
confidence level		
confounding variable		
consistency		
control		
control group		
convenience sample		
correlation coefficient		
critical value		
data		
degrees of freedom (df)		
distribution		
dotplot		
double blind		
duality		
empirical estimate		
empirical rule		
expected count		
expected value		
experiment		
explanatory variable		
extrapolation		
fitted value		
five-number summary (FNS)		
frequency		
histogram		
independence		
influential		
intercept		
interquartile range		
labeled scatterplot		

Key Words	Page Ref.	Notes
least squares line		
lower quartile		
lurking variable		
margin-of-error		
marginal distribution		
matched-pairs design		
mean		
mean absolute deviation (MAD)		
median		
mode		
modified boxplot		
normal distribution		
normal probability plot		
nonresponse		
null hypothesis		
observational study		
observational unit		
one-sided alternative		
outlier		
outlier test		
p-value		
paired *t*-procedures		
parameter		
peaks/clusters		
placebo		
placebo effect		
population		
power		
practical significance		
precision		
prediction		

(*continued*)

Key Words	Page Ref.	Notes
probability		
probability distribution		
proportion of variability		
quantitative		
random assignment		
randomization		
randomized comparative experiment		
range		
rate		
regression line		
relative frequency		
relative risk		
representative		
residual		
residual plot		
resistant		
response variable		
robust		
sample		
sample size		
sample space		
sampling		
sampling bias		
sampling distribution		
sampling frame		
sampling variability		
scatterplot		
segmented bar graph		
shape		
side-by-side stemplot		
significance level		

Key Words	Page Ref.	Notes
simple random sampling		
Simpson's paradox		
simulation		
skewed to the left		
skewed to the right		
slope coefficient		
standard deviation (SD)		
standard error		
standard error of the sample mean		
standard normal distribution		
standard normal table		
standardization		
statistic		
statistical significance		
statistical tendency		
stemplot		
symmetric		
t-distribution		
t-interval		
t-test		
table of random digits		
tally		
technical conditions		
test decision		
test of significance		
test statistic		
transformation		
treatment		
two-sample *t*-interval		
two-sample *t*-test		
two-sample *z*-interval		

(*continued*)

Key Words	Page Ref.	Notes
two-sample z-test		
two-sided alternative		
two-sided test		
two-way table		
Type I error		
Type II error		
unbiased		
upper quartile		
variability		
variable		
voluntary response		
$y = x$ line		
z-score		

Appendix B: Sources for Studies and Datasets

Unit 1: Collecting Data and Drawing Conclusions

Topic 1: Data and Variables

Activity 1-3, Cell Phone Fraud: This study is described in "To Catch a Thief: Detecting Cell Phone Fraud," by D. Lambert and J. Pinheiro, *Statistics: A Guide to the Unknown,* 4th edition, edited by R. Peck et al., Duxbury Press, 2006, pp. 293–306.

Activity 1-4, Studies from Blink: These studies are described in *Blink: The Power of Thinking Without Thinking,* by M. Gladwell, Little, Brown, and Company, 2005.

Activity 1-6, A Nurse Accused: This study is described in "Statistics in the Courtroom: United States v. Kristen Gilbert," by G. Cobb and S. Gerlach, *Statistics: A Guide to the Unknown,* 4th edition, edited by R. Peck et al., Duxbury Press, 2006, pp. 3–18.

Activity 1-8, Top 100 Films: The list from the American Film Institute can be downloaded from `www.afi.com/Docs/tvevents/pdf/movies100.pdf`.

Activity 1-9, Credit Card Usage: This study is described in a report titled "Undergraduate Students and Credit Cards in 2004: An Analysis of Usage Rates and Trends," developed and produced by Nellie Mae, 2005. The report can be downloaded from `www.nelliemae.com/library/research_12.html`.

Activity 1-14, Natural Light and Achievement: This study is described in a report titled "Windows and Classrooms: A Study of Student Performance and the Indoor Environment," prepared by the Heschong Mahone Group, Inc., for the California Energy Commission, 2003. The report can be downloaded from `www.h-m-g.com/downloads/Daylighting/A-7_Windows_Classrooms_2.4.10.pdf`.

Activity 1-15, Children's Television Viewing: This study is described in "Reducing Children's Television Viewing to Prevent Obesity: A Randomized Controlled Trial," by T. Robinson, *Journal of the American Medical Association,* vol. 282, 1999, pp. 1561–1567.

Activity 1-16, Nicotine Lozenge: This study is described in "Efficacy of a Nicotine Lozenge for Smoking Cessation," by S. Shiffman, C. Dresler, P. Hajek, S. Gilburt, D. Targett, and K. Strahs, *Archives of Internal Medicine,* vol. 162, June 10, 2002, pp. 1267–1276.

Topic 2: Data and Distributions

Activity 2-2, Hand Washing: This study is described in a report titled "A Survey of Hand Washing Behavior (2005 Findings)," prepared by Harris Interactive for the American Society for Microbiology and the Soap and Detergent Association. The report can be downloaded from `www.cleaning101.com/newsroom/2005_survey/handhygiene/`.

Activity 2-4, Buckle Up!: These data are from *Traffic Safety Facts Research Notes,* National Highway Traffic Safety Administration, March 2006. This report can be downloaded from `www.nhtsa.dot.gov`.

Activity 2-5, February Temperatures: These data are from `www.weather.com`, March 2006.

Activity 2-6, Sporting Examples: This study is described in "Teaching an Introductory Statistics Class Based on Sports Examples," by R. Lock, *Proceedings of the Seventh International Conference on Teaching Statistics,* edited by A. Rossman and B. Chance, International Association for Statistical Education, 2006. Data were supplied by the author.

Activity 2-12, Responding to Katrina: These data are from a poll conducted by CNN, *USA Today,* and Gallup on September 8–11, 2005. The survey results were downloaded from `www.cnn.com`, September 2005.

Activity 2-13, Backpack Weights: These data are from "Oh, My Aching Back! A Statistical Analysis of Backpack Weights," by J. Mintz, J. Mintz, K. Moore, and K. Schuh, *Stats: The Magazine for Students of Statistics,* vol. 32, 2002, pp. 17–19.

Activity 2-14, Broadway Shows: These data were presented by the League of American Theaters and Producers and downloaded from `www.livebroadway.com`, July 2006.

Activity 2-15, Highest Peaks: These data were from `http://geology.com/state-high-points.shtml`, January 2007.

Activity 2-16, Pursuit of Happiness: This study is described in a report titled "Are We Happy Yet?" from the Pew Research Center and released on February 13, 2006. The report can be downloaded from `http://pewresearch.org/pubs/301/are-we-happy-yet`, February 2006.

Activity 2-17, Roller Coasters: These data are from the Roller Coaster DataBase and were downloaded from `www.rcdb.com`.

Topic 3: Drawing Conclusions from Studies

Activity 3-1, Elvis Presley and Alf Landon: The Elvis poll was reported in the *Harrisburg Patriot-News,* August 18, 1989. The *Literary Digest* poll is discussed in "Why the *Literary Digest* Poll Failed," by P. Squire, *Public Opinion Quarterly,* vol. 52, 1988, pp. 125–133.

Activity 3-2, Self-Injuries: This study is described in "Self-injurious Behaviors in a College Population," by J. Whitlock, J. Eckenrode, and D. Silverman, *Pediatrics,* vol. 117, June 2006, pp. 1939–1948.

Activity 3-3, Candy and Longevity: This study is described in "Life Is Sweet: Candy Consumption and Longevity," by I. Lee and R. Paffenbarger, *British Medical Journal,* vol. 317, 1998, pp. 1683–1684.

Activity 3-5, Childhood Obesity and Sleep: This study is described in "Relationship Between Short Sleeping Hours and Childhood Overweight/Obesity: Results from the 'Quebec en Forme' Project," by J. Chaput, M. Brunet, and A. Tremblay, *International Journal of Obesity,* vol. 30, March 2006, pp. 10801–10085.

Activity 3-6, Elvis Presley and Alf Landon: The data are from `www.misterpoll.com`, August 1, 2006.

Activity 3-8, Generation M: This study is described in a report titled "Generation M: Media in the Lives of 8–18 Year-Olds," by the Kaiser Family Foundation, March 2005. The report can be downloaded from `http://www.kff.org/entmedia/7251.cfm`.

Activity 3-10, Penny Thoughts: The data are from a Harris Interactive report titled "Abolish the Penny? A Majority of the Public Says 'No'," July 15, 2004. The report can be downloaded from `www.harrisinteractive.com`.

Activity 3-12, Web Addiction: This study was reported in the *Tampa Tribune,* August 23, 1999.

Activity 3-13, Alternative Medicine: This study was reported in *Self* magazine, March 1994.

Activity 3-14, Courtroom Cameras: This study was reported in the *Harrisburg Evening-News,* October 4, 1994.

Activity 3-15, Junior Golfer Survey: This study was reported in the "Golf Plus" section of *Sports Illustrated,* August 21, 2006, pages G8–G9.

Activity 3-16, Accumulating Frequent Flyer Miles: These data are from www.msnbc.com, September 28, 2005.

Activity 3-17, Foreign Language Study: One such study is "The Effects of Foreign Language Study in High School on Verbal Ability as Measured by The Scholastic Aptitude Test—Verbal," by P. Eddy, Center for Applied Linguistics, 1981.

Activity 3-18, Smoking and Lung Cancer: One such study is described in "Tobacco Smoking as a Possible Etiologic Factor in Bronchiogenic Cancer," by E. Wydner and E. Graham, *Journal of the American Medical Association,* vol. 143, 1950, pp. 329–338.

Activity 3-19, Smoking and Lung Cancer: This study is described in "Smoking and Death Rates: Report on Forty-Four Months of Follow-up of 187,783 Men," by E. Hammond and D. Horn, *Journal of the American Medical Association,* vol. 166, 1958, pp. 1159–1172, 1294–1308.

Activity 3-22, Yoga and Middle-Aged Weight Gain: This study was reported in an article titled "Can Yoga Prevent Middle-Age Spread? Yes" at www.medicalnewstoday.com, July 20, 2005.

Activity 3-24, Winter Heart Attacks: This study was reported in *USA Today,* October 12, 1999.

Activity 3-26, Televisions, Computers, and Achievement: This study is described in "Association of Television Viewing During Childhood with Poor Educational Achievement," by R. Hancox, B. Milne, and R. Poulton, *Archives of Pediatric Adolescent Medicine,* vol. 159, 2005, pp. 614–618.

Activity 3-27, Parking Meter Reliability: This study was reported in "Sixth-grade Project Inspires State Legislature," by S. Yang, *Los Angeles Times,* September 8, 1998, p. A3. More detail is given by L. Snell in edition 7.09 of the *Chance News* newsletter, available from www.dartmouth.edu/~chance.

Activity 3-28, Night Lights and Nearsightedness: This study is reported in "Myopia and Ambient Lighting at Night," by G. Quinn, C. Shin, M. Maguire, and R. Stone, *Nature,* vol. 399, May 13, 1999, p. 113.

Topic 4: Random Sampling

Activity 4-5, Back to Sleep: This study is described in "Factors Associated with the Transition to Nonprone Sleep Positions of Infants in the United States: The National Infant Sleep Position Study," by M. Willinger et al., *Journal of the American Medical Association,* vol. 280, 1998, pp. 329–335.

Activity 4-6, Rating Chain Restaurants: This study is described in *Consumer Reports,* July 2006, pp. 12–17.

Activity 4-11, Rose-y Opinions: A similar poll was conducted by the Gallup Organization, January 9–11, 2004.

Activity 4-15, Emotional Support: This study is reported in *The Superpollsters: How They Measure and Manipulate Public Opinion in America,* by D. Moore, Four Walls Eight Windows Publishers, 1995, p. 19.

Activity 4-19, Voter Turnout: Data from the 1998 General Social Survey can be downloaded from `http://sda.berkeley.edu/archive.htm`.

Topic 5: Designing Experiments

Activity 5-1, Testing Strength Shoes: This study is described in "Development of Lower Leg Strength and Flexibility with the Strength Shoe," by S. Cook, G. Schultz, M. Omey, M. Wolfe, and M. Brunet, *The American Journal of Sports Medicine,* vol. 21, 1993, pp. 445–448.

Activity 5-4, Botox for Back Pain: This study is described in "Botulinum Toxin A and Chronic Low Back Pain: A Randomized, Double-Blind Study," by L. Foster, L. Clapp, M. Erickson, and B. Jabbari, *Neurology,* vol. 56, 2001, pp. 1290–1293.

Activity 5-8, Treating Parkinson's Disease: The announcement for this study was downloaded from `www.clinicaltrials.gov`, a service of the National Institutes of Health, September 2006.

Activity 5-9, Ice Cream Servings: This study is described in "Ice Cream Illusions: Bowls, Spoons, and Self-Served Portion Sizes," by B. Wansink, K. van Ittersum, and J. Painter, *American Journal of Preventive Medicine,* vol. 31, 2006, pp. 240–243.

Activity 5-12, AZT and HIV: This study is described in "Reduction of Maternal-Infant Transmission of Human Immunodeficiency Virus Type 1 with Zidovudine Treatment," by E. Connor et al., *New England Journal of Medicine,* vol. 331, 1994, pp. 1173–1180.

Activity 5-15, Reducing Cold Durations: This study was reported in *USA Today,* November 1, 1999.

Activity 5-16, Religious Lifetimes: This study was reported in *USA Today,* August 9, 1999.

Activity 5-20, Literature for Parolees: This study was reported in the *New York Times,* October 6, 1993, p. B10.

Activity 5-21, Therapeutic Touch: This study is described in "A Closer Look at Standards for Therapeutic Touch," by L. Rosa et al., *Journal of the American Medical Association,* vol. 279, 1998, pp. 1005–1010.

Activity 5-25, Dolphin Therapy: This study is described in "Randomized Controlled Trial of Animal Facilitated Therapy with Dolphone in the Treatment of Depression," by C. Antonioli and M. Reveley, *British Medical Journal,* vol. 331, 2005, pp. 1231–1234.

Activity 5-26, Cold Attitudes: This study is described in "Emotional Style and Susceptibility to the Common Cold," by S. Cohen et al., *Psychosomatic Medicine,* vol. 65, 2003, pp. 652–657.

Activity 5-27, Friendly Observers: This study is described in "The Trouble with Friendly Faces: Skilled Performance with a Supportive Audience," by J. Butler and R. Baumeister, *Journal of Personality and Social Psychology,* vol. 75, 1998, pp. 1213–1230.

Unit 2: Summarizing Data

Topic 6: Two-Way Tables

Activity 6-1, Government Spending: Data from the 2004 General Social Survey (GSS) can be downloaded from the SDA Archive at http://sda.berkeley.edu/archive.htm.

Activity 6-7, "Hella" Project: These data were collected by Cal Poly students O. Mead, L. Olerich, and J. Selenkow in the fall of 2005.

Activity 6-8, Suitability for Politics: Data from the 2004 General Social Survey (GSS) can be downloaded from the SDA Archive at http://sda.berkeley.edu/archive.htm.

Activity 6-11, Children's Television Advertisements: This study is described in "A Content Analysis of Health and Physical Activity Messages Marketed to African American Children During After-School Television Programming," by C. Outley and A. Taddese, *Archives of Pediatrics and Adolescent Medicine,* vol. 160, 2006, pp. 432–435.

Activity 6-12, Female Senators: These data are from www.senate.gov, March 2007.

Activity 6-13, Weighty Feelings: Data from the 2003–2004 National Health and Nutrition Examination Survey (NHANES) can be downloaded from www.cdc.gov/nchs/about/major/nhanes/datalink.htm.

Activity 6-14, Preventing Breast Cancer: These data are from a press release titled "Initial Results of the Study of Tamoxifen and Raloxifene (STAR) Released: Osteoporosis Drug Raloxifene Shown to be as Effective as Tamoxifen in Preventing Invasive Breast Cancer" and downloaded from www.cancer.gov/newscenter/, April 17, 2006.

Activity 6-16, Flu Vaccine: These data are from a report titled "Preliminary Assessment of the Effectiveness of the 2003–04 Inactivated Influenza Vaccine—Colorado, December 2003," *Morbidity and Mortality Weekly Report,* Centers for Disease Control and Prevention, and can be downloaded from www.cdc.gov/mmwr/.

Activity 6-21, Gender-Stereotypical Toy Advertising: These data were supplied by Dr. Pamela Rosenberg, Gettysburg College, 1994.

Activity 6-23, Baldness and Heart Disease: These data are from "A Case-Control Study of Baldness in Relation to Myocardial Infarction in Men," by S. Lasko et al., *Journal of the American Medical Association,* vol. 269, 1993, pp. 998–1003.

Activity 6-24, Gender and Lung Cancer: This study was reported in *USA Today,* November 1, 1999.

Activity 6-26, Graduate Admissions Discrimination: These data are described in "Sex Bias in Graduate Admission: Data from Berkeley," by P. Bickel, E. Hammel, and J. O'Connell, *Science,* vol. 187, 1975, pp. 398–404.

Topic 7: Displaying and Describing Distributions

Activity 7-1, Matching Game: A. These data are from the Scrabble Brand Crossword Game, by Selchow and Righter. B. These data are from the Monopoly Real Estate Trading Game, by Parker Brothers. C. These data are from the *2006 Cal Poly Football Official Game Program.* D. These data are from the *2004 Olympic Games Media Guide* and can be downloaded from www.usoc.com. E. These data are a sample from the

2003–04 National Health and Nutrition Examination Survey (NHANES) and can be downloaded from `www.cdc.gov/nchs/about/major/nhanes/datalink.htm`. F. These data were supplied by one of the authors. G. These data are from the *Statistical Abstract of the United States 1992.*

Activity 7-2, Rowers' Weights: These data are from the *2004 Olympic Games Media Guide.*

Activity 7-3, British Monarchs' Reigns: These data are from *The World Almanac and Book of Facts 1995,* World Almanac Books, pp. 534–535.

Activity 7-4, Population Growth: These data are from the U.S. Census Bureau and can be downloaded from `www.npg.org/popfacts.htm`.

Activity 7-5, Diabetes Diagnosis: These data are from the 2003–04 National Health and Nutrition Examination Survey (NHANES) and can be downloaded from `www.cdc.gov/nchs/about/major/nhanes/datalink.htm`.

Activity 7-6, Go Take a Hike!: These data are from *Day Hikes in San Luis Obispo County,* by R. Stone, Day Hike Books, Inc., 2000.

Activity 7-11, College Football Scores: These data were obtained from scores listed at `http://sportsillustrated.cnn.com/`, September 2006.

Activity 7-16, Placement Exam Scores: These data were obtained from the Dickinson College Department of Mathematics and Computer Science in 1992.

Activity 7-18, Hitchcock Films: These data were gathered at a Blockbuster Video store in Carlisle, Pennsylvania, in January 1995.

Activity 7-19, Jurassic Park *Dinosaur Heights:* These data are from *Jurrasic Park,* by M. Crichton, Ballantine Books, 1990, p. 165.

Activity 7-22, Tennis Simulations: These data are a random sample taken from the data analyzed in the study titled "Computer Simulation of a Handicap Scoring System for Tennis," by A. Rossman and M. Parks, *Stats: The Magazine for Students of Statistics,* vol. 10, 1993, pp. 14–18.

Activity 7-23, Blood Pressures: These data are a sample from the 2003–04 National Health and Nutrition Examination Survey (NHANES) and can be downloaded from `www.cdc.gov/nchs/about/major/nhanes/datalink.htm`.

Topic 8: Measures of Center

Activity 8-1, Sleeping Times: These data were collected from students at Cal Poly in the spring of 2002.

Activity 8-2, Game Show Prizes: These data can be downloaded from `www.nbc.com/Deal_or_No_Deal/game/flash.shtml`.

Activity 8-7, Readability of Cancer Pamphlets: These data are from the study titled "Readability of Educational Materials for Patients with Cancer," by T. Short, H. Moriarty, and M. Cooley, *Journal of Statistics Education,* vol. 3, no. 2, 1995, and can be downloaded from `www.amstat.org/publications/jse/v3n2/datasets.short.html`.

Activity 8-10, House Prices: This statistic is from "County Home Sales Drop Nearly a Third," by J. Lynem, *The Tribune,* www.sanluisobispo.com, August 25, 2006. The original source was cited as the California Association of Realtors.

Activity 8-11, Supreme Court Justices: These data were gathered from www.supremecourtus.gov, October 2006.

Activity 8-12, Planetary Measurements: These data are from *The World Almanac and Book of Facts 1993,* p. 251.

Activity 8-17, Marriage Ages: These data were collected by Matthew Parks at the Cumberland County (PA) courthouse in June–July, 1993.

Topic 9: Measures of Spread

Activity 9-1, Baseball Lineups: These data were downloaded from www.sportsline.com/mlb/scoreboard, August 31, 2006.

Activity 9-8, Social Acquaintances: This activity is based on an exercise described in *The Tipping Point: How Little Things Can Make a Big Difference,* by M. Gladwell, Little, Brown, and Company, 2000, pp. 38–41.

Activity 9-11, Baby Weights: The national values are based on the 2000 CDC Growth Charts downloaded from the National Center for Health Statistics, www.cdc.gov/growthcharts/.

Activity 9-14, Pregnancy Durations: The values are based on Table 43 of the *National Vital Statistics Report,* vol. 52, December 17, 2003.

Activity 9-20, Monthly Temperatures: These data are from *The World Almanac and Book of Facts 1999,* p. 220.

Topic 10: More Summary Measures and Graphs

Activity 10-1, Natural Selection: These data are from "The Elimination of the Unfit as Illustrated by the Introduced Sparrow, Passer Domesticus," by H. Bumpus, *Biological Lectures: Woods Hole Marine Biological Laboratory,* 1898, pp. 209–225. This data can be downloaded from The Fields Museum Web site as posted by P. Lowther at http://fm1.fieldmuseum.org/aa/staff_page.cgi?staff=lowther&id=432.

Activity 10-3, Ice Cream Calories: These data were downloaded from Ben & Jerry's at www.benandjerrys.com, Cold Stone Creamery at www.coldstonecreamery.com, and Dreyers Grand Ice Cream at www.dreyers.com, August 2006.

Activity 10-4, Fan Cost Index: These data were downloaded from the *Team Marketing Report* at www.teammarketing.com/fci.cfm, August 2006.

Activity 10-5, Digital Cameras: These data are from *Consumer Reports,* July 2006, pp. 26–31.

Activity 10-8, Welfare Reform: These data are reported in "How Welfare Reform Changed America," by R. Wolf, *USA Today,* July 18, 2006.

Activity 10-16, Hazardousness of Sports: These data are from *Injury Facts,* presented by the National Safety Council, 1999.

Activity 10-17, Gender of Physicians: These data are from *The World Almanac and Book of Facts 2006,* pp. 186–187.

Activity 10-18, Draft Lottery: This study was described in "Randomization and Social Affairs: The 1970 Draft Lottery," by S. Fienberg, *Science,* vol. 171, 1971, pp. 255–261. The data can be downloaded from www.landscaper.net/draft.htm.

Activity 10-19, Honda Prices: These data were gathered from www.cars.com, July 27, 2006.

Unit 3: Randomness in Data

Topic 11: Probability

Activity 11-4, Jury Selection: This value was obtained from the U.S. Census Bureau, State and County QuickFacts, and downloaded in December 2006 from http://quickfacts.census.gov/qfd/states/06/06079.html.

Activity 11-8, Interpreting Probabilities: Part e: This value was obtained from the National Climatic Data Center Web site, U.S. Department of Commerce, and downloaded in September 2006 from www.ncdc.noaa.gov/oa/climate/extremes/christmas.html. Part f: This value is from "Restaurant Failure Rate Much Lower Than Commonly Assumed, Study Finds," *Ohio State Research News,* September 8, 2003. This article was downloaded from http://researchnews.osu.edu/archive/restfail.htm, September 2006.

Activity 11-19, Hospital Births: This value is from *The World Almanac and Book of Facts 1999.*

Activity 11-24, AIDS Testing: These values are from "The Statistical Precision of Medical Screening Procedures: Application to Polygraph and AIDS Antibodies Test Data," by J. Gastwirth, *Statistical Science,* vol. 2, 1987, pp. 213–238.

Topic 12: Normal Distributions

Activity 12-1, Body Temperatures and Jury Selection: These data are from "What's Normal? Body Temperature, Gender, and Heart Rate," by A. Shoemaker, *Journal of Statistics Education,* vol. 4, no. 2, 1996, www.amstat.org/publications/jse/v4n2/datasets.shoemaker.html.

Activity 12-2, Birth Weights: The values are based on data from the National Vital Statistics System, U.S. Department of Health and Human services, and were downloaded in January 2007 from http://209.217.72.34/VitalStats/TableViewer/tableView.aspx?ReportId=318.

Activity 12-3, Blood Pressure and Pulse Rate Measurements: These data are a sample from the 2003–04 National Health and Nutrition Examination Survey (NHANES), National Center for Health Statistics, and can be downloaded from http://www.cdc.gov/nchs/about/major/nhanes/datalink.htm.

Activity 12-12, Heights: These values are from Table 240 of the *Statistical Abstract of the United States 1998,* National Center for Health Statistics.

Activity 12-13, Weights: These values are from Table 241 of the *Statistical Abstract of the United States 1998,* National Center for Health Statistics.

Topic 13: Sampling Distributions: Proportions

Activity 13-3, Kissing Couples: This study is described in "Adult Persistence of Head-Turning Asymmetry," by O. Güntürkün, *Nature,* vol. 421, 2003, p. 711.

Activity 13-10, Calling Heads or Tails: The claim is from *Statistics You Can't Trust: A Friendly Guide to Clear Thinking About Statistics in Everyday Life,* by S. Campbell, Think Twice Publishing, 1998, p. 158.

Activity 13-12, Halloween Practices: These data are from a Gallup Poll conducted on October 21–24, 1999, and released on October 29, 1999, at `www.gallup.com`.

Activity 13-14, Pet Ownership: These data are from Table 1232 of the *Statistical Abstract of the United States 2006,* National Center for Health Statistics. The original source for the data is the *U.S. Pet Ownership and Demographics Sourcebook,* American Veterinary Medical Association.

Activity 13-16, Volunteerism: These data are from a news release titled "Volunteering in the United States, 2005," Bureau of Labor Statistics, December 9, 2005, and can be downloaded from `www.bls.gov/news.release/volun.nr0.htm`.

Activity 13-18, Cursive Writing: This value is from "The Handwriting Is on the Wall," by M. Pressler, *Washington Post,* October 11, 2006, and can be downloaded from `www.washingtonpost.com`.

Topic 14: Sampling Distributions: Means

Activity 14-3, Christmas Shopping: These data are from a Gallup Poll conducted on November 18–21, 1999, and released on November 26, 1999, at `www.gallup.com`.

Activity 14-5, Heart Rates: These data are from the 2003–04 National Health and Nutrition Examination Survey (NHANES), National Center for Health Statistic, and can be downloaded from `http://www.cdc.gov/nchs/about/major/nhanes/datalink.htm`.

Activity 14-11, Cars' Fuel Efficiency: This value is from *Consumer Reports 1999 New Car Buying Guide.*

Topic 15: Central Limit Theorem

Activity 15-1, Smoking Rates: These values are from the November 11, 2005, issue of *Morbidity and Mortality Weekly Report,* Centers for Disease Control and Prevention, and can be downloaded from `www.cdc.gov/mmwr/`.

Activity 15-5, Capsized Boat Tour: Information on this incident is from Wikipedia, Ethan Allan Boating Accident, and can be downloaded from `http://en.wikipedia.org/wiki/Ethan_Allen_Boating_Accident`.

Activity 15-13, Non-English Speakers: This value is from *The World Almanac and Book of Facts 1995,* p. 600.

Unit 4: Inference from Data: Principles

Topic 16: Confidence Intervals: Proportions

Activity 16-14, West Wing *Debate:* This MSNBC/Zogby poll is described in a November 7, 2005, press release from Zogby International, available at www.zogby.com.

Activity 16-15, Magazine Advertisements: These data were gathered by the authors in September 1999.

Activity 16-18, Charitable Contributions: Data from the 2004 General Social Survey (GSS) can be downloaded from the SDA Archive at http://sda.berkeley.edu/archive.htm.

Topic 17: Tests of Significance: Proportions

Activity 17-1, Flat Tires: This campus legend is recounted in an "Ask Marilyn" column, by Marilyn vos Savant, *Parade,* March 3, 1996. Laurie Snell elaborates on the story in edition 5.04 (1996) of *Chance News,* which can be downloaded from www.dartmouth.edu/~chance.

Activity 17-5, Baseball "Big Bang": This "Ask Marilyn" column appeared in *Parade,* May 3, 1998. The data were gathered from www.sportsline.com/mlb/scoreboard.

Activity 17-13, Political Viewpoints: Data from the 2004 General Social Survey (GSS) can be downloaded from the SDA Archive at http://sda.berkeley.edu/archive.htm.

Activity 17-17, Baseball "Big Bang": These data are reported in edition 7.05 (1998) of *Chance News,* available at www.dartmouth.edu/~chance.

Activity 17-21, Hiring Discrimination: These data are reported in *Statistics for Lawyers,* by M. Finkelstein and B. Levin, Springer-Verlag, 1990, pp. 161–162.

Activity 17-23, Veterans' Marital Problems: These data are from "Why Does Military Combat Experience Adversely Affect Marital Relations?" by C. Gimbel and A. Booth, *Journal of Marriage and the Family,* vol. 56, 1994, pp. 691–703.

Topic 18: More Inference Considerations

Activity 18-10: Voter Turnout: These data are from the 1998 General Social Survey, which can be downloaded from the SDA Archive at http://sda.berkeley.edu/archive.htm.

Activity 18-15, Penny Activities: These data are reported in edition 11.02 (2002) of *Chance News,* available at www.dartmouth.edu/~chance.

Topic 19: Confidence Intervals: Means

Activity 19-14, Sentence Lengths: These data were gathered from the opening sentences of Chapter 3 of *The Testament,* by J. Grisham, Doubleday, 1999.

Activity 19-17, Close Friends: These data are from the 2004 General Social Survey (GSS) and can be downloaded from the SDA Archive at http://sda.berkeley.edu/archive.htm.

Topic 20: Tests of Significance: Means

Activity 20-1, Basketball Scoring: These data were gathered from www.espn.com, December 1999.

Activity 20-3, Golden Ratio: These data are from *Handbook of Small Data Sets,* by D. Hand, F. Daly, K. McConway, and D. Lunn, Chapman and Hall, 1993, p. 118.

Activity 20-8, UFO Sighters' Personalities: These data are from "Close Encounters: An Examination of UFO Experiences," by N. Spanos, P. Cross, K. Dickson, and S. DuBreuil, *Journal of Abnormal Psychology,* vol. 102, 1993, pp. 624–632.

Activity 20-9, Basketball Scoring: These data are from *Sports Illustrated,* November 29, 1999.

Unit 5: Inference from Data: Comparisons

Topic 21: Comparing Two Proportions

Activity 21-4, Perceptions of Self-Attractiveness: These data are from a Gallup poll conducted on July 22–25, 1999, and released on September 15, 1999, at www.gallup.com.

Activity 21-19, Underhanded Free Throws: These data are from the column "Paging Dr. Barry," by Rick Reilly, *Sports Illustrated,* December 11, 2006.

Activity 21-22, Wording of Surveys: These data are from Tables 11.2, 11.3, and 11.4 of *Questions and Answers in Attitude Surveys,* by H. Schuman and S. Presser, Academic Press, 1981.

Activity 21-24, Questioning Smoking Policies: This study was conducted by Janet Meyer at Dickinson College.

Activity 21-25, Teen Smoking: This study was reported in *The Harrisburg Patriot-News,* October 4, 1994.

Topic 22: Comparing Two Means

Activity 22-4, Got a Tip?: These data are from "Effect of Server Introduction on Restaurant Tipping," by K. Garrity and D. Degelman, *Journal of Applied Social Psychology,* vol. 20, 1990, pp. 168–172.

Activity 22-9, Got a Tip?: These data are from "Sweetening the Till: The Use of Candy to Increase Restaurant Tipping," by D. Strohmetz, B. Rind, R. Fisher, and M. Lynn, *Journal of Applied Social Psychology,* vol. 32, 2002, pp. 300–309.

Activity 22-11, Ideal Age: These data are from a Harris Poll conducted on September 16–23, 2003, and reported on October 22, 2003, at www.harrisinteractive.com.

Activity 22-12, Editorial Styles: These data were collected from the editorials "Hillary is In," *Washington Post,* January 27, 2007, p. A18, and "Our View on Size of the Military: Strains on U.S. Forces Justify Increasing Troop Strength," *USA Today,* January 27, 2007.

Activity 22-16, Classroom Attention: These data are from "Learning What's Taught: Sex Differences in Instruction," by G. Leinhardt, A. Seewald, and M. Engel, *Journal of Educational Psychology,* vol. 71, 1979, pp. 432–439.

Topic 23: Analyzing Paired Data

Activity 23-4, Alarming Wake-Up: This study is described in "Comparison of a Personalized Parent Voice Smoke Alarm with a Conventional Residential Tone Smoke Alarm for Awakening Children," by G. Smith, M. Splaingard, J. Hayes, and H. Xiang, *Pediatrics,* vol. 118, 2006, pp. 1623–1632.

Activity 23-5, Muscle Fatigue: These data are from "Men Are More Fatigable Than Strength-Matched Women When Performing Intermittent Submaximal Contractions," by S. Hunter, A. Critchlow, I. Shin, and M. Enoka, *Journal of Applied Physiology,* vol. 96, 2004, pp. 2125–2132.

Activity 23-9, Sickle Cell Anemia and Child Development: This study is described in "Growth and Development in Children with Sickle-Cell Trait: A Prospective Study of Matched Pairs," by M. Kramer, Y. Rooks, and H. Pearson, *New England Journal of Medicine,* vol. 299, 1978, pp. 686–689.

Activity 23-13, Catnip Aggression: These data are from "Catnip Bonanza," by S. Jovan, *Stats: The Magazine for Students of Statistics,* issue 27, 2000, pp. 25–27.

Activity 23-16, Maternal Oxygenation: These data are from "The Effect of Maternal Oxygen Administration on Fetal Pulse Oximetry During Labor in Fetuses with Nonreassuring Fetal Heart Rate Patterns," by M. Haydon et al., *American Journal of Obstetrics and Gynecology,* vol. 195, 2006, pp. 735–738.

Activity 23-17, Mice Cooling: These data are from "Calirometric Determination of Average Body Temperature of Small Mammals and Its Variation with Environmental Conditions," by J. Hart, *Canadian Journal of Zoology,* vol. 29, 1951, pp. 224–233.

Activity 23-18, Clean Country Air: These data are from "Urban Factor and Tracheobronchial Clearance," by P. Camner and K. Phillipson, *Archives of Environmental Health,* vol. 27, 1973, p. 82.

Activity 23-22, Comparison Shopping: These data were collected by students at Cal Poly in San Luis Obispo, California, November 1999.

Unit 6: Inferences with Categorical Data

Topic 24: Goodness-of-Fit Tests

Activity 24-1, Birthdays of the Week: These data are from *The World Almanac and Book of Facts 2000,* p. 352.

Activity 24-5, Leading Digits: These data are from *The World Almanac and Book of Facts 2006,* pp. 851–853.

Activity 24-6, Mendel's Peas: These data are from "Experiments in Plant Hybridization," by G. Mendel (1865) and can be downloaded from `www.mendelweb.org/Mendel.html`.

Activity 24-8, Poohsticks: These data are from "The Poohsticks Phenomenon," by R. Knight, *British Medical Journal,* vol. 329, 2004, pp. 1432–1434.

Activity 24-9, Leading Digits: These data are from *The World Almanac and Book of Facts 2006,* p. 526.

Activity 24-10, Political Viewpoints: Data from the 2004 General Social Survey (GSS) can be downloaded from the SDA Archive at `http://sda.berkeley.edu/archive.htm`.

Activity 24-12, Halloween Treats: These data are from "Trick, Treat, or Toy: Children Are Just as Likely to Choose Toys as Candy on Halloween," by M. Schwartz, E. Chen, and K. Brownell, *Journal of Nutrition Education and Behavior,* vol. 35, 2003, pp. 207–209.

Topic 25: Inferences for Two-Way Tables

Activity 25-3, Got a Tip?: These data are from "The Effects of a Joke on Tipping When It Is Delivered at the Same Time as the Bill," by N. Gueguen, *Journal of Applied Social Psychology,* vol. 32, 2002, pp. 1955–1963.

Activity 25-5, Pets as Family: These data are from a report titled "Gauging Family Intimacy: Dogs Edge Cats (Dads Trail Both)" from the Pew Research Center and released on March 7, 2006. The report was downloaded from `http://pewresearch.org/pubs/303/gauging-family-intimacy`, February 2007.

Activity 25-6, Newspaper Reading: Data from the 2004 General Social Survey (GSS) can be downloaded from the SDA Archive at `http://sda.berkeley.edu/archive.htm`.

Activity 25-10, Removing Gingivitis: These data are from "Adjustment of Frequently Used Chi-square Procedures for the Effect of Site-to-Site Dependencies in the Analysis of Dental Data," by A. Donner and D. Banting, *Journal of Dental Research,* vol. 68, 1989, pp. 1350–1354.

Activity 25-11, Asleep at the Wheel: These data are from "Driver Sleepiness and Risk of Serious Injury to Car Occupants: Population Based Control Study," by J. Connor, R. Norton, S. Ameratunga, E. Robinson, I. Civil, R. Dunn, J. Bailey, and R. Jackson, *British Medical Journal,* vol. 324, 2002, pp. 1125–1129.

Unit 7: Relationships in Data

Topic 26: Graphical Displays of Association

Activity 26-1, House Prices: These data were downloaded from `www.zillow.com`, February 7, 2007.

Activity 26-2, Birth and Death Rates: These data are from *The World Almanac and Book of Facts 1999,* p. 874.

Activity 26-3, Car Data: These data are from *Consumer Reports' 1999 New Car Buying Guide.*

Activity 26-6, Televisions and Life Expectancy: These data are from *The World Almanac and Book of Facts 2006,* pp. 750–850.

Activity 26-7, Airline Maintenance: These data are from the March 2007 issue of *Consumer Reports,* p. 18.

Activity 26-9, Challenger *Disaster:* These data are from "Lessons Learned from *Challenger:* A Statistical Perspective," by S. Dalal, E. Folkes, and B. Hoadley, *Stats: The Magazine for Students of Statistics,* vol. 2, 1989, pp. 14–18.

Activity 26-14, Fast-Food Sandwiches: These data were downloaded from Arby's, `www.arbys.com`, June 2006.

Activity 26-16, College Alumni Donations: These data are from Harvey Mudd College Alumni Office promotional material, September 1999.

Activity 26-17, Peanut Butter: These data are from the September 1990 issue of *Consumer Reports.*

Topic 27: Correlation Coefficient

Activity 27-2, Governors' Salaries: The salary data are from *The World Almanac and Book of Facts 2006,* p. 65. The housing data are from the U.S. Census Bureau, State and County QuickFacts and can be downloaded from `http://quickfacts.census.gov/qfd/index.html`.

Activity 27-10, Monopoly: These data are from the Parker Brothers board game.

Activity 27-19, Climatic Conditions: These data are from Tables 368–375 of the *Statistical Abstract of the United States 1992.*

Topic 28: Least Squares Regression

Activity 28-3, Animal Trotting Speeds: These data are from "Speed, Stride Frequency, and Energy Cost per Stride: How Do They Change with Body Size and Gait?" by N. Heglund and C. Taylor, *Journal of Experimental Biology,* vol. 138, 1988, pp. 301–318.

Activity 28-4, Textbook Prices: These data were collected by Cal Poly students Michelle Shaffer and Andrew Kaplan in November 2006.

Activity 28-6, Airfares: The airfare data are from the January 8, 1995, issue of the *Harrisburg Sunday Patriot-News.* The distance data are from the Delta Air Lines Worldwide Timetable Guide, effective December 15, 1994.

Activity 28-17, Box Office Blockbusters: These data were downloaded from `www.boxofficemojo.com`, February 2007.

Activity 28-22, Gestation and Longevity: These data are from *The World Almanac and Book of Facts 2006,* p. 305.

Topic 29: Inference for Correlation and Regression

Activity 29-1, Studying and Grades: These data were collected by students at the University of the Pacific.

Activity 29-4, Plasma and Romance: These data are from "Raised Plasma Nerve Growth Factor Levels Associated with Early-Stage Romantic Love," by E. Emanuele, P. Politi, M. Bianchi, P. Minoretti, M. Bertona, and D. Geroldi, *Psychoneuroendocrinology,* vol. 31, 2006, pp. 288–294.

Appendix C: List of Data Files and Applets

The following data files and applets, listed alphabetically, are needed for the indicated activities. They can be found on the Web-based Student Resource Center (see page xxii for how to access online resources) and on the CD that accompanies this book.

Data Files	Activity Numbers
ACQUAINTANCESCP	9-9, 19-10
AIRFARES	28-7, 28-8, 29-16
ARBYS06	26-14, 26-15, 28-16
BACKPACK	10-12, 19-6, 20-17
BLOODPRESSURES	7-23, 12-3
BODYTEMPS	12-19, 15-19, 19-3, 19-7, 20-11, 22-10
BOXOFFICE05	28-17, 28-18
BROADWAY06	26-11, 27-15
BUMPUS	10-7, 12-20, 22-21
CARS99	27-1
CHALLENGER	26-9, 27-17
CLIMATE	27-19
CLOSEFRIENDS	22-22
COOLMICE	23-17
DAYHIKES	7-14
DEALAMOUNTS	8-2
DIABETES	7-5
DIGITALCAMERAS	10-5, 27-21
DRAFTLOTTERY	10-18, 27-7
EDITORIALS07	22-12, 22-13
ELECTRICBILL	28-11
EXAMSCORES	23-19, 23-20, 27-6
FANCOST06	10-4
FEBTEMPS	8-19, 9-7
GESTATION06	28-22
GOVERNORS05	27-2
HEARTRATE	14-5
HMCDONORS	26-16, 27-16
HONDAPRICES	12-21, 28-14, 28-15, 29-10, 29-11
HOUSEPRICESAG	28-2, 28-12, 28-13, 29-3
HYPOATM	9-24, 19-21, 22-25
HYPOCOMMUTE	22-2
HYPOEXAMS	27-8
HYPOSLEEP	20-2, 20-7
HYPOSTDDEV	9-25
ICECREAMCALORIES	10-3
MARRIAGEAGES	8-17, 9-6, 23-1, 29-17, 29-18
MARRIAGEAGES100	23-12
MATCHING	8-3
MATERNALOXYGENATION	23-16
MATHPLACEMENT	7-16
MONOPOLY	27-10, 27-11
NBAPOINTS	20-1
PEANUTBUTTER	26-17

Data Files *(continued)*	Activity Numbers
PHYSICIANS	10-17
PLANETS	8-12, 28-20, 28-21
QUIZPCT	8-18
ROWERS04	8-4
SEATBELTUSAGE05	8-5
SHOPPING	23-22, 23-23, 26-21
SOLITAIRE	27-18
SPORTSEXAMPLES	8-14, 10-11, 22-26
SPORTSHAZARDS	10-16
TENNISSIM	8-21, 9-16, 22-18
TEXTBOOKPRICES	28-4, 29-5
TROTSPEEDS	28-3
TVLIFE	26-6, 27-3, 28-19
UOPGPA	29-1, 29-2
VALUESFJ	9-3
WELFAREREFORM	10-8

TI-83 and TI-84 Plus Programs

CELL	25-2
CHANGE	14-2
CHIGOF	24-4
CORR	27-1, 27-2, 27-3
DOTPLOT	2-3, 8-3, 13-13
HEART	14-5
ICPRICE	11-18
LBLSCAT	26-5
SIMSAMP	11-19, 13-9, 13-10, 13-11, 13-13
STDER	29-1
STREAK	11-21

Applets

Friendly Observer Simulation	21-1
Guess the Correlation	27-4
Least Squares Regression	28-1
Normal Probability Calculator	12-2, 15-5
Power Simulation	18-5, 18-16, 18-17, 18-18
Random Babies	11-1, 11-2
Randomization of Subjects	5-3
Reese's Pieces	13-2
Sampling Pennies	14-2
Sampling Regression Lines	29-1, 29-8
Sampling Words	4-3, 4-4, 4-8, 9-15, 14-6
Simulating Confidence Intervals	16-4, 16-22, 19-1
Test of Significance Calculator	17-2, 20-1, 21-2, 22-1
Two-Way Table Simulation	21-8

Index

D

Q

R

S

U

V

W

Y

Z